RESIDENTIAL CONSTRUCTION ACADEMY

Facilities Maintenance

MAINTAINING, REPAIRING, AND REMODELING

Third Edition

Kevin Standiford

CENGAGE
Learning·

Australia • Brazil • Japan • Korea • Mexico • Singapore • Spain • United Kingdom • United States

Residential Construction Academy Facilities Maintenance: Maintaining, Repairing, and Remodeling, Third Edition
Kevin Standiford

Vice President and General Manager-Skills and Planning: Dawn Gerrain

Senior Product Manager: James Devoe

Senior Director, Development-Skills and Planning: Marah Bellegarde

Senior Product Development Manager: Larry Main

Development Editor: Brooke Wilson, Ohlinger Publishing Services

Senior Content Developer: Jennifer Starr

Product Assistant: Aviva Ariel

Brand Manager: Kay Stefanski

Market Development Manager: Erin Brennan

Senior Production Director: Wendy A. Troeger

Production Manager: Mark Bernard

Content Project Manager: David S. Barnes

Art Director: Bethany Casey

Media Developer: Deborah Bordeaux

Cover images: ©iStockphoto.com/ kkgas
©iStockphoto.com/Gencay M. Emin

For product information and technology assistance, contact us at
Cengage Learning Customer & Sales Support, 1-800-354-9706
For permission to use material from this text or product,
submit all requests online at **www.cengage.com/permissions.**
Further permissions questions can be e-mailed to
permissionrequest@cengage.com

Library of Congress Control Number: 2013937531

ISBN-13: 978-1-133-28243-3

Cengage Learning
200 First Stamford Place, 4th Floor
Stamford, CT 06902
USA

Cengage Learning is a leading provider of customized learning solutions with office locations around the globe, including Singapore, the United Kingdom, Australia, Mexico, Brazil, and Japan. Locate your local office at: **www.cengage.com/global**

Cengage Learning products are represented in Canada by Nelson Education, Ltd.

To learn more about Cengage Learning, visit **www.cengage.com**
Purchase any of our products at your local college store or at our preferred online store **www.cengagebrain.com**

Notice to the Reader

Publisher does not warrant or guarantee any of the products described herein or perform any independent analysis in connection with any of the product information contained herein. Publisher does not assume, and expressly disclaims, any obligation to obtain and include information other than that provided to it by the manufacturer. The reader is expressly warned to consider and adopt all safety precautions that might be indicated by the activities described herein and to avoid all potential hazards. By following the instructions contained herein, the reader willingly assumes all risks in connection with such instructions. The reader is notified that this text is an educational tool, not a practice book. Since the law is in constant change, no rule or statement of law in this book should be relied upon for any service to any client. The reader should always refer to standard legal sources for the current rule or law. If legal advice or other expert assistance is required, the services of the appropriate professional should be sought. The publisher makes no representations or warranties of any kind, including but not limited to, the warranties of fitness for particular purpose or merchantability, nor are any such representations implied with respect to the material set forth herein, and the publisher takes no responsibility with respect to such material. The publisher shall not be liable for any special, consequential, or exemplary damages resulting, in whole or part, from the readers' use of, or reliance upon, this material.

Printed in the United States of America
3 4 5 6 7   23 22 21 20 19

Brief Contents

Table of Contents

© iStockphoto.com/Gencay M. Emin

v

Preface

About the Residential Construction Academy Series

One of the most pressing problems confronting the building industry today is the shortage of skilled labor. The construction industry must recruit hundreds of thousands of new craft workers each year to meet future needs. This shortage is expected to continue because of projected job growth and a decline in the number of available workers. At the same time, the training of available labor is an increasing concern throughout the country and is affecting construction trades, threatening the ability of builders to construct quality homes.

These challenges led to the creation of the innovative *Residential Construction Academy Series*. The *Residential Construction Academy Series* is the perfect way to introduce people of all ages to the building trades while guiding them in the development of essential workplace skills, including carpentry, electrical wiring, HVAC, plumbing, masonry, and facilities maintenance. The products and services offered through the Residential Construction Academy are the result of cooperative planning and rigorous joint efforts between industry and education. The program was originally conceived by the National Association of Home Builders (NAHB)—the premier association in the residential construction industry—and HBI (Home Builders Institute) – a national leader for career training in the building industry and the NAHB Federation's workforce development partner.

For the first time, construction professionals and educators created national skills standards for the construction trades. In the summer of 2001, NAHB, through the HBI, began the process of developing residential craft standards in six trades: carpentry, electrical wiring, HVAC, plumbing, masonry, and facilities maintenance. Groups of employers from across the country met with an independent research and measurement organization to begin the development of new craft training standards. Builders and remodelers, residential and light commercial, custom single family and high production or volume builders are represented. The guidelines from the National Skills Standards Board were followed in developing the new standards. In addition, the process met or exceeded American Psychological Association standards for occupational credentialing.

Next, through a partnership between HBI and Delmar / Cengage Learning, learning materials—textbooks, videos, and instructor's curriculum and teaching tools—were created to teach these standards effectively. A foundational tenet of this series is that students *learn by doing*. Integrated into this colorful, highly illustrated text are Procedure sections designed to help students apply information through hands-on, active application. A constant focus of the *Residential Construction Academy* is teaching the skills needed to be successful in the construction industry and constantly applying the learning to real-world applications.

An enhancement to the Residential Construction Academy Series is industry Program Credentialing and Certification for both instructors and students by HBI. National Instructor Certification ensures consistency in instructor teaching/training methodologies and knowledge competency when teaching to the industry's national skills standards. Student Certification is offered for each trade area of the Residential Construction Academy Series in the form of rigorous testing. Student Certification is tied to a national database that will provide an opportunity for easy access for potential employers to verify skills and competencies. Instructor and Student Certification serve the basis for Program Credentialing offered by HBI. For more information on HBI Program Credentialing and Instructor and Student Certification, please visit HBI.org.

About This Book

A facility maintenance technician is responsible for the day-to-day maintenance and operational tasks that support a commercial facility. Duties often include but are not limited to:

- Responsibility for various activities related to the repair and maintenance of the electrical, plumbing, heating, and ventilation systems
- Painting and minor repair to walls, ceilings, and floors
- Preventive/predictive maintenance per requirements and unscheduled or emergency maintenance when required to support operations
- Recommendations to modify or replace equipment when necessary to support demand or improve building efficiency
- Snow removal and other groundskeeping duties as assigned
- Installation and repairs of locks
- Inspection and maintenance of lights and fire extinguishers
- Inspection and maintenance of swimming pool areas
- Management of remodeling jobs, including obtaining the proper building permits
- Performing energy audits as well as recommending weatherization updates to the facility

Facilities Maintenance, Maintaining, Repairing, and Remodeling, Third Edition, provides coverage for the areas in residential wiring that are required of an entry-level facility maintenance technician, including the basic hands-on skills and the more advanced theoretical knowledge needed to gain job proficiency. In addition to the electrical area, other topics covered include customer service skills, carpentry, surface painting, plumbing, appliance repair, pest prevention, groundskeeping, HVAC systems, remodeling, and weatherization. The format of the text is designed to be easy to learn and easy to teach.

Maintaining, Repairing, and Remodeling

A facility maintenance technician is responsible for a variety of maintenance and repair duties in order to maintain the present state of a facility. When appliances, equipment, and building materials wear out or malfunction, it is the job of the facility maintenance technician to restore these building elements to their original condition. Remodeling duties like obtaining permits, securing financing, and selecting contractors can be duties of the facilities maintenance technician as well. Remodeling and demolition information, safety tips, and tricks of the trade are included throughout the third edition.

New to This Edition

Most of the chapters in the third edition have been updated to include advances and changes in the HBI standards for facility maintenance technicians. A significant addition is the inclusion of new information on remodeling and demolition, which is clearly highlighted throughout the chapters where applicable, and three new chapters covering weatherization and solar systems.

The third edition includes the following revisions:

New! Introduction to the Construction Industry and Facilities Maintenance

- New Introduction to *Facilities Maintenance*

Chapter 3: Applied Safety Rules

- New information related to remodeling and demolition safety

Chapter 4: Fasteners, Tools, and Equipment

- New information covering specialty hand and power tools used in demolition

Chapter 6: Electrical Facilities Maintenance

- New information on remodeling as it relates to electrical systems

Chapter 7: Carpentry

- New information on demolition involved in remodeling various rooms in a residential structure
- Expanded content on suspended ceilings
- Added content on steel framing
- Expanded content on drywall and fastening techniques and methods
- Added content for cricket and saddle

Chapter 8: Surface Treatments

- New information on painting when remodeling

Chapter 9: Plumbing

- New information on remodeling as it relates to plumbing systems

Chapter 11: Appliance Repair and Replacement

- New information on demolition of the kitchen, including safety information related to utilities, appliances, and plumbing fixtures

Chapter 12: Trash Compactors

- Expanded coverage of troubleshooting

Chapter 15: Landscaping and Groundskeeping

- New information on demolishing an in-ground pool

Chapter 17: Blueprint Reading, Building Codes, and Permits

- New information on permits and inspections required for remodeling and demolition
- Added content regarding permits and zoning
- Added content for specifications
- Added content for quantity takeoff

New! Chapter 18—Weatherization Concepts

- New Chapter in *Facilities Maintenance*

New! Chapter 19—Weatherization Installation, Maintenance, and Repair

- New Chapter in *Facilities Maintenance*

New! Chapter 20—Solar Systems: Maintenance and Repair

- New Chapter in *Facilities Maintenance*

New! Appendix A—Remodeling and Demolition Tips

- New Appendix in *Facilities Maintenance*

Features of This Book

This innovative series was designed with input from educators and industry professionals and informed by the curriculum and training objectives established by the National Skills Standards Committee. The following features aid learning.

Learning features such as the **Objectives, Glossary of Terms**, and **Introduction** set the stage for the coming body of knowledge and help learners identify key concepts and information. These learning features serve as a road map for continuing through the chapter. Learners also may use them as an on-the-job reference. In addition, an end-of-book glossary containing a complete list of terms and definitions from the chapters offers an important reference.

From Experience boxes provide tricks of the trade and mentoring wisdom that make a particular task a little easier for the novice to accomplish.

Safety is featured throughout the text to instill safety as an attitude among learners. Safe job site practices by all workers are essential; if one person acts in an unsafe manner, all workers on the job are at risk of being injured. Learners will come to know and appreciate that adherence to safety practices requires a blend of ability, skill, and knowledge that should be continuously applied to all tasks they perform in the construction industry.

Caution boxes highlight safety issues and urgent safety reminders for the trade.

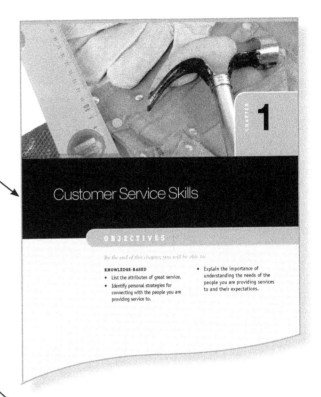

© iStockphoto.com/kkgas

CHAPTER 1

Customer Service Skills

OBJECTIVES

By the end of this chapter, you will be able to:

KNOWLEDGE-BASED
- List the attributes of great service.
- Identify personal strategies for connecting with the people you are providing service to.
- Explain the importance of understanding the needs of the people you are providing services to and their expectations.

FROM EXPERIENCE

In most cases when a carpet is being replaced, the underlayment will be worn out. Therefore, even if the underlayment appears to be in good shape, it is recommended that you always replace it when installing a new carpet.

CAUTION

Only use pipe insulation that is specifically designed for the size pipe you are intending to insulate. Failure to use the correct size will diminish the effectiveness of the insulation.

Sealing Switches and Outlets

As stated earlier switches and outlets along exterior walls are one source of transfer. This source of heat transfer can be easily eliminated by applying outlet and switch sealers. The following steps outline how to install switch and outlet sealers.

Step 1: Disconnect the power to the outlet(s) and/or switches.
Step 2: Using a screwdriver remove the device cover (see Figure 19-22).
Step 3: Place the precut sealer around the device.
Step 4: Replace the device cover (see Figure 19-23).
Step 5: Reenergize the circuit.

Remodeling and demolition information, safety tips, and tricks of the trade highlight special considerations for these situations.

REMODELING

Remodeling and Landscaping

As mentioned in the introduction, when a structure is remodeled that contains a swimming pool, most locations require that a permit be obtained before the remodeling begins. The general procedure for remodeling a structure that contains swimming pool is as follows.

Demolishing an In-Ground Pool

Most locations require two inspections to remove an in-ground pool. The first inspection ensures that either the bottom of the pool has been removed or that proper drainage has been applied (holes punched in the bottom of the pool). The second is typically conducted after the pool has been backfilled.

- Once the proper permits and paper work have been completed, if the pool is constructed from steel or fiberglass, then the materials must be removed and properly disposed of before the pool area can be backfilled.
- Some areas allow for pools constructed from concrete/gunite to be crushed and used as part of the fill. (Check with the local building department.)
- A section of the bottom of the pool must be broken and removed to allow the water table in the area of the pool to rise and lower naturally.
- Some areas allow for the concrete pavers and decking surrounding the pool to be crushed as used as backfill also. (Check with the local building department.)
- Crushed stone can also be used to backfill the pool area within 16 inches from the top.
- The remaining pool area can be backfilled using top soil, compacting as more and more soil is added. Compacting the soil will help prevent settling.

Always check with your local and state agencies before starting a remodeling project. Before beginning the demolition or deconstruction process, the proper permits must be obtained.

Procedures provide step-by-step coverage of typical facilities maintenance tasks, explaining the work in detail.

PROCEDURE 7-1

Constructing the Grid Ceiling System

- Locate the height of the ceiling, marking elevations of the ceiling at the ends of all wall sections. Snap chalk lines on all walls around the room to the height of the top edge of the wall angle. If a laser is used, the chalk line is not needed since the ceiling is built to the light beam.

- Fasten wall angles around the room with their top edge lined up with the line. Fasten tem into the framing wherever possible, not more than 24 inches apart. If available, power nailers can be used for efficient fastening.

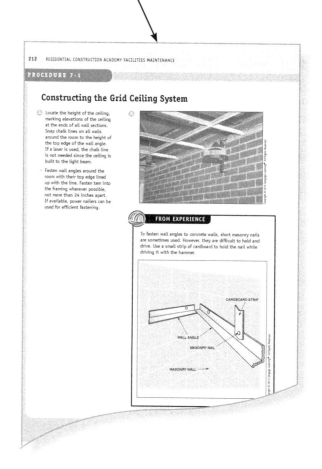

FROM EXPERIENCE

To fasten wall angles to concrete walls, short masonry nails are sometimes used. However, they are difficult to hold and drive. Use a small strip of cardboard to hold the nail while driving it with the hammer.

CARDBOARD STRIP

WALL ANGLE

MASONRY NAIL

MASONRY WALL →

Chapter Review Questions and **Job Sheets** enable the reader to assess the knowledge and skills obtained from reading the chapter.

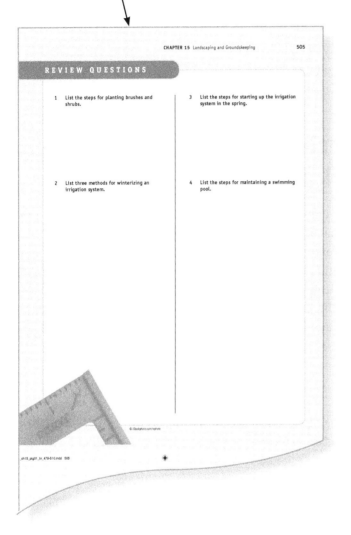

REVIEW QUESTIONS

1 List the steps for planting brushes and shrubs.

2 List three methods for winterizing an irrigation system.

3 List the steps for starting up the irrigation system in the spring.

4 List the steps for maintaining a swimming pool.

Name: _____

Date: _____

TOOLS AND
EQUIPMENT
JOB SHEET

1

Tools and Equipment

FASTENERS

Upon completion of this job sheet, you should be able to demonstrate your ability to select the correct fastener for the appropriate repair project.

1. Take an inventory and list the different types of fasteners used at your facilities. Write a brief description of where they would be used.
    _____
    _____

2. Are the correct fasteners used for the appropriate jobs at your facilities?
    a. If no, what, if anything, should be done?
    _____

3. What are the different types of screws available and what determines which ones are used?
    _____
    _____

4. Write a brief explanation of what determines whether a nail or a screw should be used.
    _____
    _____

INSTRUCTOR'S RESPONSE:
    _____
    _____
    _____

In addition, key topic and skill sets are addressed in the following areas:

- Basic concepts of remodeling: Introduction
- Introduction to customer service skills: Chapter 1
- Introduction to task management: Chapter 2
- Introduction to OSHA safety: Chapter 3
- Basic concepts of electrical theory: Chapter 5
- Basic concepts of weatherization: Chapter 19
- Basic concepts of solar energy: Chapter 20

Turnkey Curriculum and Teaching Material Package

We understand that a text is only one part of a complete, turnkey educational program. We also understand that instructors want to spend their time teaching, not preparing to teach. The *Residential Construction Academy* series is committed to providing thorough curriculum and preparatory materials to aid instructors and alleviate some of their heavy preparation commitments. An integrated teaching solution is provided with the text, including the Instructors Resources CD, a printed Instructor's Resource Guide, a Workbook, and Companion web site.

Workbook

Designed to accompany *Residential Construction Academy Facilities Maintenance, Third Edition*, the Workbook is an extension of the core text and provides additional review questions and problems designed to challenge and reinforce the student's comprehension of the content presented in the text.

Instructor Resources CD

Designed as an integrated teaching package, Cengage Learning's Instructor Resources CD is a complete guide to classroom management. The CD-ROM contains lesson plans, instructor tips, answers to review questions and workbook questions, and other supporting materials for instructors using this series.

Features contained in the Instructor Resources CD include:

- Lesson Plans: These include goals, discussion topics, suggested reading, and suggested homework assignments. You have the option of using these lesson plans with your own course information.
- Instructor Tips: These are hints that provide direction on how to present the material and coordinate the subject matter with student projects.
- Answers to Review Questions: These are solutions that enable you to grade and evaluate end-of-chapter tests and exercises.

- Answers to Workbook Questions: These are solutions that enable you to track and evaluate student learning
- PowerPoint Presentations: These provide the basis for a lecture outline that helps you present concepts and material. These are key points and concepts can be graphically highlighted for student retention.
- Test Questions: Multiple-choice questions of varying levels of difficulty are provided in these chapter-by-chapter testbanks. These questions can be used to assess student comprehension, and they can be edited to meet the needs of the course.
- Correlation Grid: This maps the book content to the HBI Skill Standards.

Companion Site

The Companion site is an excellent supplement for students and instructors. It features many useful resources to support the text.

About the Author

Kevin Standiford, author, contributor, and consultant, has been in the technology fields of manufacturing processes, HVACR, process piping, and robotics for more than 20 years. While attending college to obtain his bachelor of science degree in mechanical engineering technology, he worked for McClelland Consulting Engineers as a mechanical designer, designing HVAC, complex processing piping, and cogeneration systems for commercial and industrial applications. During his college years, he became a student member of the American Society of Heating, Refrigerating and Air-Conditioning Engineers, where he developed and later wrote a paper on a computer application that enabled the user to simulate, design, and draw heating and cooling systems by using AutoCAD. The paper was entered into a student design competition and won the first prize for the region and state. After graduation, Kevin worked for Pettit and Pettit Consulting Engineers, one of the leading HVAC engineering firms in the state of Arkansas, as a mechanical design engineer. While working for Pettit and Pettit, Kevin designed and selected equipment for large commercial and government projects by using manual design techniques and computer simulations.

In addition to working at Pettit and Pettit, Kevin started teaching part-time evening engineering and design courses for Garland County Community College in Hot Springs, Arkansas.

Subsequently, he stopped working full time in the engineering field and started teaching technology classes, which included heat transfer, duct design, and properties of air. It was also at this time that Kevin started writing textbooks for Cengage Learning. The first textbook was a descriptive geometry book, which included a section on sheet metal design. Today Kevin is a full-time consultant, working for both the publishing and engineering industries, and a part-time instructor. In the publishing industry, Kevin has worked on numerous e-resource products, mapping, and custom publications for Cengage Learnings's HVAC, CAD, and plumbing titles.

Acknowledgments

The NAHB and HBI would like to thank the many individuals, members, and companies that participated in the creation of the Facilities Maintenance National Skill Standards. These standards helped guide us in the creation of this text. Thanks also to Debbie Standiford for her work on Chapter 8.

In addition, we thank the following people who provided important feedback throughout the development of the book:

Jim Eichenlaub
Executive Director
Builders Association of Metropolitan Pittsburgh
Pittsburgh, PA

Michael Frank
Facilities Maintenance Instructor
Quentin M. Burdick Job Corps
Minot, ND

Kevin Fry
Facilities Maintenance Instructor
Excelsior Springs Job Corp
Excelsior Springs, MO

Rodney Gaugh
Facilities Maintenance Instructor
Home Builders Institute
St. Louis, MO

Sean M. Kelly
Facilities Maintenance Instructor
Old Dominion Job Corps
Monroe, VA

Mark Martin
Carpentry Instructor
Penobscot Job Corps
Bangor, ME

Daryl Martinez
Facilities Maintenance Instructor
Talking Leaves Job Corps Center
Tahlequah, OK

Pat McHale
Building Trades Instructor
Monroe Career and Technical Institute
Bartonsville, PA

Shannon Pfeiffer
Program Administrator/
Project Coordinator
Building Trades Academy
Chesapeake, VA

George Vick
Building Trades Instructor
Building Trades Academy
Chesapeake, VA

Nathan Wilson
Facilities Maintenance Instructor
Hubert H. Humphrey Job Corps
St. Paul, MN

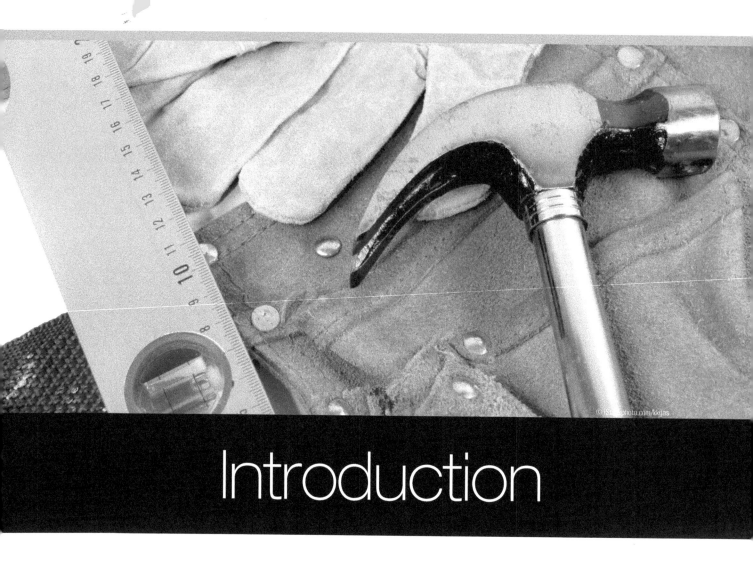

Introduction

Organization of the Construction Industry

The residential construction industry is one of the biggest sectors of the American economy. According to the U.S. Department of Labor, construction is one of the nation's largest industries. About 64 percent of wage and salary jobs in construction are in the specialty trade contractors sector, primarily plumbing, heating and air-conditioning, electrical, and masonry. The National Association of Home Builders (NAHB) reports that home building traditionally accounts for 50–55 percent of the construction industry. Opportunities are available for people to work at all levels in the construction industry, from those who handle the tools and materials on the job site to the senior engineers and architects, who spend most of their time in offices. Few people spend their entire lives in a single occupation, and even fewer spend their lives working for one employer. You should be aware of all the opportunities in the construction industry so that you can make career decisions in the future, even if you are sure of what you want to do at this time.

Construction Personnel

The occupations in the construction industry can be divided into four categories:

- Unskilled or semiskilled labor
- Skilled trades or crafts
- Technicians
- Design and management

Unskilled or Semiskilled Labor

Construction is labor intensive. That means it requires a lot of labor to produce the same dollar value of end products in comparison with other industries, where labor may be a smaller part of the picture. Construction workers with limited skills are called *laborers* or *pre-apprentices*. Laborers or pre-apprentices are sometimes assigned the tasks of moving materials, running errands, and working under the close supervision of a skilled worker. Their work is strenuous, and so construction laborers must be in excellent physical condition.

Construction laborers or pre-apprentices are those workers who have not reached a high level of skill in a particular trade and are not registered in an apprenticeship program. These laborers often specialize in working with a particular trade, such as mason's tenders or carpenter's helpers or pre-apprentices (Figure I-1).

FIGURE I-1: Mason.

Although the mason's tender may not have the skill of a bricklayer, the mason's tender knows how to mix mortar for particular conditions, can erect scaffolding, and is familiar with the bricklayer's tools. Many laborers or pre-apprentices go on to acquire skills and become skilled workers. Laborers who specialize in a particular trade are often paid slightly more than completely unskilled laborers.

Skilled Trades

A *craft* or *skilled trade* is an occupation in which one works with tools and materials and building structures. The building trades are the crafts that deal most directly with building construction (see Figure I-2).

The building trades are among the highest paying of all skilled occupations. However, work in the building trades can involve working in cold conditions in winter or blistering sun in the summer. Also, job opportunities will be best in an area where construction is being done in abundance. The construction industry usually is growing at a high rate nationwide. Generally, plenty of work is available to provide a comfortable living for a good worker.

Carpenter
  Framing carpenter
  Finish carpenter
  Cabinetmaker
Plumber
  New construction
  Maintenance and repair
Roofer
Electrician
  Construction electrician
  Maintenance electrician
Mason
  Bricklayer (also lays concrete blocks)
  Cement finisher
HVAC technician
Plasterer
  Finish plaster
  Stucco plaster
Tile setter
Equipment operator
Drywall installer
  Installer
  Taper
Painter

FIGURE I-2: Building trades.

Apprenticeship

The skills needed to gain employment in the building trades are often learned in an apprentice program. Apprenticeships are usually offered by trade unions, trade associations, community or technical colleges, and large employers. *Apprentices* attend class a few hours a week to learn the necessary theory. The rest of the week they work on a job site under the supervision of a *journeyman* (a skilled worker who has completed the apprenticeship and has experience on the job). The term "journeyman" is a gender neutral term that has been used for decades. It is worth noting that many highly skilled building trades workers are women. Apprentices receive a lower salary than journeymen, often about 50 percent of what a journeyman receives. The apprentice wage usually increases as stages of the apprenticeship are successfully completed. By the time the apprenticeship is completed, the apprentice can be earning as much as 95 percent of what a journeyman earns. Many apprentices receive college credit for their training. Some journeymen receive their training through school or community or technical college and on-the-job training. In one way or another, some classroom training and on-the-job supervised experience are usually necessary to reach journeyman status. Not all apprentice programs are the same, but a typical apprenticeship lasts 4 or 5 years and requires between 100 and 200 hours per year of classroom training along with 1,200 to 1,500 hours per year of supervised work experience.

Technical Career	Some Common Jobs
Surveyor	Measures land, draws maps, lays out building lines, and lays out Roadways
Estimator	Calculates time and materials necessary for project
Drafter	Draws plans and construction details in conjunction with architects and engineers
Expeditor	Ensures that labor and materials are scheduled properly
Superintendent	Supervises all activities at one or more job sites
Inspector	Inspects project for compliance with local building codes at various stages completion
Planner	Plans for best land and community development

FIGURE I-3: Technicians.

Technicians

Technicians provide a link between the skilled trades and the professions. Technicians often work in offices, but their work also takes them to construction sites. Technicians use mathematics, computer skills, specialized equipment, and knowledge of construction to perform a variety of jobs. Figure I-3 lists several technical occupations.

Most technicians have some type of college education, often combined with on-the-job experience, to prepare them for their technical jobs. Community and technical colleges often have programs aimed at preparing people to work at the technician level in construction. Some community or technical college programs are intended especially for preparing workers for the building trades, while others have more of a construction management focus. Construction management courses give the graduate a good overview of the business of construction. The starting salary for a construction technician is about the same as that for a skilled trade, but the technician can be more certain of regular work and will have increased opportunities for advancement.

Design and Management

Architecture, engineering, and contracting are the design and management professions. These *professions* are occupations that require more than four years of college and a license to practice. Many contractors have less than four years of college, but they often operate at a very high level of business, influencing millions of dollars, and so they are included with the professions here. These construction professionals spend most of their time in offices and are not frequently seen on the job site.

Architects usually have a strong background in art and aesthetics, so they are well prepared to design attractive, functional buildings. A typical architect's education includes a four-year degree in fine art, followed by a master's degree in architecture. Most of their construction education comes during the final years of work on the architecture degree.

Engineers generally have more background in math and science, so they are prepared to analyze conditions and calculate structural characteristics. There are many specialties within engineering, but civil engineers are most commonly found in construction. Some civil engineers work mostly in road layout and building. Other civil engineers work mostly with structures in buildings. They are sometimes referred to as structural engineers.

Contractors are the owners of the businesses that do most of the building. In larger construction firms, the principal (the owner) may be more concerned with running the business than with supervising construction. Some contractors are referred to as general contractors and others as *subcontractors* or *trade contractors* (Figure I-4). The general contractor is the principal construction company hired by the owner to construct the building. A general contractor might have only a skeleton crew, relying on subcontractors or trade contractors for most of the actual construction. The general contractor's site or field superintendent coordinates the work of all the subcontractors or trade contractors.

It is quite common for a successful journeyman to start his or her own business as a contractor, specializing in the field in which he or she was a journeyman. These are the subcontractors or trade contractors who sign on to do a specific part of the construction, such as framing or plumbing. As the contractor's company grows and the company works on several projects at one time, the skilled workers with the best ability to lead others may become foremen. A foreman is a working supervisor of a small crew of workers in a specific trade. All contractors have to be concerned with business management. For this reason, many successful contractors attend college and get a degree in construction management. Most states require contractors to have a license to do contracting in their state. Requirements vary from state to state, but a contractor's license usually requires several years of experience in the trade and a test on both trade information and the contracting business.

An Overview of Design and Construction

To understand the relationships between some of the design and construction occupations, we shall look at a scenario for a typical housing development. The first

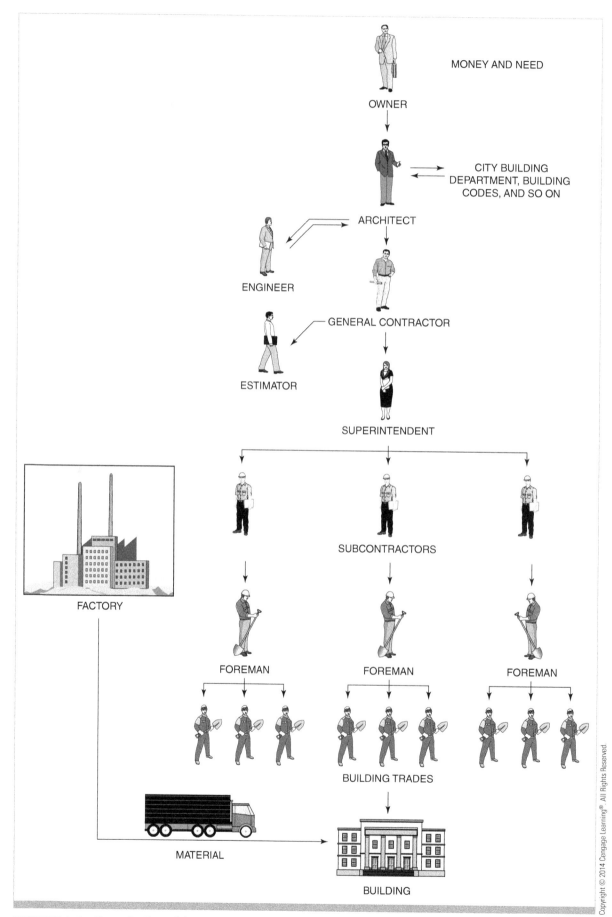

FIGURE I-4: Organization of the construction industry.

people to be involved are the community planners and the *real estate developer*. The real estate developer has identified a 300-acre tract on which he would like to build nearly 1,000 homes, which he will later sell at a good profit. The developer must work with the city planners to ensure that the use of the site he has planned is acceptable to the city. The city planner is responsible for ensuring that all building in the city fits the city's development plan and zoning ordinances. On a project this big, the developer might even bring in a planner of his own to help decide where parks and community buildings should be located and how much parking space they will need.

As the plans for development begin to take shape, it becomes necessary to plan streets and to start designing houses to be built throughout the development. A civil engineer is hired to plan and design the streets. The civil engineer will first work with the developer and planners to lay out the locations of the streets, their widths, and drainage provisions to get rid of storm water. The civil engineer also considers soil conditions and expected traffic to design the foundation for the roadway.

An architectural firm, or perhaps a single architect, will design the houses. Typically several standard or stock plans are used throughout a development, but many homeowners wish to pay extra to have a custom home designed and built. In a custom home, everything is designed for that particular house. Usually the homeowner, who will eventually live in the house, works with the architect to specify the sizes, shapes, and locations of rooms, interior and exterior trim, type of roof, built-in cabinets and appliances, the use of outdoor spaces, and other special features. Architects specialize in use of space, aesthetics (attractive appearance), and livability features. Most architectural features do not involve special structural considerations, but when they do, a structural engineer is employed to analyze the structural requirements and help ensure that the structure will adequately support the architectural features.

One part of construction that almost always involves an engineer is the design of roof trusses. Roof trusses are the assemblies that make up the frame of the roof. Trusses are made up of the top chords, bottom chords, web members, and gussets. The engineer considers the weight of the framing materials, the weight of the roof covering, the anticipated weight of any snow that will fall on the roof in winter, and the span (the distance between supports) of the truss to design trusses for a particular purpose. The architect usually hires the engineer for this work, and so the end product is one set of construction drawings that includes all the architectural and engineering specifications for the building. Even though the drawings are sometimes referred to as architectural drawings, they include work done by architects, engineers, and their technicians. Building codes require an architect's seal on the drawings before work can begin. The architect will require an engineer to certify certain aspects of the drawings before putting the architect's seal on them.

Forms of Ownership

Construction companies vary in size from small, one-person companies to very large international organizations that engage in many kinds of construction. However, the size of the company does not necessarily indicate the form of ownership. Three types of ownership and the advantages and disadvantages of each are shown in Figure I-5.

Forms of Ownership	What it Means	Advantages	Disadvantages
Sole Proprietorship	A sole proprietorship is a business whose owner and operator are the same person.	The owner has complete control over the business and there is a minimum of government regulation. If the company is successful, the owner receives high profits.	If the business goes into debt the owner is responsible for that debt. The owner can be sued for the company, and the owner suffers all the losses of the company.
Partnership (*General* and *Limited Liability Partnership* (*LLP*))	A partnership is similar to a sole proprietorship, but there are two or more owners. *General*: In a general partnership, each partner shares the profits and losses of the company in proportion to the partner's share of investment in the company. *LLP*: A limited liability partner is one who invests in the business, receives a proportional share of the profit or loss, but has limited liability.	*General Partnership*: The advantage is that the partners share the expense of starting the business and partnerships are not controlled by extensive government regulations. *LLP*: A limited liability partner can only lose his or her investment.	*General Partnership*: Each partner can be held responsible for all the debts of the company. *LLP*: Every LLP must have one or more general partners who run the business. The general partners in an LLP have unlimited liability and they can be personally sued for any debts of the company.
Corporation	In a corporation a group of people own the company. Another, usually smaller, group of people manage the business. The owners buy shares of stock. A share of stock is a share or a part of the business. The value of each share increases or decreases according to the success of the company.	In a corporation, no person has unlimited liability. The owners can only lose the amount of money they invested in stock. The owners of a corporation are not responsible for the debts of the corporation. The corporation itself is the legal body and is responsible for its own debts.	The government has stricter regulations for corporations than for the other forms of ownership. Also, corporations are more expensive to form and to operate than are proprietorships and partnerships.

FIGURE I-5: The three types of ownership.

Unions and Contractor's Associations

The construction industry contains thousands of organizations of people with common interests and goals. Whole directories of these organizations are available in libraries and on the Internet. Two categories of construction organizations are of particular importance to construction students: craft unions and contractors' associations.

Unions

A *craft union*, usually just called a "union," is an organization of workers in a particular building trade. Workers' unions were first formed in the 1800s, when factory workers were being forced to work extreme hours under unsafe conditions and for very low wages. Although working conditions in both factories and construction have improved dramatically, unions continue to serve a valuable role in the construction industry. Figure I-6 lists several national construction craft unions.

Individuals pay dues to be members of the union. Dues pay for the benefits the union provides to its members. Most unions have an apprenticeship program that includes both classroom instruction and on-the-job supervised work experience. Some of the members' dues pay for instructors, classroom space, and training supplies. Unions usually provide a pension for members who have worked in the trade. Because they represent a large block of members, unions can be a powerful force in influencing government to do such things as pass worker safety laws, encourage more construction, and support technology that is good for construction. Unions negotiate with employers (contractors) to establish both a pay rate and working conditions for their members. It is quite typical to find that union members enjoy a higher hourly pay rate than nonunion workers in the same trade.

Contractors' Associations

Associations of contractors include just about every imaginable type of construction contractor. Figure I-7 lists only a small number of the largest associations that have apprenticeship programs. Some contractors' associations are formed to represent only nonunion contractors, a few represent only union contractors, and others represent both. Many associations of nonunion contractors were originally formed because the

International Association of Bridge, Structural, Ornamental and Reinforcing Iron Workers (www.ironworkers.org/)

International Association of Heat and Frost Insulators and Asbestos Workers (www.insulators.org/)

International Brotherhood of Boilermakers, Iron Ship Builders, Blacksmiths, Forgers and Helpers (www.boilermakers.org/)

International Brotherhood of Electrical Workers (www.ibew.org/)

International Brotherhood of Teamsters (www.teamster.org/)

International Union of Bricklayers and Allied Craftworkers (www.bacweb.org/)

International Union of Elevator Constructors (www.iuec.org/)

International Union of Operating Engineers (www.iuoe.org/)

International Union of Painters and Allied Trades (www.iupat.org/)

Laborers' International Union of North America (www.liuna.org/)

Operative Plasterers' and Cement Masons' International Association of the United States and Canada (www.opcmia.org/)

Sheet Metal Workers' International Association (www.smwia.org/)

United Association of Journeymen and Apprentices of the Plumbing and Pipefitting Industry of the United States and Canada (www.ua.org/)

United Brotherhood of Carpenters and Joiners of America (www.carpenters.org/)

United Union of Roofers, Waterproofers and Allied Workers (www.unionroofers.com/)

Utility Workers Union of America (www.uwua.org/)

FIGURE I-6: Construction craft unions.

Air Conditioning Contractors of America (http://www.acca.org)
Air Conditioning Heating and Refrigeration Institute (http://www.ahrinet.org/))
Associated Builders and Contractors (http://www.abc.org)
National Association of Home Builders (http://www.nahb.org)
Home Builder's Institute (http://www.hbi.org)
Independent Electrical Contractors Association (http://www.ieci.org)
National Electrical Contractors Association (http://www.necanet.org)
National Utility Contractors Association (http://www.nuca.com)
Plumbing-Heating-Cooling Contractors Association (http://www.phccweb.org)
The Associated General Contractors (AGC) of America (http://www.agc.org)

FIGURE I-7: These are only a few of the largest construction associations.

contractor members felt a need to work together to provide some of the benefits that union contractors receive—such as apprentice training and a lobbying voice in Washington, DC.

Building Codes

Most towns, cities, and counties have building codes. A *building code* is a set of regulations (usually in the form of a book) that ensure that all buildings in that jurisdiction (area covered by a certain government agency) are of safe construction. Building codes specify such things as minimum size and spacing of lumber for wall framing, steepness of stairs, and fire rating of critical components. The local building department enforces the local building codes. States usually have their own building codes, and state codes often require local building codes to be at least as strict as the state code. Most small cities and counties adopt the state code as their own, meaning that the state building code is the one enforced by the local building department.

Until recently three major model codes were published by independent organizations. (A model code is a suggested building code that is intended to be adopted as is or with revisions to become a government's official code.) Each model code was widely used in a different region of the United States. By themselves model codes have no authority. They are simply a model that a government agency can choose to adopt as their own or modify as they see fit. In 2013, the International Code Council published a new model code called the *International Building Code*. They also published the *International Residential Code* to cover home construction (Figure I-8 A and B). Since publication of the first *International Building Code*, states have increasingly adopted it as their building code.

Other than the building code, many codes govern the safe construction of buildings: plumbing codes, fire protection codes, and electrical codes. Most workers on the job site do not need to refer to the codes much during construction. It is the architects and engineers who design the buildings that usually see that the code requirements are covered by their designs. Plumbers and electricians do, however, need to refer to their respective codes frequently. Especially in residential construction, it is common for the plans to indicate where fixtures and outlets are to be located, but the plumbers and electricians must calculate loads and plan their work so it

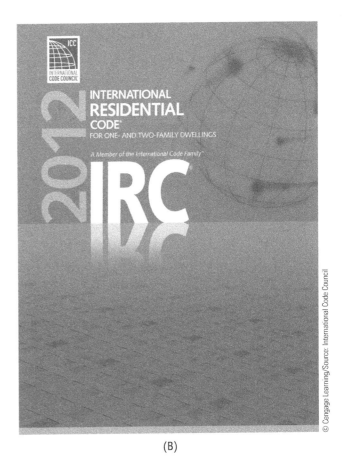

(A)   (B)

FIGURE I-8: (A) 2012 International building code. (B) 2012 International residential code.

meets the requirements of their codes. The electrical and plumbing codes are updated frequently, so the workers in those trades spend a certain amount of their time learning what is new in their codes.

Construction Trends

Every industry has innovations, and construction is no exception. As a construction professional, it is important to be aware of new technologies, new methods, and new ways of thinking about your work. This is as important for a worker's future employment as being aware of safety and ethical business practices. Some of the key technological trends include disaster mitigation, maintenance, building modeling, and energy-efficient or green building.

Disaster Mitigation

Both new and existing buildings need to be strengthened and improved to deal with earthquakes, floods, hurricanes, and tornados. Actions like improving wall bracing or preparing moisture management reduce damage and improve safety when these events occur. These actions are increasingly required by building regulations (especially in disaster-prone areas) and requested by property owners and insurers.

Maintenance

Regardless of the situation or the circumstances, the preservation and upkeep of a structure is the top priority with any facility owner. Besides the initial construction cost, maintenance is the single most expensive reoccurring expense that a facility owner can have through the life cycle of the facility (not counting remodeling). However, when the facility maintenance technician performs preventive maintenance or even scheduled maintenance, the return on investment will more than pay for the cost of the maintenance and the technician. For example, changing the air filter in the air-conditioning systems will not only improve the indoor environment for the occupant but will also extend the life of the air-conditioning equipment. For a commercial facility, the replacement of the air-conditioning equipment could be a significant investment. Because preventive maintenance is such an important industry trend and the primary responsibility of the facility maintenance technician, property owners are more concerned about the costs, effort, and time required to repair and maintain their structures. This has led to improved preventive maintenance techniques as well as significant research into materials that are more durable, construction assemblies that manage moisture, air and elements better, and overall higher-quality construction work. Because maintenance is the primary responsibility of the facility maintenance technician, this has led to the development of more high-quality resources, such as training videos, training classes and programs, manufacturer-based training opportunities, publications, and certification programs.

Building Modeling

One of the newest trends in construction technology doesn't include construction materials at all; it includes being able to design, simulate, and manage buildings with the use of computer and information technology. Some of these tools, such as computer-aided drafting (CAD) and computer-aided manufacturing (CAM), have been around for decades. Others, such as energy modeling and simulation software or project management tools, are being used more often. Still others, such as building information modeling (BIM), are gathering many of these previous tools into single computing platforms. In all cases, the ability to use computers and professional software is becoming mandatory among workers.

Energy-Efficient or Green Building

Probably the most significant trend in the construction industry over the last decade has been energy efficient or *green building*—that is, planning, design, construction, and maintenance practices that try to minimize a building's impact on the environment throughout its use. Although definitions of green building are always evolving, most agree on key concepts that are important.

Occupant Health and Safety

The quality of indoor air is influenced by the kinds of surface paints and sealants that are used as well as the management of moisture in plumbing lines, HVAC equipment, and fixtures. Long-term maintenance and care by homeowners and remodelers also can shape the prevalence of pests, damage, and mold. Builders and remodelers are becoming more aware of the products and assemblies they use that could have an effect on indoor environments.

Water Conservation and Efficiency

Many builders and property owners are attempting to collect, efficiently use, and reuse water in ways that all save the overall amount of water use. From using rainwater collectors to irrigate lawns, to installing low-flow toilets and water-conserving appliances, to feeding used "graywater" from sinks and showers into secondary non-occupant water needs, water efficiency is a trend in energy efficient or green building, but especially where water shortages or droughts are prevalent.

Low-Impact Development

Builders concerned with the effect of the construction site on the land, soils, and underground water are incorporating storm water techniques, foundation and pavement treatments, and landscaping preservation methods to minimize disturbances to the land and surrounding natural environments.

Material Efficiency

Builders are becoming more aware of the amount of waste coming from construction sites, and inefficiency in the amount of materials (like structural members) that they install in buildings. Many of the materials that are used in construction also do not come from naturally renewable sources or from recycled content materials. Using materials from preferred sources, using them wisely, and then appropriately recycling what is left is an industry trend (Figure I-9).

FIGURE I-9: Construction site waste recycling.

Energy Efficiency and Renewable Energy Sources

The most widely known of all energy-efficient or green building trends involves the kind and amount of energy that buildings use. Often, builders can incorporate the use of renewable energy sources (such as solar photovoltaics) or passive solar orientation into their designs. Then, the combination of good building envelope construction and efficient equipment and appliances can reduce utility costs for property owners, much like the maintenance trend reduces repair costs (Figure I-10).

There are many ways to keep track of the latest trends in the construction industry. Trade or company's magazines, online resources and blogs, and the latest research coming out of government and university laboratories are several ways to keep informed and up-to-date on the latest industry trends.

FIGURE I-10: Duct insulation increases system energy efficiency.

Organization of the Facility Maintenance Industry

The general maintenance and repair industry is a large sector of the American economy. According to the U.S. Department of Labor, general maintenance and repair is one of the nation's largest industries, with 1.2 million wage and salary jobs. Around 22.59 percent of these workers are currently employed by the lessors of real estate; 17.67 percent are currently employed with other activities related to real estate. According to the Bureau of Labor Statics, general maintenance and repair workers are employed in almost every industry. Opportunities are available for people to work at all levels in this industry, from those who perform lawn maintenance to the senior technicians who possible have specialty training and licenses. You should be aware of all the opportunities in the facility maintenance industry so that you can make career decisions in the future, even if you are sure of what you want to do at this time.

The Nature of the Facility Maintenance Business

In construction, a craft or trade worker is someone who specializes in a particular building trade (electrical, HVAC, plumbing, etc.); however, a general maintenance and repair worker must have skills in wide range of construction fields. These technicians are responsible for the repair and maintenance of mechanical equipment, and plumbing and electrical systems, as well as the air-conditioning systems. These technicians' duties include, but are not limited to:

- Responding to emergency situations as required
- Inspecting and diagnosing problems and determining the best possible method to correct them
- Troubleshooting and repairing electrical switching and air-conditioning motors
- Adjusting building computer-control settings
- Unclogging drains
- Reading, interpreting, and working from blueprints, plans, drawings, and specifications
- Repairing drywall and building partition walls
- Collaborating with the facility's management and construction contractors to ensure terms of agreements are met and work is completed satisfactorily
- Participating in the development and administration of the building maintenance operating budget
- Participating in the estimation of cost, time and materials of maintenance, repair and renovation work
- Ordering parts and supplies from manufacturers and building supply houses
- Maintaining security and control of the facility tools

- Monitoring the safety and accessibility of the facility and developing work requests to address safety and accessibility issues such as those related to The Americans with Disabilities Act ADA, Occupational Safety and Health Administration (OSHA, city, and local code requirements).

Facilities Maintenance, Repair, and Remodeling

The primary function of a facility maintenance technician is the upkeep of the facility. Normally, this involves maintaining the present state of the facility. As appliances, equipment, and building materials wear, break, and malfunction, it is the job of the facility maintenance technician to restore these building elements to their original condition.

Repair and Facilities Maintenance

Repair can be defined as the act of troubleshooting an issue that relates to a facility and resolving the issue by repairing or replacing the defective building component or appliance. A repair project will not usually involve obtaining a permit; however, you should always verify that a permit is not required. Repair will not have an effect on the property value.

Repair issues that are commonly addressed in facilities maintenance are:

- Dealing with amenities that are worn, consumed, dull, dirty, and/or clogged
- Dealing with amenities that are broken or damaged
- General upkeep and maintenance

New Construction

Typically, the facility maintenance technician is responsible for preservation of the facility. If the scope of the project requires enhancing the facility by adding additional space or changing a particular area, then it is no longer considered repair, but instead it is considered either new construction or remodeling. New construction involves either adding on to the existing structure or building a new structure in place of or next to the existing structure. New construction will require a building permit and will affect the value of the structure.

Remodeling and Facilities Maintenance

Remodeling is typically a renovation that improves the existing structure, but it can sometimes involve new construction. Remodeling will often affect the value of a structure. It is important to note that remodeling projects will sometimes require a remodeling permit. Additional information about remodeling permits can be found in Appendix A: Remodeling and Demolition Tips.

Typically, there are two classifications of remodeling: *commercial* and *residential*. Commercial remodeling projects usually require hiring a general contractor and allowing them to act as the project manager. The general contractor is responsible for obtaining the proper permits and hiring the proper technicians to perform any specialized task that might be necessary.

For residential remodeling projects, there are three different managing techniques that can be employed—hiring a general contractor and allowing that person to act as the project manager, having the facility maintenance technician act as the general contractor and directly hire the specialized contractors, having the facility maintenance technician complete the remodeling project him- or herself (in the areas that do not require special licenses).

Demolition and Deconstruction

Regardless of the type of remodeling that is underway (with the exception of constructing a stand-alone structure), part of the existing structure will have to be removed or demolished. This is typically done by either demolishing or deconstructing part of the area to be remodeled. When *demolition* is involved, little or no attention is given to preserving the elements that are being removed from the structure. This is in contrast to *deconstruction* in which the overall goal is to preserve the elements that are being removed for possible reuse.

Working in the Facility Maintenance Industry

Often success in a career depends more on how people act or how they present themselves to the world than it does on how skilled they are at their job. Most employers would prefer to have a person with modest skills but a great work ethic than a person with great skills but a weak ethic.

Ethics

Ethics are principles of conduct that determine which behaviors are right and wrong. The two aspects of ethics are values and actions. *Values* have to do with what we believe to be right or wrong. We can have a very strong sense of values, knowing the difference between right and wrong, but not act on those values. If we know what is right but we act otherwise, we lack ethics. To be ethical, we must have good values and act accordingly.

We often hear that someone has a great work ethic. That simply means that the person has good ethics in matters pertaining to work. Work ethic is the quality of putting your full effort into your job and striving to do the best job you can. Good work ethics become habits, and the easiest way to develop good work ethics is to consciously practice them.

Working on a Team

Teamwork is critical, as facility maintenance technicians often have to work with other building trades. Regardless of the type of work that is being performed, interaction with other skilled technicians is inevitable. This is especially true during new construction.

Constructing a building is not a job for one person acting alone. The work at the site requires cooperative effort by carpenters, masons, plumbers, painters, electricians, and others. Usually several workers from each of these trades collaborate. A construction project without teamwork can experience many problems. For example, one carpenter's work might not match up with another carpenter's work. There could be too much of some materials and not enough of others. Walls may be enclosed before the electrician runs the wiring in them.

Teamwork is very important on a construction site, but what does being a team player on a construction team mean? Effective team members have the best interests of the whole team at heart. Each team member has to carry his or her own load, but it goes beyond that. Sometimes a team member might have to carry more than his or her own load, just because that is what is best for the team. If you are installing electrical boxes and the plumber says one of your boxes is in the way of a pipe, it might be in the best interests of the project to move the electrical box. That would mean you would have to undo work you had just completed and then redo it. It is, after all, a lot easier to relocate an outlet box than to reroute a sink drain.

The following are six traits of an effective team:

- *Listening*: Team members listen to one another's ideas. They build on teammates' ideas.
- *Questioning*: Team members ask one another sincere questions.
- *Respect*: Team members respect one another's opinions. They encourage and support the ideas of others.
- *Helping*: Team members help one another.
- *Sharing*: Team members offer ideas to one another and tell one another what they have learned.
- *Participation*: Team members contribute ideas, discuss them, and play an active role together in projects.

Education and Training for Facility Maintenance

Whereas in the past many general maintenance and repair technicians acquired their skills from on the job training, today facility managers and owners are requiring their facility maintenance technicians to complete training in either a certification program and/or postsecondary educational program. In addition to participating in postsecondary education, facility maintenance technicians can obtain resources through the Environmental Protection Agency (EPA) as well as from other government agencies.

Lifelong Learning

Lifelong learning refers to the idea that we all need to continue to learn throughout our entire lives. We have greater opportunities to learn and greater opportunities to move up a career ladder today. Our lives are filled with technology, innovative new materials, and new opportunities. People change not only jobs, but entire careers several times during their working life. Those workers who do not understand the new technology in the workplace, along with those who do not keep up with the changes in how their company is managed, are destined to fall behind economically. There is little room in a fast-paced company of this century for a person whose knowledge and skills are not growing as fast as the company or the industry. To keep up with new information and to develop new skills for the changing workplace, everyone must continue to learn throughout life.

Job Outlook

Employment for facility maintenance technicians is expected to increase approximately 11 percent from now through the year 2018, according to the U.S. Department of Labor Bureau of Labor Statistics. As the population and the economy grow, more facility maintenance technicians will be needed to maintain office and apartment buildings, stores, schools, hospitals, hotels, and factories.

The average hourly wage for all facility maintenance technician as of 2011 (the newest data available) was $17.75. The lowest 10 percent earned $10.01, and the highest 10 percent earned $27.30. Facility maintenance technicians may earn more or less per hour, depending on a number of factors such as experience, education, licenses, and worker demand.

Looking Forward

Before you can start a career in the residential construction industry as facility maintenance technician, you must acquire the knowledge, skills, and work attitudes required for the job. This textbook covers all of the areas needed for you to become proficient as facility maintenance. Your job now is to be a student and study the material presented in this textbook to the best of your ability. Your instructor will guide you through the process. To a large degree, how well you do your job now as a student will determine how well you eventually do your job as a facilities maintenance technician in the residential construction industry.

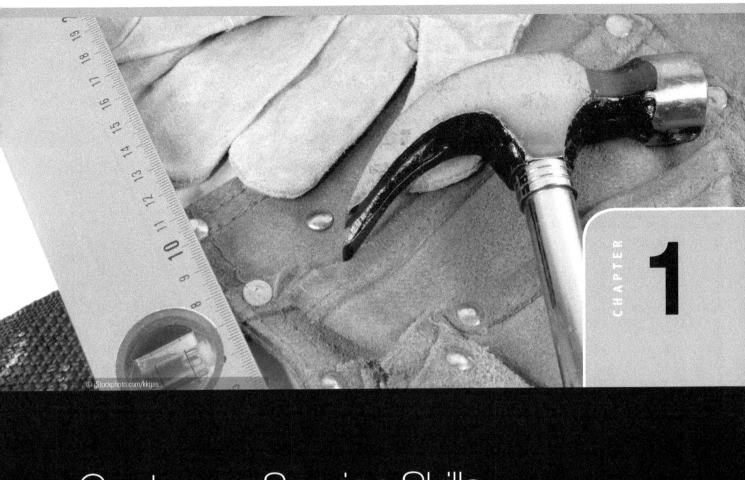

Customer Service Skills

OBJECTIVES

By the end of this chapter, you will be able to:

KNOWLEDGE-BASED

- List the attributes of great service.
- Identify personal strategies for connecting with the people you are providing service to.

- Explain the importance of understanding the needs of the people you are providing services to and their expectations.

Confidence having a belief in yourself and your abilities

Competence having the skills, knowledge, ability, or qualifications to complete a task

Appreciation the expression of gratitude toward your customers

Self-talk what you say silently to yourself as you go through the day or when you are faced with difficult situations

Empathy the capacity to understand your customers' state of mind or emotion

Honesty acting truthfully with your customers

Reliability the quality of being dependable

Courtesy acting respectful toward your customers

Introduction

Great service starts with a great attitude about your job, yourself, and the people you are providing service to. The purpose of this chapter is to highlight some important characteristics that lead to an excellent customer service attitude. The facilities maintenance technician will also be introduced to the importance of treating work projects as team projects and exploring the many characteristics required to be an effective team member. Excellent customer service allows service workers to meet the needs of the people they are providing service to in the most pleasant and efficient way possible.

An Excellent Customer Service Attitude

Attitude really is *everything* when providing service. It sets the stage for all other actions. A positive and service-oriented attitude can help you overcome many problems, dissatisfactions, and mistakes. It can also help in building long-term and loyal relationships with coworkers and customers.

Attitude is also a key factor in developing first impressions. It tells how you feel about yourself, your job, and the customers. It reveals confidence in doing an excellent job or reflects a sense of competence.

Finally, attitude is contagious, not only to other employees or contractors with whom you work, but also to the customers themselves. Remember that customers can have their own problems (including attitude problems), and you have an opportunity to improve the customer's day with your positive outlook.

Attitude constitutes several elements: *confidence, competence, appreciation, empathy, honesty, reliability, responsiveness, patience, open-mindedness,* and *courtesy*. If presented properly, these elements ensure others that you are capable of doing what you say you can do. Let's look at each of these elements independently.

Confidence

Do you project confidence that you can do your job, solve a problem, or find the information necessary to achieve the customer's goals? What do the people you work with think? Do they have confidence that you can do the job?

A high level of **confidence** lowers the stress and anxiety in those around you. Lowered stress and anxiety leads to calm and rational thinking, better relationships, and improved patience, all of which are key ingredients to a pleasant and productive work environment.

Confidence is fostered by several behaviors:

- Know how to do your job well.
- Know that the services you provide and represent will meet the customer's needs and expectations.
- If you do not know how to do something or do not have sufficient knowledge, learn it and practice your skills often.
- Believe in your own abilities to complete the task, solve the problem, or find needed information.
- Control your self-talk so that you treat yourself (and others) in a positive way.

Competence

Are you skilled in the tasks you are asked to do? Do you have the appropriate knowledge? Are you efficient so that you can do the tasks quickly and with ease? If you have skills that need improvement, or do you know where to go for help? How will you improve your **competence** throughout your career? If you were asked to develop new skills, how would you go about doing this?

You must know how to perform well the tasks of your job. You must understand fully the services you are presented with. You must keep current and know where to find information quickly and how to effectively communicate that information to others. Being highly competent tends to increase confidence. The more you know about what you are doing, the better service you can offer your customers.

Appreciation

Do you appreciate your customers? How do you show your appreciation to others? Do you appreciate the skills and know-how you have developed in yourself?

Appreciation is a mindset established through self-talk. **Self-talk** is what you say silently to yourself as you go through the day or when you are faced with difficult situations. Self-talk includes all the things you are saying to yourself as you work with a customer and complete your tasks. What you are saying to yourself is always reflected in some type of behavior—a tone of voice, an action, or an easily perceived attitude—to your customers. You choose what you think about your customer.

> **Negative or condescending self-talk:** *She has no clue about all the things this system can do! What an idiot!*

Positive self-talk: *This is a great opportunity to help her learn about all the things this system can do! Part of what she bought was the training; she is paying for it every bit as much as she paid for the TV. We owe it to her. She is going to be so glad she bought it.*

The kind of self-talk you choose is just that—your choice. Positive self-talk sets you up for a positive attitude followed by positive words and actions. Negative self-talk does just the opposite. It sets you up for a negative attitude followed by damaging words and ineffective actions.

Appreciation also includes how you feel about yourself and the efforts you have made to be successful. Self-talk applies here, too. With negative self-talk, you set yourself up for failure. Positive self-talk can provide just the right amount of confidence and motivation to succeed.

Negative self-talk: *I have no clue how to fix this problem. I might as well give up now before I waste any more time.*

Positive self-talk: *I haven't seen this problem before. This will be a good opportunity to test my skills. Let's see what I can find out.*

Once you make a habit of believing in yourself and your abilities, words and actions will follow accordingly. You will see that you actually find ways to succeed.

Empathy

Do you listen to your customers with empathy and truly want to help them meet their needs? How do you show empathy to those around you? What are the effects when you show empathy?

Empathy does not mean agreeing with everything your customer says. Nor does it mean promising to do everything your customer wants. **Empathy** does mean understanding what your customer's needs are and where they are coming from. Empathy is putting yourself in the other person's shoes, so to speak. Whether you agree with them, customers do have a right to express their opinions and points of view. They also have a right to be heard. Empathy is realizing this fact and acting and communicating accordingly.

The first step in empathy is to listen carefully to the customer without forming an opinion or making a judgment based on your own point of view. Taking turns to speak is not really listening. It is just being polite. Listening means that you do whatever you need to understand the customer's concerns, problems, questions, and points of view. Listening to a customer does two important things: (1) it gives you the information you need to solve the problem or meet the need and (2) it affirms the customer that he or she has been heard. This affirmation is critical and will go a long way toward building trust in you. Once the customer trusts you, you then have the opportunity to address the issue or complete your job successfully. You might be able to provide more information to help the customer fix the problem. You may change the customer's viewpoint by showing that the customer didn't really want what he or she thought in the first place. You may simplify what was initially perceived as a complicated issue. Basically, you must listen to the other person before you can expect him or her to listen to you.

Empathy is important in everyday dealings with customers, but it is especially important when the customer is frustrated, angry, or upset. When people are upset, their ability to think clearly or logically is often diminished. They often say irrational things, demand unreasonable responses, or behave inappropriately. Empathy diffuses emotion. When an angry customer experiences empathy from an employee, the typical response is to calm down and return to rational and reasonable thinking. With rational and reasonable thinking, most problems can be solved satisfactorily.

Some common phrases that show empathy include:

Wow! I would be mad too if the product broke in the first week.

This problem is definitely an inconvenience. Let's see how to fix this as quickly as possible.

I know it is frustrating when there is this much of a delay. Let me explain why our process is so important.

Be specific in your statements and try to avoid being general. Identify the emotion that you think the customer is feeling. Try to also identify the source of the emotion specifically.

A word of warning: If you try to show empathy without sincerity, you only make matters worse. Your customers are not stupid. They will see through your insincerity and feel patronized.

Honesty

Are you honest with your customers, and do your words and actions convey honesty? Do you strive to build trusting relationships with your customers? How specifically do you try to be honest with your customers? What are the consequences of dishonesty?

Honesty is an essential element to success whether you are the CEO or the lowest rung of the ladder. Also trusting relationships with customers foster customer loyalty and ongoing business.

Customers need to trust that you will be honest with them and that you are providing them with the right information. Dishonesty is quite easy to detect. It comes through clearly in words, tone, and body language, not to mention in customer dissatisfaction eventually when the dishonesty is found out—and dishonesty is *always* found out sooner or later.

To be honest, you must say the truth, follow through on promises, and state that you do not know the answer when appropriate. Customers will quickly determine that you are trustworthy and will continue to give you their business, often even when you cannot fix a problem or meet the immediate need. Also they will tell others about how they were treated.

You can get into "honesty" problems in many different ways. The following list reflects only a few examples that you might have experienced yourself.

- Promising that a product will meet a need when you know it won't
- Underestimating a wait time for service when you know it will be longer
- Stating that a product will be available on time when you know delivery has been delayed
- Underestimating the cost of a repair when you know that it will cost more

There are some basic tips that will help to ensure that the customer sees you as trustworthy:

- Overestimate wait times. If you serve the customer sooner than expected, the customer will be thrilled.

- Never promise you can deliver anything that you are not sure. If necessary, tell the customer you need to check your facts with others and get back to them (and then get back to them when you say you will).

- Don't tell a customer what he or she wants to hear, unless it is the truth. Instead, tell him or her the facts and work on solving any problems.

- Explain your actions thoroughly and in terms your customer can understand. Sometimes a customer may distrust you simply because he or she does not have enough information to know that you are doing, what you are supposed to be doing, or even if the information you are dispensing is factual and accurate. Explain yourself in terms the customer can understand. But never talk down to them.

- Present yourself as working on behalf of the customer. For example, address the customer's needs and offer information or make recommendations related to those needs.

- Provide advantages and disadvantages of a product or service option so that the customer can make informed decisions.

Reliability

How do you relate yourself in terms of reliability? Do you *always* do what you say you will do? Can customers depend on you and your products and services to meet their needs? Are you regularly late, or do you typically run on time? How important is being punctual to you?

Reliability is essential to your customers. Customers want services they can depend on. Basically, **reliability** means doing what you say you will do and when you will do it. If you have scheduled a service call for 9 AM, then show up at 9 AM (not before, unless you call to ensure it is convenient, and not after). Obviously, things happen that can be out of your control. In these cases, it is important to notify customers accordingly. It is also important to do everything in your power to control the situations. Good planning, accurate estimation of a project's time needs and travel time, and organization all help in this control.

Reliability is also reflected in availability, prompt replies, quick follow-up, and fast work. It implies accountability for actions and any potential problems. In order to be accountable, you must be available. Can customers contact you? Can your coworkers or office contact you easily? If a message must be left, do you respond quickly?

Keys to reliability are listed here. Add your own keys as they relate to your specific job.

- Ensure that customers and coworkers can contact you.

- If a message must be left, check your messages often and respond immediately.

- If you are out of touch for a specific time, let people know, so they won't be disappointed by your lack of response.

- Use an alarm to remind you of important meetings. (Some watches and cell phones have this feature.)
- Show up on time—not before or after the agreed-upon time, unless it has been approved ahead of time.
- Learn to accurately estimate how long it will take to complete a task, receive a part, schedule an appointment, and so on. Don't guess; instead, wait until you have all the information before estimating time.
- Learn how to do the job right to begin with. Reliability also means being able to depend on the quality of work.

Responsiveness

Do you respond to your customers quickly, accurately, and with the goal of meeting their needs and answering their questions? How do you show your customers that you are responsive? How you respond to the customer typically determines how well your service is received, which ultimately translates to either customer satisfaction and loyalty or customer dissatisfaction.

There is nothing more frustrating than being passed from one person to another without getting what you need. Yet, this is often the experience of customers in many businesses. For good service, responsiveness is on the top of the list of key elements.

Patience

Do you have patience when dealing with the customer and/or solving a problem? People that are considered to be good natured or tolerance of delay or incompetence is considered to have patience. It is extremely important to exercise patience when dealing with a customer. Remember without the customer you would not have a job.

Open-Mindedness

Do you have open-mindedness? People that have the ability to consider different opinions or ideals are considered to be open-minded. Being open-minded does not, however, mean that all opinions or ideals are correct for a particular problem, but having the ability to listen to the customers ideals and opinions will help strengthen the technician/customer relationship.

Courtesy

What does it mean when someone is courteous? Are you courteous when you deal with others? How specifically do you demonstrate courtesy?

Courtesy sends a positive and powerful message to customers, whether they are your external paying customers or those internal customers with whom you work on a daily basis. Courtesy is also a habit that, once formed, becomes second nature.

Characteristics of a courteous employee are reflected in the following behaviors:

- Saying "please," "thank you," and "you're welcome"
- Responding with "yes, ma'am" or "no, sir"

- Saying "I'm sorry" or "excuse me"
- Addressing people by their names and using Mr., Ms., or Miss as appropriate (e.g., if you do not know them well)
- Saying "yes" instead of "yeah"
- Being friendly
- Smiling often
- Opening doors and allowing others to go through first
- Introducing yourself to new people
- Being attentive and focusing on the person in front of you without being distracted
- Using appropriate language

Courtesy also implies sincerity. Show that you sincerely appreciate your customers by thanking them for their business and being specific.

> For example: *Thank you for buying our entertainment system. I know you are going to love it. We really appreciate your business.*

Remember your customers' names because you honestly feel that they are important enough to do so. Be sincere in all of your actions. Your sincerity—or lack of sincerity—will be obvious to your customers.

It is more natural for people to forget common courtesies when stressed—for example, when a customer is angry or when you are frustrated because of a problem situation. These are the times when you need to be exceptionally courteous. You should be courteous even when you feel that the person does not deserve it and when customers are not being courteous themselves. This is why developing the habit of being courteous is so important. If courtesy is a habit, you are less likely to forget about it when stressed.

Tone of Voice

Sometimes how you say something means as much as what you say. The wrong tone can cause a misinterpretation of your words. Saying "thank you" in an angry tone serves only to agitate your customer. Asking about the problem in a disinterested tone shows the customer that you are not sincere. Using sarcasm typically causes a customer to become angry and feel disrespected.

Combining a positive, friendly, and confident tone with positive and confident words such as "absolutely," "definitely," "not a problem," and so forth can be very effective. Say the following phrases with a positive, friendly, and confident tone to get the point:

> *Absolutely. I can have this fixed in no time!*
> *Definitely. I will order the part for you today.*
> *Not a problem. I will reschedule the service call for that date.*
> *Yes! I will be happy to move this for you.*

Try to match your tone to the customer's needs. If the customer is in a hurry, then make your tone urgent and energetic. If the customer is frustrated, use a confident and helpful tone. If your customer is doubtful or has many questions, use a reassuring and confident tone.

Listening

There is a significant difference between listening and hearing. To listen means that you truly attempt to understand what the speaker is saying. Real listening is a highly active process. Without listening, there is no communication—but only speaking and hearing.

Effective listening does several things for the relationship:

- It shows that you sincerely care about the customer and the customer's needs.
- It demonstrates attentiveness to the customer.
- It allows you to gain critical information with which to complete your task successfully or to solve the problem.
- For frustrated customers, it reduces irritation by ensuring them that they are being heard.
- It fosters an effective and productive relationship between you and the customer.

Avoid the Words "I Can't"

Focus on what you can do rather than what you cannot do. If you cannot do exactly what the customer wants, explain what you *can* do for the customer that either comes close to the customer's request or meets the same need in a different way. Be a problem solver.

To connect with customers, develop your skills in the following strategies and then practice them consistently:

- Evaluate your body language and use it to convey the appropriate messages.
- Focus on how you say things and your tone of voice.
- Be attentive to customers.
- Develop effective listening skills and practice these with customers.
- Respond positively to customers.

SUMMARY

- Great service starts with a great attitude about your job, yourself, and the people you are providing service to.
- Attitude includes several elements: confidence, competence, appreciation, empathy, honesty, reliability, responsiveness, patience, open-mindedness, and courtesy.
- Confidence is having a belief in yourself and your abilities
- Appreciation is a mindset established through what you say silently to yourself as you go through the day or when you are faced with difficult situations.

- Empathy is identifying with and understanding what your customer's needs are and where they are coming from.
- Honesty is an essential element to success. Building a trusting relationship with customers fosters customer loyalty and ongoing business.
- Reliability is essential to your customers. How you respond to the customer typically determines how well your service is received.
- It is extremely important to exercise patience when dealing with a customer.
- Courtesy sends a positive and powerful message to customers, whether they are your external paying customers or those internal customers with whom you work on a daily basis.
- The wrong tone can cause a misinterpretation of your words. Asking about the problem in a disinterested tone shows the customer that you are not sincere.
- Listening means that you truly attempt to understand what the speaker is saying.

REVIEW QUESTIONS

Define the following attitudes:

1   **Confidence**

2   **Competence**

3   **Appreciation**

4   **Empathy**

5   **Honesty**

6   **Reliability**

© iStockphoto.com/icphoto

REVIEW QUESTIONS

7 **Responsiveness**

9 **Open-mindedness**

8 **Courtesy**

10 **Patience**

Customer Service

CUSTOMER SERVICE CHECKLIST

One of the best ways to improve your customer service skills is to practice. No one is born with great customer service skills; instead one must acquire them over time through practice. In this job sheet, you and several other classmates will practice customer service skills on one another. Using the checklist provided below, evaluate each other's performance as one student acts as a technician and another acts as the customer.

In this simulation you are dealing with a disgruntled customer who needs a dishwasher repaired immediately and can't wait.

Simulation #1

Customer Service Checklist		
Customer Service Skill	Effectiveness (Scale 1–10)	Comment
Responsiveness		
Open-mindedness		
Patience		
Reliability		
Honesty		
Empathy		
Appreciation		
Competence		
Confidence		
Courtesy		
Tone of Voice		
Listening		

In this simulation you are running late for an appointment and must communicate to the customer that you are running late.

Simulation #2

Customer Service Checklist		
Customer Service Skill	Effectiveness (Scale 1–10)	Comment
Responsiveness		
Open-mindedness		
Patience		
Reliability		
Honesty		
Empathy		
Appreciation		
Competence		
Confidence		
Courtesy		
Tone of Voice		
Listening		

In this simulation you have to explain to the customer that the reason the dishwasher is clogged is that the customer is not scraping off their dishes before loading and starting the dishwasher.

Simulation #3

Customer Service Checklist		
Customer Service Skill	Effectiveness (Scale 1–10)	Comment
Responsiveness		
Open-mindedness		
Patience		
Reliability		
Honesty		
Empathy		
Appreciation		
Competence		
Confidence		
Courtesy		
Tone of Voice		
Listening		

In this simulation you have to explain to the customer that running over an automatic sprinkler with a mower is not covered under warranty.

Simulation #4

Customer Service Checklist		
Customer Service Skill	**Effectiveness (Scale 1–10)**	**Comment**
Responsiveness		
Open-mindedness		
Patience		
Reliability		
Honesty		
Empathy		
Appreciation		
Competence		
Confidence		
Courtesy		
Tone of Voice		
Listening		

INSTRUCTOR'S RESPONSE:

_____
_____
_____

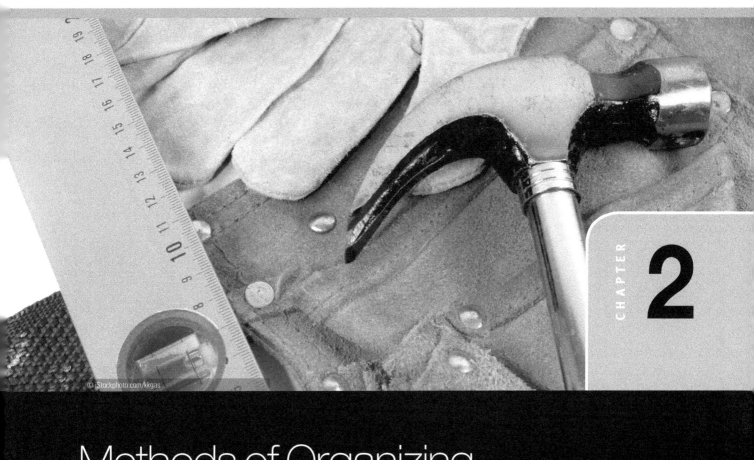
© iStockphoto.com/kkgas

Methods of Organizing, Troubleshooting, and Problem Solving

OBJECTIVES

By the end of this chapter, you will be able to:

SKILL-BASED

- Establish priority of work tasks.
- Assign tasks.

- Carry out work order systems.
- Using the steps outlined in the text to properly trouble a technical issue.

Task an activity that needs to be performed to complete a project

Priority giving a task precedence over others

Assigning tasks giving a task to someone to complete

Troubleshooting the process of performing a systematic search for a resolution to a technical problem

Diagnostics tests used in the process of determining a malfunction

Introduction

To successfully complete a project regardless of its complexity or nature, it must first be divided into tasks that can be assigned and the results measured. This is irrespective of whether it is a new installation or replacement or repair of a defective piece of equipment. The only difference is that, if the project consists of replacing or repairing equipment, then removal and/or troubleshooting will be included as a task to be performed.

Time Management

Time management is the act of controlling the amount of time spent on a particular activity for the sake of improving productivity. When time management is properly executed, stress levels are reduced and productivity is increased. Effective time management for most people does not come naturally, but instead it is a skill that is learned. In fact, time management can be achieved by:

- Planning each day
- Prioritizing your tasks
- Not taking on unnecessary tasks
- Taking the time necessary to perform a high-quality task
- Breaking up larger tasks into smaller tasks
- Limiting unnecessary distractions

Establish Priority of Work Tasks

Traditional wisdom about setting priorities promises you higher productivity and a greater sense of accomplishment. All you have to do is write out a to-do list, prioritize it by the order of importance and urgency (using the ubiquitous A, B, and C labels), and then tackle it, right? Then why, after a period time, is that same C item still on your to-do list? Also why, after a busy day of completing **tasks**, do you still find yourself saying, "I didn't get anything done today"?

Two reasons may exist for this:

1. Priorities changed during the day, but for good reason. You may not have accomplished A, B, or C on your to-do list, but you did respond appropriately to the additional tasks that you were presented with that day.

2. You may have fallen into the "ACT, then THINK" method of setting priorities. To prevent yourself from falling into this method of setting **priorities**, understand three common priority-setting traps and how you can avoid them.

   a. **Whatever hits first**—Do you "choose" your priorities simply by responding to things as they happen? If so, your priorities are really choosing you. Think about how this general lack of control over your day contributes to your stress level. You need to clarify your priorities by determining each task's importance and level of urgency (i.e., "THINK, then ACT"). This means negotiating with people to respond in a time frame that's convenient to you and agreeable to them.

   b. **Path of least resistance**—When was the last time you heard yourself say, "It's just easier to do it myself"? This is not always an incorrect assumption, but if you're saying it too often, you're probably not giving the other workers enough credit or you have the wrong person working for you. Ask yourself these questions: Am I trying to avoid conflict? Does the task at hand require more expertise than the other workers have? Should time or money be invested to train someone to take on some of the lower-priority tasks I am currently performing? Answers will help you determine what alternative action you need to take.

   c. **The squeaky wheel**—In most situations, it is not hard to identify who the squeaky wheels are. Their requests are always urgent and need to be done right away. Usually, you do the work on their time frame. Unless the request is really urgent, give them a specific time or date when they can expect you to complete the task. Eventually they'll understand that their requests to complete tasks need to be prioritized with all of your own prioritized tasks.

There are numerous systems and methods that can be used to assist individuals with setting priorities; however, one approach taken by most facility maintenance technicians is the "must do, should do, and nice to do" method. In this method prioritization, priorities are places task into one of the three different categories and addressed accordingly.

Assign Tasks

Assigning tasks isn't just a matter of telling someone else what to do. There is a wide range of responsibilities that you can assign to a person along with a task. The more experienced and reliable the person is, the more unsupervised tasks that can be assigned to the person. The more critical the task is, the more cautious you need to be when assigning tasks. It is important that each worker understand his or her part in a job and can perform the assigned task. If a worker does not have the ability to complete an assigned task, as a supervisor or manager, you have to assign that task to someone else who can do it.

Before assigning a task to a worker, consider the following aspects of the job the worker will do:

- What hazards are in the workplace environment or around the worker?
- Are there special work situations that come up which could lead to new risks for this worker? For example, are there risks that might be encountered outside the normal work area? Just once a week? During a task to fetch materials?
- Are there occasional risks from coworkers, as in welding or machining, that could affect the workers nearby?
- In slow periods, workers might be asked to "help out" other employees. Ensure that any hazards associated with those jobs are reviewed with the worker, by both you and the coworker who will supervise those tasks.

Ensure that you communicate with the worker about the job tasks clearly and frequently, repeating and confirming this training over the first few weeks of work. Some workers are overwhelmed with instructions at first and may need to hear this information repeated more than once. Also:

- Inform workers not to perform any task until they have been properly trained.
- Inform workers that if they don't know or if they are unsure of something, they need to ask someone first. Get them to think in a safety-minded way about all their work.

Use Work Order Systems

Once tasks have been assigned to the appropriate workers, the tasks will need to be completed. Following are suggestions on how to document the completed tasks:

1. Develop a progress report for the current week. This report should contain the following information:
   a. Work accomplished: Document the tasks accomplished during the previous week. They should be specific in nature.
   b. Major findings: Document any issues that were encountered when a specific task was dealt with.
   c. Worker: In the documentation, record the name of the person who completed the specific task.
   d. Estimated hours to complete: This is used to identify the worker's effectiveness in task estimation. The goal is to compare the predicted value (estimated the week before) with the actual number of hours it took to accomplish a specific task.
   e. Actual hours to complete: Record the actual number of hours the worker took to complete a specific task.

2. Plan items for the following week. The report should contain the following information:
   a. Work items for next week: List all the tasks that the worker plans to accomplish for the following week.
   b. Worker: List the worker who is responsible for completing the listed task.
   c. Estimated hours to complete: For planning purposes, the estimated hours to completion should be identified to show how long it is expected to take to actually complete the task.

The Troubleshooting Process

Some consider **troubleshooting** to be an art form; whereas, others consider it to be science. Actually it is a combination of art and science—trying to solve a problem using a pure scientific approach may not always work. In any case, some steps can be followed (especially by new technicians) to assist in the development of troubleshooting skills. One of the most important of these steps—establishing a good rapport with the customer—was briefly discussed in Chapter 1. Often what the customer perceives as the ultimate problem is nothing more than a symptom. Often when the technician establishes a good rapport with the customer, he or she can easily get the actual information needed to determine the actual problem. However, as mentioned earlier, rapport is only a small portion of the troubleshooting process. Actually two major phases can be associated with the troubleshooting process: the identification and the repair processes. The identification process is more than just shining a flashlight into a boiler or heat exchanger and trying to spot a defect or a malfunctioning part. This process can be divided into several key phases or steps as follows:

1. Gathering information
2. Verifying the issue
3. Looking for quick fixes
4. Performing the appropriate **diagnostics** (tests used in the process of determining a malfunction)
5. Using additional resources to research the issue (if necessary)
6. Escalating the issue (if necessary)
7. Completing the repair process

Gathering Information

When starting to troubleshoot a system, the first step is to gather the information necessary to correctly identify the problem. This is often done by simply asking the customer a few simple questions. However, when questioning the customer, keep two general rules in mind. They are:

1. Start with open questions such as "What is the issue?" Open questions cannot be answered with a "yes" or "no."
   a. When did the issue first occur?
   b. Is the issue reoccurring?
   c. What exactly where you doing when the issue first started?
2. Let the customer explain in his or her own words what he or she has experienced/ is experiencing. *Never interrupt* a customer or add comments to what he or she is telling you.

Verifying the Issue

As stated earlier, the situation or problem that the customer describes is often not the actual problem. Therefore, always verify whether the problem described by the customer is the actual problem of the system and not just a symptom.

Looking for Quick Fixes

Although in many cases the actual fix is more involved than simply resetting or changing the battery in a thermostat, there are still cases in which the simplest and/or most obvious fix corrects the problem. For example, suppose that you were called to look at a customer's gas central heating unit because it would not ignite. If the customer was using an electronically controlled thermostat, it might be useful to check its battery before starting to break the furnace down.

Performing the Appropriate Diagnostics

If the quick fix does not resolve the issue, then you will need to perform a more thorough diagnostics. Often equipment manufacturers will supply troubleshooting charts and information to help diagnose the equipment.

Using Additional Resources to Research the Issue

If you have never encountered a problem like the one that is currently before you, and you are having trouble locating the issue, don't be afraid to go to the Internet, a distributor, technical support, or even a colleague to help resolve the issue.

Escalating the Issue

If you are working for a large company and continue to have trouble locating and correcting the problem, the issue can often be escalated to a service manager. If you are self-employed or working for a small company and you encounter a problem that you cannot resolve, the equipment manufacturer can often be of assistance. In any case, though, you should never escalate an issue unless you are truly stumped.

Complete the Repair Process

Once the issue has been correctly identified, the repair process can proceed. Like the identification process, the repair process involves several steps. They are:

1. Repairing or replacing the faulty item and/or equipment
2. Testing the system thoroughly to verify that the repair actually corrected the issue
3. Educating the homeowner about the nature of the problem and the action(s) taken to correct it
4. Completing all administrative paperwork

Verifying that the Repair Actually Corrected the Issue

Verifying that the repair actually corrected the issue is one of the most critical steps in the repair process. Never leave a customer site without first testing the repair and/or installation to confirm that you actually corrected the problem.

Educating the Customer about the Nature of the Problem and Actions Taken to Correct It

Always show the customer the worn and replaced parts and explain to him or her why the old parts are defective. If a customer understands the issue and the corrective action taken, he or she is less likely to become dissatisfied with the repair job.

Completing All Administrative Paperwork

Completing the paperwork is especially important when dealing with warranty work. If the necessary paperwork is not completed correctly and on time, then there will be a delay in the service company receiving its payment. In some cases, the claim may even be denied.

Solving a Technical Problem

When solving a technical problem, the facility maintenance technician doesn't start by pulling out a calculator and entering numbers; there is a systemic approach that must be taken. This approach starts by defining and then researching the problem. Once sufficient information has been ascertained, the facility maintenance technician can start determining all possible solutions to the problem. The steps required for successful problem solving are listed next and can be executed in order with the exception of Steps 1, 2, 7, and 8.

Step 1. State the problem
Step 2. List unknown variables
Step 3. List what is given
Step 4. Create diagrams
Step 5. List all formulas
Step 6. List assumptions
Step 7. Perform all necessary calculations
Step 8. Check answers

The following example illustrates each step of this process. Note that not every step may apply to all situations; however, the overall concept is still the same. In this example, you will calculate the amount of concrete necessary to pour a slab for the condenser shown in Figure 2-1. The cooling tower is 8 ft × 5 ft and requires a 1 ft × 6 in. overhang around the perimeter.

Step 1 State the Problem

In this step a statement of the problem is created. The statement should be kept as simple and direct as possible. There is no need to list every detail about the problem, just the key points. Additional information can be added later if necessary. A statement for the problem shown in Figure 2-1 might read:

Statement:
For a 8 ft × 5 ft condensing unit, calculate the amount of concrete required for a slab having a 1 ft × 6 in. overhang and 6 in. thick.

FIGURE 2-1: Cooling tower.

Step 2 List Unknown Variables

Create a list of all the unknown attributes for the stated problem. Even though the problem is in its infancy, some calculations might have to be performed to determine some of the missing information. Be sure to provide ample space for any future elements that might require calculations. Therefore, the list should look like this:

Find:
Amount of concrete needed in cubic feet.
Amount of concrete needed in cubic yards.

Step 3 List What Is Given

A list of all known parameters associated with the stated problem is next created:

Given:
Condensing unit width = 5 ft
Condensing unit length = 8 ft
Required overhang = 1 ft × 6 in.

Step 4 Create Diagrams

Often the best way to determine exactly what is going on in a problem is to make a simple sketch. The sketch should be void of any unnecessary details that might hinder the interpretation of the actual problem (see Figure 2-2).

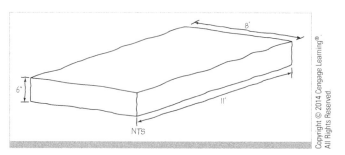

FIGURE 2-2: Rough hand sketch of concrete pad.

Step 5 List All Formulas

In this step, a list is created of all the formulas that will be used to solve the problem, as well as a source reference for each formula. This list will serve as a guide that will facilitate checking the final answer. When a technical solution is validated, all stages of the solution are verified, including formulas and their connotation. To solve the sample problem, the following formulas will be used:

Formulas:
Feet = inches/12
Volume = length $\times$ width $\times$ thickness (height)
Cubic yards = cubic feet/27

Step 6 List Assumptions

Often when solving a technical problem it is necessary to assume some of the details of the project. Suppose that an engineer is calculating the amount of heat that is transferred via conduction through the exterior walls of a building to its surroundings. Although the average outside temperature can be obtained from the *ASHRAE Fundamentals Handbook* for all major cities in the United States, it may be necessary to assume a temperature if the area where the building is located is not listed. In the case of the sample problem, we do not have any assumptions. Therefore, we would list N/A for our assumptions or leave this completely off.

Assumptions:
N/A

Step 7 Perform All Necessary Calculations

The necessary calculations are carried out. Solving the sample problem would yield the following calculations.

Converting the thickness from inches to feet

Feet = inches/12
Feet = 6 in./12
Feet = 0.5

Determining the volume of the slab (cubic feet)

$Volume_{cubic\ feet}$ = length $\times$ width $\times$ thickness
$Volume_{cubic\ feet}$ = 11 ft $\times$ 8 ft $\times$ 0.5 ft
$Volume_{cubic\ feet}$ = 44 cubic feet

Converting the volume from cubic feet into cubic yards

$Volume_{cubic\ yards}$ = $volume_{cubic\ feet}$/27
$Volume_{cubic\ yards}$ = $44_{cubic\ feet}$/27
$Volume_{cubic\ yards}$ = 1.63

Step 8 Check Answers

Any time a calculation is made it must be checked for accuracy.

Check:
To check our work for this sample problem, we will work the problem in reverse.

Convert from cubic yards to cubic feet

$1.62_{\text{cubic yards}} \times 27 = \text{volume}_{\text{cubic feet}}$
$44 = \text{volume cubic feet}$

Find the thickness of the slab If we use two of the known dimensions of the slab, we should be able to find the remaining dimension. For example, if we divide the volume$_{\text{cubic feet}}$ by the length and then again by the width, the remaining portion should be the thickness of the slab.

First:

      Unknown = total cubic feet/length of slab

      Unknown = 44 cubic feet/11 feet

      Unknown = 4 square feet

Second:

      Thickness = unknown (from step #1)/width of slab

      Thickness = 4 square feet/8 ft

      Thickness = 0.5 ft

Or

      Thickness (in.) = 0.5 ft $\times$ 12 in.

      Thickness (in.) = 6 in.

SUMMARY

- To successfully complete a project, regardless of its complexity or nature, it must first be divided into tasks that can be assigned and the results measured.

- Time management is the act of controlling the amount of time spent on a particular activity for the sake of improving productivity.

- To assist individuals in setting priorities, one approach many facility maintenance technicians use is "must do, should do, and nice to do."

- When assigning a task, there is a wide range of responsibilities that you can assign to a person along with a task. After a task has been assigned, you should develop a progress report for the current week.

- The troubleshooting process can be divided into seven phases:
  - Gathering information
  - Verifying the issue
  - Looking for quick fixes
  - Performing the appropriate diagnostics (tests used in the process of determining a malfunction)
  - Using additional resources to research the issue (if necessary)
  - Escalating the issue (if necessary)
  - Completing the repair process

- The steps required for successful problem solving are:
    - State the problem
    - List unknown variables
    - List what is given
    - Create diagrams
    - List all formulas
    - List assumptions
    - Perform all necessary calculations
    - Check answers

REVIEW QUESTIONS

1   Why is it important to assign priority to work tasks?

2   How is priority assigned to a task?

3   List four aspects that should be considered before a task is assigned.

4   Why is it important to estimate the amount of time to complete a task before starting it?

5   When the amount of time to complete a task spans multiple days or weeks, why is it important to plan the event in consecutive days?

Name: _____

Date: _____

Assigning a Task

ASSIGNING A TASK CHECKLIST

Completing this checklist will help you continuously improve your skills in assigning tasks. As you are dividing a project into tasks to be assigned and completed, use the comments section to record special notes and/or concerns related to that task. In addition, a typical work order is shown below that can be used to complete a project.

Considerations	Yes/No	Comment
Are there hazards in the workplace environment or around the worker?		
Are there special work situations that come up that could lead to new risks for this worker? For example: Are there risks that might be encountered outside the normal work area? Just once a week? Can the materials needed to complete the project?		
Are there occasional risks from coworkers, such as those from welding or machining, that could affect the workers nearby?		
If hazards are associated with a task, have those hazards been reviewed with the worker?		

Work Order

CUSTOMER INFORMATION

Date: _____ Time: _____ Call Taken By: _____

Name: _____

Address: _____

City _____ State: _____ Zip: _____

Service Requested:

_____

_____

_____

_____

_____

_____

_____

_____

_____

_____

_____

_____

_____

_____

_____

_____

Work Completed:

_____

_____

_____

_____

_____

_____

_____

_____

_____

_____

_____

_____

_____

_____

_____

Completed By: _____

Date Completed: _____

Time Completed: _____

INSTRUCTOR'S RESPONSE:

_____

_____

_____

Applied Safety Rules

OBJECTIVES

By the end of this chapter, you will be able to:

KNOWLEDGE-BASED

- Explain the purpose of OSHA.
- Explain the basic safety guidelines and rules for general workplace safety.
- Explain the basic safety guidelines and rules for working with and around an electrical power tool and circuit.
- Explain the basic safety guidelines and rules for demolition and remodeling.

SKILL-BASED

- Create a basic fall protection plan.
- Work safely with ladders and extension ladders.
- Correctly identify and select the proper fire extinguisher for a particular application.

Occupational Safety and Health Administration (OSHA) branch of the U.S. Department of Labor that strives to reduce injuries and deaths in the workplace

Cardiopulmonary resuscitation (CPR) an emergency first aid procedure used to maintain circulation of blood to the brain

Asphyxiation loss of consciousness caused by a lack of oxygen or excessive carbon dioxide in the blood

Frostnip the first stage of frostbite, which causes whitening of the skin, itching, tingling, and loss of feeling

Frostbite injury to the skin resulting from prolonged exposure to freezing temperatures

Ground fault circuit interrupter (GFCI) electrical device designed to sense small current leaks to ground and deenergize the circuit before injury can result

Personal protective equipment (PPE) any equipment that will provide personal protection from a possible injury

Class A fire extinguishers fire extinguishers used on fires that result from burning wood, paper, or other ordinary combustibles

Class B fire extinguishers fire extinguishers used on fires that involve flammable liquids such as grease, gasoline, or oil

Class C fire extinguishers fire extinguishers used on electrically energized fires

Class D fire extinguishers fire extinguishers typically used on flammable metals

Introduction

The Occupational Safety and Health Act (OSHA) of 1970 was passed by Congress "to assure so far as possible every working man and woman in the Nation safe and healthful working conditions and to preserve our human resources." Under the act, OSHA was established within the Department of Labor and was authorized to regulate health and safety conditions for all employers with few exceptions. This chapter is designed to provide the facilities technician with the knowledge to ensure safety for themselves and their coworkers when performing maintenance duties at the facilities where they work.

Purpose of OSHA

The **Occupational Safety and Health Administration (OSHA)** was created to:

- Encourage employers and employees to reduce workplace hazards and implement new or improve existing safety and health standards
- Provide for research in occupational safety and health and develop innovative ways of dealing with occupational safety and health problems
- Establish "separate but dependent responsibilities and rights" for employers and employees for achieving better safety and health conditions
- Maintain a reporting and recordkeeping system to monitor job-related injuries and illnesses

- Establish training programs to increase the number and competence of occupational safety and health personnel
- Develop mandatory job safety and health standards and enforce them effectively

Basic Fall Protection Safety Procedures

Maintaining written fall protection procedures protects not only workers from falls but also management from charges of incompetence. Having individual workers or supervisors decide as to when fall protection is required and what kinds of fall protection equipment to use is an acceptable practice only where workers are routinely exposed to simple hazards, such as homebuilders on a roof. However, when workers are involved with lots of nonroutine jobs, such as removing a branch from a roof, safety is enhanced if management puts in writing the fall protection and rescue procedures that employees are required to use.

The written plan must describe how workers will be protected when working 10 feet or more above the ground, on other work surfaces, or in water.

The plan should:

1. Identify all fall hazards in the work area
2. Describe the method of fall arrest or fall restraint to be provided
3. Outline the correct procedures for assembly, maintenance, inspection, and disassembly of the fall protection system to be used
4. Explain the method of providing overhead protection for workers who may be in or pass through the area below the work site
5. Communicate the method for prompt and safe removal of injured workers

Before a fall protection plan can be developed, it is necessary to understand two important definitions:

- **Fall arrest system**—Equipment that protects someone from falling more than 6 feet or from striking a lower object in the event of a fall, whichever distance is less. This equipment includes approved full-body harnesses and lanyards properly secured to anchorage points or to lifelines, safety nets, or catch platforms.
- **Fall restraint system**—An apparatus that keeps a person from reaching a fall point; for example, it allows someone to work up to the edge of a roof but not fall. This equipment includes standard guardrails, a warning line system, a warning line and monitor system, and approved safety belts (or harnesses) and lanyards attached to secure anchorage points.

Developing a Fall Protection Work Plan

To develop a fall protection work plan, you must identify the responsibilities of your company and the work areas to which the plan applies. This information should be listed as the first item in your plan. After listing your company's responsibilities and the work areas in the plan:

1. Identify all fall hazards in the work area. To determine fall hazards, you must review all jobs and tasks to be done. After all fall hazards have been identified, list the employees required to work 10 feet or more above the ground, on other work surfaces, or in water.

2. Determine the method of fall arrest or fall restraint to be provided for each job and task that is to be done 10 feet or more above the ground, on another work surface, or in water.

3. Describe the procedures for assembly, maintenance, inspection, and disassembly of the fall protection system to be used.

4. Describe the correct procedures for handling, storage, and security of tools and materials.

5. Describe the method of providing overhead protection for workers who may be in or pass through the area below the work site.

6. Describe the method for prompt and safe removal of injured workers.

7. Identify where a copy of this plan has to be posted.

8. Train and instruct all personnel in all of these items.

9. Keep a record of employee training and maintain it on the job.

First Aid

No matter how careful the service technicians are, accidents and mishaps do happen, which require immediate medical attention. Although not all injuries require a visit to the doctor or hospital, at least some treatment is necessary to prevent further injury or infection. All service vehicles should be equipped with a first aid kit, which contains the basic medical supplies, such as burn cream, bandages, alcohol pads, eye wash, eye pads, tweezers, antiseptic spray, gauze bandages, and **cardiopulmonary resuscitation (CPR)** face shields. Note that the following sections are not intended to provide medical advice, but to provide basic information regarding immediate treatment for a number of situations commonly encountered in the field.

Bleeding

If a cut results in bleeding, place a clean folded cloth over the area and apply firm pressure. If blood soaks through the cloth, do not remove it. Simply cover the cloth with another and continue to apply pressure until the bleeding stops. If at all possible, elevate the cut area to a level above the heart to help stop the bleeding. If the cut is relatively small, the injury can be washed with soap and warm water and then bandaged.

Asphyxiation

Asphyxiation is loss of consciousness caused by a lack of oxygen or excessive carbon dioxide in the blood. An oxygen level below 19 percent may result in unconsciousness. As a result of electric shock or inhalation of refrigerant, the victim may stop breathing. When a victim's respiratory system fails, the flow of oxygen through the body may stop within a matter of minutes. If the victim stops breathing, CPR should be administered. CPR is an emergency first aid procedure used to maintain circulation of blood to the brain.

Exposure

Exposure results when a refrigerant comes in contact with the skin. Frostbite can occur any time a technician is exposed to prolonged freezing temperatures. The two main stages of exposure are **frostnip** and **frostbite**. The first stage of exposure, called "frostnip," causes whitening of the skin, itching, tingling, and loss of feeling. In the final stage, called "frostbite," the skin turns purple and blisters form on the skin. In rare situations, the exposure can result in gangrene, which usually requires amputation of the affected area. The area can be treated by covering the area with something warm and dry and then obtaining professional medical attention. Never rub, massage, poke, or squeeze the affected area, as this can result in tissue damage. A warm bottle of water can be placed gently against the affected area to warm it slightly.

Environmental Protection Agency (EPA) and Department of Transportation (DOT) Hazardous Materials Safety Procedures

Hazardous materials or chemicals are those substances regulated by federal, state, and local laws, regulations, and ordinances.

Safety Procedures

When dealing with hazardous materials or chemicals, be sure to follow these general guidelines:

- Make sure that the names on container labels match the substance names on the corresponding material safety data sheets (MSDSs). If a label is missing or the MSDS is unavailable, notify your supervisor; do not use the chemical until the correct MSDS is obtained. Never remove a manufacturer-affixed label from any container.

- Be familiar with the hazards associated with the chemicals intended to be used, and ensure that all required hazard controls are in place.

- Handle and store hazardous materials only in the areas designated by your supervisor.

- Use an appropriate fume hood or other containment device for procedures that involve the generation of aerosols, gases, or vapors containing hazardous substances.

- When working with materials of high or unknown toxicity, remain in visual and auditory contact with a second person who understands the work being performed and all pertinent emergency procedures.

- Avoid skin contact by wearing gloves, long sleeves, and other protective apparel as appropriate. Upon leaving the work area, remove any protective apparel; place it in an appropriate labeled container; and thoroughly wash your hands, forearms, face, and neck.

- Be prepared for accidents and spills. If a major spill occurs, evacuate the area and dial 911.

Electrical Safety Procedures

Electrical accidents can occur when electricity is present in faulty wiring and equipment or when poor work practices are followed. Accidents involving electricity can lead to burns and tissue damage and, in some cases, cardiac arrest and death when the body forms part of the electric circuit. Electric shock can be unsettling to the victim even if there is no apparent injury.

Other possible consequences of electrical accidents are fire and explosion (because sparking can be a source of ignition) and damage to equipment. Many of these accidents can be traced back to faults such as frayed or broken insulation or practices such as inappropriate work on live equipment.

Confined Spaces

When working in confined spaces, some general rules include the following:

- Station a person outside the confined space to watch the person or persons working inside.

- The outside person should never enter the space, even in an emergency, but should contact the proper emergency personnel.

- Use only electrical equipment and tools that are approved for the atmosphere found inside the confined area.

- As a general rule, a person working in a confined space should wear a harness with a lanyard that extends to the outside person, so the outside person could pull him or her to safety if necessary.

Trenches

When working in or around trenches some general safety rules should be followed:

- Unless absolutely necessary don't walk near a trench. This can cause the dirt to loosen and increase the possibility of a cave-in.

- Never step over or jump over a trench.

- Place barriers around trenches.

- Ladders should be used to exit and enter trenches.

- When working in trenches greater than 5 feet, shoring is required.

General Safety Precautions

- Never work on "hot" or energized equipment unless it is necessary to conduct equipment troubleshooting.
  - Install a lock out/tag out at the point of disconnection so people will not restore power to the circuit.
- Try not to ever work alone.
  - Have someone with you who can turn off the power or give artificial respiration and/or CPR.
- Do not connect too many pieces of equipment to the same circuit or outlet, as the circuit or outlet could become overloaded.
- Be sure that **ground fault circuit interrupters (GFCIs)** are used in high-risk areas such as wet locations. (GFCIs are electrical devices designed to sense small current leaks to ground and deenergize the circuit within as little as $\frac{1}{40}$ of a second, before injury can result.)
- Test the meter on a known live circuit to make sure that it is operating.
- Test the circuit that is to become the deenergized circuit with the meter.
- Inspect all equipment periodically for defects or damage.
- Replace all cords that are worn, frayed, abraded, corroded, or otherwise damaged.
- Always follow the manufacturer's instructions for use and maintenance of all electrical tools and appliances.
- Keep equipment operating instructions on file.
- Always unplug electrical appliances before attempting any repair or maintenance.
- All electrical equipment used on campus should be UL or FM approved.
- Keep cords out of the way of foot traffic so that they don't become tripping hazards and are not damaged by traffic.
- Never use electrical equipment in wet areas or run cords across wet floors.

Safety and Maintenance Procedures for Power Tools and Cords

Hand and power tools are a common part of our everyday lives and are present in nearly every industry. These tools help us easily perform tasks that would otherwise be difficult or impossible. However, these simple tools can be hazardous and have the potential to cause severe injuries when used or maintained improperly. Paying special attention to hand and power tool safety is necessary to reduce or eliminate these hazards:

- Do not use electric-powered tools in damp or wet locations.
- Keep guards in place, in working order, and properly adjusted. Safety guards must never be removed while using a tool.
- Avoid accidental starting. Do not hold a finger on the switch button while carrying a power tool.
- Safety switches must be kept in working order and must not be modified. If you feel it necessary to modify a safety switch for a job you're doing, use another tool.

- Work areas should have adequate lighting and be free of clutter.
- Observers should remain a safe distance away from the work area.
- Be sure to keep good footing and maintain good balance.
- Do not wear loose clothing, ties, or jewelry when operating tools.
- Wear appropriate gloves and footwear while using tools.

Ladder Safety and Maintenance Procedures

Ladders can be divided into two main types: straight and step. Straight ladders are constructed by placing rungs between two parallel rails. They generally contain safety feet on one end that help prevent the ladder from slipping (see Figure 3-1).

Step ladders are self-supporting, constructed of two sections hinged at the top. The front section has two tails and steps; the rear portion has two rails and braces (see Figure 3-2).

Safe Ladder Placement

- Ladders, including step ladders, should be placed in such a way that each side rail (or stile) is on a level and firm footing and that the ladder is rigid, stable, and secure.
- The side rails (or stiles) should not be supported by boxes, loose bricks, or other loose packing.
- No ladder should be placed in front of a door opening toward the ladder unless the door is fastened open, locked, or guarded.
- According to OSHA Standard CFR 1926.1053(b)(5)(i), "Non-self-supporting ladders shall be used at an angle such that the horizontal distance from the top support to the foot of the ladder approximately one-quarter of the working length of the ladder (the distance along the ladder between the foot and the top support)." In other words, for every 4 feet of working length, the ladder is extended upward, the base must be moved out 1 foot (as indicated in Figure 3-3).
- Where a ladder passes through an opening in the floor of a landing place, the opening should be as small as is reasonably practicable.
- A ladder placed in such a way that its top end rests against a window frame should have a board fixed to its top end. The size and position of this board should ensure that the load to be carried by the ladder is evenly distributed over the window frame.

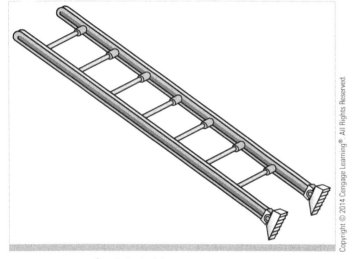

FIGURE 3-1: Straight ladder.

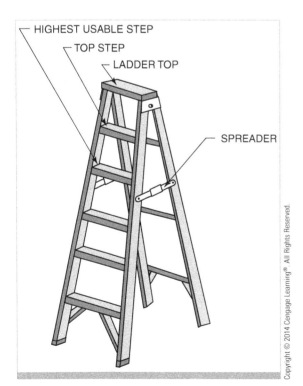

HIGHEST USABLE STEP

TOP STEP

LADDER TOP

SPREADER

FIGURE 3-2: Typical step ladder.

Safely Securing Ladders

- The ladder should be securely fixed at the top and foot so that it cannot move either from its top or from its bottom points of rest. If this is not possible, then it should be securely fixed at the base. If this is also not possible, then a person should stand at the base of the ladder and secure it manually against slipping.

- Ladders set up in public thoroughfares or other places (where there is potential for accidental collision with them) must be provided with effective means to prevent the displacement of the ladders due to collisions, for example, through the use of barricades.

Safe Use of Ladders

- Your body should be centered in relation to the ladder.

- When working from a ladder, if at all possible, use one hand to hold onto the ladder while using the other to perform the necessary task.

- If at all possible avoid pushing and pulling while on the ladder.

- Move materials with extreme caution.

- If the ladder is to heavy or long to move by yourself, get help before attempting to move it.

- If another technician is available have him or her hold the ladder when in use.

- Always move one step at a time.

- When possible, transport materials using a line instead of carrying them up the ladder.

- Only one person at a time should use or work from a single ladder.

- Always face the ladder when ascending or descending it.

- Carry tools in a tool belt, pouch, or holster, not in your hands, so you can keep hold of the ladder.

- Wear fully enclosed slip-resistant footwear when using the ladder.

- Do not climb higher than the third rung from the top of the ladder.

- Do not use ladders made by fastening cleats across a single rail or stile.

- When there is significant traffic on ladders used for building work, separate ladders for ascent and descent should be provided, designated, and used.

- Make sure the weight your ladder is supporting does not exceed its maximum load rating (user and materials). Use a ladder that has proper length for the job. Proper length is a minimum of 3 feet extending over the roofline or working surface. The three top rungs of a straight, single, or extension ladder should not be stood on.

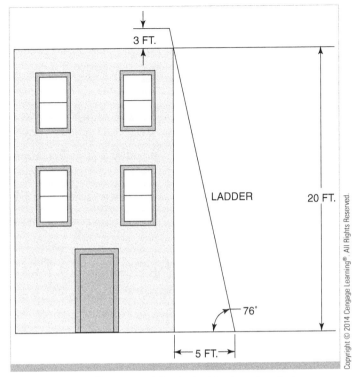

FIGURE 3-3: For every 4 feet a ladder is extended upward, the base must be moved out 1 foot.

CAUTION

Accidents can be prevented by using the right ladder for the job and not trying to make the ladder you are using do more than it is made to do.

- When setting up straight, single, or extension ladders, for every 4 feet of working length the ladder is extended upward, the base must be moved out 1 foot.
- Metal ladders will conduct electricity. Use a wooden or fiberglass ladder in the vicinity of power lines or electrical equipment. Do not let a ladder, made from any material, come in contact with live electric wires.
- Be sure all locks on extension ladders are properly engaged.
- Make sure that the ground under the ladder is level and firm. Large, flat wooden boards braced under the ladder can level a ladder on uneven or soft ground. A good practice is to have a helper hold the bottom of the ladder.
- Follow the instruction labels on ladders.

   Figures 3-4 and 3-5 illustrate these safety guidelines.

Appropriate Personal Protective Equipment

Personal protective equipment (PPE) is defined as all equipment, including clothing for shielding against weather, intended to be worn or held by people at work and

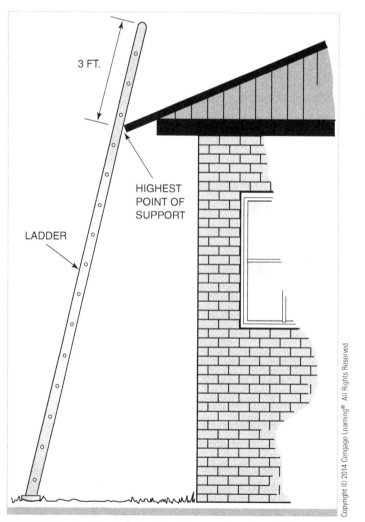

FIGURE 3-4: The ladder should extend at least 3 feet above the highest point of support.

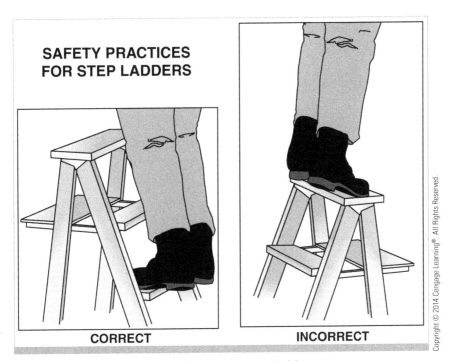

FIGURE 3-5: Safe practices for using a step ladder.

that protects them against one or more risks to their health or safety. This equipment includes, but is not limited to, the following:

- Hard hats
- Gloves
- Eye protection
- High-visibility clothing
- Safety footwear
- Safety harnesses
- Ear protection

To choose the right type of PPE, carefully consider the different hazards in the workplace. This will enable you to assess which types of PPE are suitable to protect against the hazard and for the job to be done. Figures 3-6 through 3-11 show examples of PPE.

FIGURE 3-6: Typical hard hat with attached safety googles.

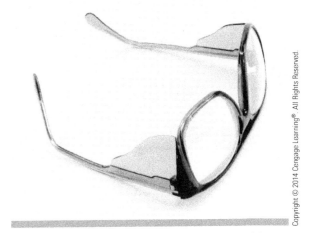

FIGURE 3-7: Safety glasses with side shields.

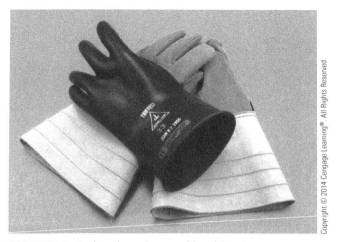

FIGURE 3-8: Leather gloves with rubber inserts.

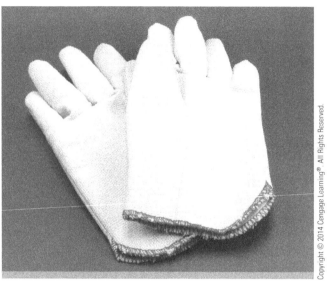

FIGURE 3-9: Kevlar gloves protect against cuts.

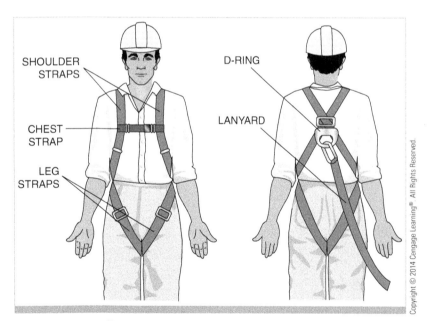

SHOULDER
STRAPS

CHEST
STRAP

LEG
STRAPS

D-RING

LANYARD

FIGURE 3-10: Typical safety harness.

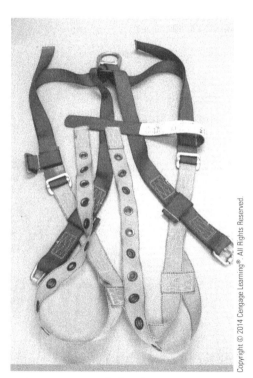

FIGURE 3-11: Safety harness.

Safe Methods for Lifting and Moving Materials and Equipment to Prevent Personal Injury and Property Damage

General safety principles can help reduce workplace accidents. These include work practices, ergonomic principles, and training and education. Whether moving materials manually or mechanically, employees should be aware of the potential hazards associated with the task at hand and know how to exercise control over their workplaces to minimize the danger.

Proper methods of lifting and handling protect against injury and make work easier. You need to think about what you are going to do before bending to pick up an object. Over time, safe lifting technique should become a habit.

Learn the correct way to lift: Get solid footing, stand close to the load, bend your knees, and lift with your legs, not your back (see Figure 3-12).

1	2	3	4
APPROACH THE LOAD AND SIZE IT UP AS TO WEIGHT, SIZE, AND SHAPE. CONSIDER YOUR PHYSICAL ABILITY TO HANDLE THE LOAD.	PLACE FEET CLOSE TO THE OBJECT TO BE LIFTED AND 8 TO 12 INCHES APART FOR GOOD BALANCE.	BEND THE KNEES TO THE DEGREE THAT IS COMFORTABLE AND GET A HANDHOLD. THEN USING BOTH LEG AND BACK MUSCLES . . .	LIFT THE LOAD STRAIGHT UP, SMOOTHLY AND EVENLY. PUSH WITH YOUR LEGS AND KEEP THE LOAD CLOSE TO YOUR BODY.

5	6	7	
LIFT THE OBJECT INTO CARRYING POSITION, MAKING NO TURNING OR TWISTING MOVEMENTS UNTIL THE LIFT IS COMPLETED.	TURN YOUR BODY WITH CHANGES OF FOOT POSITION AFTER LOOKING OVER YOUR PATH OF TRAVEL, MAKING SURE IT IS CLEAR.	SETTING THE LOAD DOWN IS JUST AS IMPORTANT AS PICKING IT UP. USING LEG AND BACK MUSCLES, COMFORTABLY LOWER LOAD BY BENDING YOUR KNEES. WHEN LOAD IS SECURELY POSITIONED, RELEASE YOUR GRIP.	

FIGURE 3-12: How to safely lift using your knees.

Procedures to Prevent and Respond to Fires and Other Hazards

For a fire to burn, three things are needed: fuel, heat, and oxygen. Fuel is anything that can burn, including materials such as wood, paper, cloth, combustible dusts, and even some metals. Fires are divided into four classes: A, B, C, and D (see Figure 3-13).

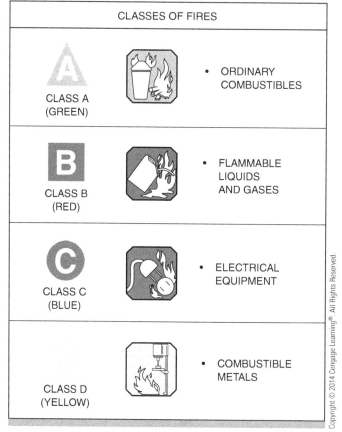

FIGURE 3-13: Four classes of fires.

Class A Fires

This class involves common combustible materials such as wood or paper. **Class A fire extinguishers** often use water to extinguish a fire (see Figure 3-14).

Class B Fires

This class involves fuels such as grease, combustible liquids, or gases. **Class B fire extinguishers** generally employ carbon dioxide (CO_2).

Class C Fires

This class involves energized electrical equipment. **Class C fire extinguishers** usually use a dry powder to smother the fire.

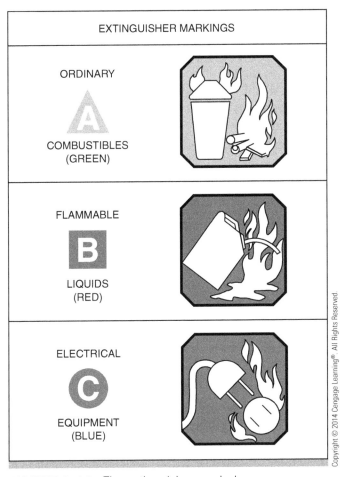

EXTINGUISHER MARKINGS

ORDINARY

COMBUSTIBLES
(GREEN)

FLAMMABLE

LIQUIDS
(RED)

ELECTRICAL

EQUIPMENT
(BLUE)

FIGURE 3-14: Fire extinguisher symbols.

Class D Fires

This class consists of burning metal. **Class D fire extinguishers** place a powder on top of the burning metal that forms a crust to cut off the oxygen supply to the metal.

Prevent a Fire from Starting in Your Home

The most common causes of residential fires are careless cooking and faulty heating equipment. When cooking, never leave food on a stove or in an oven unattended. Avoid wearing clothes with long, loose-fitting sleeves. Have your heating system checked annually, and follow manufacturer's instructions when using portable heaters.

Smoking is the leading cause of fire deaths and the second most common cause of residential fires. If you are a smoker, do not smoke in bed, never leave burning cigarettes unattended, do not empty smoldering ashes in the trash, and keep ashtrays away from upholstered furniture and curtains. In addition, keep matches and lighters away from children. Safely store flammable substances used throughout the home. Finally, never leave burning candles unattended.

Procedure to Prevent Uncontrolled Chemical Reactions

A chemical reactivity hazard is a situation with the potential for an uncontrolled chemical reaction that can result directly or indirectly in serious harm to people, property, or the environment.

To maintain a safe and healthful working environment, the Department of Energy recommends the following practices wherever chemicals are stored. These practices are based on regulations, rules, and guidelines designed to reduce or eliminate hazardous incidents associated with the improper storage of chemicals.

1. Adhere to the manufacturer's recommendations for each chemical stored, noting any precautions on the label.

2. Label all chemicals. The name and address of the manufacturer or other responsible party must be listed on the label. Chemicals with a shelf life should be labeled with the date received.

3. Store chemicals in the locations recommended (e.g., where the temperature range, vibration, or amount of light does not exceed the manufacturer's recommendations).

4. Inspect annually all chemicals in stock and storage. Hazardous chemicals should be inspected every six months. Some hazardous chemicals may require more frequent inspections. Any outdated materials should be properly disposed of or replaced if necessary.

5. Ensure that provisions are made for liaison with local planning committees, the state emergency planning commission, and local fire departments in the event of a chemical emergency.

6. Keep only enough inventory necessary for uninterrupted operation. Chemical inventory should be maintained at a minimum to reduce fire, exposure, and disposal hazards.

7. Rotate new shipments of chemicals with existing stock so that the oldest stock is available first.

REMODELING

Remodeling and Demolition Safety

The most important aspect of any project is safety. Regardless of the type of project, safety should be the primary concern of anyone involved in construction, maintenance, remodeling and/or repair. In addition to preventing accidents that lead to injury and possible death, safety has a direct effect on the bottom line of a project by avoiding increases in worker compensation rates, lost productivity from hours not worked, and increases in the cost of insurance, as well as possible fines from regulatory agencies.

Before starting any remodeling project, always talk to the occupants of the structure. It is important to find out whether any of the occupants have allergies or health conditions that could be affected by new products introduced into the structure or dust and fumes caused by the demolition process. To minimize the exposure to the occupants of the structure, the area(s) being remodeled should be sealed off using a plastic barrier. This plastic should be kept in place until the remodeling process is complete and the cleanup has been performed. All ventilation ducts leading into the affected areas should be sealed off to prevent the spread of contamination from the remodeling process.

Typical safety equipment needed for demolition includes:

- **Safety glasses**—Safety glasses should meet ANSI Z87.1-2003 safety standard or OSHA safety guidelines.

- **Hard hat**—Hard hats that are used to protect employees against impact and penetration of falling objects must meet the specifications listed in the ANSI (American National Standards Institute) Z89.1-1969, safety requirements for industrial head protection.

- **Hearing protection**—The OSHA requirements for hearing protection are currently outlined in 29 CFR 1910.95.

- **Demolition work gloves**—Consult OSHA standard 29 CFR 1926.28 requirements for PPE when selecting demolition gloves.

- **Safety shoes**—Foot protection is outlined in 1910.136(a) general requirements.

- **Appropriate clothing**—Typically consists of long-sleeved shirts and long pants.

SUMMARY

- OSHA was created to encourage employers and employees to reduce workplace hazards and implement new safety and health standards or improve existing ones, as well as establish training programs to increase the number and competence of occupational safety and health personnel.

- Maintaining written fall protection procedures protects not only workers from falls but also management from charges of incompetence.

- The written plan must describe how workers will be protected when working 10 feet or more above the ground, on other work surfaces, or in water.

- To develop a fall protection work plan, you must identify the responsibilities of your company and the work areas to which the plan applies.

- If a cut results in bleeding, place a clean folded cloth over the area and apply firm pressure.

- Asphyxiation is loss of consciousness caused by a lack of oxygen or excessive carbon dioxide in the blood. An oxygen level below 19 percent may result in unconsciousness.

- Exposure results when a refrigerant comes in contact with the skin.

- Make sure that the names on container labels match the substance names on the corresponding material safety data sheets (MSDSs). If a label is missing or the MSDS is unavailable, notify your supervisor; do not use the chemical until the correct MSDS is obtained. Never remove a manufacturer-affixed label from any container.

- Electrical accidents can occur when electricity is present in faulty wiring and equipment or when poor work practices are followed.

- When working in confined spaces, station a person outside the confined space to watch the person or persons working inside.

- When working in or around trenches unless it absolutely necessary stay away from the trench.

- Never work on "hot" or energized equipment unless it is necessary to conduct equipment troubleshooting.

- Do not use electric-powered tools in damp or wet locations.

- Keep guards in place, in working order, and properly adjusted. Safety guards must never be removed while using the tool.

- According to OSHA Standard CFR 1926.1053(b)(5)(i), "Non-self-supporting ladders shall be used at an angle such that the horizontal distance from the top support to the foot of the ladder approximately one-quarter of the working length of the ladder (the distance along the ladder between the foot and the top support)."

- Personal protective equipment (PPE) is all equipment, including clothing for shielding against weather, intended to be worn or held by people at work and that protects them against one or more risks to their health or safety.

- Always talk to the occupants of the structure before starting any remodeling project to find out whether any of the occupants have allergies or health conditions that could be affected by the demolition process.

- Typical safety equipment needed for demolition includes safety glasses, a hard hat, hearing protection, demolition work gloves, safety shoes, and appropriate clothing.

REVIEW QUESTIONS

1 What is the most common cause of residential fires?

2 What is the leading cause of fire deaths in residential buildings?

3 Define the following:

Class A fires

Class B fires

Class C fires

Class D fires

4 Define the term *personal protective equipment (PPE)*.

5 True or false? Ladders should be securely fixed at the top and foot in such a way that they cannot move either from their top or from their bottom points of rest.

6 True or false? If it is not possible to secure a ladder at both the top and the bottom, then it shall be securely fixed at the top.

REVIEW QUESTIONS

7 What is the purpose of OSHA?

9 List two items that should be addressed in a fall protection plan.

8 True or false? To develop a fall protection work plan, you must identify the responsibilities of your company and the work areas to which the plan applies.

10 True or false? Always follow the manufacturer's instructions for use and maintenance of all electrical tools and appliances.

Name: _____

Date: _____

Applied Safety Rules

WORKPLACE SAFETY

Upon completion of this job sheet, you should be able to demonstrate your awareness of work area safety items. As you survey your work area and answer the following questions, you should learn how to evaluate the safety of your area. Evaluate your work area and how you fit into it.

1. Are you properly dressed for work?
   a. If yes, describe how you are dressed.

   _____
   _____

   b. If no, explain why you are not properly dressed.

   _____
   _____

2. Do you have the necessary safety equipment?
   Hard hat Yes____ No____
   Gloves Yes____ No____
   Eye protection Yes____ No____
   High-visibility clothing Yes____ No____
   Safety footwear Yes____ No____
   Safety harnesses Yes____ No____
   Ear protection Yes____ No____

3. Carefully inspect your work area, note any potential hazards.

4. Where are tools stored at your facility?
   a. Are they clean and neatly stored? Yes____ No____

5. Explain how you could improve tool storage.

6. Where is the first aid kit at your facility?

INSTRUCTOR'S RESPONSE:

_____
_____
_____

Applied Safety Rules

IDENTIFYING AND HANDLING HAZARDOUS MATERIALS

Upon completion of this job sheet, you should be able to demonstrate your ability to identify hazardous materials and explain how to handle them.

1. Inspect your facility. Identify and list all hazardous materials found.
   a. Solvents

   _____

   _____

   b. Gasoline

   _____

   _____

   c. Cleaners

   _____

   _____

   d. Others

   _____

   _____

2. Check the containers in which hazardous materials are stored. Are they clearly marked? **YES___ NO___**

3. Check to see if your facility has a material safety data sheet (MSDS) file. Is it located near the hazardous materials? **YES___ NO___**

4. Make sure your facility has an MSDS list posted on a bulletin board where everyone can read it.

5. Read the MSDS bulletins on each of the materials you find at the facility and explain to the instructor how you would handle a spill of each material.

INSTRUCTOR'S RESPONSE:

_____

_____

_____

Name: _____

Date: _____

3

Applied Safety Rules

BASIC FALL PROTECTION SAFETY

Upon completion of this job sheet, you should be able to demonstrate your understanding of basic fall protection safety procedures.

1. Does your facility have written fall protection and rescue procedures?
   a. If yes, are the procedures readily accessible to the workers?

   _____

   _____

   b. If no, explain why this should be brought to the attention of your supervisor.

   _____

   _____

2. When working higher than 6 feet off the ground, do workers use a fall arrest system?
   a. If no, should this be brought to the attention of your supervisor?

   _____

   _____

3. Identify all fall hazards at your facility.

4. Explain the method of providing overhead protection for workers who may be in or pass through the area below a work site.

   _____

   _____

INSTRUCTOR'S RESPONSE:

   _____

   _____

   _____

Name: _____

Date: _____

Applied Safety Rules

LADDER SAFETY AND MAINTENANCE PROCEDURES

Upon completion of this job sheet, you should be able to demonstrate your understanding of ladder safety and maintenance procedures.

1. Name the two main types of ladders and explain the difference between the two.

    _____

    _____

2. Explain why it is important for straight, single, or extension ladders to be set up for every four feet of working length the ladder is extended upward, the base must be moved out 1 foot.

    _____

    _____

3. Explain why it is important for the ground under the ladder be level and firm.

    _____

    _____

4. List the ladders at your facility and write down the weight limitations for each.

    _____

    _____

    _____

INSTRUCTOR'S RESPONSE:

_____

_____

_____

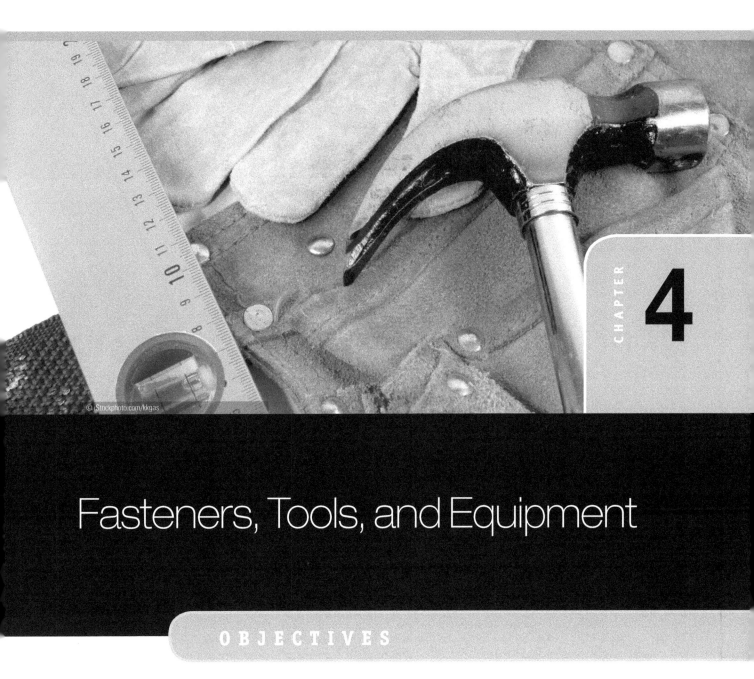

CHAPTER 4

Fasteners, Tools, and Equipment

OBJECTIVES

By the end of this chapter, you will be able to:

KNOWLEDGE-BASED

- Describe the safe use of tools, including power tools used by facilities maintenance technicians.
- Describe the proper anchors, fasteners, and adhesives necessary for a specific project.
- Describe the basic hand tools used in demolition.
- Describe the basic power tools used in demolition.

SKILL-BASED

- Select, and install the proper anchors, fasteners, and adhesives necessary for a specific project.
- Select and properly use the appropriate hand tool for a specific project.
- Select and properly use the appropriate power or stationary tool for a specific project.

Power tool a tool that contains a motor

Screws used when stronger joining power is needed

Caulk used to fill cracks as well as other defects in interior wall as well as exterior walls and foundation cracks

Introduction

Choosing the appropriate tool or equipment for a project is an important element in maintaining any facility. Many types of tools are available for handling many of the problems that may occur. There are times when using tools is not the appropriate method for repair and other equipment may be necessary. This chapter introduces a variety of fasteners, solvents, and adhesives.

Fasteners

A fastener is used when it is necessary to mechanically join two or more mating surfaces or objects. Today, there is a fastener on the market for just about any situation or application. Fasteners are often employed in situations where wood meets wood, concrete, or brick, and most are approved by the Uniform Building Code requirements. However, you should always consult your local building code before selecting a particular type of fastener to incorporate.

Nails

There is a huge difference in nails. Some are specialty nails:

- **Common nails**—Most often used nails; used for most applications in which the special features of the other nail types are not needed (see Figure 4-1)
- **Box nails**—Used for boxes and crates
- **Finishing nails**—Can be driven below the surface of the wood and concealed with putty so that it is completely hidden
- **Casing nails**—Used for installing exterior doors and windows
- **Duplex nails**—Used for temporary structures, such as locally built scaffolds
- **Roofing nails**—Used for installing asphalt and fiberglass roofing shingles
- **Masonry nails**—Used when nailing into concrete or masonry
- **Brad**—A thin, short, finishing nail used for nailing trim

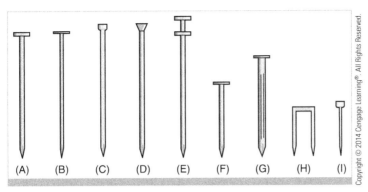

FIGURE 4-1: Some of the most common kinds of nails: (A) common nail, (B) box nail, (C) finishing nail, (D) casing nail, (E) duplex nail, (F) roofing nail, (G) masonry nail, (H) staple, and (I) brad.

Screws

Screws are used when stronger joints are needed, or when other materials must be secured to wood. The screw is tapered to help draw the wood together as the screw is inserted. Screw heads are designed for a specific purpose. Their final seating and appearance and are usually flat, oval, or round (see Figure 4-2).

Types of screws include the following:

- **Drywall screws**—Used to attach drywall to wall studs
- **Sheet metal screws**—Used to secure metal to wood, metal, plastic, or other materials. They are threaded entirely from the point to the head, and have sharper threads than wood screws.
- **Particleboard and deck screws**—Corrosion-resistant screws used for installing deck materials and/or particleboard
- **Lag screws**—Are driven in with a wrench and are typically used when heavy holding power is needed
- **Wood Screws**—Are similar in function to lag screws and are obtainable in a wide range of head styles, sizes, and materials.

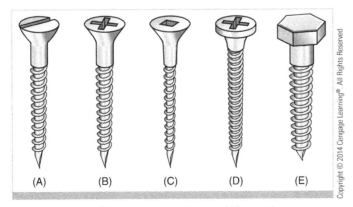

FIGURE 4-2: Common screw types: (A) wood screw, (B) drywall screw, (C) particleboard screw, (D) panhead sheet metal screw, and (E) lag screw.

Screws should penetrate two-thirds of the combined thickness of the materials being joined. For example, if two 1.5-inch-thick pieces of wood are being joined using screws, then the combined thickness of the material would be 3 inches. Therefore, a 2-inch screw is needed. Use galvanized or other rust-resistant screws where moisture and rust could be a problem.

Screw head shapes are usually determined by the screw types and the pitch and/or depth of the threads. The most common shapes are oval, pan, bugle, flat, round, and hex (Figure 4-3). Different types of slots are also available for these screws (Figure 4-4).

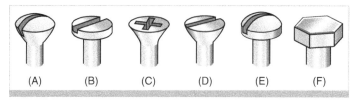

FIGURE 4-3: Screw head shapes: (A) oval, (B) pan, (C) bugle, (D) flat, (E) round, and (F) hex.

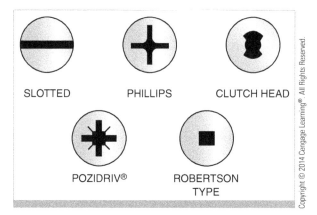

FIGURE 4-4: Common types of screw slots.

Nuts and Bolts

Nuts and bolts are usually used together: The bolt is inserted through a hole drilled through the materials to be fastened together, while the nut is threaded onto the bolt from the other side and tightened. Using nuts and bolts also allows for the disassembly of parts.

Types of bolts include the following:

- **Cap screws**—Available with hex heads, slotted heads, Phillips head, and Allen drive (Figure 4-5)
- **Stove bolts**—Either round or flat heads and threaded all the way to the head; used to join sheet metal parts (Figure 4-6)
- **Carriage bolts**—A bolt designed to fasten together timber as well as other dimensional lumber. Carriage bolts have a rounded head and contain thread all along their shank (Figure 4-7).

See Appendix B, page 000, for a complete list of screws, bolts, washers, nuts, nails, and other fasteners.

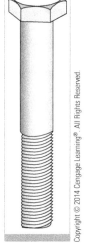

FIGURE 4-5: Cap screw.

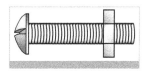

FIGURE 4-6: Stove bolt.

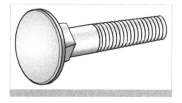

FIGURE 4-7: Carriage bolt.

Adhesives

Sometimes nails and screws need a little help in holding two or more pieces of material together. When this happens adhesives are often used. There are several different types of adhesives available on the market today. However, when selecting an adhesive, always read the directions and the intended usage of the adhesive being considered.

Some of the more common adhesives available on the market today are:

- **Carpenter's wood glue**—Used primary in the furniture and craft industries, carpenter wood glue is an adhesive used to fasten two pieces of wood together. Vulnerable to moisture, polyvinyl sets in an hour, dries clear, and will not stain.

- **Epoxy**—In general, epoxies offer the greatest holding capability. They generally provide a holding power greater than the material that they are fastening. In addition, epoxies resist almost anything from water to solvents. They are extremely useful when filling cavities that otherwise would be difficult, if not impossible, to fill otherwise. Normally, they are designed to be used in warm temperature; however, always read the manufacturer's instructions before attempting to use an epoxy. Not all epoxies dry in the same amount of time or have the same resin to hardener ratio. Tips for using epoxy include the following:
  - One gallon of epoxy can be used to cover 12.8 square feet at $\frac{1}{8}$ in. thickness or 6.4 square feet at $\frac{1}{4}$ in. thickness.
  - While epoxies are normally able to endure high temperatures for short periods, they are not recommended for applications in which the temperature exceeds 200°F.
  - To ensure maximum bond strength remove all oil, dirt, rust, paint, and water. Use a degreasing solution (alcohol or acetone) to eliminate oil and grease. Sand and/or wipe away paint, dirt, or oxidation.
  - In general, epoxy cures enough to provide significant strength after 3 hours; however, see the manufacturer's instructions and specification for exact drying times. As a rule, at 70°F the working time is 15 minutes. However, it is still possible to relocate work up to 45 minutes after application.
  - Uncured epoxy can be cleaned up with soap and water or denatured alcohol; however, always follow the manufacturer's instruction for cleanup. Cured epoxy can be removed by scraping, cutting, or removing in layers with a good paint remover.

Contact Cements

Laminates are usually bonded to countertop and tabletops using contact cement. Before using contact cement, always read and follow the manufacturer's instructions. Also, only use contact cement in a well-ventilated area, and never use it around open flames. When using contact cement, apply a thin, even coat of cement on the countertop and the back of the laminate. Allow the cement to somewhat dry before bonding; for the exact drying time, consult the manufacturer's instructions. Because contact cement will not pull apart, before pressing the laminate to the countertop, be sure to perfectly align to the surfaces.

Caulks

Caulk is used to fill cracks around exterior window and door frames, and to fill exterior wall and foundation.

Types of caulk include:

- **Painter's caulk**—Used by painters to plug and patch holes, cracks, and imperfections before paint is applied. Painter's caulk is an inexpensive latex-based caulk. Painter's caulk is often used to provide a smooth joint in a corner where textured materials meet, thus, allowing for a straight line in the corner.

- **Acrylic latex**—Can be used for both interior and exterior applications, acrylic latex is paintable. Acrylic latex is soluble in water.

- **Siliconized latex**—Available in a variety of different colors, this form of latex caulk is durable multipurpose caulk with silicone. It is soluble in water.

- **100 percent silicone**—This is most often used in the sealing of glass, ceramic tiles, and metallic surfaces. It is ideal for nonporous substances but is less appropriate for porous surfaces such as wood and masonry. Silicone caulk remains flexible and is impervious to water. It is not soluble in water and cannot be painted, and it is usually available in clear and white. Read the manufacturer's instructions and precautions before attempting to use it. Proper ventilation is required when using silicone caulk.

- **Tub and tile**—This is an acrylic caulk that provides a flexible, watertight seal. This type of caulk is designed to resist mildew and is soluble in water.

- **100 percent silicone kitchen and bath sealant**—This has the same characteristics as plain 100 percent silicone sealant.

- **Gutter and foundation sealant (butyl rubber)**—This can be used on metal, wood, or concrete. It is appropriate for use in areas that experience extreme temperature variations. It is not soluble in water and, therefore, requires solvent cleanup. It is often used on metal flashing and around skylights.

- **Roof repair caulk**—This is a convenient butyl rubber/asphalt formulation for sealing flashing, roofing, skylights, and so on. It is soluble with mineral spirits.

- **Adhesive caulk**—This is used in the installation of sinks, countertops, and similar components. It dries hard and is not flexible.

- **Concrete and mortar repair**—This form of caulk is used to repair concrete and masonry projects. It retains some elasticity thus permitting it to remain in mortar and concrete cracks. It is soluble in water.

Applying Caulk

Step 1. Caulk comes either in a squeeze tube or as a cartridge (see Figure 4-8). When using a cartridge and a caulk gun, always cut the cartridge nozzle at a 45° angle. The diameter of the opening should equal to the size of the gap being caulked. Using a piece of wire, break the inner seal of the cartridge by poking it through the opening in the tip.

Step 2. Hold the caulk gun at a 45° angle and apply even pressure when applying caulk (Figure 4-9).

Step 3. Once the caulk starts to flow, move the tip at an even pace along the joint.

When caulking around a tub keep in mind that filling the tub with water will cause the tub to slightly sink. Therefore, always fill the tub with water before starting a caulk job. If you caulk an empty tub, you will need to supply enough caulk to compensate for the sinking. Otherwise, you'll end up with cracked caulk the next time someone takes a bath.

FIGURE 4-8: Caulk is made of a variety of materials.

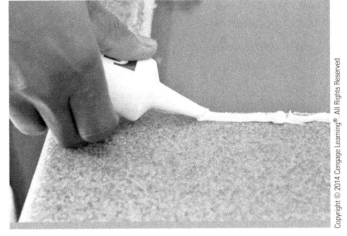

FIGURE 4-9: Applying caulk.

Step 4. Once all the seams have been completely caulked, use a caulk smoother to even out the finish. If a caulk smoother is not available, then use your finger or a Popsicle stick to smooth the caulk. If you are using your finger or a Popsicle stick, wet it first before using it.

Step 5. When application of all of your caulk is done, give it time to thoroughly set. Be sure to read and follow the manufacturer's instructions regarding drying times to ensure that the caulk is adequately set before attempting to paint it.

Solvents

Solvents are used in hundreds of products to improve cleaning efficiency. The solvents in cleaners help make counters, showers, toilets, tubs, carpets, and other items clean.

- **All-purpose cleaner**—Nonhazardous, biodegradable, nontoxic, super-concentrated cleaner; good for cleaning everyday spills and messes

- **Spot remover**—Used in the removal of chewing gum, tar, grease, makeup, crayon, candle wax, and adhesives from most surfaces

- **Basin, tub, and tile cleaner**—Removes and clean soap scum, lime scale, body oils, and other deposits from shower walls and doors, sinks, ceramic tile, fiberglass, vinyl, porcelain, and stainless steel

Hand Tools

Hand tools, such as hammers, screwdrivers, levels, and tapes, do not use a motor.

As a general rule of thumb, more adhesive does not make for a stronger bond—with the exception of epoxy. Too much adhesive will weaken a joint. Therefore, it is critical to always follow the manufacturer's instructions when using an adhesive. Always clean the surfaces to be bonded before applying the adhesive, by slightly sanding smooth surfaces, thus permitting them to

(continued)

(continued)

form a better bond. In accordance with the manufacturer's instructions, coat the surface with a thin layer of adhesive. Securely hold the mating surfaces together using clamps while the adhesive is drying. Drying time will vary. Wipe away any excess adhesive immediately after clamping.

Measuring, Marking, Leveling, and Layout Tools

All construction projects require some form of measurements to be taken in order for the project to be completed. Regardless of the project, accuracy and attention to detail when making a measurement will result in either an extremely well-built project or one that will not stay together. Tools that fall in this category are tape measures, squares, combination squares, chalk line reels, levels, and plumb bobs.

Tape Measures

Tape measures are available in lengths ranging from 6 to 100 feet. The wider the blade of the tape measure, the easier the tape measure is to use. In addition, wide blades are also safer than narrow blades. Blades 1 inch or wider are recommended. Tape measures that employ cushioned bumpers offer a degree of protection to the hook, protecting it from the damage that usually happens from repeated usage (extending and retracting) of the blade. It is the play in the tapes measure's hook (see Figure 4-10) that enables the technician to take either inside or outside measurements without having to adjust or compensate measurement for the hook (see Figure 4-11). Rounded objects, odd shapes, and contours can be measured because of the tape's flexibility. For inside measurements, the tape measure's case is added to the measurement obtained. The length of the tapes measure's case is typically marked on the outside of the case (see Figure 4-12).

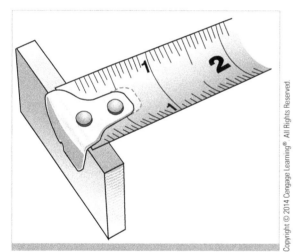

FIGURE 4-10: The hook or fitting on the end of a tape measure will slide to adjust for the thickness of the fitting.

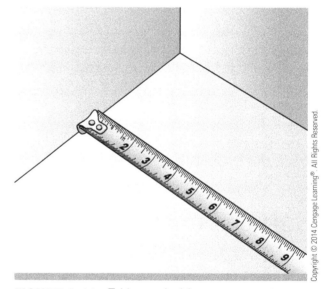

FIGURE 4-11: Taking an inside measurement.

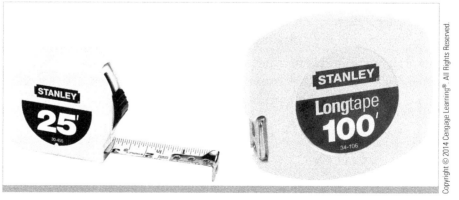

FIGURE 4-12: Tape measure.

Squares

Ensuring the squareness of an assembly, as well as determining the proper angles of various assembly components, is typically done using a square. Squares are also used for layout work. True perpendicular lines can be obtained by using a framing square, also known as a carpenter's square. Framing squares are also used for squaring corners, laying out steps, as well as other tasks requiring the use of a square (see Figure 4-13).

Combination Squares

One of the most flexible tools available is the combination square (Figure 4-14). A combination square consists of a movable handle that locks in place around a 12-inch steel rule. Combination squares have a variety of different uses. Some of these uses include squaring the ends of boards, marking various angles, and checking for level, using the spirit level. Combination squares are also used to transfer lengths, as well as mark a continuous line along the length of a board.

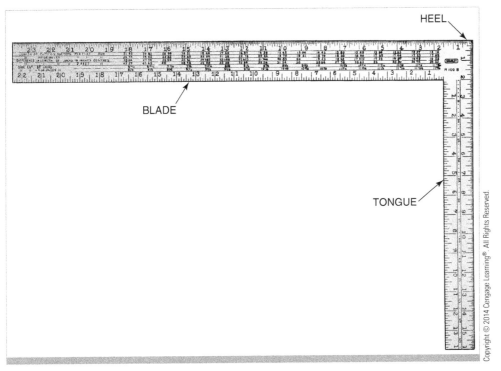

FIGURE 4-13: Framing square.

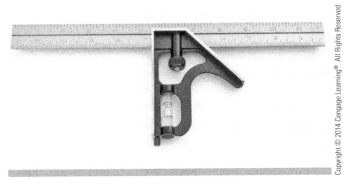

FIGURE 4-14: Combination square.

Chalk Line Reels

When it is necessary to create a straight line on a work surface, the chalk line is typically used. A chalk line consists of a chalk-coated string or line contained within a holder (see Figure 4-15). Using the chalk line is rather easy. Start by pulling the line out and positioning it between the two points you are trying to connect. Once the line is in position, pull the line tight. While keeping the line pulled tight, with one hand snap the chalk line. This will leave a chalk mark on the work surface, thus connecting the two points (shown in Figure 4-16). Some chalk lines are equipped with a point on one end to allow the chalk line to be used as a plumb bob.

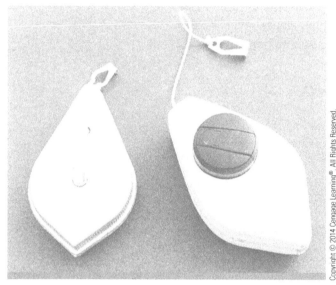

FIGURE 4-15: Chalk line reel.

FIGURE 4-16: Snapping a chalk line.

Levels

When it is absolutely necessary for a work piece to be aligned to true horizontal plane (level) or aligned to a true vertical plane (plumb), a level is often used (Figure 4-17).

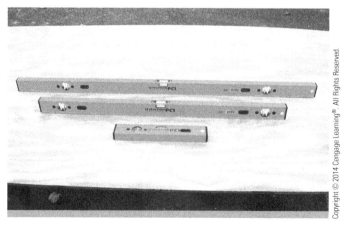
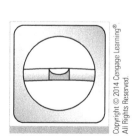

FIGURE 4-17: A spirit level has one or more transparent vials containing a liquid and a bubble.

Plumb Bobs

A plumb bob (Figure 4-18) is a heavy, balanced weight on a string, which you drop from a specific point to locate another point exactly below it or to determine true vertical.

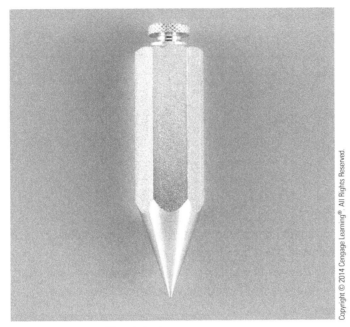

FIGURE 4-18: Plumb bob.

Boring and Cutting Tools

Boring and cutting tools are a category of tools used for making holes or cutting wood, metals, and plastic. They include drill bits, saws, and so on.

Drill Bits

When drilling a hole, be sure to use the correct type of drill bit for the type of material being drilled and the type of hole being produced. The drill bit must be used correctly and be sharpened as appropriate. A set of high-speed steel (HSS) twist drills and some masonry bits will probably be adequate for the average facilities maintenance technician.

Types of drill bits include:

- **Twisted bit**—For hand and electric drills, this is the most common drilling tool used (see Figure 4-19). Twisted bits can be used on lumber, metal, plastic, and similar materials. Most twist bits are made from either of the following:
  - **HSS**—Suitable for most types of material, HSS is a type of tool steel that is harder and able to withstand higher temperatures than high carbon steel.
  - **Carbon steel**—This is specially designed for wood and should not be used for drilling metals. These bits have a tendency to be less flexible and more brittle than HSS bits.
- **Masonry bit**—Designed especially for drilling into brick, block, stone, quarry tiles, or concrete, these bits are normally used in power drills (see Figure 4-19).
- **Forstner bit**—Used to drill flat-bottomed holes
- **Spade bit**—Used for boring holes in rough carpentry (see Figure 4-19)

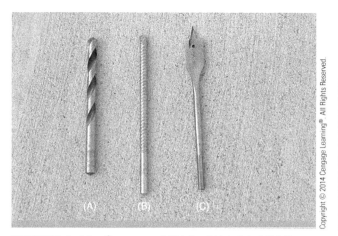

FIGURE 4-19: Common drill bits: (A) twist drill bit, (B) masonry bit, and (C) spade bit.

FIGURE 4-20: Hole saw.

- **Hole saw**—Designed for cutting large, fixed-diameter holes in wood, plastic, or metal (see Figure 4-20). Tips for using a drill:
  - Always read the manufacturer's instructions and safety guidelines before attempting to use a drill, and follow them.
  - Always wear eye protection.
  - Don't apply too much pressure on small-diameter drill bits.
  - When drilling through material, always ease up on pressure.
  - The larger the drill bit diameter, the slower the drill speed and the material's feed rate.
  - Use a vise or clamp to hold the material to prevent it from spinning if the drill bit catches.

Saws

Various types of saws are available, such as rip, crosscut, and wallboard saws. All of them look basically the same, and their primary purpose is cutting of timber from boards, and sometimes making larger joints.

Types of saws include:

- **Crosscut saw**—Available in many sizes and configurations, the crosscut saw is a good general-purpose saw, typically used for cutting wood across its grain (see Figure 4-21). The kerf, or the actual cut made by the saw, is as wide as the set of teeth in the saw blade.

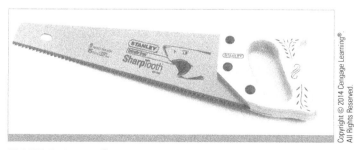

KERF

FIGURE 4-21: Crosscut saw.

- **Ripsaw**—This is used to cut with the grain of the wood (see Figure 4-22). Ripsaw teeth are filed straight across the blade, so each tooth is shaped like a little chisel.

FIGURE 4-22: Ripsaw.

- **Hacksaw**—This is a basic hand saw used mostly for cutting metals (see Figure 4-23). Some have pistol grips, which make the hacksaw easy to grip. Hacksaws cut in straight lines.

- **Wallboard saw**—This is used for cutting electrical outlet holes and other small rough sawing where a powered saber saw will not fit (see Figure 4-24). A self-starting keyhole saw (a type of wallboard saw) is very handy and comfortable to use.

- **Coping saw**—Used to cut the profile of one piece of molding on the end of molding (see Figure 4-25)

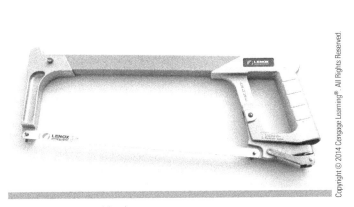

FIGURE 4-23: Hacksaw.

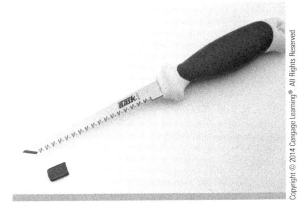

FIGURE 4-24: Wallboard saw.

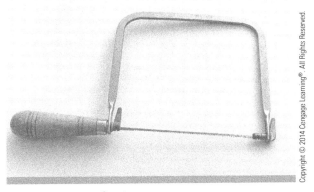

FIGURE 4-25: Coping saw.

Hammers

Hammers are used in construction and demolition for driving nails, fitting parts, and breaking up objects. Various types of hammers include:

- **Sledge hammer**—Mainly used on exterior and demolition projects, these are designed to deliver heavy force (see Figure 4-26).

- **Mason's hammer**—Designed for working on brick, concrete, or mortar (see Figure 4-27), mason's hammers are used for cutting and setting brick.

- **Claw hammer**—A general-use hammer (see Figure 4-28). Tips for using a hammer:
  - When working with hard wood, splitting can be prevented by first drilling a small pilot hole in the material.
  - To begin hammering, grip the hammer firmly (but not too tightly) in the middle of the handle and move your hand in a shaking motion, similarly to shaking hands with someone. In other words, shake hands with your hammer.
  - Using the other hand, hold the nail between your thumb while placing the nail where it is to be driven.
  - Strike the head of the nail, using the center of the hammer's face, and drive the nail in with firm, smooth blows.

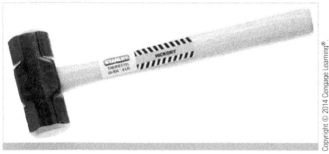

FIGURE 4-26: Sledge hammers.

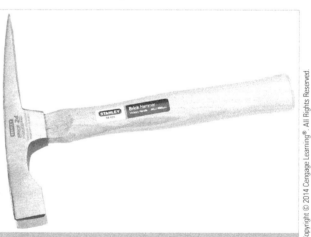

FIGURE 4-27: Mason's hammer.

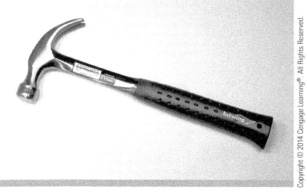
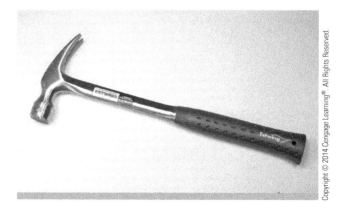

FIGURE 4-28: Curved-claw hammer and framing hammer.

- Make sure that the striking face of the hammer is parallel with the surface being hit.
- Striking the nail with a sideways or glancing blow can result in bending the nail as well as leaving hammer marks in the work surface.

Screwdrivers

Screwdrivers (Figure 4-29) are used to tighten and loosen as well as insert and remove screws. To select the right screwdriver for a particular task, match the appropriate type and size of screwdriver to the screw head (see Figure 4-30).

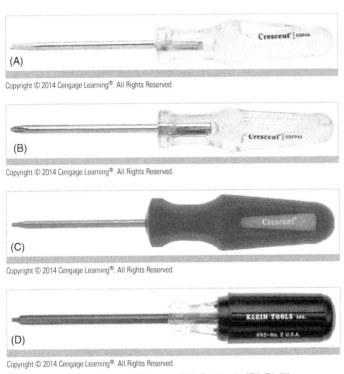

(A)

(B)

(C)

(D)

FIGURE 4-29: Screwdrivers: (A) Slotted, (B) Phillips, (C) Torx, (D) Square or Robertson.

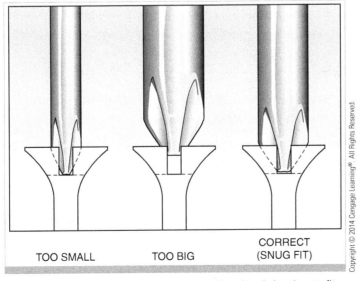

TOO SMALL   TOO BIG   CORRECT (SNUG FIT)

FIGURE 4-30: The screwdriver must be the right size to fit the fastener.

Types of screwdrivers include:

- Slotted
- Torx
- Phillips
- Square (Robertson)

Pliers

Pliers are used for leverage in gripping. They are designed for various tasks and require different jaw configurations for gripping, turning, pulling, and crimping (see Figure 4-31).

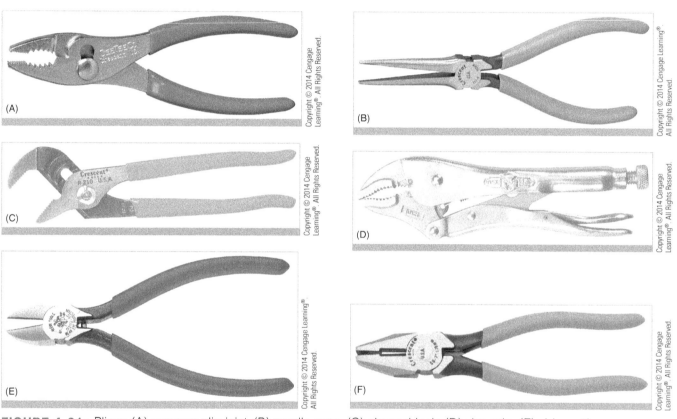

FIGURE 4-31: Pliers: (A) common slip-joint, (B) needle nose, (C) channel-lock, (D) vise grip, (E) side-cutting, (F) electrician's.

Wrenches

Wrenches are used to turn nut, bolts, and other hard-to-turn parts.

Some common types of wrenches include:

- **Socket wrench**—This has a ratchet handle, which allows the user to move the handle back and forth without having to take the socket off the nut and reposition it. This type of wrench uses detachable sockets designed to fit a variety of different nuts and bolts (see Figure 4-32).

- **Open-end wrench**—Usually has different sizes at each end (see Figure 4-33)

- **Box-end wrench**—Usually has different sizes at each end that form a complete circle and either 6 or 12 points (see Figure 4-34)

- **Adjustable wrench**—Has jaws that can be adjusted by turning the adjusting screw (see Figure 4-35)

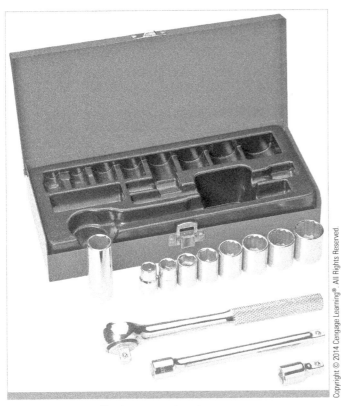

FIGURE 4-32: Socket wrench set.

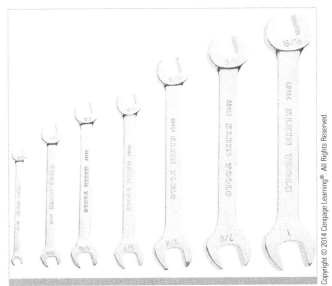

FIGURE 4-33: Open-ended wrenches.

FIGURE 4-34: Box-end wrench.

FIGURE 4-35: Adjustable wrench.

Portable Power Tools

The term **power tool** is used to describe a motorized tool. Power tools are often used for specific applications that cannot be easily done using hand tools. In many instances, a particular task cannot be performed without using a power tool. On a jobsite, portable tools are preferable because of the ease of transporting them from one location or site to another.

Power Saws

Power saws are commonly used in construction because they provide a means of cutting a material quickly and more efficiently than a hand saw.

Types of power saws include:

- **Circular saw**—The most widely used portable power saw (see Figure 4-36), most circular saws are designed to incorporate a rig guide, thus allowing for the technician to maintain a uniform cut on long passes.

FIGURE 4-36: Circular saw.

- **Reciprocating saw**—This is similar to a hand saw except that the blade moves back and forth (see Figure 4-37). These saws are ideal for ripping and crosscutting; however, these saws are not easy to control because of platform design of the saber saw. These saws are mostly used for demolition during construction.

- **Saber saw**—Equipped with a small, thin reciprocating blade, these are ideal for irregular and scroll work, as well as for ripping and crosscutting (see Figure 4-38). Tips for using saws:
  - Read the owner's manual carefully before operating your power saw.
  - Observe all of the safety precautions discussed in the owner's manual.
  - Make sure that the guards and safety devices designed for your saw are working properly and in place.
  - Keep your hands away from the blade.
  - When the saw is in use always stand to one side or the other of the rotating blade in case of wood is kicked back.
  - Before attempting to make any adjustments to the saw or the blade, unplug the saw.
  - To prevent binding and burning always ensure that the blades used on the saw are clean and sharp.
  - Never use power saws in damp or wet conditions.

FIGURE 4-37: Reciprocating saw.

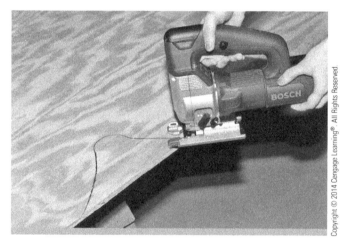

FIGURE 4-38: Saber saw.

Drills and Drivers

The most widely used drill today is the rotary drill. Of all the electric drills, the rotary drill is the most basic and is available in both corded and cordless forms. Rotary drills are used mainly for boring holes in various materials. These drills can be either corded or cordless. Drivers are akin to the rotary drill but generally have a greater torque, thus allowing the technician to drive and remove screws as well as drill through materials at a greater rate (see Figure 4-39).

Planes

Often it is necessary to trim a door or a door frame for a perfect fit. These operations are ideal for a plane. In addition, planes are perfect for planing rough boards (see Figure 4-40).

FIGURE 4-39: Portable power drill.

Tips on how to plane:

- When operating any power tool always read, understand, and follow all safety instructions.

- The end of the wood should be placed on the front of the planer, but ensure that the cutter block is not touching the lumber. Gently press down on the front handle while turning on the plane. Move the tool at an even rate along the wood. When the end has been reached, transfer pressure to the rear handle and glide off the wood to avoid gouging out the last few inches of the lumber. To plane small pieces of lumber, choose a portable planer that can be inverted in its own accessory stand.

FIGURE 4-40: Portable electric plane.

- To plane a wide surface, adjust the plane to its finest cut and plane diagonally to the grain in overlapping strokes. You will still need to use a hand plane to get rid of slight machining marks.

- Instead of taking the timber down in one pass, make several light passes.

- Never use the plane in damp or wet conditions.

Routers

Raised panel doors, rounded edges, trimming laminate are just a few jobs that are well suited for routers.

There are two types of routers: the plunge and fixed (or standard) routers. Both types offer the same results, although each type is better for particular jobs.

1. **Fixed router**—Used to make many different cuts including grooves, dadoes, rabbets, and dovetails, it is also used to shape edges and make cutouts (see Figure 4-41).

2. **Plunge router**—This is used for various applications in which a fixed router cannot be used. A typical application for a plunge router in construction is to cut a groove in the middle of stock. The base is in constant contact with the stock, while the cutting bit is pushed up and down in a plunging motion (see Figure 4-42).

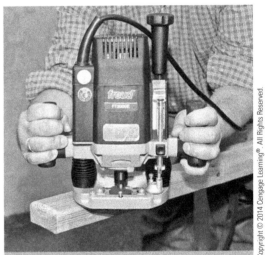

FIGURE 4-41: Portable electric router.

Finishing plastic laminates is done using special router blades. Flush-cut blades and beveled cutting blades are the most popular.

Four basic types of router bits (Figure 4-44) are:

1. **Grooving bits**—Used to cut a groove into a piece of wood, grooving bits are most commonly used in the production of decorative street and address signs. Different types of grooving bits include the V-groove, the round-nose, and the straight bit.

2. **Joinery bits**—Used extensively in the furniture and cabinet industries, these bits are used in the production of several different types of joints. Joinery router bits include the finger joint, the drawer lock, the rail and stile, and the dovetail bits.

3. **Edge bits**—These are used to produce numerous shaped edges in woodwork. Examples of edge bits include the beading, flush, and round-over bits.

4. **Specialized bits**—These bits are designed for specific purposes such as the key hole, raised panel, and T-slot bits.

A tip for using the router: Adjusting the edge guide so that it is running against a straight edge of the workpiece will help ensure that the router is going in a straight line (see Figure 4-43).

FIGURE 4-42: Plunge router.

FIGURE 4-43: A guide attached to the base of the router rides along the edge of the stock and controls the sideways motion of the router.

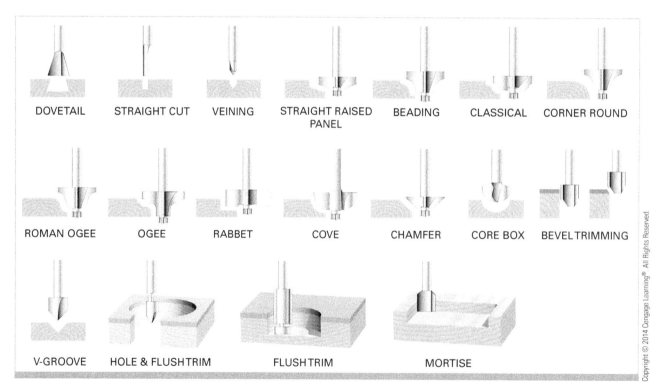

FIGURE 4-44: Router bit selection guide.

Sanders

Sanders are ideal for removing large amounts of material quickly while putting a glass-smooth surface on your projects.

Types of sanders include:

- **Detail sanders**—Designed for sanding odd shapes and small nooks in woodwork, these are small handheld sanders (see Figure 4-45).

- **Random orbit sanders**—These types of sander allow the technician to move the sander in any direction without leaving scarring in the surface of the material being sanded (see Figure 4-46).

- **Belt sanders**—These are designed to smooth out rough wood, removing large amount of material, removing paint and varnishes as well as reducing the thickness of a piece of wood (see Figure 4-47).

- **Disk sanders**—A stationary tool mounted onto a bench, disk sanders are equipped with a circular pad that accepts specially made sanding sheets. Most disk sanders also have a belt mounted vertically or horizontally on their frame (see Figure 4-48).

- **Spindle sanders**—These are stationary tools mounted on a bench, equipped with a cylindrical spindle located in the center of a large worktable. The spindle holds special sanding tubes of various grit sandpapers. Some spindle sanders are equipped with an oscillating feature that causes the spindle to raises and as it rotates. The oscillating feature increases the rate at which the sander removes stock. Spindle sanders are good for edge sanding, especially around curves and circles (see Figure 4-49).

Sandpaper is a sheet coated with abrasive particles of flint, garnet, emery, aluminum oxide, or silicon carbide. These particles are mounted on paper or cloth in "open coat" or "closed coat" density. Types of sandpaper include:

- **Flint**—The least expensive sandpaper available today; flint is a gray material that wears down quickly.

- **Garnet**—A slightly more expensive sandpaper that the flint, garnet a much harder grit than flint and more suitable for woodworking.

FIGURE 4-45: Detail sander.

FIGURE 4-46: Random orbital sander.

FIGURE 4-47: Belt sander.

FIGURE 4-48: Disk sander.

FIGURE 4-49: Spindle sander.

- **Emery**—This type is generally used on metal.
- **Aluminum oxide**—Used on either wood or metal, aluminum oxide sand paper is a reddish-colored, very sharp grit sandpaper.
- **Silicon carbide**—Commonly used for finishing metal or glass, this bluish-black material is the hardest of all.

Sandpaper grit is specified by numbers from 1500 to 12; the smaller the number, the more coarse the grit (see Figure 4-50).

Grit	Common Name	Uses
40–60	Coarse	Heavy sanding and stripping, roughing up the surface
80–120	Medium	Smoothing of the surface, removing smaller imperfections and marks
150–180	Fine	Final sanding pass before finishing the wood
220–240	Very Fine	Sanding between coats of stain or sealer
280–320	Extra Fine	Removing dust spots or marks between finish coats
360–600	Super Fine	Fine sanding of the finish to remove some luster or surface blemishes and scratches

FIGURE 4-50: Sample sandpaper grit table.

Stationary Tools

Stationary tools are often larger, more powerful, and more complex than hand tools. These include:

FIGURE 4-51: Miter saw.

- **Miter Saw**—When a precise angle is needed a miter saw is typically used. These saws normally replace the miter block and hand saw as well as the various miter saw mechanisms employed to support a manual saw in a framework. Typical applications include cutting studwork, cutting miters for picture frames, dados, skirting, architraves, and coves. They are also used in general-purpose carpentry and joinery (see Figure 4-51).
- **Chop Saw**—A lightweight circular saw attached to a spring-loaded pivoting arm and supported by a metal base. When it is necessary to have an exact or square cut, these saw are the best (see Figure 4-52).
- **Band saw**—Capable of performing a whole range of cuts, such as ripping, crosscutting, beveled cuts, and curves, the band saw is also capable of resawing and

FIGURE 4-52: Chop saw.

cutting a thick board into several thinner boards (see Figure 4-53). Tips for using the band saw:

- When operating a band saw, always stand to the left of the band. This will help prevent injury in the event of the band breaking. When bands break, they have a tendency to fly off to the right. In the case of a band failure, shut off the power to the band saw and wait until the band saw has completely stopped before attempting to repair the saw.
- Use gloves when working with (uncoiling, removing, and installing) a band saw blade.
- Always keep your hand and fingers away from unprotected parts of the blade.
- Always read and follow the manufacturer's instructions and guidelines when making adjustments to the sliding bar or post. When the guide is positioned too high, the blade will not have the proper support.
- If at all possible, never back out of a cut. This could push the blade off the wheels.
- Never use band saws in damp or wet conditions.

FIGURE 4-53: Band saw.

- **Table saws**—The most widely used stationary woodworking power tools, table saws are typically used to rip, miter, bevels edges, crosscuts, and mills nearly every imaginable woodworking joint. The diameter of the largest blade that can be used on the saw is the decisive factor in determining the table saw size. A 10-inch table saw can use up to a 10-inch saw blade (see Figure 4-54). Tips for using the table saw include:
  - Never operate a saw if the guards are not in the correct position and operating properly.

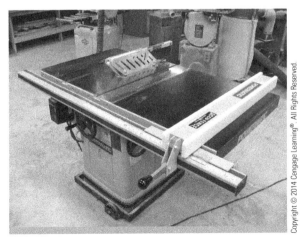

FIGURE 4-54: Table saw.

▨ If a piece of scrap needs to be removed or a workpiece needs extra support, never do this work by reaching across the saw blade.

▨ When operating a table saw, never stand directly in line with the blade—always stand to one side. Material that is caught by the saw will be ejected in line with the blade. Standing in line with the blade can result in serious injury or even death.

▨ When cutting, NEVER PULL the workpiece through the saw. Start and finish the cut from the front of the saw.

▨ When crosscutting, ensure that the workpiece is held firmly against the miter gauge. Ensure that the miter gauge works freely in the slot and that it clears both sides of the blade when tilted. Note that on some saws the miter gauge can be used only on one side when the blade is tilted.

▨ Use a push stick in accordance with the manufacturer's guidelines.

▨ Never use the miter gage and fence simultaneously.

▨ Never use a table saw in damp or wet conditions.

- **Drill press**—More precise than any portable drill, a drill press is composed of a drilling head mounted above an adjustable workbench, both being secured to a sturdy base. Most models provide the user with greater control while drilling by including a clamp and a guide (see Figure 4-55). Tips for using the drill press include:

  ▨ Always secure the material being drilled.

  ▨ Only use drill bits that are designed specifically for drill presses.

  ▨ After the drill press has been powered off, never attempt to stop a moving chuck by grabbing it.

  ▨ Always check to ensure that the chuck is properly secured before attempting to turn the drill press on.

Safety tips for using stationary tools include:

- Never operate a stationary tool unless all the safety devices and guards are in position and operating.

- Never use a cutting edge or blade that is not sharp.

- Never perform maintenance, accessory changes, and adjustments unless the tool is off and unplugged.

- Never wear loose fitting clothing. High-powered stationary tools can catch clothing and draw the operator's body into the tool.

- Never wear gloves when operating any type of stationary saw. Gloves can get caught in the tool.

- Never place your fingers and hands in front of the saw blades and other cutting tools.

- Read and understand the owner's manual and safety precautions before using the tools.

- Know the location of emergency start/stop switches as well as the tool's main power switch.

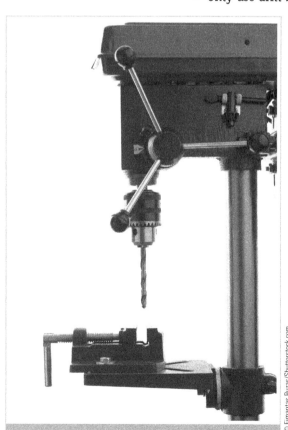

© Eimantas Buzas/Shutterstock.com

FIGURE 4-55: Drill press.

- Before operating a tool, always ensure that all blades, bits, and accessories are properly mounted as well as all locking handles and clamps are tight.
- Never use them in damp or wet conditions.

Tools, whether hand or power tools, can be dangerous if proper safety precautions are not followed. When working on projects that require the use of these tools, kindly ask the client to stay a safe distance from you while you are working and use all of the possible safety precautions.

REMODELING

Specialty Hand Tools for Demolition

Just as with any other project or type of construction, the proper tools make a big difference in the quality and ease with which a demolition project can be accomplished. In addition to the standard hand tools typically used by a facility maintenance technician, demolition sometimes requires additional specialty tools. These specialty tools include:

- **FuBar forcible entry tool**—Provides the facility maintenance technician with a demolition head, gas shut-off feature, pry bar, nail slot for pulling nails, and board jaw for gripping dimensional lumber
- **Nail jack**—Extracts nails and brads
- **Prybar and nail puller**—For pulling nails and prying apart structural components
- **Dead-on annihilator**—For breaking apart hard surfaces, pulling nails, ripping up tile, dry wall, shingles, and the like
- **Wonder bar (flat pry bar)**—For breaking apart hard surfaces, pulling nails, ripping up tile, dry wall, shingles, and the like
- **Lee Valley box tool**—For opening crates, chopping holes, and making crude openings
- **Demo Dawg**—For pulling nails, shingles, flooring, and drywall (glued and nailed)
- **Gutster**—For pulling nails, siding, roofing, flooring, joist, beams, cabinets, wire mesh, and drywall
- **10-lb Double-face sledge hammer, 36-inch fiberglass handle**—For applying more blunt force over a wide area than a framing hammer
- **Demolition screwdrivers**—Typically have a rubber grip and are made from cold-formed steel
- **Claw hammer**—For pulling nails and the like
- **Hacksaw (with demolition blade)**—For cutting through wood, nails, and metal
- **6-lb engineer hammer, 16-inch handle**—For striking metal, concrete, stone, and wood
- **Demolition axe**—For cutting through wood and siding.

Specialty Power Tools for Demolition

Specialty power tools for demolition are typically used for removing windows, doors, trim, drywall, flooring, and shingles and ceiling materials. Typical specialty power tools used for demolition include:

- **Breaker hammer**—For breaking up concrete, asphalt, rock, and other masonry materials
- **Demolition hammer drill**—For breaking up and/or drilling and chipping concrete, asphalt, rock, and other masonry materials
- **Reciprocating saw with demolition blade**—For cutting metal, plastic, and wood
- **Power floor stripper**—For removing glued-down flooring (padding, carpeting, etc.)
- **Skill saw (demolition blade)**—For cutting metal, plastic, and wood, and concrete, rock, and other masonry materials
- **Chainsaw with demolition/rescue chain**—For cutting metal, plastic, timber, and wood

SUMMARY

- Screws are employed when strong joints are needed or when other materials must be fastened to wood.
- Nuts and bolts are usually used together: The bolt is inserted through a hole drilled through the materials to be fastened together, while the nut is threaded onto the bolt from the other side and tightened.
- Sometimes when nails and screws just aren't holding an item together by themselves, adhesives are needed.
- Caulk is used to fill cracks as well as other defects in interior walls as well as exterior walls and foundation cracks.
- Solvents are used in hundreds of products to improve cleaning efficiency.
- Hand tools, such as hammers, screwdrivers, levels, and tape measures, do not use a motor.
- Tape measures are available in lengths ranging from 6 to 100 feet.
- All saws look basically the same, and their primary purpose is cutting of timber from boards, and sometimes making larger joints.
- Hammers are used for driving nails, fitting parts, and breaking up objects.
- Screwdrivers are used to tighten and loosen, as well as insert and remove, screws.
- Wrenches are used to turn nuts, bolts, and other hard-to-turn parts.

- Power saws are commonly used in construction because they provide a means of cutting a material quickly and more efficiently than a hand saw.

- The most widely used drill today is the rotary drill. Of all the electric drills the rotary drill is the most basic and is available in both corded and cordless forms.

- Demolition sometimes requires additional specialty hand tools in addition to the standard hand tools typically used by a facility maintenance technician.

- Specialty power tools for demolition are typically used for removing windows, doors, trim, drywall, flooring, and shingles and ceiling materials.

REVIEW QUESTIONS

1   What are fasteners commonly used for?

2   Define the following:
    a. Common nail
    b. Box nail
    c. Finishing nail

3   When is a screw used?

4   When are adhesives used?

5   What is a solvent?

6   List three types of saws commonly used today in facility maintenance.

7   How does a claw hammer differ from a mason's hammer?

8   How does an open-end wrench differ from a box-end wrench?

© iStockphoto.com/icphoto

9   List two types of routers commonly used today.

10 List the four basic types of router bits commonly used today.

    Circle the letter that indicates the correct answer.

11 The term _____ is used to indicate the size of a common nail.

    a. Inch

    b. Diameter

    c. Length

    d. Penny

12 Using nuts and bolts also allows for the disassembly of parts.

    a. True

    b. False

13 Wood screws generally have ___.

    a. A flat head

    b. A round head

    c. An oval head

    d. Any of the above

Name: _____

Date: _____

Tools and Equipment

FASTENERS

Upon completion of this job sheet, you should be able to demonstrate your
ability to select the correct fastener for the appropriate repair project.

1. Take an inventory and list the different types of fasteners used at your facilities.
   Write a brief description of where they would be used.

   _____

   _____

2. Are the correct fasteners used for the appropriate jobs at your facilities?
   a. If no, what, if anything, should be done?

   _____

   _____

3. What are the different types of screws available and what determines which ones
   are used?

   _____

   _____

4. Write a brief explanation of what determines whether a nail or a screw should be used.

   _____

   _____

INSTRUCTOR'S RESPONSE:

   _____

   _____

   _____

Tools and Equipment

ADHESIVES

Upon completion of this job sheet, you should be able to demonstrate your ability to select the correct adhesive for the appropriate repair project.

1. Write a brief explanation of what determines when to use an adhesive.

   _____

   _____

2. Take an inventory of the types of adhesives used at your facility. Briefly describe where the different adhesives should be used.

   _____

   _____

INSTRUCTOR'S RESPONSE:

_____

_____

_____

Name: _____

Date: _____

3

Tools and Equipment

HAND TOOLS

Upon completion of this job sheet, you should be able to demonstrate your ability to safely use hand tools.

1. Compare four different brands of tape measures.
   a. How are the symbols different from one tape measure to another and what do the symbols mean?

   _____

   _____

   b. What are the increments of the scales?

   _____

   _____

2. Check your combination square for accuracy.
   a. Using a board that has a perfectly straight edge attach a sheet of.
   b. Draw a line along the outer edge of the blade while holding the square against the edge of the board.
   c. Draw a second line about $\frac{1}{16}$ of an inch from the first line by, flip the square over so the opposite side of the blade faces up, align the square on the edge of the stock.
   d. If the square is true, the lines will be parallel.

3. Investigate the different types of drill bits available at your facility. Write a brief description of the differences and the materials they are used on.

   _____

   _____

4. Investigate the different types of hand saws available at your local home improvement center and write a brief description of the physical differences of the saws and how each one should be used.

   _____

   _____

5. Compare and contrast the uses of the portable power tools discussed in this course with their comparable hand tools and/or stationary tools. Discuss how you decide which tool is most appropriate for the project.

INSTRUCTOR'S RESPONSE:

_____

_____

_____

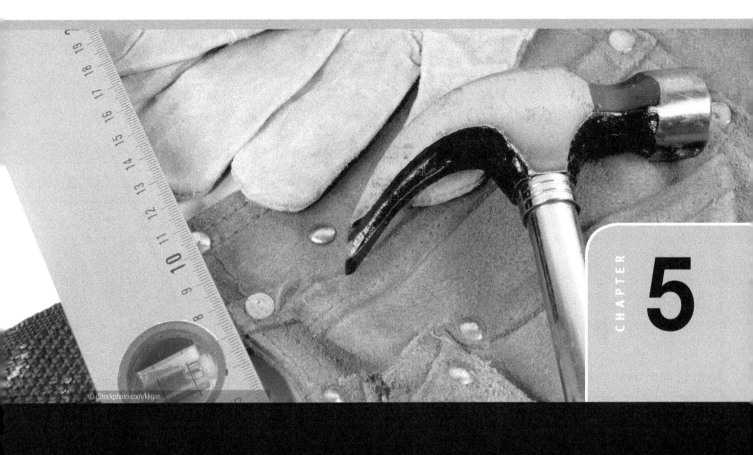

© iStockphoto.com/kkgas

CHAPTER 5

Practical Electrical Theory

OBJECTIVES

By the end of this chapter, you will be able to:

KNOWLEDGE-BASED

- Understand the principles of basic electricity.
- Describe the difference between AC and DC currents.
- Understand the properties of common electrical wires used by facilities maintenance technicians and understand and correctly measure wire size and load-carrying capacity.

- Understand the operation and functions of emergency circuits.
- Describe different types of emergency backup electrical power systems.

SKILL-BASED

- Calculate electrical load using Ohm's law.

Atom the smallest particle of an element

Matter a substance that takes up space and has weight

Element any of the known substances (of which 92 occur naturally) that cannot be separated into simpler compounds

Law of charges rule that states that like charges repel and unlike charges attract

Law of centrifugal force rule that states that a spinning object has a tendency to pull away from its center point and that the faster it spins, the greater the centrifugal force

Valence shell the outermost shell of an atom

Coulomb one coulomb is the amount of electric charge transported in one second by one ampere of current

Ampere (amp) unit of current flow

Electron theory states that the current in a circuit flows from the negative terminal to the positive terminal

Conventional current flow theory theory that states that the current in a circuit flows from the positive terminal to the negative terminal

Direct current (DC) electron flow that occurs in only one direction; used in the industry only for special applications such as solid-state modules and electronic air filters

Alternating current (AC) electron flow that occurs in one direction and then reverses direction at regular intervals

Voltage the potential electrical difference for electron flow from one line to another in an electrical circuit

Ohm's law rule that deals with the association between voltage and current and a material's ability to conduct electricity

Watt a unit of power applied to electron flow; it is the rate at which work is done when one amp of current flows through an electrical potential difference of one volt (Watt = Amp × Voltage); one watt equals 3.414 Btu

Continuous load when the maximum load current of a load is expected to continue for at least 3 hours then it is known as a continuous load

---

> *Do not make any electrical measurements without specific instructions from a qualified person. Use only electrical conductors of proper size to avoid overheating and possibly fire. Electrical circuits must be protected from current overloads.*

Introduction

Electricity is the driving force that provides most of the power for the industrialized world. It is used to light homes, cook meals, heat and cool buildings, and drive motors, and serves as the ignition for most automobiles. This chapter provides the technician with a basic understanding of practical electrical theory. Without a good understanding of the basic principles of electricity, maintaining electrical applications and equipment would be difficult if not impossible.

Basic Electrical Theory

Within the last hundred years, many practical uses of electricity have become common; however, it has been known as a force for much longer. Around 2,500 years ago, the Greeks discovered electricity. They observed that when amber was rubbed with other materials, the amber received an unknown force that had the power to attract objects such as dried leaves, feathers, bits of cloth, and other lightweight materials. The Greeks referred to this as *elektron*. Later the term "electric" was derived from this word, which means "to be like amber" or to have the capability of attracting other objects.

The Structure of Matter

Before the facility maintenance technicians can grasp the fundamentals of basic electricity, they must understand basic atomic theory. The **atom** is the basic building block of the universe. All matter is made by combining atoms into groups (see Figure 5-1). **Matter** is any substance that has mass and occupies space. An object's weight comes from the earth's gravitational pull. Matter constitutes atoms, which are small parts of a substance and may combine to form molecules. Atoms of one substance may be combined chemically with those of another to form a new substance. When molecules are formed, they cannot be broken down. They can exist in any further without changing

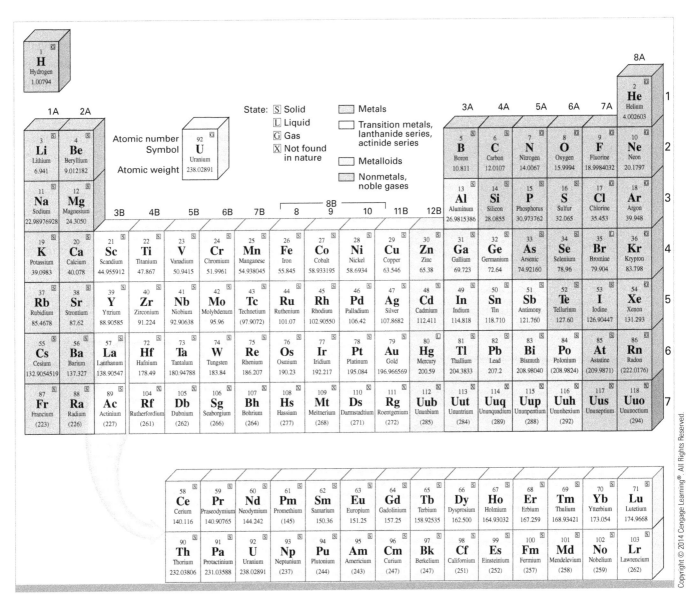

FIGURE 5-1: Periodic table of elements.

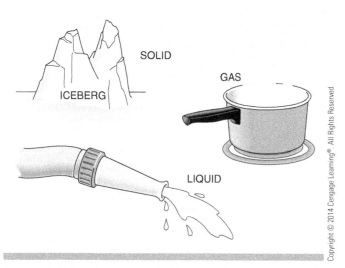

FIGURE 5-2: Using water to illustrate the three states of matter.

FIGURE 5-3: The orbital path of electrons.

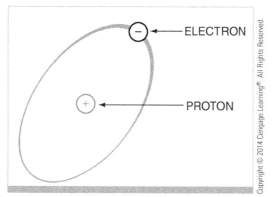

FIGURE 5-4: A hydrogen atom contains one electron and one proton.

the chemical nature of the substance. Matter exists in three states: solid, liquid, and gas. Water, for example, can exist as a solid in the form of ice, as a liquid, or as a gas in the form of steam (see Figure 5-2).

An atom is the smallest part of an element that still contains the characteristics of the element. An **element** is any of the known substances (of which 92 occur naturally) that cannot be separated into simpler compounds. The atom is composed of three principal parts: protons, neutrons, and electrons. The protons and the neutrons are located in the center of the atom (also known as the nucleus). Protons have a positive charge, whereas neutrons are electrically neutral and, therefore, have little or no effect when considering electrical characteristics. Because the neutron has no charge, the nucleus will have a net positive charge. Electrons have a negative charge and travel around the nucleus in orbits. The number of electrons in an atom will equal the number of protons in an atom. Electrons in the same orbit travel at the same distance from the nucleus but do not follow the same orbital path (see Figure 5-3).

The hydrogen atom is the simplest atom to illustrate because it has only one proton and one electron (see Figure 5-4). It is also the lightest of the elements. Not all atoms are as simple as the hydrogen atom. Most wiring used to conduct an electrical current is made of copper. Figure 5-5 shows a copper atom, which has 29 protons and 29 electrons. Some electron orbits are farther away from the nucleus than others. As can be seen, 2 travel in an inner orbit, 8 in the

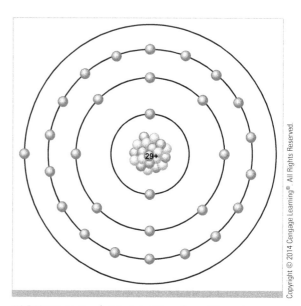

FIGURE 5-5: A copper atom has 29 protons and 29 electrons.

next, 18 in the next, and 1 in the outer orbit. It is this single electron in the outer orbit that makes copper a good conductor.

The Law of Charges

To understand atoms and atomic theory you must first have a good grasps of two basic laws of physics. These are the laws of charges and the law of centrifugal force. The **law of charges** states that opposite charges attract, whereas like charges repel (see Figure 5-6). If two objects with unlike charges come close to each other, the lines of force attract (Figure 5-7). Likewise, if two objects with like charges come close to each other, the lines of force repel (Figure 5-8). For example, consider what happens if you try to connect the north poles of two magnets. It is this principle that helps the electrons to maintain their orbit around the nucleus.

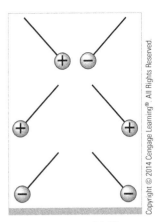

FIGURE 5-6: Unlike charges attract and like charges repel.

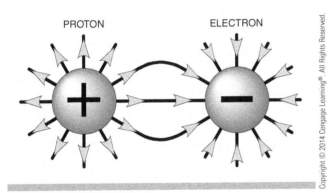

FIGURE 5-7: Unlike charges are attracted to each other.

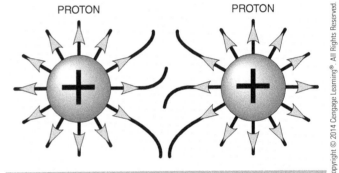

FIGURE 5-8: Like charges repel each other.

The Law of Centrifugal Force

Another law that is important to understanding the behavior of an atom is the **law of centrifugal force**, which states that a spinning object will pull away from its center point and that the faster it spins, the greater the centrifugal force will be (see Figure 5-9). If you tie an object to a string and spin it around, it will try to pull away from you. The faster the object spins, the greater will be the force that tries to pull the object away.

Although atoms are often drawn flat, electrons orbit the nucleus in a spherical fashion (see Figure 5-10). Electrons travel at such a high rate of speed that they form a shell around the nucleus. For this reason, electron orbits are often referred to as shells.

The number of electrons that any one orbit, or shell, can contain can be determined using the formula ($2N^2$), where N represents the number of the shells or orbits (see Figure 5-11).

FIGURE 5-9: Centrifugal force causes an object to pull away from its axis point.

FIGURE 5-10: Electron orbits.

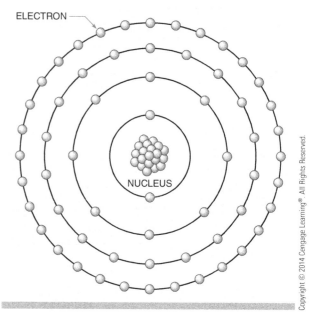

ELECTRON

NUCLEUS

FIGURE 5-11: Electrons orbit the nucleus in a circular fashion.

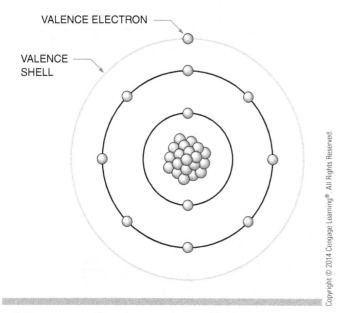

VALENCE ELECTRON

VALENCE SHELL

FIGURE 5-12: The electrons located in the outer orbit of an atom are valence electrons.

The outer shell of an atom is known as the **valence shell**, and electrons located there are known as valence electrons (Figure 5-12). It is the valence shell and the valence electrons contained within that shell that is the primary focus as far as the study of electricity is concerned. The valence shell of an atom cannot contain more than eight electrons. A conductor, for instance, is generally made from a material that contains one or two valence electrons. Atoms with one or two valence electrons are unstable and can be made to give up these electrons with little effort. Conductors are materials that permit electrons to flow through them easily. Examples of good conductors are gold, silver, copper, and aluminum.

Conductors

Good conductors contain atoms with few electrons in the outer orbit. Three common metals—copper, silver, and gold—are good conductors, and the atom of each has only one electron in the outer orbit. These electrons are considered to be free electrons because they move easily from one atom to another.

Insulators

Atoms with several electrons in their outer orbits are poor conductors. It is difficult to free these electrons, and materials composed of such atoms are considered to be insulators. Glass, rubber, and plastic are examples of good insulators.

Quantity Measurement for Electrons

A **coulomb** is a quantity measurement for electrons. One coulomb contains 6.25×10^{18}, or 6,250,000,000,000,000,000, electrons. Coulomb's law of electrostatic charges describes the electrostatic interaction between two electrically charged particles. It states that the force of electrostatic attraction or repulsion is directly proportional to the product of the two charges and inversely proportional to the square of the distance between them.

The amp, or ampere, is named for André Ampére, a scientist who lived from the late 1700s to the early 1800s. The amp (A) is defined as one coulomb per second. One amp of current flows through a wire when one coulomb flows past a point in one second (see Figure 5-13).

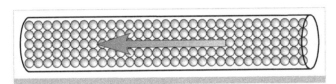

FIGURE 5-13: One ampere equals one coulomb per second.

The amount of electricity that flows through a circuit is measured in **amperes (amps)**. The two theories that describe current flow are: the electron theory and the conventional current flow theory. The **electron theory** states that because electrons are negative charged particles, current flows from the most negative point in the circuit to the most positive. The **conventional current flow theory** is older than the electron theory and states that current flows from the most positive point to the most negative (see Figure 5-14).

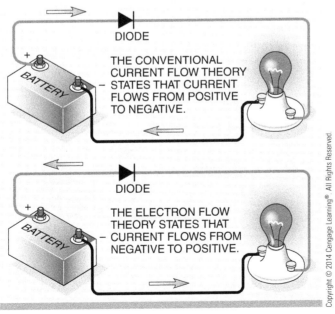

FIGURE 5-14: Conventional current flow and electron flow theories.

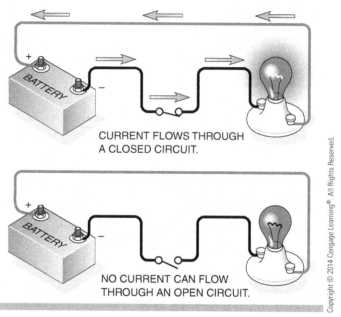

FIGURE 5-15: Current flows only through a closed circuit.

Before current can flow through a circuit, there must first be a complete path for the current to take. A complete circuit is often referred to as a closed circuit because the power source, conductors, and load form a closed loop (see Figure 5-15). A short circuit, which has very little or no resistance, generally occurs when the conductors leading from and back to the power source become connected (see Figure 5-16). Another type of circuit, one that is often confused with a short circuit, is a grounded circuit (see Figure 5-17). Grounded circuits can also cause an excessive amount of current flow. When an unintended path is established to the ground, a grounded circuit has occurred. Finally, an open circuit is one in which a path cannot be established between the positive and negative terminals of the power source. Many circuits contain an extra conductor called the grounding conductor. The grounding conductor helps prevent a shock hazard when the ungrounded, or hot, conductor comes in contact with the case or frame of the appliance.

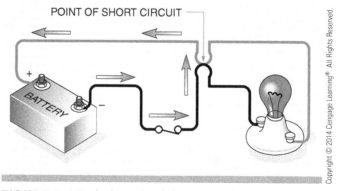

FIGURE 5-16: A short circuit bypasses the load and permits too much to flow.

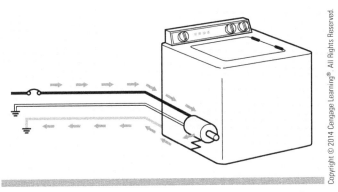

FIGURE 5-17: The grounding conductor provides a low resistance path to ground.

Direct Current

Direct current (DC) travels in one direction. Because electrons have a negative charge and travel to atoms with a positive charge, DC is considered to flow from negative to positive.

Alternating Current

Alternating current (AC) is continually and rapidly reversing current. The charge at the power source (generator) is continually changing direction; thus the current continually reverses itself. For several reasons, most electrical energy generated for public use is AC. It is much more economical to transmit electrical energy long distances in the form of AC. The voltage of this type can be readily changed so that it has many more uses. DC still has many applications, but it is usually obtained by changing AC to DC or by producing the DC locally where it is to be used.

Probably the single greatest advantage of alternating current is that is can be easily transmitted over long distances where direct current cannot. Voltage can be stepped up for long-distance transmission and then stepped back down when it is to be used by some device. It is this fact that makes high-voltage transmission line possible. The advantage of high-voltage transmission is that less current is required to produce the same amount of power. Reducing current allows smaller wires to be used, which results in a savings of material.

Electrical Units of Measurement

Electromotive force (emf) or voltage (V) is used to indicate the difference of potential in two charges. **Voltage** is defined as the force that causes electrons to move from atom to atom in a conductor. When an electron surplus builds up on one side of a circuit and a shortage of electrons exists on the other side, a difference of potential or emf is created. The unit used to measure this force is the volt.

All materials oppose or resist the flow of electrical current to some extent. In good conductors this opposition or resistance is very low, whereas in poor conductors it is high. The unit used to measure resistance is ohm. A conductor has a resistance of 1 ohm when a force of 1 volt causes a current of 1 amp to flow.

Volt = Electrical force or Pressure (V)
Ampere = Quantity of electron flow rate (A)
Ohm = Resistance to electron flow (Ω)

The Electrical Circuit

An electrical circuit must have a power source, a conductor to carry the current, and a load or device to use the current. Generally, there is also a means for turning the electrical current flow on and off. Figure 5-18 shows an electrical generator for the source, a wire for the conductor, a light bulb for the load, and a switch for opening and closing the circuit.

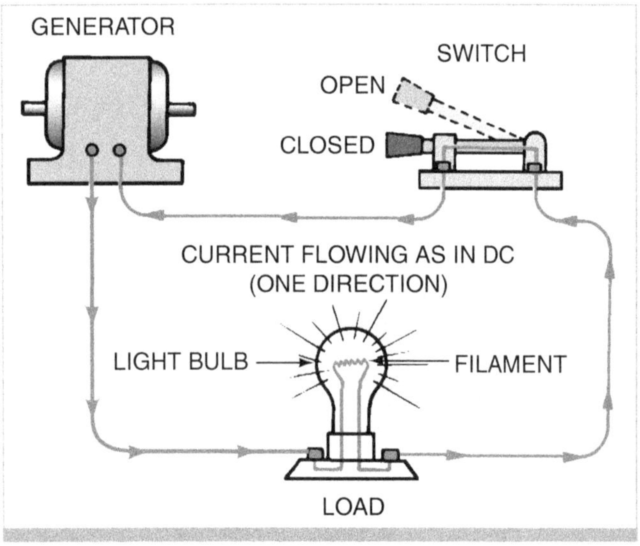

The generator produces the current by passing many turns of wire through a magnetic field. If it is a DC generator, the current will flow in one direction. If it is an AC generator, the current will continually reverse itself. However, the effect on this circuit will generally be the same for both AC and DC. The wire or conductor provides the path for the electricity to flow to the bulb and complete the circuit. The electrical energy is converted to heat and light energy at the bulb element.

The switch is used to open and close the circuit. When the switch is open, no current will flow. When it is closed, the bulb element will produce heat and light because current is flowing through it.

FIGURE 5-18: An electric circuit.

Calculate Electrical Load Using Ohm's Law

Ohm's law deals with the relationship between voltage and current and a material's ability to conduct electricity. Typically, this relationship is written as Voltage = Current × Resistance. In its simplest form, Ohm's law states that it takes 1 volt to push 1 amp through 1 ohm.

In a DC circuit, the voltage is inversely proportional to the resistance and directly proportional to the current.

The resistance, voltage, and/or current of a device (load) can be calculated by rearranging Ohm's law (Figure 5-19). The three basic forms of Ohm's law formulas are as follows:

$$E = I \times R$$

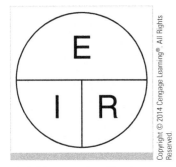

FIGURE 5-19: Ohm's law chart for finding values of voltage, current, and resistance.

If the current and the resistance is known then the voltage can be determined.

I = E/R

Current can be found if the voltage and resistance are known.

R = E/I

Resistance can be found if the voltage and current are known.

Where

E = Emf or voltage

I = Intensity of current or amperage

R = Resistance

The first formula states that if the current and the resistance are known then the voltage can be determined. The second formula states that the current can be found if the voltage and resistance are known. The third formula states that if the voltage and current are known, the resistance can be found.

For example, determine the voltage for a load that draws 30 amps and has a resistance of 500 ohms (see Figures 5-20 and 5-21).

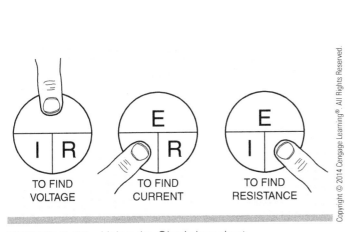

FIGURE 5-20: Using the Ohm's law chart.

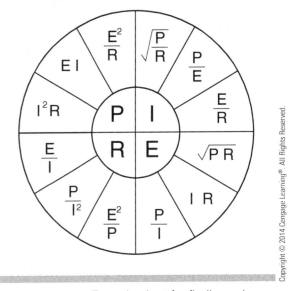

FIGURE 5-21: Formula chart for finding values or voltage, current, resistance, and power.

E = I × R

E = 30 amps × 500 ohms

E = 15,000 volts

Determine the current for a 120-volt load that has a resistance of 0.5 ohms.

I = E/R

I = 120 volts/0.5 ohms

I = 60 amps

Determine the resistance for a 120-volt load that draws 15 amps.

R = E/I
R = 120 volts/15 amps
R = 8 ohms

Wattage is proportional to the amounts of voltage and current flow. It can be calculated using the formula: Watt = Amps × Voltage
Where:

Watt = The rate at which work is done when 1 amp of current flows through an electrical potential difference of 1 volt

Amp = Ampere (amp) unit of current flow

Voltage = The potential electrical difference for electron flow from one line to another in an electrical circuit

For example: determine the wattage of a 120-volt load that draws 15 amps.

R = E/I
R = 120 volts/15 amps
R = 8 ohms
Watts = E × I
Watts = 120 volts × 15 amps
Watts = 1,800 watts

Characteristics of Circuits

There are two kinds of circuits: series and parallel. Three characteristics make a series circuit different from a parallel circuit:

1. The voltage is divided across the electrical loads in a series circuit. For example, if there are three resistances of equal value in a circuit, and the voltage applied to the circuit is 120 volts, the voltage is equally divided across each resistance (see Figure 5-22). If the resistances are not equal, the voltage is divided across each according to its resistance (Figure 5-23).

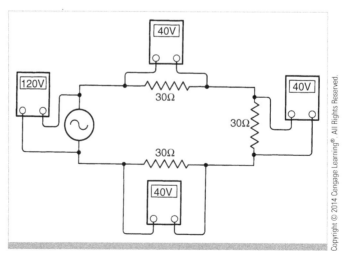

FIGURE 5-22: Three resistors of equal value divide the voltage equally in a series circuit.

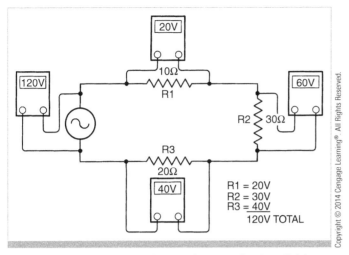

FIGURE 5-23: Three resistors of unequal value divide the voltage according to their values.

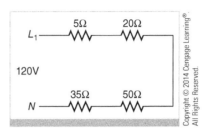

FIGURE 5-24: Series circuit.

For example:

In the circuit above (Figure 5-24), find the total resistance, the voltage drop for each element, and the current for the circuit.

1. Find the total resistance of the circuit.

   The total resistance of the circuit can be found by using the following formula.

   Rt = R1 + R2 + R3 + R4 + R . . .
   Rt = 5 ohms + 20 ohms + 35 ohms + 50 ohms
   Rt = 110 ohms

2. Find the amperage draw of the circuit.

   The amperage draw of the circuit can be calculated by using Ohm's Law.

   I = E/R
   I = 120 volts/110 ohms
   I = 1.09 amps

3. Find the voltage drop across each element in the circuit.

   The voltage drop across each element in the circuit can be calculated by rewriting Ohm's law.

   E = IR

   For the 5-ohm resistor, the voltage drop would be:

   E = IR
   E = 1.09 amps × 5 ohms
   E = 5.45 volts

   For the 20-ohm resistor, the voltage drop would be:

   E = IR
   E = 1.09 amps × 20 ohms
   E = 21.8 volts

For the 50-ohm resistor, the voltage drop would be:

$E = IR$
$E = 1.09 \text{ amps} \times 50 \text{ ohms}$
$E = 54.5 \text{ volts}$

For the 50-ohm resistor, the voltage drop would be:

$E = IR$
$E = 1.09 \text{ amps} \times 35 \text{ ohms}$
$E = 38.15 \text{ volts}$

2. The total current for the circuit flows through each electrical load in the circuit. With one power supply and three resistances, the current must flow through each to reach the others.

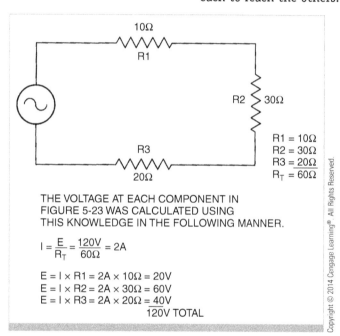

THE VOLTAGE AT EACH COMPONENT IN FIGURE 5-23 WAS CALCULATED USING THIS KNOWLEDGE IN THE FOLLOWING MANNER.

$I = \dfrac{E}{R_T} = \dfrac{120V}{60\Omega} = 2A$

$E = I \times R1 = 2A \times 10\Omega = 20V$
$E = I \times R2 = 2A \times 30\Omega = 60V$
$E = I \times R3 = 2A \times 20\Omega = \underline{40V}$
$120V \text{ TOTAL}$

FIGURE 5-25: The sum of three resistors is the total resistance of the circuit.

3. The total resistance in the circuit is equal to the sum of the resistances in the circuit. For example, the total resistance of a circuit containing three resistances is the sum of the three resistances (see Figure 5-25).

Three characteristics of a parallel circuit make it different from a series circuit:

1. The total voltage for the circuit is applied across each circuit resistance. The power supply feeds each power consuming device (load) directly (see Figure 5-26).

2. The current is divided among the different loads, or the total current is equal to the sum of the currents in each branch (see Figure 5-27).

3. The total resistance is less than the value of the smallest resistance in the circuit. Calculating the resistances in a parallel circuit requires a procedure that is different from simply adding them.

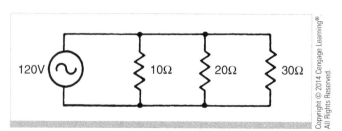

FIGURE 5-26: The total voltage for a parallel circuit is applied across each component.

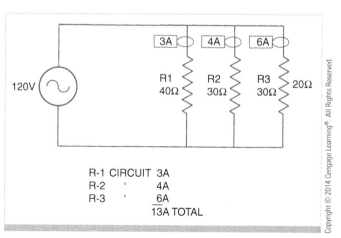

R-1 CIRCUIT 3A
R-2   "   4A
R-3   "   6A
          _____
          13A TOTAL

FIGURE 5-27: The current is divided between the different loads in a parallel circuit.

A parallel circuit allows current along two or more paths at the same time. This type of circuit applies equal voltage to all loads. The general formula used to calculate total resistance in a parallel circuit is as follows:

$$R_{total} = \cfrac{1}{\cfrac{1}{R_1} + \cfrac{1}{R_2} + \cfrac{1}{R_3} + \ldots}$$

For example, the total resistance of the circuit in Figure 5-28 is determined as follows:

$$R_{total} = \cfrac{1}{\cfrac{1}{R_1} + \cfrac{1}{R_2} + \cfrac{1}{R_3}}$$

$$= \cfrac{1}{\cfrac{1}{40} + \cfrac{1}{30} + \cfrac{1}{20}}$$

$$= \cfrac{1}{0.025 + 0.0333 + 0.05}$$

$$= \cfrac{1}{0.1083}$$

$$= 9.234 \text{ ohms}$$

In the following parallel circuit (Figure 5-29) calculate the total resistance of the circuit, the total current draw of the circuit, and the current draw for each resistor.

1. Calculate the current draw for each individual load.

   The current for each individual resistor can be calculated by using Ohm's law.

   I = E/R

   For the 35.7-ohm resistor, the current draw would be:

   I = 120 volts/35.7 ohms
   I = 3.36 amps

   For the 15.5-ohm resistor, the current draw would be:

   I = 120 volts/15.5 ohms
   I = 7.74 amps

2. Determine the total resistance for the circuit. The total resistance of the parallel circuit can be obtained by using the formula:

$$RT = \frac{(R1 \times R2)}{(R1 = R2)}$$

   Rt = (15.5 × 35.7)/(15.5 + 35.7)
   Rt = (553.35)/(51.2)
   Rt = 10.80 ohms

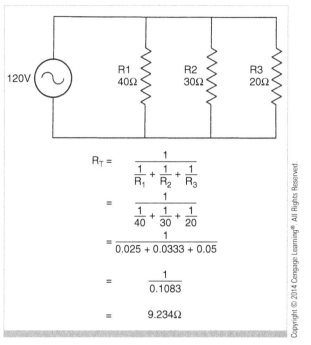

$$R_T = \cfrac{1}{\cfrac{1}{R_1} + \cfrac{1}{R_2} + \cfrac{1}{R_3}}$$

$$= \cfrac{1}{\cfrac{1}{40} + \cfrac{1}{30} + \cfrac{1}{20}}$$

$$= \cfrac{1}{0.025 + 0.0333 + 0.05}$$

$$= \cfrac{1}{0.1083}$$

$$= 9.234\Omega$$

FIGURE 5-28: The total resistance for a parallel circuit must be calculated.

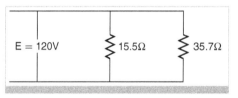

FIGURE 5-29: Parallel circuit.

R1	R2	R3
40Ω	30Ω	20Ω

CALCULATED TOTAL RESISTANCE IS 9.234Ω
CALCULATED AMPERAGE:

$$I = \frac{E}{R}$$

$$= \frac{120}{9.234}$$

$$= 12.995 \text{ OR } 13A$$

FIGURE 5-30: The current in Figure 5-28 can be calculated from the total resistance when the applied voltage is known.

3. Determine the total current draw for the circuit. The total current draw for the circuit can be calculated by using Ohm's law.

$$I = E/R$$
$$I = 120 \text{ volts}/10.80 \text{ ohms}$$
$$I = 11.11 \text{ amps}$$

To determine the total current draw, Ohm's law can be used. For example, in the previous problem, the voltage is known and the current flow for each component can be calculated from the total resistance, as shown in Figure 5-30. Again, notice that the total resistance is less than that of the smallest resistor.

A complex circuit is also known as a series-parallel circuit; it is a circuit containing loads in both series and parallel (see Figure 5-31). Complex circuits are the most widely used circuits in residential wiring today.

To solve a complex circuit:

- Start with a small, simple chunk of the circuit that you know how to work with that is either all series or all parallel.
- Then keep redrawing the circuit at each step as it simplifies.

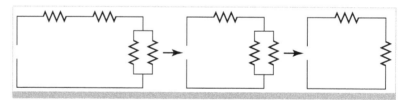

FIGURE 5-31: Complex circuit.

Electrical Power

Electrical power (P) is measured in watts. A **watt** (W) is the power used when 1 amp flows with a potential difference of 1 volt. Therefore, power can be calculated by multiplying the voltage and the current flowing in a circuit.

Watts = volts × amperes

or

$$P = E \times I$$

The consumer of electrical power pays the electrical utility company according to the number of kilowatts (kW) used for a certain time span usually billed as kilowatt hours (kW h). A kilowatt is equal to 1000 watts. To determine the power being consumed, divide the number of watts by 1000:

$$P(\text{in kW}) = \frac{E \times I}{1000}$$

AC vs. DC

As mentioned earlier, there are two types of electrical current used to provide power to electrical circuits today: alternating current (AC) and direct current (DC). AC is electrical current in which the magnitude and direction will vary cyclically, whereas that of DC does not modulate and will, therefore, remain constant. AC is primarily used in residential, commercial, and industrial applications as the primary source of power. DC is used primarily in electronic, low-voltage applications, batteries, and so on.

Square Waves

Alternating current differs from direct current in that it reverses its direction of flow at set intervals. Depending upon how it is produced, alternating current wave forms will vary. One of the wave form frequently encountered is the square wave. A square wave is produced by electronic devices called oscillators.

Triangle Waves

Another common AC wave form is the triangle wave. The triangle wave is a linear wave. A linear wave is one in which the voltage rises at a constant rate with respect to time.

Sine Waves

As stated earlier alternating current is continually reverses direction. This can be seen using an oscilloscope. An oscilloscope is an instrument that measures the amount of voltage over a given period.

The sine wave illustrates the voltage of one cycle through 360°. In the United States and Canada, this is repeated 60 times in one second. The rate at which the cycle is repeated is known as the frequency and is measured in hertz (Hz). In the United States and Canada, the standard frequency is 60 Hz. Figure 5-32 is a sine wave as it would be displayed on an oscilloscope. At the 90° point, the voltage reaches its peak (positive); at 180° it is back to 0; at 270° it reaches its negative peak; and at 360° it is back to 0.

Figure 5-33 shows the peak and peak-to-peak values. As illustrated in figure 5-33, the voltage is at its peak for a very short amount of time in each cycle. Because of

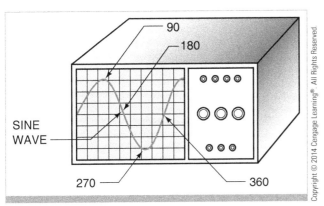

FIGURE 5-32: A sine wave displayed on an oscilloscope.

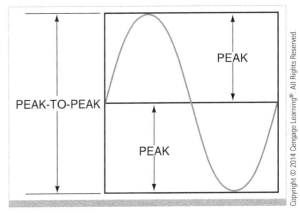

FIGURE 5-33: Peak and peak-to-peak values AC voltage values.

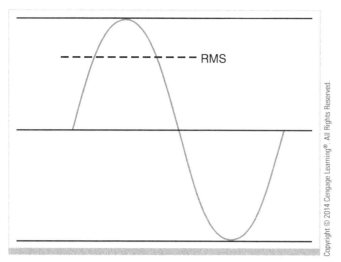

FIGURE 5-34: The RMS or effective voltage.

this, the peaks of the voltages are not the effective voltage values. The effective voltage is the root-mean-square (RMS) voltage (Figure 5-34). The RMS voltage is 0.707 × the peak voltage. If the peak voltage were 170 volts, the effective voltage measured by a voltmeter would be 120 volts (170 volts = 0.707 × 120.19 volts).

Single-Phase AC vs. Three-Phase AC

Single-phase AC is the most commonly used electrical supply for single-family and multifamily dwellings. Single-phase AC consists of two ungrounded conductors (hot wires) and one grounded conductor (neutral wire). When measuring voltage between the two hot wires of this type of system, you will read approximately 240 volts. If you read from any one hot wire to the neutral, you should read approximately 120 volts.

Three-phase AC is most often used for commercial buildings, healthcare facilities, and industrial facilities. There are always three ungrounded conductors (hot wires) in this type of system. The three-phase system may or may not have a neutral wire. The common voltage levels for the three-phase system will be 208, 230, 240, 277, 480, 575, or 600 when measured between the two hot wires. This is referred to as the line voltage. If a neutral is present, then voltage read from any one hot wire to the neutral wire will equal the line voltage divided by the square root of 3 (1.732).

Wire Sizes and Load-Carrying Capacity

The National Electric Code (NEC) governs the type and size of wire that can be used for a particular application and certain amperage. It provides the basic minimum size for a conductor to prevent the conductor from overheating and causing a fire. When selecting a conductor size, NEC® Article 310 should be referred.

This article contains information about the following:

- Insulation types
- Conductors in parallel
- Wet and dry locations
- Marking, maximum operating temperatures
- Permitted use
- Direct burial
- Ampacity tables
- Adjustment and correction factors

Typically, in a residential application, copper is the most popular conductor. Copper wire bends easily, has good mechanical strength, resists corrosion, and can easily be joined together.

In an electrical circuit, the conductors and overload protection are the key components. The conductors must be selected to meet the requirements of the equipment that they service. Conductors must be selected to conform to the requirement outlined in NEC as well as the local building codes. Overload protection is vital not only for people that might come in contact with the equipment but also for the equipment itself.

The National Electrical Code (NEC) establishes important fundamentals that weave their way through the decision-making process for an electrical installation. The NEC defines **continuous load** as "a load where the maximum current is expected to continue for three hours or more." General lighting outlets and receptacle outlets in residences are not considered to be continuous loads.

When wiring a house, it is all but impossible to know which appliances, lighting, heating, and other loads will be turned on at the same time. Different families have different lifestyles. The rules for doing the calculations are found in Article 220 of the NEC. For lighting and receptacles, the computations are based on volt-amperes per square foot. For small appliance circuits, such as those in kitchens and dining rooms, the basis is 1,500 volt-amperes per circuit. For large appliances such as dryers, electric ranges, ovens, cooktops, water heaters, air conditioners, heat pumps, and so on, all of which are not used continuously or at the same time, demand factors are used in the calculations. Following the requirements in the NEC, the various calculations roll together in steps that result in the proper sizing of branch circuits, feeders, and service equipment.

The American Wire Gauge

The American Wire Gauge (AWG) is used mainly in the United States for the diameters of round, solid, nonferrous electrical wire. The gauge size is important for determining the current carrying capacity of a conductor. Gauge sizes are determined by the number of draws necessary to produce a given diameter or wire. When electrical wire is manufactured it pulled through a series of dies with each die having a smaller diameter that the last. Therefore, a 12 AWG wire would have been pulled (drawn) through 12 dies. As the AWG number of a conductor increases, its diameter decreases. For example, a 12 AWG (0.0808 inches in diameter) wire will have a smaller diameter than a 4 AWG (0.2043 inches in diameter) wire.

It is the cross-sectional area of particular gauge (diameter) wire that is an important factor in determining the amount of current that a conductor can safely carry.

Emergency Circuits

An emergency circuit is intended to supply illumination and power automatically to designated areas and equipment when the normal source of power fails. For the emergency circuit, a portable or temporary alternate source must be available whenever the emergency generator is out of service for major maintenance or repair.

Ground fault circuit interrupts (GFCIs) are used to prevent people from being electrocuted. They work by sensing the amount of current flow on both the

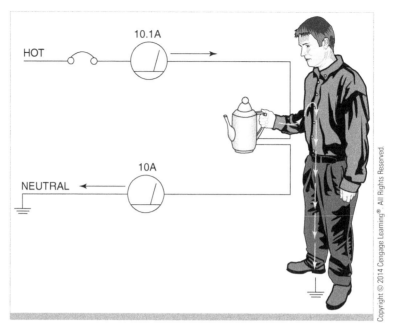

FIGURE 5-35: A ground fault.

ungrounded (hot) and grounded (neutral) conductors supplying power to a device. A ground fault occurs when a path to ground, other than the intended path, is established (see Figure 5-35).

Ground fault indication is required for emergency systems operating at more than 150 volts to ground and overcurrent devices rated at 1,000 amps or more. Wiring for emergency circuits must be kept entirely independent of all other wiring unless required to be associated with normal source wiring.

Emergency Backup Electrical Power Systems

Emergency backup electrical power systems are used to provide continued access to electrical services during power outages. Currently there are a variety of different systems on the market today that range in capacity as well as price that can supply enough power to run a single device or an entire facility until the power is restored. These systems can be based on fossil fuel–powered generators, battery-based storage systems, or propane or natural gas supply, or can be directly wired into the facility's circuit.

An emergency generator can be set up to power the structure during an outage or just the essential loads, such as the furnace, security systems, and various appliances.

Transformers

Transformers are electrical devices that produce an electrical current in a second circuit through electromagnetic induction. Transformers have a primary winding, a core usually made of thin plates of steel laminated together, and a secondary winding. There are step-up and step-down transformers. When the windings of a transformer are physically and electrically isolated from one another the transformer is known as an isolation transformer.

Electric Motors

Electric motors are used in almost every aspect of life. For a facility maintenance technician, electric motors are used to drive compressors, fans, pumps, dampers, and any other device that needs energy to power its movement. The electric motor is a device that converts electrical energy into mechanical energy (motion). Regardless of the application or the type of equipment, electric motors play an important part in the operation of a facility. An electric motor works on the principle of a rotating magnetic field is responsible for the continuous rotation of an electric motor. Currently, there are many different types of electric motors available today with each having different running and starting characteristics. Most single-phase motors are designed and used according to their running and starting torque. Some of the commonly used electric motors are:

- **Split-phase motors**—Have a medium amount of starting torque and good running efficiency
- **Capacitor start motors**—Use start capacitors to increase the starting torque of the motor
- **Capacitor start, capacitor run motors**—Use both start and run capacitors
- **The PSC motor**—Uses only a run capacitor
- **The shaded-pole motor**—Has very low starting torque
- **Three-phase motors**—Are used for commercial and industrial applications
- **Variable-speed motors**—Ramp up and down, often using DC converters, inverters, and rectifiers
- **ECM motors**—Are commutate with permanent magnets

SUMMARY

- The atom is the basic building block of the universe; is composed of three principal parts: protons, neutrons, and electrons; and is the smallest part of an element that still contains the characteristics of the element.
- An element is any of the known substances (of which 92 occur naturally) that cannot be separated into simpler compounds.
- Matter is any substance that has mass and occupies space.
- Protons have a positive charge; neutrons have no charge and have little or no effect when considering electrical characteristics; electrons have a negative charge and travel around the nucleus in orbits.

- The law of charges states that opposite charges attract and like charges repel.

  The law of centrifugal force states that a spinning object will pull away from its center point and that the faster the object spins, the greater the force will be that tries to pull the object away.

- The valence shield is the outer shield of an atom. The electrons located there are known as valence electrons.

- Conductors are materials that permit electrons to flow through them easily. Atoms with several electrons in their outer orbits are poor conductors and, therefore, are good insulators.

- A coulomb is a quantity measurement for electrons. One coulomb contains 6.25×10^{18}, or 6,250,000,000,000,000,000 electrons.

- The amp (A) is defined as one coulomb per second. One amp of current flows through a wire when one coulomb flows past a point in one second.

- The electron theory states that because electrons are a negative particle, current flows from the negative terminal to the positive terminal.

- The conventional current flow theory is older than the electron theory and states that current flows from the positive terminal to the negative terminal.

- Direct current (DC) travels in one direction and alternating current (AC) is continually and rapidly reversing.

- Electromotive force (emf) or voltage (V) is used to indicate the difference of potential in two charges. Voltage is defined as the force that causes electrons to move from atom to atom in a conductor.

- Ohm's law deals with the association between voltage and current and a material's ability to conduct electricity.

- There are two kinds of circuits: series and parallel.

- The characteristics of a series circuit
  - The voltage is divided across the electrical loads in a series circuit.
  - The total current for the circuit flows through each electrical load in the circuit.
  - The total resistance in the circuit is equal to the sum of the resistances in the circuit.

- The characteristics of a parallel circuit
  - The total voltage for the circuit is applied across each circuit resistance.
  - The current is divided among the different loads, or the total current is equal to the sum of the currents in each branch.

- The total resistance is less than the value of the smallest resistance in the circuit. Calculating the resistances in a parallel circuit requires a procedure that is different from simply adding them.

- A complex circuit is also known as a series-parallel circuit; it is a circuit containing loads in both series and parallel.

- Electrical power (P) is measured in watts. A watt (W) is the power used when 1 amp flows with a potential difference of 1 volt.

- When selecting a conductor size, NEC® Article 310 should be referred to.

- The American Wire Gauge is used mainly in the United States for the diameters of round, solid, nonferrous electrical wire.

REVIEW QUESTIONS

1   What are the three states of matter?

2   What are the three principal parts of an atom?

3   State the law of charges.

4   What is a coulomb?

5   What is an amp?

6   What is electricity?

REVIEW QUESTIONS

7   What is a watt?

8   True or false? Direct current flows in one direction only.

9   True or false? Alternating current can be transformed and direct current cannot.

10   True or false? Alternating current reverses its direction of flow at periodic intervals.

11   True or false? The most common AC wave form is the square wave.

12   True or false? There are 180° in one complete sine wave.

13   True or false? One complete wave form is
     called a cycle.

14   The number of complete cycles
     that occur in one second is called
     _____.

15   _____ is measured in
     hertz (Hz).

16   The _____is the maximum
     amount of voltage attained by the
     wave form.

17   The _____ value of voltage
     will produce as much power as a like
     amount of DC voltage.

18   What are the three principle parts that
     make up an atom?

REVIEW QUESTIONS

19 _____ have a positive charge; _____ have no charge and have little or no effect when considering electrical characteristics; electrons have a negative charge and travel around the nucleus in orbits.

20 Most of the electrical power generated in the world is _____.

21 Define "valence shell."

22 Explain the difference between AC and DC voltage.

23 Explain the difference between single-phase and three-phase AC power.

24 A _____ is a circuit that contains both series and parallel circuits.

© iStockphoto.com/icphoto

Name: _____

Date: _____

Practical Electrical Theory

UNDERSTANDING OHM'S LAW

Upon completion of this job sheet, you should have a basic understanding of Ohm's law.

1. Determine the voltage of a load drawing 20 amps and having a resistance of 25 ohms.

2. Determine the wattage of a device drawing 10 amps and having a resistance of 75 ohms.

INSTRUCTOR'S RESPONSE:

_____

_____

_____

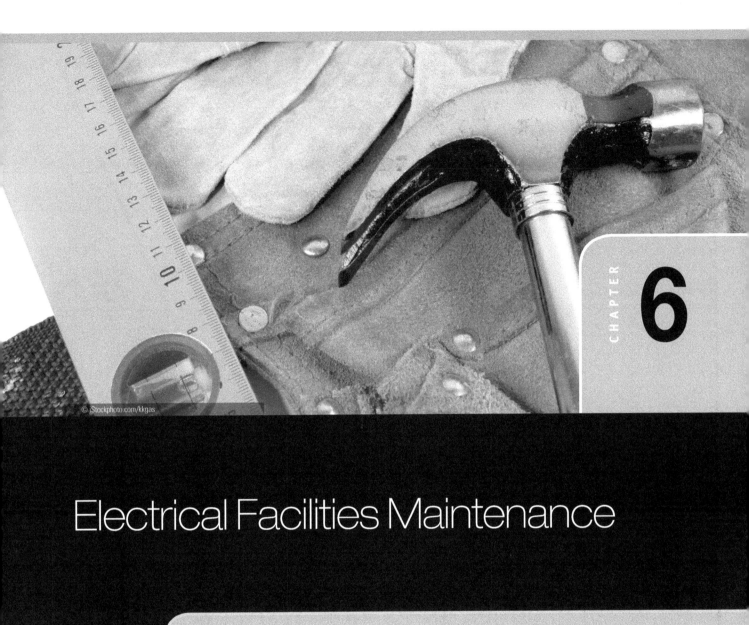

Electrical Facilities Maintenance

OBJECTIVES

By the end of this chapter, you should be able to:

KNOWLEDGE-BASED

- Understand and apply OSHA regulations that cover electrical installations.
- Describe the difference between AC and DC.
- Correctly identify and select the boxes most commonly used in electrical installations.
- Correctly identify and select different types of electrical devices and fixtures.
- Describe the different types of emergency backup systems.
- Describe the basic steps for performing demolition on an electrical system in a kitchen, bathroom, and utility room.

SKILL-BASED

- Follow systematic, diagnostic, and troubleshooting practices.
- Perform tests on smoke alarms, fire alarms, medical alert systems, and emergency exit lighting.
- Perform tests on GFCI receptacles.
- Repair and/or replace common electrical devices such as receptacles and switches.
- Repair and/or replace lighting fixtures and/or bulbs, and ballasts.

Qualified technician a person that has the necessary knowledge and skills directly related to the operation of electrical equipment

Voltage the amount of electrical pressure in a given circuit, measured in *volts*

Resistance the opposition to current flow in a given electrical circuit, which is measured in *ohms*

Current (or *amperage*) the flow of electrons through a given circuit, which is measured in *amps*

Megohmmeter an instrument used to measure high electrical resistance

Megger another name for a megohmmeter

Power the electrical work that is being done in a given circuit, which is measured in wattage (watts) for a purely resistive circuit and volt-amps (VA) for an inductive/capacitive circuit

Introduction

Once the facilities maintenance technician has a good understanding of basic electrical theory and electrical safety, the technician can attempt to troubleshoot, repair, and install basic electrical circuits and appliances. The previous chapter introduced the technician to basic electrical theory. This chapter introduces the concepts of safety and the basic procedures for troubleshooting and repairing basic electrical circuits and devices (switches, receptacles, and so on).

Safety, Tools, and Test Equipment

Safety should be the most important aspect of any assigned task or project. Completing the project on time is meaningless if it was completed under the possible risk of personal injury and/or death. This includes using tools and equipment that are properly maintained and designed for that particular task.

Safety

Many safety considerations must be adhered to while working on electrical systems, equipment, and devices. Many of the safety procedures are outlined in the OSHA standard (29 CFR Part 1926) and must be followed while working on electrical systems, equipment, and devices.

In any installation, repair, or removal of electrical equipment or devices, use the proper wiring methods and practices stipulated in the National Electrical Code, better known as the NEC® (NFPA® 70). The NEC® is written to ensure the protection of people and property from the hazards that arise from installation and/or repair of electrical systems or equipment. One must also follow all local codes and regulations that might exceed the minimum standards required by the NEC®.

Prior to beginning any work on electrical systems, equipment, or devices, always consult with the local authority having jurisdiction (building/electrical inspector) on the local codes and regulations that might affect the proper and safe installation of electrical systems, equipment, and devices in your area.

The OSHA and NEC® safety articles mandate that only highly trained and qualified technicians can work on electrical systems, equipment, or devices. This is intended to protect the occupants of a structure as well as the facility's property owner. In addition, it protects the person performing the installation and maintenance by defining a minimum skill and knowledge level that the technician must obtain. A **qualified technician** is defined as a person who has the necessary knowledge and skills required for the operation of electrical equipment. This includes the necessary safety training outlining the hazards and dangers to the individual operating the equipment as well as potential damage to the structure and its occupants.

Lockout/tag-out procedures, as stated in the OSHA standard 1926.417 and the NFPA® 70E standard, should be followed at all times to prevent electrical shock or even death by electrocution. Various forms of lockouts exist, which can be placed on safety disconnects, switches, and breakers. You must use the lockout that will prevent current from flowing through the circuit that you are working on. Figure 6-1 shows some examples of the different lockouts that are available.

The NFPA® 70E mandates that qualified personnel should never work on any part of an energized electrical system that is over 50 volts AC without the proper arc flash/arc blast personnel protective clothing (PPE) and without using 1,000-volt-rated tools. This mandate is in effect for your protection only. Also, safety glasses should be worn at all times while working on electrical systems, equipment, or devices.

Working Space around Electrical Equipment

Adequate lighting and space is required around electrical equipment so that maintenance on the electrical equipment can be performed safely. The NEC® 110.26 covers the minimum working space, access, headroom, and lighting requirements for electrical equipment such as switchboards, panelboards, and 600 volt or less motor control centers.

Tools and Test Equipment

Always ensure that all hand tools and power tools are inspected before and after use. Look for any defects or damage that may cause injury while the hand tool or power tool is in use. If you should happen to discover a tool that shows sign of excessive wear and or damage, then that tool should be taken out of use and serviced by a technician who is qualified to service the damaged tool. If the tool cannot be serviced, then the tool must remain out of use and be replaced

FIGURE 6-1: Examples of lock-out/tag-out devices.

as soon as possible. If a qualified technician is not available to repair the damaged tool, the tool should be either discarded or put away until such time as a qualified person can repair it. Regularly maintain and clean all hand tools and power tools to ensure their proper operation.

Always use calibrated test equipment when testing electrical systems, equipment, or devices to ensure accurate measurements. Never use test equipment on any energized circuit above what it is rated for. This could cause the test equipment to explode, possibly leading to serious injury or even death.

Megohmmeter

A **megohmmeter**, also referred to as a **"megger,"** is an instrument used to measure high electrical **resistance**. These types of meters are seldom used for residential structures by facility maintenance technicians or even electricians. These types of meter are often used to test the resistance of transformer windings, motor windings, and the like. Megohmmeters are designed to measure resistance in megohms (one million ohms).

Analog Meters

Analog meters are characterized by the fact that they incorporate a scale and a needle to indicate the value of the electrical property being measured. There are different types of analog meter movements. One of the most common is the d'Arsonval movement (also referred to moving coil meter).

Analog meters operate on the principle that like magnetic poles repel each other. Since the turning force of this meter depends on the repulsion of magnetic fields, it will operate on DC current only. If an AC current is connected to the moving coil, the magnetic polarity will change 60 times per second and the net. Turning force will be zero.

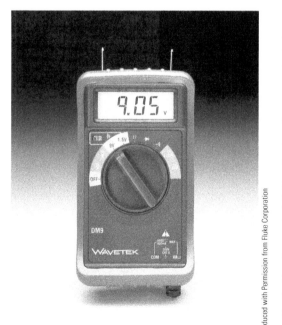

FIGURE 6-2: A typical VOM with digital readout.

The Voltmeter

The voltmeter (VOM) is constructed to connect across the power source, allowing the technician to measure the potential difference between two points. Figure 6-2 illustrates a typical multimeter with a digital readout.

Most voltmeters are multirange voltmeters, which means that they are designed to use one meter movement to measure several ranges of **voltage**. When the selector switch of this meter is turned, different resistances or steps of resistance are added to the circuit to increase the range or removed from the circuit to decrease the range. *Refer to Procedure 6-1 Using a Voltage Tester and Procedure 6-2 Using a Noncontact Voltage Tester for step-by-step instructions.*

Reading a Meter

Learning to read the scale of a multimeter takes time and practice. Most people use meters every day without thinking about it. Learning to read the scale of a multimeter is similar to learning to read a speedometer or fuel gauge.

Notice that the three voltmeter scales use the primary numbers 3, 6, and 12, and are in multiples of 10 of these numbers. Since the numbers are in multiples of 10, it is easy to multiply or divide the readings in your head by moving a decimal point. Remember that the position of a decimal in any number can be shifted to the left or right by simply multiplying or dividing it by 10. Multiplication moves the decimal point to the right, and division moves the decimal point to the left.

The Ammeter

The ammeter, unlike the voltmeter, is a very-low-impedance device. The ammeter is used to measure **current**; therefore, it is necessary to connect the ammeter in series to the load, thereby permitting the load to limit the current flow. DC ammeters are constructed by connecting a common moving coil type of meter across a shunt. An ammeter shunt is a low-resistance device used to conduct most of the circuit current away from the meter movement.

Since the meter movement is connected in parallel with the shunt, the voltage drop across the shunt is the voltage applied to the meter.

Most AC ammeters employ a current transformer instead of shunts to change scale values. Because these types of meters employ a transformer, the ammeter is typically connected to the secondary side of the transformer, while the load is connected to the primary side of the transformer. Selecting different taps on the secondary side of the current transformer will produce the various ranges required for the meter.

The Series Ohmmeter

When resistance is to be measured, the meter must first be zeroed. This is done with the ohms-adjust control, the variable resistor located on the front of the meter. To zero the meter, connect the leads and turn the ohms adjust knob until the meter indicates zero at the far-right end of the scale. An ohmmeter should always be readjusted to zero when the scale is changed.

Digital Ohmmeters

Digital ohmmeters display the resistance in digital display similar to that of a calculator instead of using a meter movement. When using a digital ohmmeter, care must be taken to notice the scale indication on the meter. The ohmmeter, whether digital or analog, must never be connected to a circuit when the **power** is turned on.

Digital Multimeters

Digital multimeters have become increasingly popular in the past few years. The most apparent difference between digital meters and analog meters is that digital meters display their reading in discrete digits instead of with a pointer and scale.

AC Clamp-On Ammeter

An AC clamp-on ammeter is a versatile instrument, which is also called clip-on, tang-type, snap-on, or other names. Some can also measure voltage or resistance or both. Unless you have an ammeter like this, you must interrupt the circuit to place the ammeter in the circuit. With this instrument you simply clamp the jaws around a single conductor, as shown in Figure 6-3. *Refer to Procedure 6-3 Using a Clamp-On Ammeter for step-by-step instructions.*

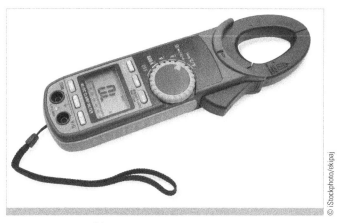

FIGURE 6-3: Clamp-on ammeter.

Megohmmeter

A megohmmeter is used for measuring very high resistances. This particular device can measure up to 4,000 megohms (Figure 6-4).

Wiring and Crimping Tools

Wiring and crimping tools are available in many designs. Figure 6-5 illustrates a combination tool for crimping solderless connectors, stripping wire, cutting wire, and cutting small bolts. This figure also illustrates an automatic wire stripper. To use this tool, insert the wire into the proper strip-die hole. The length of the strip is determined by the amount of wire extending beyond the die away from the tool. While firmly gripping the wire in one hand, apply pressure to the tool with the other hand. Release the handles and remove the stripped wire.

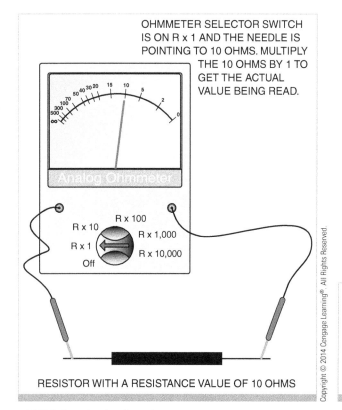

OHMMETER SELECTOR SWITCH IS ON R x 1 AND THE NEEDLE IS POINTING TO 10 OHMS. MULTIPLY THE 10 OHMS BY 1 TO GET THE ACTUAL VALUE BEING READ.

Analog Ohmmeter

R x 100
R x 10   R x 1,000
R x 1   R x 10,000
Off

RESISTOR WITH A RESISTANCE VALUE OF 10 OHMS

FIGURE 6-4: Analog ohmmeter.

FIGURE 6-5: Wire-stripping and crimping tool.

Electrical Conductors

When working with electricity, the facilities maintenance technician will be exposed to various conductors (wire, cables, etc.). Having a good understanding of different types of conductors (wires) used in an electrical circuit is essential when working with electricity. Using the wrong size or type of conductor can result in a loss of property and/or even a fatality.

> *If a wire is too small for the current passing through, it will overheat and possibly burn the insulation and could cause a fire.*

Wire Sizes

All conductors have some resistance. The resistance depends on the material, the cross-sectional area, and the length of the conductor. A conductor with low resistance carries a current more easily than that with high resistance.

The proper wire (conductor) size must always be used. The size of a wire is determined by its diameter or cross section (Figure 6-6). A larger-diameter wire has more current-carrying capacity than a smaller-diameter wire.

Standard copper wire sizes are identified by American Standard Wire Gauge numbers and measured in circular mils. A circular mil is the area of a circle $\frac{1}{1,000}$ inches. in diameter. Temperature is also considered because resistance increases as temperature increases. Increasing wire size numbers indicate smaller wire diameters and greater resistance. For example, number 12 wire size is smaller than number 10 wire size and has less current-carrying capacity. The technician should not choose and install a conductor of a particular size unless licensed to do so. The technician should, however, be able to recognize an undersized conductor and bring it to the attention of a qualified person. As mentioned previously, an undersized wire may cause voltage to drop, breakers or fuses to trip, and conductors to overheat.

The conductors are sized by their amperage-carrying capacity, which is also called ampacity. Figure 6-7 contains a small part of a chart from the NEC® and a partial footnote for one type of conductor. This chart as shown and the footnote should not be used when determining a wire size. It is shown here only to acquaint you with the way in which wire size is presented in the NEC®. The footnote actually reduces the amount of amperes for number 12 and number 14 wires listed in the NEC® table in Figure 6-7. The reason is that number 12 and number 14 wires are used in residential structures, where circuits are often overloaded unintentionally by the occupants. The footnote exception simply adds more protection.

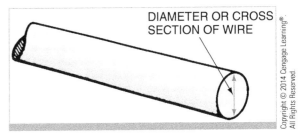

DIAMETER OR CROSS SECTION OF WIRE

FIGURE 6-6: The cross section of a wire.

WIRE SIZE	60 C (140F)
	Types TW, UF
AWG or kemil	
COPPER	
18	—
16	—
14*	20
12*	25
10*	30
8	4 0
6	5 5
4	7 0
3	8 5
2	9 5
1	110

Small Conductors. Unless specifically permitted in (e) through (g), the overcurrent protection shall not exceed 15 amperes for No. 14, 20 amperes for No. 12, and 30 amperes for No. 10 copper.

FIGURE 6-7: This section of the NEC shows an example of how a wire is sized for only one type of conductor. It is not the complete and official document but is a representative example.

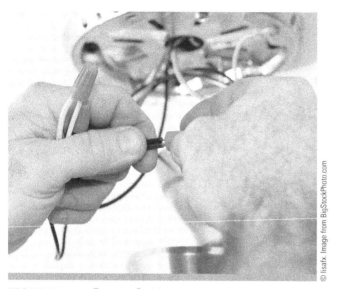

FIGURE 6-8: Romex Cable.

NM Cable

Using improper wire types and wire sizes can result in a fire! The most common type of cable used in single-family and multifamily dwellings is NM cable (commonly referred to as Romex; Figure 6-8). NM cable is covered in Article 334 of the NEC®, where you will find its proper use and installation methods.

NM cable is available in many sizes and conductor pairs. This means that an NM cable comes with two-, three-, or four-current-carrying conductors and a ground. The ground that is present in NM cable is typically bare (without insulation); all of the current-carrying conductors will have an insulation covering that is identified by the following colors:

- Black
- Red
- Blue
- White

The blue insulated conductor is found only in four-conductor NM cable. Admittedly, a four-conductor NM cable with ground is rarely used or seen, but it does exist. Most often, NM cable is sold as a two conductor with ground or a three conductor with ground.

Electrical Devices, Fixtures, and Equipment

Typically, the electrical repairs that a facilities maintenance technician does will consist of troubleshooting, repairing, and/or replacing electrical devices, fixtures, and equipment as opposed to replacing conductors. When troubleshooting, repairing, and/or replacing electrical devices, fixtures, and equipment, always follow manufacturer's recommendations, warnings, and specifications. In most cases, manufacturers of electrical devices (outlet, switch, light fixture, etc.) supply how-to wiring diagrams with the device.

Boxes

A multitude of boxes are used today for electrical installations. Only the most common of these boxes will be discussed here. One of the most common boxes in use is the single gang nail-up box (Figure 6-9). It is made of plastic or fiberglass and is typically used during new construction before the drywall is installed.

This box will have enough room for only one yoked device, such as a switch, a receptacle, or a dimmer switch. A two-gang nail-up box will have enough room for two yoked devices, and a three-gang nail-up box will have room for three yoked devices.

The "cut-in" box (see Figure 6-10) is made of plastic or fiberglass and is used when a device must be installed after the drywall is already in place. ***Refer to Procedures 6-4 Installing Old-Work Electrical Boxes in a Sheetrock Wall or Ceiling for step-by-step instructions.***

Round ceiling boxes (see Figure 6-11) are available in various diameters, depths, and shapes. Four common types of ceiling boxes are:

- Pancake (the shallow box)
- 4" round nail-up box
- 4" round cut-in box
- Fan-rated ceiling box

Ceiling fan boxes (see Figure 6-12) are used to support a lighting fixture in the ceiling. It is important to know that a ceiling fan should not be installed on a ceiling box that is not fan-rated. If a ceiling box is rated to support a fan, it will be clearly and legibly stamped into the interior of the box. If a fan is mounted to a box that is not rated for a ceiling fan, the fan could fall during operation and cause injury to someone underneath it.

Weatherproof boxes (see Figure 6-13) are used whenever a device (receptacle or switch) is installed in a wet location. This box is designed to keep the elements of the weather from affecting the electrical circuit and device that is contained within the box. If this box is installed in the direct weather and it houses a receptacle, it is to have an "in-use" cover. An in-use cover looks like a giant clear bubble.

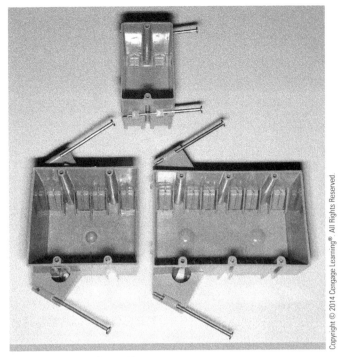

FIGURE 6-9: Gang Boxes.

FIGURE 6-10: Cut-in box.

FIGURE 6-11: Round ceiling box.

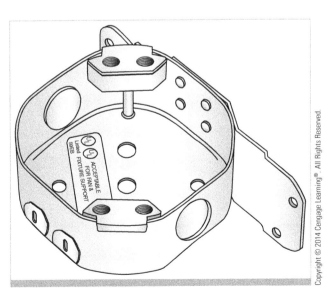

FIGURE 6-12: Ceiling fan box.

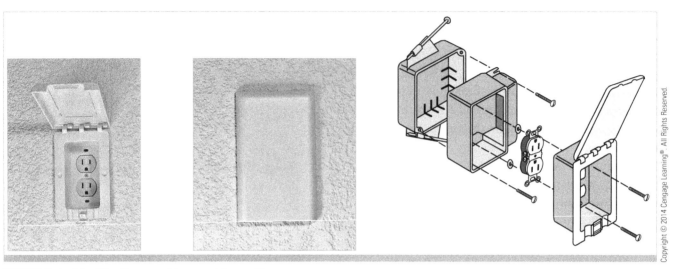

FIGURE 6-13: Weatherproof box and box with cover.

Feeder Circuits

A feeder circuit is the wires that connect the service equipment to the overload pro-tection device located on the final branch-circuit. Some commercial wiring situations may call for another load center (panel), which is called a subpanel, to be located in another part of the structure. The reason for installing a subpanel is usually to locate a load center closer to an area of the house where several circuits are required. The feeder is the wiring extending from the main panel to the subpanel. The feeder typi-cally constitutes wiring in a cable or individual wires in an electrical conduit that are large enough to feed the electrical requirements of the subpanel they are servicing.

Surface Metal Raceways

When it is impossible to conceal conductors (for example, around a desk, counter, cabinets) a raceway is often used. Surface raceways, which are governed by Articles 386 and 388 of the NEC®, are typically made of either metal or nonmetallic materials.

It should be noted that the number of conductors in a raceway is limited to the design of the raceway. Also that the combined size of the conductors, splices, and tape should not exceed 75 percent of the raceway. Also all conductors that are to be installed in a raceway should be spliced either in a junction box or within the raceway.

Multioutlet Assemblies

The NEC® defines a multioutlet assembly as a raceway that is designed to house conductors and receptacles that has been assembled in the field (on the jobsite) or assembled at the manufacturers facility. The raceway can be freestanding, surface mounted, or flush mounted. They offer a high degree of flexibility to an installa-tion by allowing for the likelihood that the installation and use requirements could change. For example, in an office building where cubicles are utilized, multioutlet assemblies offer a suitable solution. They are also utilized in heavy-use situations, for example, computer rooms and laboratories.

Floor Outlets

Floor outlets are often used in large office complexes, where desks will be positioned too far away from a wall to effectively run a power cord. Floor outlets can be contained within a under floor raceway or by installing floor boxes. The installation requirements for an underfloor raceway are set forth in the NEC® Article 390.

Panelboards

A panelboard is, according to the NEC®, a single panel or a group of panels designed to act as a single panel. Panelboards are located in either a cabinet or a cutout box positioned in such a way that only the front is accessible.

When separate feeders are to be installed from the main service equipment to each of the areas of the commercial building, each feeder will terminate in a panelboard, which is to be installed in the area to be served.

Circuit Protection Devices

Circuit protection is essential to prevent the conductors in the circuit from being overloaded. If one of the power-consuming devices were to cause an overload due to a short circuit within its coil, the circuit protector would stop the current flow before the conductor became hot and overloaded. A circuit consists of a power supply, the conductor, and the power-consuming device. The conductor must be sized large enough that it does not operate beyond its rated temperature, typically 140°F (60°C) while in an ambient temperature of 86°F (30°C). For example, a circuit may be designed to carry a load of 20 amps. As long as the circuit is carrying up to its amperage, overheating is not a potential hazard. If the amperage in the circuit is gradually increased, the conductor will become hot (Figure 6-14A and 14B). Proper understanding of circuit protection is a lengthy process. More details can be obtained from the NEC® and further study of electricity.

Fuses

A fuse is a simple device used to protect circuits from overloading and overheating. Most fuses contain a strip of metal that has a higher resistance than the conductors in the circuit. This strip also has a relatively low melting point. Because of its higher resistance, it will heat up faster than the conductor. When the current exceeds the rating on the fuse, the strip melts and opens the circuit.

Plug Fuses

Plug fuses have either an Edison base or a type S base (Figure 6-15(A)). Edison-base fuses are used in older installations and can be used only for replacement. Type S fuses can be used only in a type S fuse holder specifically designed for the fuse; otherwise, an adapter must be used

(A)   (B)

FIGURE 6-14: (A) A circuit breaker. (B) A cutaway.

FIGURE 6-15: (A) A type S base plug fuse. (B) Type S fuse adaptor.

Electrical circuits must be protected from current overloads. If too much current flows through the circuit, the wires and components will overheat, resulting in damage and possible fire. Circuits are normally protected with fuses or circuit breakers.

(Figure 6-15(B)). Each adapter is designed for a specific ampere rating, and these fuses cannot be interchanged. The amperage rating determines the size of the adapter. Plug fuses are rated up to 125 volts and 30 amps.

Dual-Element Plug Fuses

Many circuits have electric motors as the load or part of the load. Motors draw more current when starting and can cause a plain (single element) fuse to burn out or open the circuit. Dual-element fuses are frequently used in this situation (Figure 6-16). One element in the fuse will melt when there is a large overload such as a short circuit. The other element will melt and open the circuit when there is a smaller current overload lasting more than a few seconds. This allows for the larger starting current of an electric motor.

Cartridge Fuses

For 230-volt to 600-volt service up to 60 amps, the ferrule cartridge fuse is used (Figure 6-17A). From 60 amps to 600 amps, knife-blade cartridge fuses can be used (Figure 6-17B). A cartridge fuse is sized according to its ampere rating to prevent a fuse with an inadequate rating from being used. Many cartridge fuses have an arc-quenching material around the element to prevent damage from arcing in severe short-circuit situations (Figure 6-18).

Circuit Breakers

A circuit breaker can function as a switch as well as a means for opening a circuit when a current overload occurs. Most modern installations in houses and many commercial and industrial installations use circuit breakers rather than fuses for circuit protection. Circuit breakers use two methods to protect the circuit. One is a bimetal strip that heats up with a current overload and trips the breaker, opening the circuit. The other is a magnetic coil that causes the breaker to trip and open the circuit when there is a short circuit or other excessive current overload in a short time (Figure 6-19).

FIGURE 6-16: A dual-element plug fuse.

(A)

(B)

FIGURE 6-17: (A) A ferrule-type cartridge fuse. (B) A knife-blade cartridge.

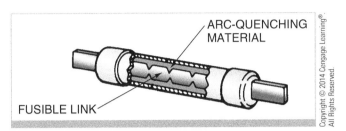

FIGURE 6-18: A knife-blade cartridge fuse with arc quenching.

FIGURE 6-19: Single-pole circuit breaker (left). Double-pole circuit breaker (right).

Emergency Backup Systems

Emergency backup systems are designed to keep critical parts of an electrical system energized in the event of power loss. Hospitals and assisted living facilities have emergency backup systems in order that life support systems will not be interrupted during the loss of power. These types of emergency backup systems will typically use a diesel engine that drives a generator. The facility will switch over to the generator only in the event of power loss.

Another form of an emergency backup system is the uninterruptible power supply (UPS) (see Figure 6-20). This system uses a battery or batteries to supply constant power to the circuits connected to it in the event of power loss. This system is available in various sizes. A UPS can be small enough to fit on a desk, or it can be so large that it may require its own room or even its own building on the facility grounds. The larger UPS systems require a lot of maintenance to maintain the reliability of the batteries. Many safety issues have to be considered when working on or around these large battery backup systems, and it is for this reason that only qualified personnel work on these systems.

FIGURE 6-20: Uninterruptible power supply (UPS).

Electrical Switches and Receptacles

Switches and receptacles are electrical devices. There are many variations of each, so we will discuss only the most common of each.

Switches

Four of the most commonly used switches are the single-pole, double-pole, three-way, and four-way switches. These switches are available in 15-amp and 20-amp ratings. Be sure to use the correct amp rating when installing a switch in a lighting branch circuit.

- **Single-pole switch** (see Figure 6-21)—This switch is used when a light or fan is turned on or off from only one location.

- **Double-pole switch** (see Figure 6-22)—This switch is used when two separate circuits must be controlled with one switch. They are used to control 240-volt loads, such as electric heat, motors, and electric clothes dryers.

- **Three-way switch** (see Figure 6-23)—This switch is used when a light or fan can be turned on or off from two different locations.

- **Four-way switch** (see Figure 6-24)—This switch is used when a light or fan can be turned on or off from three or more different locations. This switch must be used with two three-way switches.

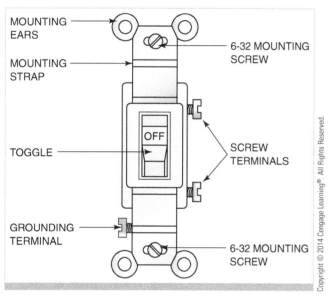

FIGURE 6-21: Single-pole switch.

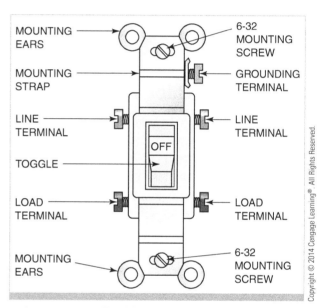

FIGURE 6-22: Double-pole switch

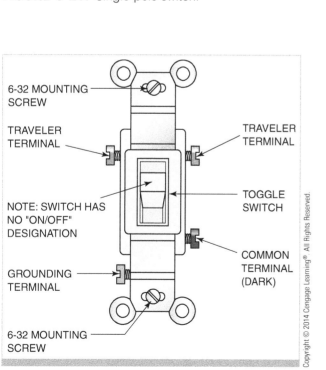

FIGURE 6-23: Three-way switch.

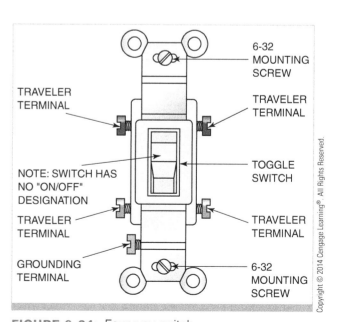

FIGURE 6-24: Four-way switch.

Receptacles

There are numerous types of receptacles available on the market today that are designed for residential structures. The most common receptacles are:

- 240-volt 30-amp or 50-amp single receptacle
- 240-volt 20-amp single receptacle
- 120-volt 15-amp or 20-amp single receptacle
- 120-volt 15-amp or 20-amp duplex receptacle
- 120-volt 15-amp or 20-amp ground fault circuit interrupter (GFCI) receptacle

The *240-volt 30-amp single receptacle* is generally used as a clothes dryer receptacle. It has four wires that are connected to it: two ungrounded conductors (hot wires), one grounded conductor (neutral), and one grounding conductor (bare ground wire).

The *240-volt 50-amp single receptacle* is generally used as an electric range (stove) receptacle. It also has four wires that are connected to it, as mentioned above.

The *240-volt 20-amp single receptacle* is generally used as an air conditioner receptacle. As the previous receptacles, it also has four wires that are connected to it. Notice the slots on the front of the receptacle.

The *120-volt $\frac{15}{20}$-amp single receptacle* (see Figure 6-25) is used most often for electrical equipment that requires a 20-amp circuit. This is so that either a 15 or 20 amp cord can be plugged into it. Any receptacle devices that are 120-volt, 20-amp rated will have this feature.

The *120-volt 15-amp duplex receptacle* (see Figure 6-25) is the most common receptacle in use today. The *120-volt $\frac{15}{20}$-amp duplex receptacle* is similar to the previous receptacle, except that it has the "T" slot to accommodate a 20-amp load. ***Refer to Procedure 6-5 Installing Duplex Receptacles in a Nonmetallic Electrical Outlet Box and Procedure 6-6 Installing Duplex Receptacles in a Metal Electrical Outlet Box for step-by-step instructions.***

The *GFCI duplex receptacle* (see Figure 6-26) is designed to trip when there is a difference between the current going to the load and the current returning from the load. This device will trip if there is a difference of 4 mA between the two. This device is easily recognizable because of the Trip and Reset buttons that are on the face of the receptacle. Once this receptacle is properly installed on a branch circuit, every device and fixture connected to the branch circuit past the GFCI receptacle is protected. This receptacle is to be used above countertops in bathrooms and kitchens, in wet or damp locations (such as basements), and outside. This receptacle is also available in 15 amps and 20 amps.

Refer to Procedure 6-7 Installing Feed-Through GFCI and AFCI Duplex Receptacles in Nonmetallic Electrical Outlet Boxes for step-by-step instructions.

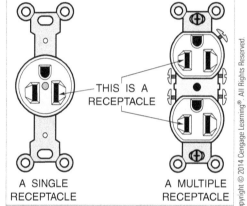

THIS IS A RECEPTACLE

A SINGLE RECEPTACLE   A MULTIPLE RECEPTACLE

FIGURE 6-25: Single and duplex receptacle.

FIGURE 6-26: Ground fault circuit interrupter receptacles (GFCI).

Fixtures

Various types of lighting fixtures are available today. Only the most common types will be discussed in this section.

Fixtures are described by the way they mount and by the type of bulb that is used within them. For example, a surface-mount, incandescent, ceiling fixture is one that mounts against the ceiling surface and has an incandescent bulb, the most commonly used light bulb. It is the standard frosted or clear light bulb that we use in our homes. It tends to give off a yellowish light when compared to a fluorescent light bulb. Fluorescent bulbs, most commonly referred to as "tubes," tend to give off a white light. ***Refer to Procedure 6-8 Installing a Light Fixture Directly to an Outlet Box for step-by-step instructions.***

Here is a list of the most common light fixtures:

- Surface-mount incandescent fixture: mounts against ceiling or wall (see Figure 6-27). ***Refer to Procedure 6-9 Installing a Cable-Connected Fluorescent Lighting Fixture Directly to the Ceiling for step-by-step instructions.***

- Surface-mount fluorescent fixture: mounts against ceiling or wall (see Figure 6-28). ***Refer to Procedure 6-12 Installing a Fluorescent Fixture (Troffer) in a Dropped Ceiling for step-by-step instructions.***

- Recessed-can incandescent lighting fixture: mounts in the ceiling (see Figure 6-29). ***Refer to Procedure 6-10 Installing a Strap on a Lighting Outlet Box Lighting Fixture for step-by-step instructions.***

- Pendant-type incandescent lighting fixture: hangs from the ceiling on a chain or cable (see Figure 6-30).

- Chandelier-type lighting fixture: multiple lamp fixtures hang from the ceiling on a chain or cable (see Figure 6-31). ***Refer to Procedure 6-11 Installing a Chandelier-Type Light Fixture Using the Stud and Strap Connection to a Lighting Outlet Box for step-by-step instructions.***

FIGURE 6-27: Surface-mount incandescent fixture.

FIGURE 6-28: Surface-mount fluorescent fixture.

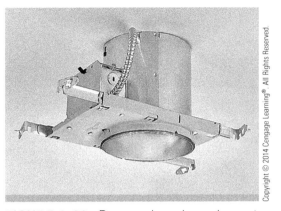

FIGURE 6-29: Recessed-can incandescent lighting fixture.

FIGURE 6-30: Pendant-type incandescent lighting fixture.

FIGURE 6-31: Chandelier-type lighting fixtures.

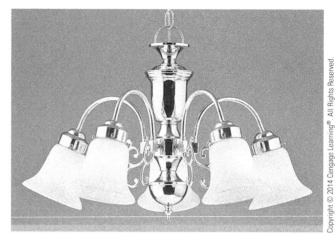

FIGURE 6-31: (*Continued*)

Remodeling and Electrical Systems

When a structure is remodeled that contains an electrical service, the fixtures are most often removed and either replaced or reused. In either case, the general procedures for remodeling a structure that contains electrical fixtures are as follows.

Remodeling a Kitchen

One of the areas in a residential structure that is often remodeled is the kitchen. Typically, this is completed when the appliances are either replaced or updated. The general procedure for demolishing a kitchen is:

• Disconnect all utilities servicing the kitchen.

• Install a plastic barrier separating the affected area from the rest of the structure.

• Cover any appliance that is not going to be removed from the area with a protective covering (cardboard).

• Disconnect electrical wiring to any appliance that is to be removed. Any exposed wire should be capped using wire nuts to prevent exposure to the bare conductors.

• Disconnect any plumbing servicing appliances and/or fixtures to be removed. Any pipes that are exposed should be capped (including vents).

• Remove appliances that are in the affected area. For example, if the technician is demolishing kitchen counter and/or cabinets, then at this stage the vent hood, as well as a drop in stove and/or cook top should be removed. Even if these components are not going to be reused, they could be recycled at an architectural salvage yard.

Remodeling a Bathroom

Another area in a residential structure that is often remodeled is the bathroom. This is typically done when the plumbing fixtures are updated or replaced or when the cabinetry is updated. The general procedure for demolishing a bathroom is:

• Disconnect all utilities servicing the bathroom.

• Install a plastic barrier separating the affected area from the rest of the structure.

- Disconnect electrical wiring to any appliance that is to be removed. Any exposed wire should be capped using wire nuts to prevent exposure to the bare conductors.

Remodeling a Utility Room

This area in a residential structure is not remodeled as often as a kitchen or a bathroom, but the demolition process is similar. Utility rooms are typically remodeled when the cabinetry is updated or when appliances are updated and/or replaced. The general procedure for demolishing a utility room is:

- Disconnect all utilities servicing the utility room.
- Install a plastic barrier separating the affected area from the rest of the structure.
- Disconnect electrical wiring to any appliance that is to be removed. Any exposed wire should be capped using wire nuts to prevent exposure to the bare conductors.

CAUTION

Always check with your local and state agencies before starting a remodeling project.

Before beginning the demolition or deconstruction process the proper permits must be obtained.

Electrical Maintenance Procedures

As mentioned earlier, facilities maintenance technicians are more likely to maintain an existing electrical system as opposed to installing a new system. Therefore, the technician should have a good understanding of the troubleshooting process as well as the proper technique for maintaining electrical systems.

Troubleshooting

Troubleshooting is a process in which a person gathers information and forms a logical conclusion as to the problems that may be present within the system. All problems should be looked at logically. Take time and consider the most logical problems that would cause the symptoms that are present in the faulty circuit. If you do this, you have a higher chance of success in finding and fixing the problem. All problems that occur in a system will give tell-tale signs that will help you find the problem. Training is always beneficial. This gives you a knowledge base that will help you form the logical conclusions needed to solve problems with the system.

Troubleshooting occurs while you are gathering the information about the faulty system. Diagnosing begins while the data are being collected and is completed when a decision as to what the problem may be is formulated. Once you have diagnosed a problem, you must prove the diagnosis and repair the problem.

A simplified, step-by-step guideline that could be used while troubleshooting follows. *Remember: Do not perform any of the following steps unless you are a qualified individual.*

1. You are notified of a problem.
2. Ask the person who has reported the problem as many questions as possible as to what was witnessed during the failure. This may include something that was seen, smelling a distinct or peculiar odor, or feeling heat in the general vicinity of the problem.

3. Begin troubleshooting while making sure that all safety standards are adhered to. It is a good idea to start troubleshooting at the most logical area that would cause the described symptoms. You should start diagnosing the problem as soon as you receive the descriptions given to you and as you start receiving data from troubleshooting.

4. Safely remove any covers or panels that will give you access to the part of the electrical system that you are troubleshooting.

5. Visually inspect the equipment and devices for any signs of overheating or disintegration. If signs are visible, go to step 6a. If signs are not visible, go to Step 6b.

6a. *Be sure that you are qualified to work on the device and equipment before working on any part of an electrical system.* Deenergize the circuit or system, attach your lockout/tag-out device to the disconnecting means, and place the key in *your* pocket. Go to Step 7.

6b. Take the appropriate step to safely acquire voltage and/or current readings at the suspected device or equipment. If your readings indicate that voltage is present and current is not following as it should be, the suspected device or equipment may be faulty and it may need to be replaced. If you decide to replace the faulty device or equipment, de-energize the circuit or system, attach your lockout/tag-out device to the disconnecting means, and place the key in *your* pocket; then go to Step 7.

7. Go back to the device or equipment and verify that the circuit is, in fact, deenergized.

8. Once it has been verified that the circuit is deenergized, begin working on your fault.

   In the event that you may need to repair and/or replace common electrical devices such as receptacles, switches, interior and exterior lighting fixtures, bulbs, or ballasts, follow the simple procedures listed in the last section of this chapter.

9. Once the fault is repaired, and all covers are back in place, remove the lockout/tag-out from the source of energy and reenergize the circuit.

10. Go back to the device or equipment that was replaced and verify that it is working properly. If it is, inform the person who called the job in. If it is not working, you may want to consider calling a qualified electrician to troubleshoot and diagnose the problem.

Perform Tests

Regularly perform tests on the following to ensure that they are operating properly before an emergency arises:

- Smoke alarms
- Fire alarms
- Medical alert systems
- Emergency exit lighting
- GFCI receptacles

Test Smoke Alarms and Fire Alarms

Individual smoke and fire alarms typically have a test button that can be pushed. Be aware that pushing the test button on any one alarm may set off all alarms that are on the system as the fire alarm code requires all of them to be tied together.

Some fire alarm systems may require you to put the system in test mode before testing. If this is not done and the fire alarm is activated during the test, the sprinkler systems may activate. Placing the fire alarm system in test mode would allow the electrical portion of the system to be tested without the sprinklers activating during the test. Be sure to repair or replace any defective alarm or smoke detector according to manufacturer's specifications and wiring diagrams, and report any malfunctions to your supervisor.

Many of the smoke detectors have a 9-volt battery in them so that the alarm can continue to work in the event that there is a loss of power. The batteries should be checked on a regular basis. If a detector is found to have a dead or weak battery, the battery should be changed immediately. If the smoke detector does not operate properly, it should be changed immediately (see Figure 6-32). Regardless of the condition of the battery, it should be replaced at least once a year.

FIGURE 6-32: Smoke detector.

To test a smoke detector:

1. Most smoke detectors will have a test button located on the outside of the unit that is designed to allow the owner to test the unit's battery. To test the battery in a smoke detector, press locate and press the test button.

2. If the battery in the unit is over a year old, replace it immediately.

3. Smoke detectors can also be checked by placing a lit candle approximately 6 inches from the smoke detector. This allows the heat from the candle to enter into the unit. If the heat is going to set the unit off, then it should happen within 20 seconds.

4. If the unit fails to sound off within 20 seconds, try blowing out the candle and letting the smoke rise to the unit.

5. If the unit fails to sound, open the detector and inspect its insides to make sure that the unit is clean and all the electrical connections are good.

6. If the unit is clean and all the electrical connections are good and the unit still will not sound, replace it with a new smoke detector.

To replace a smoke detector battery:

1. For most smoke detectors, replacing the battery is as simple as first removing the smoke detector's cover by gently pulling down or twisting the cover in a counterclockwise direction and then pulling down on the cover.

2. Once the cover has been removed, the battery should be easy to locate.

3. Remove the old battery and replace it with a fresh battery.

4. Close the case and test the smoke detector.

5. Properly dispose of the old battery.

Always check the manufacturer's documentation and even their website for additional troubleshooting tips and possible adjustments.

Test GFCI Receptacles

1. Go to the receptacle and locate the "Test" button.

2. Press the test button, and listen for a very light "pop" in the receptacle. You may also notice a small indicator that will light up after you have tripped the GFCI. If the GFCI tripped, then it is working correctly.

3. Press the "Reset" button, and the GFCI should reset and the indicator (if preset) should turn off.

4. If it did not trip, press the test button again.

If the GFCI still does not trip, then it is faulty and needs to be replaced, or it could have been wired incorrectly and. therefore. would need to be rewired correctly. For the correct installation procedures of a GFCI receptacle, see page 150.

Test Medical Alert Systems

Assisted living facilities may have medical alert systems and hospitals will have them. If your facility has medical alert systems, all tests that are conducted on these systems must be performed in strict accordance with the manufacturer's specifications and should be performed only by qualified personnel who have been properly trained on the proper test procedures for that system. If any defects are found in the system, they should be reported to your supervisor immediately.

Replace Detectors, Devices, Fixtures, and Bulbs

Be sure that as you replace or repair any defective equipment or devices, you do so according to the manufacturer's specifications. Also, don't forget to report any malfunctions to your supervisor.

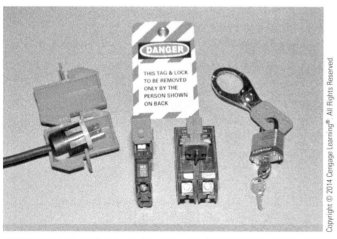

FIGURE 6-33: Lockout/tag-out devices.

Replace Smoke Detectors

1. Acquire a fiberglass or nonconductive ladder that will be tall enough for you to reach the detector once you are on it. *Do not use an aluminum ladder.* Set the ladder up under the faulty detector, making sure that all four legs of the ladder are solidly in place.

2. Go to the panel and deenergize the branch circuit that supplies the smoke detectors.

3. Lock out/tag out the breaker (see Figure 6-33).

4. Go back to the faulty detector and remove it from its mounting base by twisting the detector's body in a counterclockwise direction. The detector should release from the base (see Figure 6-34).

FIGURE 6-34: Removing faulty detector.

5. Make a note of how the smoke detector is wired. If necessary, make a sketch on a piece of paper to follow when you are reconnecting the power leads. You may find that the smoke detector you are replacing has a quick connector plug on the back of the smoke detector. If you are replacing the faulty detector with a new one that is the same model, simply unplug the quick connect from the smoke detector. If a quick connect is not present, you may have to disconnect the power leads from the smoke detector by removing the wirenuts. There should be three leads on a newer type of smoke detector (see Figure 6-35). One will be black, one white, and one orange. The black conductor is the hot wire, the white conductor is the neutral, and the orange conductor is for the repeating circuit. (This wire triggers all of the other smoke alarms.)

6. If you are replacing the faulty detector with a newer version of the same model, then it might be possible to reuse the mounting bracket and screws from the older detector. If you are replacing the faulty smoke detector with a different brand or model, you may find that the base of the old one will not work for the new one, and that the old mounting base must, therefore, be removed. This can be accomplished by loosening the two fasteners that are securing the base to the

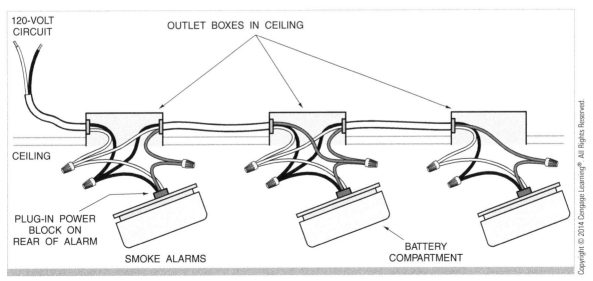

FIGURE 6-35: Interconnected smoke detectors.

FIGURE 6-36: Mounting base still attached to box.

box. Once the base is loose, simply twist it in a counterclockwise direction as you did with the detector. The mounting base should release from the box. Now replace the old base with the new base and tighten the screws (see Figure 6-36).

7. Reconnect the power leads. If you are replacing an older two-wire model with a newer three-wire model, you will not need to use the orange lead. Simply cap the orange lead and stuff it into the box.

8. Set the new detector against the mounting base, and twist it in a clockwise direction until it is securely mounted.

9. Go to the panel, remove your lockout from the breaker, and turn on the breaker. If the breaker does not trip, go to the next step. If the breaker trips, set the breaker to the off position and lock it out again. Go back to the detector, remove it from the base, and visually inspect the connections. Be sure that there is no exposed copper wire on the black lead that could be touching the bare copper ground wire. Once you think everything is correct, put the detector back on its base, go back to the panel, remove your lockout from the breaker, and turn the breaker back on. If the breaker trips again, call a qualified electrician to troubleshoot the problem. If the breaker does not trip, go to the next step.

10. Go back to the detector and verify that it is working correctly.

11. Gather all tools and ladders, and clean up the area that you were working in.

Replace a Switch

When replacing a switch first start by taking out the existing switch. This is done according to the following steps:

To remove the old switch:

1. Disconnecting the power to the existing switch by turning off the power to the circuit at the breaker panel.

2. Remove the switch's cover plate (see Figures 6-37 and 6-38). Before starting to work on the circuit or any electrical circuit, always verify that the power has been disconnected by using a circuit tester such as a neon-tester. If the circuit is disconnected, then the tester will indicate this.

3. The switch can be pulled out of its box once the screws that hold it to the box have been removed.

4. Before removing the conductor from the switch, make note of the color of the conductor's insulator in relationship switch's terminals.

5. The conductors connected to the switch can be removed by either cutting the conductors using wire cutters (typically not recommended) or loosening the terminal screws. In some cases, especially in new switches, the conductors are connected to the back of the switch. Normally, the conductors that are connected to the back of the switch can be removed by inserting a small straight slotted screw into the slot next to the conductor.

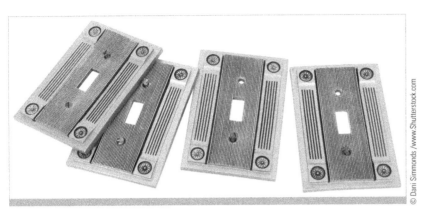

FIGURE 6-37: Cover plate.

FIGURE 6-38: Cover plate removed.

To install the new switch:

1. Start by reconnecting the conductors to the new switch connecting the proper conductor to the proper terminal as indicated in the Step 4. If the conductor is to be connected to the switch by the terminal screws, start by making a hook-shaped bend in the conductor. Place the hook under the terminal screw of the terminal to which the conductors is to be connected. Finally, tighten the terminal screw, snugging the conductor to the switch (see Figure 6-39).

2. Replace the remaining conductors to the switch, using the technique outlined in the previous step.

3. Once all conductors have been reconnected, push the switch back into its box, replace the switch cover, and reestablish the power to the circuit (see Figure 6-40).

Replace a Receptacle

The following steps should be used to replace an existing receptacle:

1. Disconnecting the power to the existing receptacle by turning off the power to the circuit at the breaker panel 2. Before starting to work on the circuit or any electrical circuit, always verify that the power has been disconnected by using a circuit tester such as a neon-tester. If the circuit is disconnected, then the tester will indicate this.

3. Once it has been verified that the power has been discounted from the receptacle, the screws holding the switch cover the receptacle, as well as the screws that hold the receptacle to its box, can be removed. This is accomplished by first removing the screws holding the cover to the receptacle.

4. Before removing the conductor from the receptacle, make note of the color of the conductor's insulator in relationship receptacle's terminals. Also, make special note of the common terminal of the receptacle. It can be identified by its color. The common terminal is normally brass in color.

5. Once the conductors of the old receptacle have been identified, they can be removed from the old receptacle.

FIGURE 6-39:
Replacement switch.

FIGURE 6-40: New switch.

6. Attach the conductors to the new receptacle connecting the proper conductor to the proper terminal as indicated in the previous step. If the conductor is to be connected to the receptacle by the terminal screws, start by making a hook-shaped bend in the conductor. Place the hook under the terminal screw of the terminal to which the conductors is to be connected. Finally, tighten the terminal screw, snugging the conductor to the receptacle.

7. Replace the remaining conductors to the receptacle, using the technique outlined in the previous step.

8. Once all conductors have been reconnected, push the receptacle back into its box, replace the receptacle cover, and reestablish the power to the circuit.

Replace a GFCI Receptacle

A GFCI receptacle has a line side and a load side (see Figure 6-41).

When a GFCI is terminated with only one set of conductors (hot, neutral, and ground), the hot and neutral wires are to be terminated to the line side of the device. The conductors must be connected to their proper terminal screws, that is, the hot (typically the black wire) will be connected to the brass terminal, the neutral (typically the white wire) will be connected to the silver terminal, and finally the green terminal is used to connect the ground wire (typically a bare copper wire).

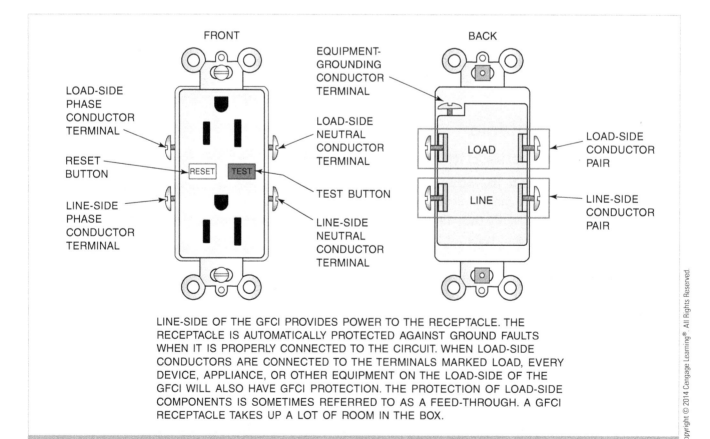

LINE-SIDE OF THE GFCI PROVIDES POWER TO THE RECEPTACLE. THE RECEPTACLE IS AUTOMATICALLY PROTECTED AGAINST GROUND FAULTS WHEN IT IS PROPERLY CONNECTED TO THE CIRCUIT. WHEN LOAD-SIDE CONDUCTORS ARE CONNECTED TO THE TERMINALS MARKED LOAD, EVERY DEVICE, APPLIANCE, OR OTHER EQUIPMENT ON THE LOAD-SIDE OF THE GFCI WILL ALSO HAVE GFCI PROTECTION. THE PROTECTION OF LOAD-SIDE COMPONENTS IS SOMETIMES REFERRED TO AS A FEED-THROUGH. A GFCI RECEPTACLE TAKES UP A LOT OF ROOM IN THE BOX.

FIGURE 6-41: Ground fault circuit interrupter (GFCI).

If other receptacles on the branch circuit are fed through the GFCI receptacle, there will be five wires on the receptacle: two black wires, two white wires, and one ground wire. If this is the case, terminate the wires that are feeding the line voltage into the GFCI receptacle to its line side. The wires that are feeding the remaining receptacles on the branch circuit should be connected to the GFCI receptacle from the load side (see the procedure on page 166).

Replace or Repair a Light Fixture (120-Volt Outlet)

1. Disconnect the power to the light fixture by turning off the power to the circuit at the breaker panel. If at all possible, use natural lighting to illuminate the work area, thereby reducing the amount of artificial light required. If this is not possible, then use adequate work lighting. Flashlights should only be used as a last resource.

2. Test the light fixture to ensure that the power is off. Once you have verified that the power has been discounted from the light fixture, remove the screws holding the plate, as well as the mounting screws (see Figure 6-42).

3. Once the screws have been removed, and with the conductors still attached to the outlet, pull the outlet out from the junction box approximately 4–6 inches.

4. Before removing the conductor from the outlet, make note of the color of the conductor's insulator in relationship to the outlet's terminals. Also, make special note of the common terminal of the outlet. It can be identified by its color. The common terminal is normally brass.

5. Once the conductors of the old outlet have been identified, they can be removed from the old outlet. Normally this is done by starting with the common wire (the common terminal is normally brass), followed by the neutral, and finally the ground.

6. When reconnecting the conductors to the outlet, it makes no difference which vertical screws the conductor is connected to.

7. If the conductor is to be connected to the outlet by the terminal screws, start by making a hook-shaped bend in the conductor. Place the hook under the terminal screw of the terminal to which the conductor is to be connected. Finally, tighten the terminal screw, snugging the conductor to the outlet.

8. Once all conductors have been reconnected, push the outlet back into its box, replace the outlet cover, and reestablish the power to the circuit.

Note: before replacing an existing outlet make sure that the new outlet is rated the same as the one being replaced. For example, do not replace a 20-amp outlet with one rated for 15 amps.

Replace a Light Bulb

1. Before attempting to change a light bulb in a lamp or a light fixture, always turn the fixture off.

2. If there is a possibility that the bulb could be hot, allow it to cool down before attempting to touch the bulb.

FIGURE 6-42: Face plate.

FIGURE 6-43: Removing a light bulb.

3. With a firm but light grip, begin turning the light bulb in a counterclockwise direction until it is free from the light socket (see Figure 6-43).

4. Using a firm but light grip on the new bulb, position it in the light socket and begin turning it in a clockwise direction until it is snug in the light fixture socket.

5. Test the light fixture by turning the light on.

6. Follow the manufacturer's instruction on the proper disposal method for the old light bulb.

Replace a Fluorescent Bulb

1. Before replacing the bulbs in a light fixture, always make sure that the switch controlling the fixture is in the off position.

2. Once you have verified that the light switch is in the off position, start by removing the fixture's lens or diffuser (see Figure 6-44). The light bulbs on a fluorescent light fixture are typically located under the fixture's lens or diffuser (see Figure 6-45). The lens or diffuser can be removed by pushing it up and tilting it at the same time.

3. Before replacing the bulbs, always ensure that the issue not something that can be easily corrected by establishing a good contact between the bulb and the socket. To check the bulbs contact, give the bulb a gentle turn in the clockwise and counterclockwise direction and then back to the lock position

4. If this does not correct the problem, the old bulb can be replaced by rotating the bulb one quarter turn in the clockwise direction (until the bulb's prongs are lined up with the loading slot on the fixture) while firmly hold one end of the bulb.

5. Finish removing the bulb by pulling it though the loading slots.

6. Once the end has cleared the loading slot, continue to lower the bulb carefully until it is completely free of the socket. The other end of bulb should come out of the fixture when you give it a slight pull.

7. Carefully place the old bulb in a location where it will not be broken while you are installing the new bulb.

8. Lift the new bulb into position, holding it so that the prongs are positioned in such a way they will easily go into the loading slots on the fixture.

FIGURE 6-44: Fluorescent light bulb.

FIGURE 6-45: Lens/diffuser on a fluorescent fixture.

9. Insert the prongs into the loading slots. Once they are completely in the loading slots, rotate the bulb counterclockwise until the bulb clicks into place.

10. Once the bulb has been replaced, ensure that the problem has been resolved by testing the light. If the light still doesn't work, then the ballast could be defective.

Replacing a Fluorescent Ballast

1. Before you can replace the ballast in a fluorescent light fixture always confirm that the power to the circuit supplying the fluorescent light fixture has been disconnected. If you are disconnecting the power at the panel, always lock out/tag out the circuit.

2. Once you have confirmed that the circuit has been deenergized, you must locate the ballast in the fixture. In some fixtures, the ballast is a cylindrical object that is located in one end of the fixture. In other fixtures, it is a rectangular item that is located behind a panel under the light bulb. Regardless of the shape and the location of the ballast, the purpose of the ballast is still the same. In addition to stabilizing the current supply to the fluorescent light fixture, the ballast is responsible for providing the starting voltage of the fluorescent light fixture (see Figure 6-46).

3. To replace cylindrical ballast, start by turning the ballast a $\frac{1}{4}$ to $\frac{1}{2}$ turn in the counterclockwise direction. For rectangular ballast, the conductors connecting the ballast to the light fixture, as well as the screws securing it to the fixture, must be removed. However, before removing the conductor from the ballast, make note of the color of the conductor's insulator in relationship wires coming out of the ballast.

4. Once the ballast has been removed from the light fixture, replace it with a ballast equivalent to the original ballast. This can be done by taking the old ballast to an electrical supplier and letting them match the ballast.

FIGURE 6-46: Ballast on a fluorescent fixture.

5. For cylindrical ballast, insert the new ballast into the fixture while turning it clockwise. This will lock the ballast into position on the fluorescent light fixture. For rectangular ballast, always start by resecuring the ballast to the fixture, using the ballast's mounting screws. Once this has been accomplished, the conductors can be reconnected to the ballast. Always check manufacturer's instructions before installing any electrical parts and/or devices.

6. Replace the lens cover and reenergize the circuit supplying the light fixture.

Test and Replace Fuses

Identify and Replace Blown Fuses

1. Before a fuse can be replaced the service panel door must be opened and the fuse located. When searching for a blown fuse always look for visible signs of a break in the fuse of a discolored area in the fuse's sight glass (for plug fuses; see Figure 6-47). If the fuse cannot be easily identified, then the circuit map can be used to locate the circuit having the issue. Typically, the circuit map is located on the fuse panel door.

FIGURE 6-47: Screw-in fuse.

FIGURE 6-48: Fuse block.

2. If a circuit map can't be located, then the blown fuse can be located by using trial and error. In this method, each of the fuses is removed one at a time, and a new fuse is inserted in its place. If the circuit does not reenergize after replacing the fuse, then you still have not located the blown fuse. The process should be repeated until the blown fuse has been identified and replaced. The fuses can also be tested using an ohm meter.

3. Some electrical devices are protected using an overload protection device called a fuse block (see Figure 6-48). To check the fuses in a fuse block, start by pulling out on the handle and testing each individual fuse. Each cartridge fuse can be removed from the fuse block by using a cartridge fuse puller.

Install a New Fuse

1. When replacing a fuse, regardless of the type, always use a fuse having the same rating as the fuse that you are replacing. Once you have located and removed the old fuse, insert the new fuse in its place.

2. Occasionally, all the circuits serviced by the fuse panel will stop working. If this happens, start by removing the fuses in the main fuse block. Normally, the main fuse block is located in the top-left position of the fuse panel. However, sometimes the main fuse block is located in the top-right position in the fuse panel.

Reset a Circuit Breaker

Often when there is an electrical failure with a particular piece of electrical equipment the result is a tripped circuit breaker. To reset a tripped circuit breaker, always follow the manufacturer's instructions. The following are the general steps for resetting a circuit breaker:

1. Locate the breaker panel containing the tripped circuit breaker.

2. After opening the breaker panel, determine which breaker has been tripped. This can be accomplished by finding the breaker in the panel in which the handle is located in the off position or is between the on and off position.

3. If the breaker in not already in the off position, push the handle until it is in the off position. Once this has been accomplished, the breaker can be reset by returning it to the on position.

4. There is always a reason why a circuit breaker tripped. Once the breaker has been reset, if it should happen to trip again, the exact cause for the circuit breaker tripping should be determined and corrected. Failure to correct the issue can lead to property damage.

SUMMARY

- Many of the safety procedures are outlined in the OSHA standard (29 CFR Part 1926) and must be followed while working on electrical systems, equipment, and devices.

- The NEC® is written to ensure the protection of people and property from the hazards that arise from installation and/or repair of electrical systems or equipment.

- Prior to beginning any work on electrical systems, equipment, or devices, always consult with the local authority that has jurisdiction (building/electrical inspector) on the local codes and regulations that might affect the proper and safe installation of electrical systems, equipment, and devices in your area.

- Lockout/tag-out procedures, as stated in the OSHA standard 1926.417 and the NFPA® 70E standard, should be followed at all times to prevent electrical shock or even death by electrocution.

- The NFPA® 70E mandates that qualified personnel should never work on any part of an energized electrical system that is over 50 volts AC without the proper arc flash/arc blast personnel protective clothing (PPE) and without using 1,000-volt-rated tools.

- Adequate lighting and space is required around electrical equipment so that maintenance on the electrical equipment can be performed safely.

- Always ensure that all hand tools and power tools are inspected before and after use.

- Always use calibrated test equipment when testing electrical systems, equipment, or devices to ensure accurate measurements.

- A megohmmeter, also referred to as a "megger," is an instrument used to measure high electrical resistance.

- Analog meters are characterized by the fact that they incorporate a scale and a needle to indicate the value of the electrical property being measured.

- Analog meters operate on the principle that like magnetic poles repel each other.

- The voltmeter is constructed to connect across the power source, allowing the technician to measure the potential difference between two points.

- The ammeter is used to measure current; therefore, it is necessary to connect the ammeter in series to the load, thereby permitting the load to limit the current flow.
- Digital ohmmeters display the resistance in digital display similar to that of a calculator instead of using a meter movement.
- An AC clamp-on ammeter is a versatile instrument, which is also called clip-on, tang-type, snap-on, and other names.
- A megohmmeter is used for measuring very high resistances.
- All conductors have some resistance. The resistance depends on the material, the cross-sectional area, and the length of the conductor.
- If a wire is too small for the current passing through, it will overheat and possibly burn the insulation and could cause a fire.
- Standard copper wire sizes are identified by American Standard Wire Gauge numbers and measured in circular mils.
- A circular mil is the area of a circle $\frac{1}{1,000}$ inches in diameter.
- The most common type of cable used in single-family and multifamily dwellings is NM cable (commonly referred to as Romex).
- One of the most common boxes in use is the single gang nail-up box that is constructed of plastic or fiberglass and is typically placed during new construction before the drywall is installed.
- Four common types of ceiling boxes are pancake (the shallow box), 4″ round nail-up box, 4″ round cut-in box, and fan-rated ceiling box.
- Weatherproof box*es* are used whenever a device (receptacle or switch) is installed in a wet location.
- A feeder circuit is the wires that connect the service equipment to the overload protection device located on the final branch-circuit.
- When it is impossible to conceal conductors (for example, around a desk, counter, or cabinets), a raceway is often used.
- The NEC® defines a multioutlet assembly as a raceway that is designed to house conductors and receptacles that has been assembled in the field (on the jobsite) or assembled at the manufacturer's facility.
- Floor outlets are often used in large office complexes where desks will be positioned too far away from a wall to effectively run a power cord.
- A panelboard is, according to the NEC®, a single panel or a group of panels designed to act as a single panel.
- Circuit protection is essential to prevent the conductors in the circuit from being overloaded.
- A fuse is a simple device used to protect circuits from overloading and overheating.
- Plug fuses have either an Edison base or a type S base.
- Dual-element fuses are frequently used in for electric motors.
- For 230-volt to 600-volt service up to 60 amps, the ferrule cartridge fuse is used.

- A circuit breaker can function as a switch as well as a means for opening a circuit when a current overload occurs.

- Emergency backup systems are designed to keep critical parts of an electrical system energized in the event of power loss.

- Four of the most commonly used switches are the single-pole, double-pole, three-way, and four-way switches.

- The most common receptacles are:

  - 240-volt 30-amp or 50-amp single receptacle

  - 240-volt 20-amp single receptacle

  - 120-volt 15-amp or 20-amp single receptacle

  - 120-volt 15-amp or 20-amp duplex receptacle

  - 120-volt 15-amp or 20-amp ground fault circuit interrupter (GFCI) receptacle

- The GFCI duplex receptacle is designed to trip when there is a difference between the current going to the load and the current returning from the load.

- Fixtures are described by the way they mount and by the type of bulb that is used within them.

Using a Voltage Tester

- This procedure demonstrates how to determine which conductor of a circuit is grounded, using a voltage tester.

- Put on safety glasses, and observe regular safety procedures.

A Connect the tester between one circuit conductor and a well-established ground.

- If the tester indicates a voltage, the conductor being tested is not grounded.

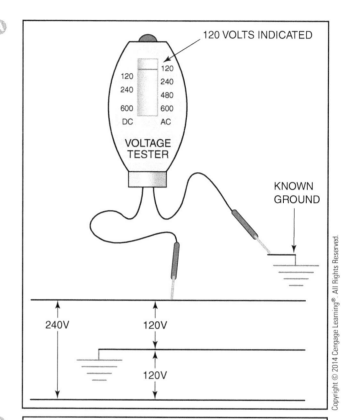

B Continue this procedure with each conductor until zero voltage is indicated between the tested conductor and the known ground. Zero voltage indicates that you have found a grounded circuit conductor.

- This procedure demonstrates how to determine the approximate voltage between two conductors using a voltage tester.

- Put on safety glasses and observe regular safety procedures.

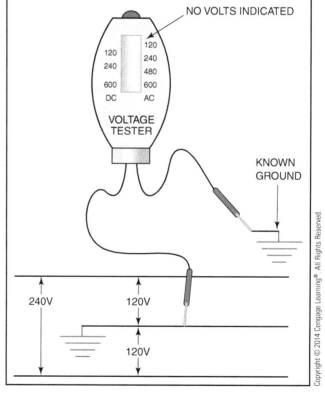

⊙ Connect the tester between the two conductors.

• Read the indicated voltage value on the meter. Note: With a solenoid type tester, you should also feel a vibration, which is another indication that voltage is present.

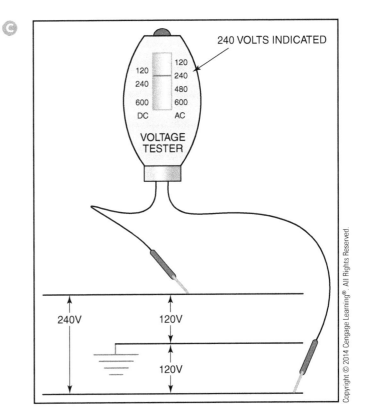

PROCEDURE 6-2

Using a Noncontact Voltage Tester

- This procedure demonstrates how to determine if an electrical conductor is energized, using a noncontact voltage tester.

- Put on safety glasses and observe regular safety procedures.

- Identify the conductor to be tested.

Ⓐ Bring the noncontact voltage tester close to the conductor. Note that some noncontact voltage testers may have to be turned on before using them.

- Listen for the audible alarm, observe a light coming on, or feel a vibration to indicate that the conductor is energized.

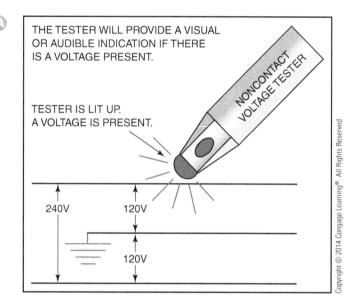

PROCEDURES 6-3

Using a Clamp-On Ammeter

- This procedure demonstrates how to measure current flow through a conductor with a clamp-on ammeter. Note that you can only take a current reading with a clamp-on ammeter clamped around a single conductor. For example, a clamp-on meter will not give a reading when clamped around a two-wire Romex cable.

- Put on safety glasses and observe regular safety procedures.

- If the meter is analog and has a scale selector switch, set it to the highest scale. Skip this step if the meter is digital and has an autoranging feature.

Ⓐ Open the clamping mechanism and clamp it around the conductor.

- Read the displayed value.

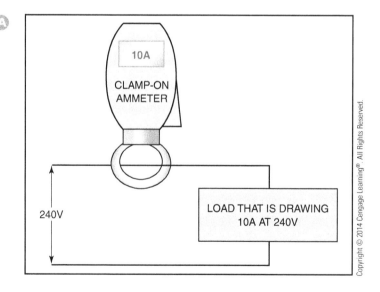

Installing Old-Work Electrical Boxes in a Sheetrock Wall or Ceiling

- Put on safety glasses and observe regular safety procedures.

- Determine the location where you want to mount the box and make a mark. Make sure there are no studs or joists directly behind the wall where you want to install the box.

- Turn the old-work box you are installing backward, and place it at the mounting location so the center of the box is centered on the box location mark. Trace around the box with a pencil. Do not trace around the plaster ears.

- Using a keyhole saw, carefully cut out the outline of the box. There are two ways to get the cut started. One way is to use a drill and a flat blade bit (say a $\frac{1}{2}$-inch size), drill out the corners, and start cutting in one of the corners. The other way often used is to simply put the tip of the keyhole saw at a good starting location and, with the heel of your hand, hit the keyhole saw handle with enough force to cause the blade to go through the sheetrock. It usually does not require much force to start the cut this way.

- Ⓐ Assuming that a cable has been run to the box opening, secure the cable to the box and insert it into the hole. Secure the box to the wall or ceiling surface with Madison hold-its, or use a metal or nonmetallic box with built-in drywall grips.

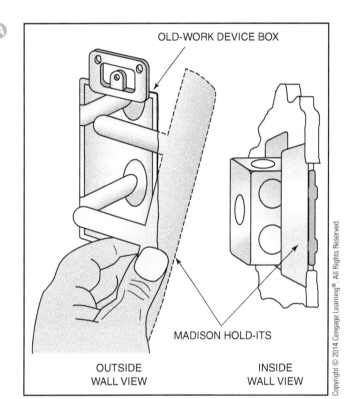

Ⓐ

OLD-WORK DEVICE BOX

MADISON HOLD-ITS

OUTSIDE WALL VIEW

INSIDE WALL VIEW

Installing Duplex Receptacles in a Nonmetallic Electrical Outlet Box

- Put on safety glasses and observe regular safety procedures.

A Using a wire stripper, remove approximately $\frac{3}{4}$ inch of insulation from the end of the insulated wires.

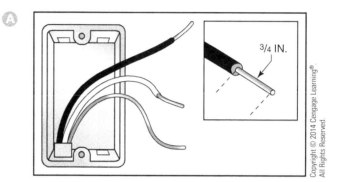

B Using long-nosed pliers or wire strippers, make a loop at the end of each of the wires.

- Take the loop formed on the hot wire (typically a black wire) and position it around the brass terminal screw. The loop should point in a clockwise direction around the screw. Positioning the loop in a counterclockwise direction will cause the loop to spread out from under the terminal screw. Once the wire has been properly positioned and pulled snug around the terminal screw, tighten the screw the correct amount.

- Take the loop formed on the neutral wire (typically a white wire) and position it around the silver terminal screw. The loop should point in a clockwise direction around the screw. Once the wire has been properly positioned and pulled snug around the terminal screw, tighten the screw the correct amount.

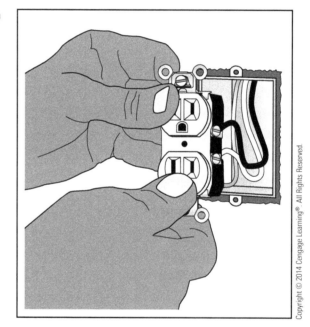

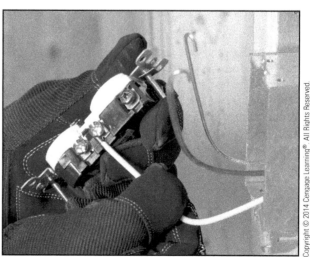

C Complete the installation by placing the loop on the bare grounding wire around the green terminal screw so that the loop is going in the clockwise direction. While pulling the loop snug around the screw terminal, tighten the screw the proper amount with a screwdriver.

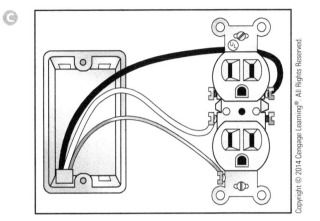

D Place the receptacle into the outlet box by carefully folding the conductors back into the device box.

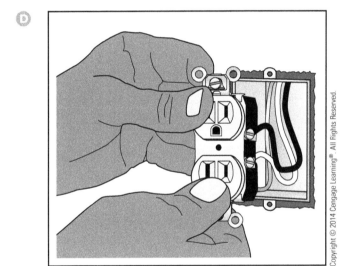

E Secure the receptacle to the device box using the 6-32 screws. Mount the receptacle so that it is vertically aligned.

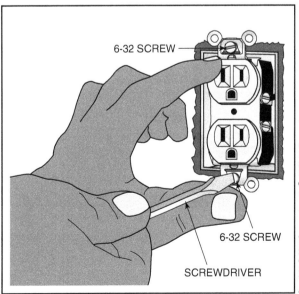

6-32 SCREW

6-32 SCREW

SCREWDRIVER

Installing Duplex Receptacles in a Nonmetallic Electrical Outlet Box (Continued)

F Attach the receptacle's cover plate to the receptacle. Be careful not to tighten the mounting screw(s) too much. Plastic faceplates tend to crack very easily.

F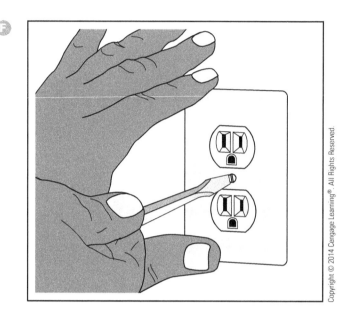

Installing Duplex Receptacles in a Metal Electrical Outlet Box

- Put on safety glasses and observe regular safety procedures.

A Attach a 6–8-inch-long grounding pigtail to the metal electrical outlet box with a 10-32 green grounding screw. The pigtail can be a bare or green insulated copper conductor.

A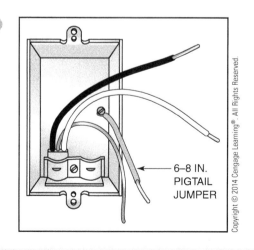

6–8 IN. PIGTAIL JUMPER

B Attach another 6–8-inch-long grounding pigtail to the green screw on the receptacle.

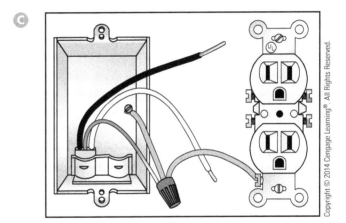

C Using a wirenut, connect the grounding pigtail attached to the box, the branch-circuit grounding conductor(s) and the grounding pigtail attached to the receptacle together.

- Using a wire stripper, remove approximately $\frac{3}{4}$ inch of insulation from the end of the insulated wires.

- Using long-nosed pliers or wire strippers, make a loop at the end of each of the wires.

- Place the loop on the black wire around a brass terminal screw and the loop on the white wire around a silver terminal screw so that the loops are going in the clockwise direction. Tighten the screws the proper amount with a screwdriver.

- Place the receptacle into the outlet box by carefully folding the conductors back into the device box.

- Secure the receptacle to the device box, using the 6-32 screws. Mount the receptacle so that it is vertically aligned.

- Attach the receptacle cover plate to the receptacle. Be careful not to tighten the mounting screw(s) too much. Plastic faceplates tend to crack very easily.

Installing Feed-Through GFCI and AFCI Duplex Receptacles in Nonmetallic Electrical Outlet Boxes

- Put on safety glasses and observe regular safety procedures.

Ⓐ At the electrical outlet box containing the GFCI or AFCI feed-through receptacle, use a wirenut to connect the branch-circuit grounding conductors and the grounding pigtail together. Connect the grounding pigtail to the receptacle's green grounding screw.

- At the electrical outlet box containing the GFCI or AFCI feed-through receptacle, identify the incoming power conductors, and connect the white (grounded) conductor to the line-side silver terminal screw and the incoming black (ungrounded) conductor to the line-side brass terminal screw.

- At the electrical outlet box containing the GFCI or AFCI feed-through receptacle, identify the outgoing conductors and connect the white (grounded) conductor to the load-side silver terminal screw and the outgoing black (ungrounded) conductor to the load-side brass terminal screw.

- Secure the GFCI or AFCI receptacle to the electrical box with the 6-32 screws provided by the manufacturer.

- A proper GFCI or AFCI cover is provided by the device manufacturer; attach it to the receptacle with the short 6-32 screws provided.

Ⓐ

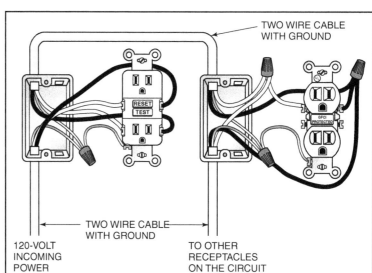

TWO WIRE CABLE WITH GROUND

RESET
TEST

GFCI
PROTECTED

TWO WIRE CABLE WITH GROUND

120-VOLT INCOMING POWER

TO OTHER RECEPTACLES ON THE CIRCUIT

- At the next "downstream" electrical outlet box containing a regular duplex receptacle, connect the white (grounded) conductor(s) to the silver terminal screw(s) and the black (ungrounded) conductor(s) to the brass terminal screw(s) in the usual way. Place a label on the receptacle that states "GFCI Protected." These labels are provided by the manufacturer.

- Continue to connect and label any other "downstream" duplex receptacles as outlined in the previous step.

PROCEDURES 6-8

Installing a Light Fixture Directly to an Outlet Box

- Put on safety glasses and observe all applicable safety rules.

- Using a voltage tester, verify that there is no electrical power at the lighting outlet where the fixture will be installed. If electrical power is present, turn off the power and lock out the circuit.

- Locate and identify the ungrounded, grounded, and grounding conductors in the lighting outlet box.

Installing a Light Fixture Directly to an Outlet Box (Continued)

Ⓐ The grounding conductor will not be connected to this fixture. If there is a grounding conductor in a nonmetallic box, simply coil it up and push it to the back or bottom of the electrical box. Do not cut it off, as it may be needed if another type of light fixture is installed at that location. If there are two or more grounding conductors in a nonmetallic box, connect them together with a wirenut and push them to the back or bottom of the lighting outlet box. If there is a grounding conductor in a metal outlet box, it must be connected to the outlet box by means of a listed grounding screw or clip. If there are two or more grounding conductors in a metal box, use a wirenut to connect them together along with a grounding pigtail. Attach the grounding pigtail to the metal box with a listed grounding screw.

• If the fixture has a white pigtail, connect it to the white conductor; otherwise, connect the white conductor to the silver fixture screw. If there is one grounded conductor in the box, strip approximately $\frac{3}{4}$ inch (19 mm) of insulation from the end of the conductor, and form a loop at the end of the conductor using an approved tool such as a T-stripper. Once the loop is made, slide it around the silver terminal screw on the fixture, so the end is pointing in a clockwise direction. Hold the conductor in place and tighten the screw. If there are two or more grounded

Ⓐ

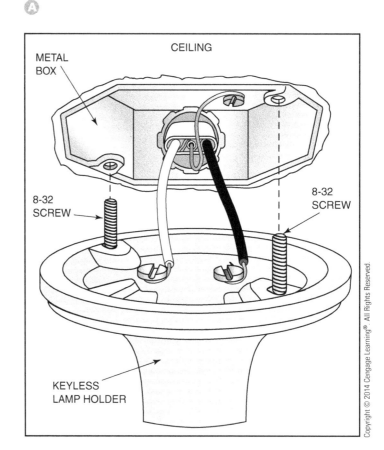

CEILING

METAL BOX

8-32 SCREW

8-32 SCREW

KEYLESS LAMP HOLDER

conductors in the box, strip the ends as described previously, and use a wirenut to connect them and a white pigtail together. Attach the pigtail to the silver grounded screw as described previously.

- The black ungrounded conductor(s) is connected to the brass terminal screw on the fixture. The connection procedure is the same as for the grounded conductor(s).

- Now the fixture is ready to be attached to the lighting outlet box. Make sure the grounding conductors are positioned so that they will not come in contact with the grounded or ungrounded terminal screws. Align the mounting holes on the fixture with the mounting holes on the lighting outlet box. Insert the 8-32 screws that are usually supplied with the fixture through the fixture holes, then thread the screws into the mounting holes located on the outlet box.

- Once finished, the fixture should make contact with either the ceiling or the wall in which the fixture is installed. This is accomplished by tightening the mounting screws until this has been accomplished. Be careful not to overtighten the screws, as you may damage the fixture.

- Install the proper lamp, remembering not to exceed the recommended wattage.

- Turn on the power and test the light fixture.

Installing a Cable-Connected Fluorescent Lighting Fixture Directly on the Ceiling

- Put on safety glasses and observe regular safety procedures. Using a voltage tester, verify that there is no electrical power at the lighting outlet where the fixture will be installed. If electrical power is present, turn off the power and lock out the circuit.

- Place the fixture on the ceiling in the correct position, making sure it is aligned and the electrical conductors have a clear path into the fixture.

- Mark on the ceiling the location of the mounting holes.

- Use a stud finder to determine if the mounting holes line up with the ceiling trusses. If they do, screws will be used to mount the fixture. If they do not, toggle bolts will be necessary. For some installations, a combination of screws and toggle bolts will be required.

- If screws are used, drill holes into the ceiling using a drill bit that has a smaller diameter than the screws to be used. This will make installing the screws easier. If toggle bolts are to be used, use a flat-bladed screwdriver to punch a hole in the sheetrock only large enough for the toggle to fit through.

- Remove a knockout from the fixture where you wish the conductors to come through. Install a cable connector in the knockout hole.

- Place the fixture in its correct position, and pull the cable through the connector and into the fixture. Tighten the cable connector to secure the cable to the fixture. This part of the process may require the assistance of a coworker. Using toggle bolts, put the bolt through the mounting hole and start the toggle on the end of the bolt.

- With a coworker holding the fixture, install the mounting screws or push the toggle through the hole until the wings spring open. This will hold the fixture in place until the fixture is secured to the ceiling.

- Make the necessary electrical connections. The grounding conductor should be properly wrapped around the fixture grounding screw and the screw tightened. The white grounded conductor is connected to the white conductor lead; then the black ungrounded conductor is connected to the black fixture conductor.

- Install the wiring cover by placing one side in the mounting clips, squeezing it, and then snapping the other side into its mounting clips.

- Install the recommended lamps. Usually, they have two contact pins on each end of the lamp. Align the pins vertically, slide them up into the lamp holders at each end of the fixture, and rotate the lamp until it snaps into place.

- Test the fixture and lamps for proper operation.

- Install the fixture lens cover.

PROCEDURES 6-10

Installing a Strap on a Lighting Outlet Box Lighting Fixture

- Put on safety glasses and observe regular safety procedures.

- Using a voltage tester, verify that there is no electrical power at the lighting outlet where the fixture will be installed. If electrical power is present, turn off the power and lock out the circuit.

- Before starting the installation process, read and follow the manufacturer's instructions as well as local building codes.

Ⓐ Mount the strap to the outlet box using the slots in the strap. With metal boxes, the screws are provided with the box. With non-metallic boxes, you must provide your own 8-32 mounting screws. Put the 8-32 screws through the slot and thread them into the mounting holes on the outlet box. Tighten the screws to secure the strap to the box.

- Identify the proper threaded holes on the strap, and install the fixture-mounting headless bolts in the holes so the end of the screw will point down.

- Make the necessary electrical connections. Make sure that all metal parts (including the outlet box), the strap, and the fixture are properly connected to the grounding conductor in the power feed cable.

Ⓐ

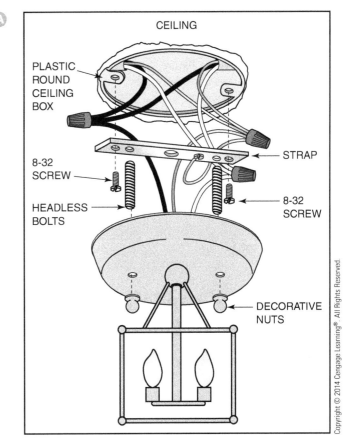

CEILING

PLASTIC ROUND CEILING BOX

STRAP

8-32 SCREW

8-32 SCREW

HEADLESS BOLTS

DECORATIVE NUTS

- Neatly fold the conductors into the outlet box. Align the headless bolts with the mounting holes on the fixture. Slide the fixture over the headless bolts until the screws stick out through the holes. Do not be alarmed if the mounting screws seem to be too long. Thread the provided decorative nuts onto the headless bolts. Keep turning the nuts until the fixture is secured to the ceiling or wall.

- Install the recommended lamp, and test the fixture operation.

- Install any provided lens or globe. They are usually held in place by three screws that thread into the fixture. Start the screws into the threaded holes, position the lens or globe so it touches the fixture, and tighten the screws until the globe or lens is snug. Do not overtighten the screws. You may return the next day and find the globe or lens cracked or broken.

PROCEDURES 6-11

Installing a Chandelier-Type Light Fixture, Using the Stud and Strap Connection to a Lighting Outlet Box

- Put on safety glasses and observe regular safety procedures.

- Using a voltage tester, verify that there is no electrical power at the lighting outlet where the fixture will be installed. If electrical power is present, turn off the power and lock out the circuit.

- Before starting the installation process, read and follow the manufacturer's instructions and specifications.

Installing a Chandelier-Type Light Fixture, Using the Stud and Strap Connection to a Lighting Outlet Box

(A) Install the mounting strap to the outlet box, using 8-32 screws.

- Thread the stud into the threaded hole in the center of the mounting strap. Make sure that enough of the stud is screwed into the strap to make a good secure connection.

- Measure the chandelier chain for the proper length, remove any unneeded links, and install one end on the light fixture.

- Thread the light fixture's chain-mounting bracket on to the stud. Remove the holding nut and slide it over the chain.

- Slide the canopy over the chain.

- Attach the free end of the chain to the chain-mounting bracket.

- Weave the fixture wires and the grounding conductor up through the chain links, being careful to keep the chain links straight. Section 410.28(F) of the NEC® states that the conductors must not bear the weight of the fixture. As long as the chain is straight and the conductors make all the bends, the chain will support the fixture properly.

- Now run the fixture wires up through the fixture stud and into the lighting outlet box.

- Make all necessary electrical connections.

- Slide the canopy up the chain until it is in the proper position. Slide the nut up the chain and thread it on to the chain-mounting bracket until the canopy is secure.

- Install the recommended lamp, and test the fixture for proper operation.

(A)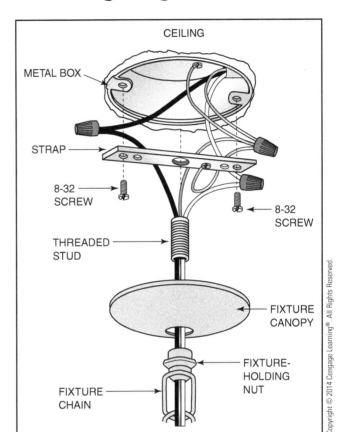

CEILING

METAL BOX

STRAP

8-32 SCREW

8-32 SCREW

THREADED STUD

FIXTURE CANOPY

FIXTURE-HOLDING NUT

FIXTURE CHAIN

Installing a Fluorescent Fixture (Troffer) in a Dropped Ceiling

- Put on safety glasses and observe regular safety procedures.

- Before starting the installation process, read and follow the manufacturer's instructions and specifications.

- During the rough-in stage, mark the position of the fixtures on the ceiling.

- Using standard wiring methods, place lighting outlet boxes on the ceiling near the marked fixture locations, and connect them to the lighting branch circuit.

- Once the dropped ceiling grid has been installed by the ceiling contractor, install the fluorescent light fixtures in the ceiling grid at the proper locations. Some electricians refer to this action as "laying in" the fixture. Once the fixture is installed, some electricians refer to the fixtures as being "laid in."

- Ⓐ Support the fixture according to NEC® requirements. Section 410.16(C) requires that all framing members used to support the ceiling grid be securely fastened the building as well as one another. The fixtures themselves must be securely attached to the grid by an approved means, such as bolts, screws, rivets, or clips. This is to prevent the fixture from falling and injuring someone.

- Using a voltage tester, verify that there is no electrical power at the lighting outlet where the fixture will be installed. If electrical power is present, turn off the power and lock out the circuit.

Ⓐ

IMPORTANT: TO PREVENT THE LUMINAIRE (FIXTURE) FROM INADVERTENTLY FALLING, *410.16(C)* OF THE CODE REQUIRES THAT (1) SUSPENDED CEILING FRAMING MEMBERS THAT SUPPORT RECESSED LUMINAIRES (FIXTURES) MUST BE SECURELY FASTENED TO EACH OTHER AND MUST BE SECURELY ATTACHED TO THE BUILDING STRUCTURE AT APPPROPRIATE INTERVALS, AND (2) RECESSED LUMINAIRES (FIXTURES) MUST BE SECURELY FASTENED TO THE SUSPENDED CEILING FRAMING MEMBERS BY BOLTS, SCREWS, RIVETS, OR SPECIAL LISTED CLIPS PROVIDED BY THE MANUFACTURER OF THE LUMINAIRE (FIXTURE) FOR THE PURPOSE OF ATTACHING THE LUMINAIRE (FIXTURE) TO THE FRAMING MEMBER.

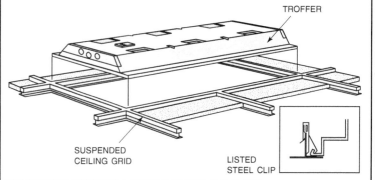

TROFFER

SUSPENDED CEILING GRID

LISTED STEEL CLIP

B Connect the fixture to the electrical system. This is done by means of a "fixture whip." A fixture whip is often a length of Type NM, Type AC, or Type MC cable. It can also be a raceway with approved conductors such as flexible metal conduit or electrical nonmetallic tubing. The fixture whip must be at least 18 inches (450 mm) long and no longer than 6 feet (1.8 m).

- Make all necessary electrical connections. The fixture whip should already be connected to the outlet box mounted in the ceiling. Using an approved connector, connect the cable or raceway to the fixture outlet box and run the conductors into the outlet box. Make sure that all metal parts are properly connected to the grounding system. Connect the white grounded conductors together and then the black ungrounded conductors together. Close the connection box.

- Install the recommended lamps, and test the fixture for proper operation.

- Install the lens on the fixture.

B

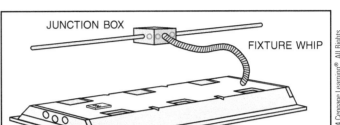

REVIEW QUESTIONS

1   What is an emergency backup system used for?

2   What is NM cable?

3   What is a single-pole switch?

4   What is a double-pole switch?

5   How do three-way switches differ from a single-pole switch?

6   What is a GFCI?

REVIEW QUESTIONS

7   What is a continuity tester?

9   How is GFCI tested?

8   How is a smoke detector tested?

10   List the steps for replacing a smoke detector.

Name: _____

Date: _____

1

Electrical Facilities Maintenance

ELECTRICAL TROUBLESHOOTING AND MAINTENANCE

Upon completion of this job sheet, you should be able to demonstrate your ability to perform basic maintenance and electrical troubleshooting.

1. Choose an area in your facility, and take an inventory of the switches being used in the area. What types are being used, and what are they made of?

   _____

   _____

2. List the types of receptacles discussed in this chapter and their uses.

   _____

   _____

3. What is the most important thing to remember when working with electricity?

   _____

   _____

4. Define troubleshooting.

   _____

   _____

5. What is a GCFI receptacle, and what is its purpose?

   _____

   _____

6. What is the purpose of a lockout/tag-out device?

   _____

   _____

INSTRUCTOR'S RESPONSE:

_____

_____

_____

© iStockphoto.com/kkgas

CHAPTER 7

Carpentry

OBJECTIVES

By the end of this chapter, you will be able to:

KNOWLEDGE-BASED

- Describe the general properties of hardwood and softwood commonly used by facilities maintenance technicians.
- Describe the effects of moisture content on different wood products.
- Perform estimating and take-off quantities for simple one-step carpentry projects.
- Estimate the amount of drywall materials needed for a particular type of installation.
- Correctly identify and select engineered products, panels, and sheet goods.
- Correctly identify framing components.

- Describe how sound is transmitted from one room to another.
- Describe the methods used to control sound from being transmitted from one room to another.
- Identify the types of suspended ceilings typically used.
- Describe the basic demolition steps for remodeling a kitchen, bathroom, and utility room.

SKILL-BASED

- Perform interior carpentry maintenance.
- Perform exterior carpentry maintenance.

Hardwood deciduous trees

Softwood coniferous, or cone-bearing, trees

Green lumber lumber that has just been cut from a log

Engineered panels human-made products in the form of large reconstituted wood sheets

Plies layers of wood used to build up a product such as plywood

Softboard a low-density fiberboard

Acoustics the ability of a material to absorb and reflect sound is referred to as acoustical

Acoustical analysis an analysis to determine the level of reverberation or reflection of sound in a space as it is influenced by the building materials used in it construction

Acoustical consultant a consultant who is experienced in providing advice on the acoustical requirements and noise control of a facility or a room

Acoustical insulation insulation that reduces the passage of sound through the section of a building

Sound transmission class (STC) a rating used to describe the resistance of a building section to the passage of sound

Impact noise sound that is caused by the dropping of object onto a floor

Back miter an angle cut starting from the end and coming back on the face of the stock

Boring jig a tool frequently used to guide bits when boring holes for locksets

Multispur bit a power-driven bit, guided by a boring jig, which is used to make a hole in a door for the lockset

Faceplate marker a tool used to lay out the mortise for the latch faceplate

Striker plate a plate installed on the door jamb against which the latch on the door engages when the door is closed

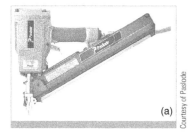

(a)

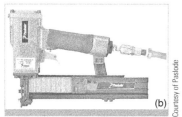

(b)

FIGURE 7-1: Pneumatic mailer and staplers are widely used to fasten building parts.

Introduction

Although facility maintenance technicians will not be responsible for new construction, they will be responsible for maintaining the integrity of the facility. This responsibility often includes performing minor construction and/or repairs using carpentry skills and techniques.

Tools

In addition to many of the tools discussed in Chapter 4, portable fastening tools are used extensively in carpentry for both construction and repair to assist in speeding up production. The two most commonly used power fastening tools are pneumatic (compressed air to drive the fastener, see Figures 7-1 through 7-6) and power-actuated (explosive charge to drive the fastener, see Figure 7-7) tools. Before operating any pneumatic tool, read and follow the manufacturer's instruction.

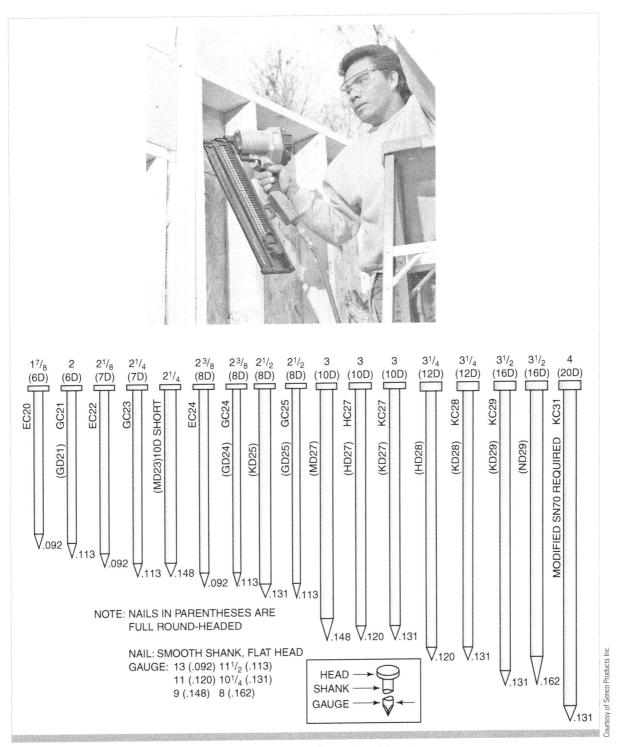

FIGURE 7-2: Heavy-duty nailers are used for floor, walls, and roof framing.

General Properties of Hardwood and Softwood

The carpenter works with wood more than any other material and must understand its characteristics to use it intelligently. Wood is a remarkable substance and is classified as hardwood or softwood. Different methods of classifying these woods exist. The most common method of classifying wood is by its source (see Figure 7-8).

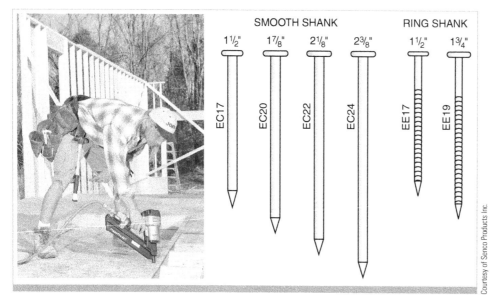

FIGURE 7-3: A light-duty nailer is used to fasten light framing, subfloors, and sheathing.

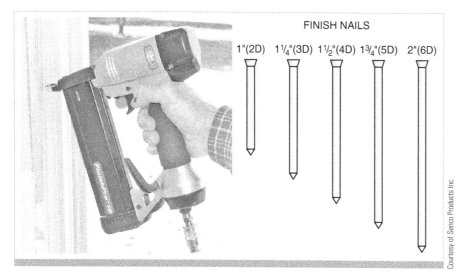

FIGURE 7-5: The finish nailer is used to fasten all kinds of interior trim.

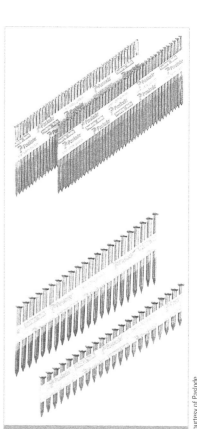

FIGURE 7-4: Both headed and finish nails used to nail guns come glued together in strips.

FIGURE 7-6: A coil roofing nailer is used to fasten asphalt roof shingles.

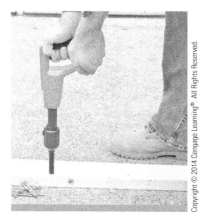

FIGURE 7-7: Power-actuated drivers are used for fastening into masonry or steel.

FIGURE 7-8: Softwood and hardwood.

Hardwood comes from deciduous trees that shed their leaves each year. **Softwood** comes from coniferous, or cone-bearing, trees, commonly known as evergreens.

Common hardwoods include:

- Ash
- Birch
- Cherry
- Hickory
- Maple
- Mahogany
- Oak
- Walnut

Common softwoods include:

- Pine
- Fir
- Hemlock
- Spruce
- Cedar
- Cypress
- Redwood

The best way to learn the different kinds of woods is to work with them and examine them.

- Look at the color and the grain.
- Feel if it is heavy or light.
- Feel if it is hard or soft.
- Smell it for a characteristic odor.

Effects of Moisture Content

When a tree is first cut down, it contains a great amount of water. Lumber, when first cut from the log, is called **green lumber** and is very heavy because most of its weight is water (see Figure 7-9). A piece 2 inches thick, 6 inches wide, and 10 feet long may contain as much as $4\frac{1}{4}$ gallons of water, weighing about 35 pounds.

Green lumber should not be used for construction. As green lumber dries, it shrinks considerably and unequally as the large amount of water leaves it. When it shrinks, it usually warps, depending on the way it was cut from the log (see Figure 7-10).

Realizing that lumber undergoes certain changes when moisture is absorbed or lost, the experienced carpenter uses techniques to deal with this characteristic of wood (see Figure 7-11).

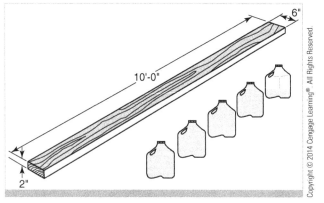

FIGURE 7-9: Green lumber contains a large amount of water.

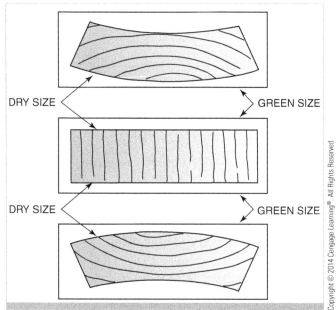

FIGURE 7-10: Lumber shrinks in the direction of the annular rings.

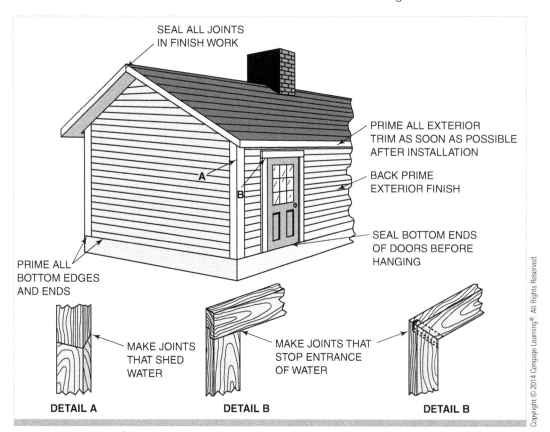

FIGURE 7-11: Techniques to prevent water from getting in behind the wood surface.

Correctly Identify and Select Engineered Products, Panels, and Sheet Goods

Before a construction project of any scope involving lumber or engineered products can be completed, the technician must correctly select the type of product that will best deliver the desired results. Engineered products are typically stronger than traditional lumber products. In addition, engineered products can be used to span greater distances than traditional lumber products (see Figure 7-12).

Engineered Panels

The term **engineered panels** refers to human-made products in the form of large reconstituted wood sheets, sometimes called panels or boards.

- **Plywood**—One of the most extensively used engineered panels (Figure 7-13). It is a sandwich of wood. Most plywood panels constitute sheets of veneer called **plies**.

- **Oriented strand board (OSB)**—A nonveneered, performance-rated structural panel composed of small oriented (lined up) strand-like wood pieces arranged in three to five layers with each layer at right angles to the other (Figure 7-14).

FIGURE 7-12: Laminated strand lumber.

Nonstructural Panels

- **Hardwood plywood**—This is available with hardwood face veneers, of which the most popular are birch, oak, and lauan.

FIGURE 7-13: APA performance-rated panels.

FIGURE 7-14: Oriented strand board (OSB).

FIGURE 7-15: Particleboard is made from wood flakes, shaving, resins, and waxes.

- **Particleboard**—This is reconstituted wood panels made of wood flakes, chips, sawdust, and planer shavings (Figure 7-15). These wood particles are mixed with an adhesive, formed into a mat, and pressed into sheet form.

- **Fiberboards**—These are manufactured as high-, medium-, and low-density boards. Medium-density fiberboard (MDF) is manufactured in a manner similar to that used to make hardboard, except that the fibers are not pressed as tightly. Low-density fiberboard, which is called **softboard**, is light and contains many tiny air spaces because the particles are not compressed tightly.

- **Hardboards**—These are high-density fiberboards, which are sometimes known by the brand name Masonite.

FIGURE 7-16: Drywall, also known as Sheetrock.

Others

- **Gypsum board**—This is used extensively for construction. It is sometimes called wallboard, plasterboard, drywall, or Sheetrock (the brand name; Figure 7-16). Gypsum board is readily available; easy to apply, decorate, or repair, and it is relatively inexpensive.

- **Plastic laminates**—These are used for surfacing kitchen cabinets and countertops. They are also used to cover walls or parts of walls in kitchens, bathrooms, and similar areas, where a durable, easy-to-clean surface is desired.

Drywall

Before the introduction of drywall in the early 1900s, home builders were left using plaster and lath for finishing interior wall. However, with the introduction of drywall, contractors could quickly construct smoother walls in a fraction of the time required to construct a plaster wall (see Figure 7-17). Drywall is a panel made of gypsum plaster that has been pressed between two thick sheets of paper. In the United States and Canada, drywall is available in widths of 48 inches (4 feet) and 54 inches (4 feet × 6 inches). Common lengths of drywall are 96 inches (8 ft), 120 inches (10 ft), 144 inches (12 ft), and 192 inches (16 ft). The most common thickness of drywall used today is $\frac{1}{2}$ inch and $\frac{5}{8}$ inch. Drywall is also available in $\frac{1}{4}$-inch and $\frac{3}{8}$-inch thicknesses.

FIGURE 7-17: Installing drywall on interior walls.

Moisture-resistant and fire-rated drywall (Type X fire-code drywall) is also available. Most building codes require the use of fire-rated drywall in areas where there is a possibility of fire. Therefore, always check your local building codes before selecting a thickness and type of drywall. In most cases, fire-rated drywall is required in utility rooms in which a furnace is used. It is also typically required on garage walls and ceilings that are adjacent to the living room.

Standard drywall (gypsum) contains approximately 21 percent moisture by weight. Therefore, when drywall is exposed to heat from a fire, the moisture content is slowly released in the form of steam, thereby helping retard the heat transfer rate. The fire-resistance property of type X fire-code drywall has been improved by adding glass-fiber reinforcement (to help it hold up longer to fire), as well as other additives, to the drywall. Before fire resistant drywall can be certified as ASTM C 1396–compliant, it must be tested and certified by an independent third-party testing facility such as Underwriters Laboratories Inc. (UL).

Standard drywall has a fire rating of approximately 30 minutes, whereas $\frac{5}{8}$-inch type X fire-code drywall has a minimum fire rating of 1 hour. The $\frac{1}{2}$-inch type X fire-resistant drywall has a minimum fire rating of 45 minutes, whereas $\frac{3}{4}$-inch type X fire-code drywall has a minimum fire rating of 2 hours (120 minutes).
When moisture is a concern, moisture-resistant drywall should be used. Moisture-resistant drywall type MR is typically used in high-humidity areas of the such as laundry rooms and bathrooms. Moisture-resistant drywall is also referred to as green or blue board because of the color of the paper that is used.

Selecting a Drywall Thickness

In most residential applications, $\frac{1}{2}$-inch drywall is used for interior walls. For ceiling applications, $\frac{1}{2}$-inch or even $\frac{5}{8}$-inch drywall is used to keep the ceiling from drooping. With thinner thicknesses of drywall, sagging can be a problem, even if the required number of drywall fasteners is used. When applying drywall to a curved surface, $\frac{1}{4}$-inch thicknesses are typically used. For commercial applications, typically $\frac{5}{8}$-inch drywall is, used because the extra thickness provides better soundproofing characteristics than $\frac{1}{2}$ inch. Again, always check the local building codes to verify the required thickness before starting a drywall application.

Drywall Fasteners

Drywall can be installed using adhesive, nails and/or screws. Each type of fastener has its advantages and disadvantages. Drywall nails will have a large head and either a barbed or ringed shanks. Of all the fasteners used with drywall, nails often have a tendency to pop out, thus destroying the wall finish. Double-nailing drywall (applying a second nail just 2 inches from the first) can help prevent this from happening. Drywall nail can only be used with stick construction and should be spaced 6 to 8 inches apart. For $\frac{1}{2}$-inch drywall, $\frac{15}{8}$ inch drywall nails are typically used. For $\frac{5}{8}$-inch drywall, $1\frac{7}{8}$-inch nails are typically used.

Drywall screws are much easier to install than drywall nails and also provide better holding capability than drywall nails. In addition, drywall screws will not pop out like drywall nails. Also, drywall screws can be used in metal stud construction. Drywall screws can be spaced 12 inches apart, thus reducing the number of fasteners required. A $1\frac{1}{4}$-inch drywall screw can be used for drywall thicknesses ranging from $\frac{1}{4}$ inch to $\frac{5}{8}$ inch. When installing drywall on metal studs, type S or type S-12 should be used. This type of drywall screw is self-tapping. Normally, type S screws are used for light-gauge metal framing members, whereas type "S-12" screws are used for heavier-gauge metal framing.

When attaching drywall to existing drywall, panel adhesive is often a better method of attachment. If the drywall is being used for a ceiling application, then consider using drywall nails or screws in addition to the adhesive.

Fastener Schedules

A schedule, on a set of construction drawings is a list added separately to the drawing plans that describe such items as windows, doors, drywall, flooring, fasteners, and so forth. The contractor working in a particular area must have a good understanding of schedules and their interpretation in order to ensure that the job is completed to the standards and specifications that have been outlined in the construction documentation. In addition, if a schedule is not followed, this could affect a manufacturer's warranty.

For windows and doors, schedules typically contain a designation of the window and/or door (duplicated on the drawing as identification), its size, and its type. In some cases, the schedule also contains a brief description.

As an example, a 30", 15 Lite French Slab Door located in the living room would be listed as shown in Table 7-1.

No.	Size	Type	Location
1	30" × 80"	30 in. Clear Pine Wood 15 Lite French Slab Door	Living Room

TABLE 7-1: Door Schedule.

For drywall fasteners a typical schedule usually contains the following information (see Table 7-2 for an example):

- **Symbols**—Shows any special symbol that might be used on the drawing to identify the fastener.

- **Location**—Describes the location in the structure in which the fastener is to be used.

- **Fastener**—The description of the fastener (thread count, type of head, length, etc.).

- **Manufacturer**—The manufacturer of the fastener.

- **Fastener Spacing**—The fastener's spacing.

- **Notes**—Any special notes associated with the fastener.

Symbol	Location	Fastener	Manufacturer	Fastener Spacing	Notes
S_c	Living Room	#6 × 1-1/4 in.	Grip-Rite	12	

TABLE 7-2: Fastener Schedule.

Estimating Drywall Materials

To estimate the amount of materials needed for a drywall application, first start by measuring the walls you will be applying the drywall to. When measuring the walls, do not subtract for windows and doors. Once the wall measurements have been established, determine the perimeter of the room. For example, if the room measures 10 feet × 14 feet, then the perimeter of the room is calculated as follows:

10 feet + 10 feet + 14 feet + 14 feet = Perimeter
20 feet + 14 feet + 14 feet = Perimeter
34 feet + 14 feet = Perimeter
Perimeter = 48 feet

Once the perimeter has been determined, the number of drywall panels can be calculated by dividing the perimeter by the width of the drywall. For example, in the preceding example, the number of 4 × 8 sheets of drywall panels required is determined as follows:

Perimeter/Width of the drywall = Number of sheets
48 feet/4 feet = Number of sheets
12 = Number of sheets

Determining the Number of Fasteners

To determine the number of fasteners needed, start by calculating the square footage of the drywall being installed. This can be done by multiplying the length of the walls by the height of the walls. In most residential applications, this can easily be accomplished by multiplying the perimeter of the room by the ceiling height. For example, in the previous illustration the square footage is determined as follows:

Perimeter × Wall Height = Area
48 feet × 8 foot = Area
384 sq. ft = area

A $5\frac{1}{4}$ pound box of drywall nails will easily fasten 1,000 square feet of drywall. Drywall screws are also sold by the pound. One pound of drywall screws is equal to approximately 150 screws. It takes approximately 350 screws to cover 500 square feet of drywall or approximately 70 screws per 100 square foot of drywall. In the preceding example, a $5\frac{1}{4}$ pound box of drywall nails will be more than adequate for the room. To determine the quantity of drywall screws, divide the total square footage by the 100. Next, multiple the quotient from the previous step by the number of drywall screws per 100 square feet of coverage. Finally, divide the product of the last step by the number of screws in a pound of drywall fasteners.

Square footage of drywall/100 square feet
384 square feet/100 square feet = 3.84
3.84 × 70 = 268.8
268.8/150 = 1.792 pounds of screws or 2 pounds

Determining the Amount of Drywall Tape

Calculating the amount of tape can be easily accomplished by take half of the square footage of drywall being installed. For example, 384 square feet of drywall would require 192 ($\frac{384}{2}$) feet of drywall tape. Drywall tape is normally sold in 60-, 250-, and 500-foot rolls. Therefore, 4 rolls of 60-foot tape or 1 roll of 250-foot drywall tape will cover the pervious example.

Determining the Amount of Drywall Compound

The following chart can be used to determine the amount of joint compound need for a particular drywall application, see Table 7-3.

Square Footage of Drywall	Amount of Compound
100–200	1 gallon
300–400	2 gallons
500–600	3 gallons
700–800	4 gallons
900–1000	5 gallons

TABLE 7-3: Amount of Compound per Square Footage of Drywall.

1 gallon of drywall compound will cover 200 square feet of drywall.

Installing a Single Layer of Drywall

To install drywall the following tools are needed:
- Measuring tape
- Chalk line
- T-square
- Straightedge
- Utility knife
- Cordless drill
- Keyhole saw
- 6-inch taping knife
- 12-inch mud knife
- 4-inch corner knife
- Dust mask
- Goggles

Before starting the installation process, plan the layout so that it will end up with as few joints as possible. If the ceiling height is less than 8 ft 1 in., then consider installing the drywall in a horizontal fashion. This will reduce the number of taping joint by 25 percent. For ceiling height greater than 8 ft 1 in., consider installing the drywall vertically.

Regardless of the application, drywall should only be delivered to the jobsite just before it is to be installed. Having drywall delivered too far in advance can lead to the drywall being damaged and/or warped. When the drywall is delivered, it should be stored in a secure dry location. It should also be stacked flat on 4 × 4 or 2 × 4 spaced ever 1–2 feet apart. When carrying drywall, never drag it or drop it. Doing so can damage the panels.

Installing Drywall on a Ceiling

When installing drywall, always start with the ceiling, followed by the walls. Panels are always applied either parallel or perpendicular to the joist. When they are applied parallel to the joist, the perimeter of the drywall must bear completely on the framing members. When applied perpendicular to the framing members, the edges are fastened where they cross framing members as well as each end (see Figure 7-18). To install drywall on a ceiling:

1. Starting in one corner of the room, carefully measure the area to be covered and cut the first panel accordingly. Any panels that require trimming should be placed so that the cut ends are against a wall. Layout lines are also used to indicate the location of fasteners in relation to the framing members. Drywall can also be attached to the ceiling using drywall adhesive.

2. Holding the panel firmly against the framing members, fasten the drywall to the framing members. Because drywall is rather heavy, it is recommended that at least two people be involved in the installation process. If this is not possible, then sheetrock jacks or deadmen be used to hold the drywall into position. Deadmen can be constructed by using 2 × 4 constructed in a "T" shape. The legs of the deadmen are typically $\frac{1}{4}$-inch to $\frac{1}{2}$ inch longer than the floor to ceiling height.

FIGURE 7-18: Installing drywall on a ceiling.

3. Continuing from where the last panel ends, install the remaining panel in that row.

4. When starting a new row, place the panels so the end joints are staggered.

Installing Drywall on a Wall

As stated earlier, when the wall height is less than 8 ft 1 in., it is recommended that the drywall be installed horizontally at right angles to the framing studs (see Figure 7-19). This is accomplished as follows:

1. Starting with the top panel, place the drywall so that it is firmly against the framing studs in one corner of the room. It can also be attached to the wall with drywall adhesive.

2. Start applying fasteners along the top edge of the panel. Make sure that the top edge is placed firmly against the ceiling before fastening the drywall to the framing studs.

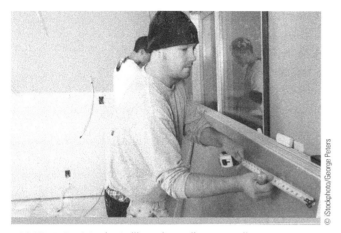

FIGURE 7-19: Installing drywall on a wall.

3. Once the top panel has been installed. Measure and cut the bottom panel. The bottom panel should be cut about $\frac{1}{4}$ inch narrower than the distance measured.

4. Laying the panel against the floor, use a drywall foot lifter to lift the drywall into place.

5. Secure the panel to the framing stud by applying fasteners.

Installing Drywall on a Curved Surface

Installing drywall on a curved surface can be easily done by slightly moistening the paper and core of the drywall. After the drywall has been moistened, it should be stacked flat on the floor and allowed to stand at the very least of one hour. When the drywall is finally installed, it should be handled with extreme caution. Moistened drywall can be easily damaged. Once the drywall has been allowed to dry, it will retain its original strength.

Multilayer Application

A multilayer drywall application is one in which one or more layers of drywall are installed over a base layer of drywall, gypsum backing board, or some other gypsum base product (see Figure 7-20). The base layer in a multilayer application is installed in the same manner as a single-layer drywall application. However, if nails are used to fasten the base layer to the framing members, then double-nailing is not necessary. For ceiling applications, the base layer is installed parallel to the joist. For wall applications, the base layer is installed parallel to the framing members.

FIGURE 7-20: Installing multiple layers of drywall.

The face layer is the final layer of drywall to be installed. It is installed either parallel or perpendicular to the framing members, whichever position will produce less taping joints. Also, the final joints should be offset from the base joints by at least 10 inches. The face layer can be attached using drywall nails, screws, and/or drywall adhesives. If nails or screws are used without drywall adhesives, then the maximum fastener spacing should be the same as for a single-layer application.

A drywall foot lifter is a special tool that can be purchased at almost any building supply company. If one is not available, one can be created by tapering a 1 × 3 or a 1 × 4.

Installing Drywall on Steel Studs

When drywall is installed on steel studs, it is attached using $1\frac{1}{4}$ inch self-tapping drywall screws and a cordless screw gun. Each fastener is started by driving the screw partially into the drywall and then pushing it forcefully with the cordless drill operated at full speed. Once the screw has penetrated the stud, it should be driven past the surface of the drywall paper without tearing the paper. When driving screws along the edge of the drywall, hold the open side of the stud with your fingers to prevent the edge from bending. The drywall screws should be spaced 10 to 12 inches apart when working in the middle of the sheet and 5 to 8 inches apart when working along the edge.

Sound Waves

Sound is a type of energy that is created by vibrations. When an object vibrates, it causes the air particles to vibrate (move). This movement causes the air particles to bump into more air particles, thus forming a sound wave. This movement continues until the wave runs out of energy. Fast vibrations produce a high-pitched sound, whereas slower vibrations produce a low-pitched sound. The speed at which the sound waves travel is known as their frequency. In other words, frequency is the number of sound waves that pass a given point in one second.

The study of sound, its production, transmission, and effects is called **acoustics**. The ability of a material to absorb and reflect sound is referred to as its *acoustical*. When a space is reviewed to determine the level of reverberation or reflection of sound in a space as it is influenced by the building materials used in it construction, this is known as an **acoustical analysis**. A consultant who is experienced in providing advice on the acoustical requirements and noise control of a facility or a room is known as an **acoustical consultant**.

Controlling Sound and Acoustical Insulation

Controlling sound and/or the transmission of sound in a facility is critical especially in a commercial environment. Aside from being annoying, excessive noise can be harmful. In addition to causing fatigue and irritability, it can cause damage to the sensitive nerves of the inner ear. Keeping unwanted noises from penetrating into a space can be accomplished by the use of acoustical insulation. **Acoustical insulation** reduces the passage of sound through the section of a building.

In a residential environment, it is especially important to prevent sound in an active area from penetrating into quiet areas. Acoustical insulation should be provided to isolate the bedroom, living areas, and bathrooms.

Sound Transmission

As mentioned earlier, sound is transmitted through the air in waves that radiate outward from the source in all directions until they strike another object (wall, ceiling, floor, etc.). When the sound waves strike these surfaces, the surfaces begin to vibrate. This vibration is the result of the pressure of the sound waves. In a structure, when walls vibrate the sound is transmitted the other side of the wall. The degree to which the sound is transmitted depends upon the construction of the wall. This also holds true for ceiling and floors.

Sound Transmission Class

The **sound transmission class (STC)** is a rating used to describe the resistance of a building section to the passage of sound. The higher the STC number, the better the building section is at providing a sound barrier. Poor construction practices are a primary cause of the transmission of sound from one location in a structure to another. For example, poorly fitting doors allow sound to pass through.

When sound waves strike a wall the framing members often act as a conductor and, therefore, aid in the transmission of sound from one location to another (one side of the wall to the other side). Therefore studs that are separated in some fashion from the wall sheathing will not conduct sound waves like studs that are attached to the sheathing. The same is true for electrical outlet boxes that are placed back to back.

Wall Construction

Before a wall can be considered to provide sufficient resistance to the transmission of sound waves, the wall must have a STC rating of 45 or greater. At one time, the only way to get a STC rating of 45 or greater was to use a double wall construction. However, doing so does increase the overall construction cost. In some cases, it made soundproofing unobtainable. Today, sound transmission is controlled by using a system of sound-deadening insulating boards with a gypsum board outer covering. In addition, resilient steel channels placed at right angles to the studs can be used to isolate the gypsum board from its stud.

Floor and Ceiling Construction

For ceiling and floor construction, insulating between an upper floor and a ceiling below provides an excellent means of increasing the STC of the ceiling and floor and provides a source of resistance to impact noise. **Impact noise** is sound that is caused by the dropping of object onto a floor. Impact noise can also be caused by moving furniture or even walking across a floor.

Impact noise is classified by its impact noise rating (INR). Like STC, the higher the INR, the more resistant a floor is to conducting impact noise.

Sound Absorption

Noise in a room can be greatly reduced by using sound-absorbing materials. The most common of the sound-absorbing materials is acoustical tile. Acoustical tiles are made of fiberboard. Acoustical tiles are often associated with ceiling tiles; however, acoustical wall tiles, also known as panels, are available for use. Acoustical tiles have a surface that consists of small holes or fissures or a combination of the two. Sound waves enter these holes, bounce back and forth, and finally die out.

Suspended Ceilings

As mentioned in preceding section, acoustical tiles are often used as a method of absorbing sound. Acoustical tiles are often used in suspended ceilings (see Figure 7-21). A suspended ceiling is an architectural element that when first introduced was used primarily commercial applications. Today, suspended ceilings are still used extensively in commercial structures, but they have also found their way into residential structures.

Suspended ceilings offer a means of concealing HVAC ductwork, electric cables, and plumbing, while still providing easy access to these utilities. In addition, suspended ceilings also can be suspended from an existing ceiling. They provide a space for recessed lighting.

Types of Suspended Ceilings

Currently, there are several types of suspended ceilings available. Some of the more common types of suspended ceiling are: exposed grid, concealed grid, Bandraster, and suspended drywall ceilings.

FIGURE 7-21: Suspended ceiling.

Exposed Grid

Exposed grid is the most common type of suspended ceiling used today. This type of suspended ceiling consists of metal strips, referred to as "mains." The mains are interconnected with "tees." These are shorter metal pieces. The mains and the "tees" form a grid system of either 2 ft × 2 ft, 2 ft × 4 ft, or 4 ft × 4 ft. The cavities are then filled with acoustical ceiling tiles.

Concealed Grid

In a concealed grid system, the acoustical ceiling tiles are used to hide the grid system. Because of the way in which these ceiling are constructed, they are more expensive than exposed grid ceilings. They are more difficult to maintain. In a concealed grid system, the main and tees are concealed by using a small grooves that are built into the perimeter of the ceiling tiles.

Bandraster

One of the most versatile types of suspended ceilings is the Bandraster suspended ceiling. This type of ceiling incorporates mains and tees that are various lengths. The various lengths associated with this type of suspended ceiling allow the facility's owner to create a variety of different patterns.

Suspended Drywall Ceilings

Another type of suspended ceiling that does not require the use of a metal grid system is the suspended drywall system. This system uses wires and hangers to suspend sheet of drywall below the ceiling. Usually drywall is attached to the suspended system runners, cross tees, and cross channels, using conventional drywall screws.

Suspended Ceiling Plans

Before you install a suspended ceiling, it is a good idea to create a suspended ceiling plan. This is accomplished by first sketching the perimeter of the room onto graph paper (see Figure 7-22). Also it is extremely important that you note the location of the ceiling joist without actually drawing them onto the actual suspended ceiling

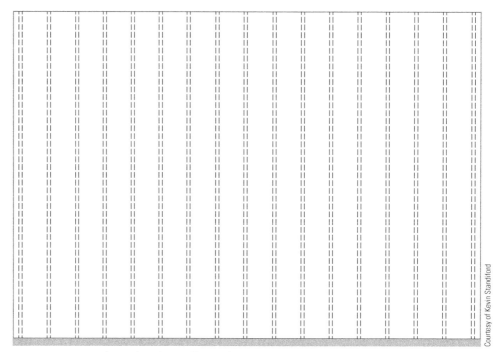

FIGURE 7-22: Locating the ceiling rafters on a sketch.

plan sketch (see Figure 7-23). After you have established the perimeter of the room onto graph paper, you will need to calculate the number of mains needed to construct the suspended ceiling. The mains of the suspended ceiling must run perpendicular to the ceiling joist. The number of mains is easily determined by dividing the length of the wall perpendicular to the ceiling joist by 4. The number 4 is the spacing

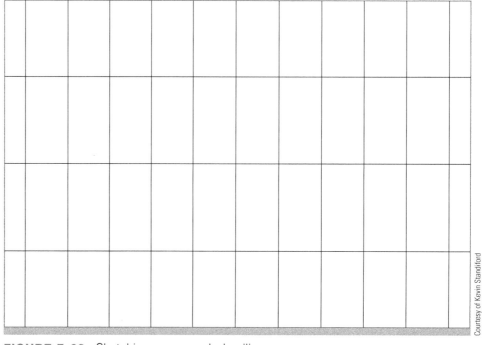

FIGURE 7-23: Sketching a suspended ceiling.

of the mains. For example, if the length of the wall perpendicular to the ceiling joist is 22 feet, then the number of mains needed can be determined as follows:

Length of wall/4 = Number of mains

22/4 = Number of mains

5 = Number of mains

Once you have determined the number of mains necessary to install the suspended ceiling, it is necessary to draw the mains on the sketch. If the number of mains needed is an odd number, then start by drawing a main in the center of the room. If the number of mains needed is an even number, then lightly draw a dashed line in the center of the room. Next draw a main on either side of the dashed line at a spacing of 2 ft.

Finally, calculate the number of tees. This is accomplished by dividing the length of the mains by 2. As with the mains, if the number of tees is an odd number, then place the first tee in the center of the main.

Framing Components

The majority of residential and light commercial constructions consist of wood framing. This is because in most cases wood framing construction is generally less expensive than other types of construction. Although many of the different framing methods are used today, the components are the same.

Wall-Framing Components

The wall frame consists of a number of different parts. An exterior wall frame consists of the following components:

- **Plates**—Top and bottom horizontal members of the wall frame
- **Studs**—Vertical members of the wall frame
- **Headers**—Run at right angles to the studs (Figure 7-24)

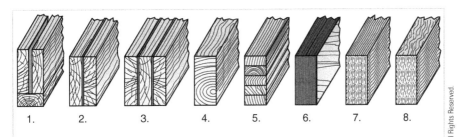

1. A BUILT-UP HEADER WITH A 2 X 4 OR 2 X 6 LAID FLAT ON THE BOTTOM.
2. A BUILT-UP HEADER WITH A 1/2" SPACER SANDWICHED IN BETWEEN.
3. A BUILT-UP HEADER FOR A 6" WALL.
4. A HEADER OF SOLID SAWN LUMBER.
5. GLULAM BEAMS ARE OFTEN USED FOR HEADERS.
6. A BUILT-UP HEADER OF LAMINATED VENEER LUMBER.
7. PARALLEL STRAND LUMBER MAKES EXCELLENT HEADERS.
8. LAMINATED STRAND LUMBER IS USED FOR LIGHT DUTY HEADERS.

FIGURE 7-24: Types of solid and built-up headers.

- **Rough sills**—Form the bottom of a window opening at right angles to the studs (Figures 7-25 and 7-26).

- **Trimmers (jacks)**—Shortened studs that support the headers.

- **Corner posts**—Same length as studs (Figure 7-27).

- **Partition intersections**—Framing needed when interior partitions meet an exterior partition (Figure 7-28).

- **Ribbons**—Horizontal members of the exterior wall frame in balloon construction (Figure 7-29).

- **Corner braces**—Used to brace walls (Figures 7-30 and 7-31).

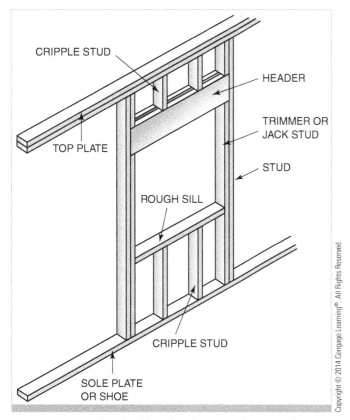

FIGURE 7-25: Typical framing for a window opening.

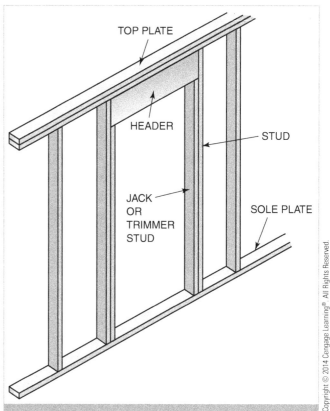

FIGURE 7-26: Typical framing for a door opening.

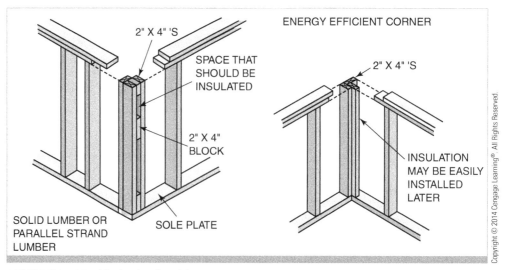

FIGURE 7-27: Methods of making corner posts.

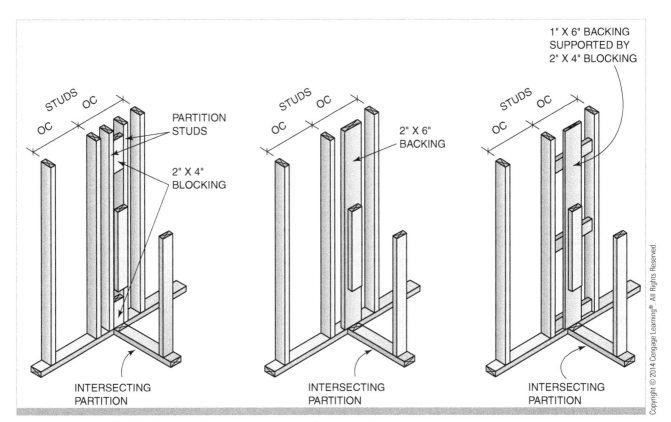

FIGURE 7-28: Partition intersections are constructed in several ways.

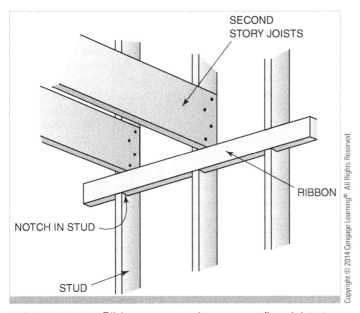

FIGURE 7-29: Ribbons are used to support floor joists to balloon frame.

Steel Framing

Originally used more for commercial structures, steel framing has become more and more popular in residential construction. It can be used for structural framing and/or interior non-load-bearing partitions and walls. Steel framing can be used for floor and ceiling joists, wall studs, and even roof trusses (see Figure 7-32). The sizing

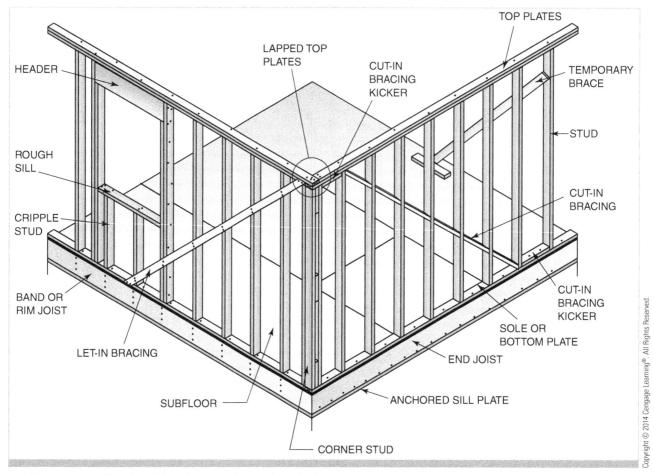

FIGURE 7-30: Wood wall bracing may be cut in or let in.

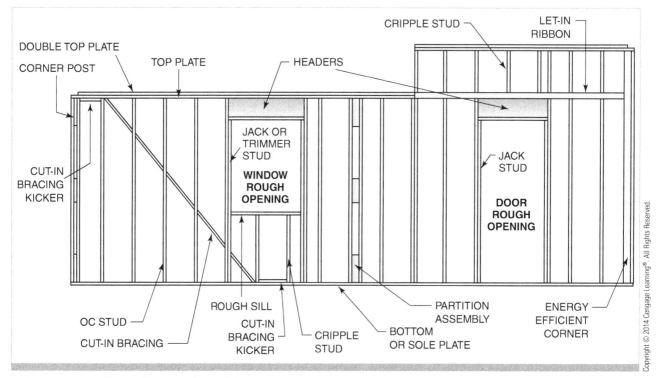

FIGURE 7.31: Parts of an exterior wall frame.

and spacing of steel structural members should only be done by trained professionals. On the jobsite, they will be listed on the structural drawing. The methods used to construct a steel frame are similar to those used for framing wood partitions. All steel framing members are coated in zinc to keep them from corroding due to moisture.

FIGURE 7-32: Steel framing studs.

Steel Framing Components

The components of a steel framing system are:

- **Stud**—For interior use, non-load-bearing wall studs are typically made of 25-, 22-, and 20-gauge steel. They have punch-outs place at regular intervals for running utilities through the partition wall. They are available in widths ranging of $3\frac{1}{2}$, $5\frac{1}{2}$, 8, 10, and 12 inches. They are available in thicknesses of $1\frac{1}{4}$ and $1\frac{5}{8}$ inches. They are also available in standard stock lengths of 8, 9, 10, 12, and 16 feet.

- **Tracks**—Unlike wood framing, the top and bottom plates of a steel framing system are referred to as "tracks" or "runners." They are fastened to the floor and ceiling for the purpose of receiving studs. They are available in standard stock lengths of 10 feet and are constructed of the same materials, gauges, width, and thicknesses as framing studs.

- **Channels**—These are created from cold-rolled 54-mil steel. They are available in several different widths and in lengths of 10, 16, and 20 feet. They are typically used in suspended ceiling and walls. When the channel is used in conjunction with steel studs as lateral bracing, the channels are inserted through the punch-outs of the studs and fastened using clip angles and welds.

- **Furring channels**—These are created from 18 and 33 mil steel with an overall cross sections size of $\frac{7}{8}$ inch $\times$ $2\frac{9}{16}$ inches. They are available in standard stock lengths of 12 feet. They are used as screw attachments for drywall.

Framing Steel Partitions

A steel frame partition wall layout is done basically the same way as for wood partition walls. Steel framing members are typically cut using a chop saw with a metal-cutting saw blade. They are connected using a $\frac{3}{8}$-inch self-tapping panhead screws or crimped to the track using a crimping tool specifically designed from steel framing studs.

Tracks are typically secured to concrete slabs by using powder-driven fasteners, concrete nails, or masonry screws. Tracks that are fastened to wood are attached with $1\frac{1}{4}$-inch oval-head screws. The track is then fastened using two fasteners about 2 inches from the end and a maximum of 24 inches apart on center.

With the open side of the studs all facing in the same direction, place all full-length studs into position within the track. All punch-outs should be vertically aligned for lateral bracings as well as utilities. All studs should be fastened to the top and bottom of the track.

Framing Wall Openings

Wall openings are framed according the type of window or door that is to be installed. For the three-piece, knocked-down door, the frame is applied after the wall covering has been installed. For a one-piece metal frame door, the frame is installed before the wall cover has been applied. Windows are framed in a similar fashion. However, the openings are generally lined using wood; jack studs, sills, and header liners are used to make attachment of the window much easier.

Metal furring is often used on ceilings and walls by attaching them at right angles to the joist and/or studs. Their spacing should not exceed 24 inches on center.

Wall furring, when used for horizontal applications, should be placed no more than 4 inches from the floor and/or ceiling. The fasteners used to install wall furrings should be staggered from one side to the other no more than 24 inches on center.

FIGURE 7-33: Roof trusses.

Trusses

A truss is an assembly of structural members joined to form a rigid framework. Roof trusses are connected in such a way that they form a triangle (see Figure 7-33). Because of their design, they can support a roof over wide spans that can reach over 100 feet. However, because of the overall design of a truss, the typical attic space is lost.

A roof truss consists of upper and lower chords and diagonals, called "web members." The upper chords act as rafters, and the lower chords serve as ceiling joists. Joints are fastened securely with metal or wood gusset plates. A Truss-framed system is an assembled truss unit that is made up of a floor truss, wall studs, and a roof truss.

Cricket

Roof penetrations can lead to problems if they are not properly sealed and, in some cases, if a method for diverting rainwater and/or snow is not provided. When snow is allowed to build up behind a roof penetration, the roof and/or penetration can be damaged. To alleviate this problem for larger roof penetration such as chimneys, a drainage diverter, also known as a "cricket" can be easily installed (see Figure 7-34). An effective cricket can be constructed using metal flashing, plywood, roofing felt, and roofing materials.

Creating a Cricket

The following steps can be used to create an effective cricket:

1. Using a tape measure, locate the chimney's center point by determining the width of the chimney and dividing it by 2.

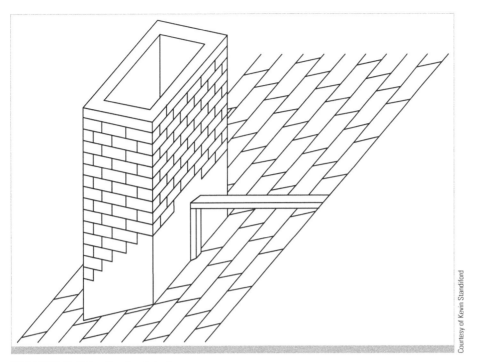

Courtesy of Kevin Standiford

FIGURE 7-34: Roof cricket.

2. Mark the center of the chimney next to the roof. Using a level, extend the centerline from the base of the chimney up approximately 8 inches. This will indicate the reference line of the cricket's ridge.

3. Using a 2 × 4, create a horizontal ridge board that is level and matches the roof angle on one end, and sits on a vertical support (2 × 4) that is mounted next to the chimney

4. Cut $\frac{3}{4}$-inch plywood so that it will extend from the edge of the chimney (along the width of the chimney) and run along the edge of the horizontal ridge board.

5. Secure the cricket to the roof and the ridge, using metal flashing.

6. The outer edges and ridge of the cricket is sealed using an exterior-grade silicon caulking.

7. Before installing the roofing onto the cricket, cover cricket decking with roofing felt.

Estimating Construction Materials

Construction materials can be estimated using a variety of different computer applications and techniques. A quick way to determine the number of 2 × 4s needed to construct a wall is to allow one 2 × 4 stud per foot. If the wall does not contain any openings (windows, door, etc.), then multiply the length of the wall by $\frac{3}{4}$ and rounding the answer to the nearest whole number. However, for a regular wall, using 1 stud per foot will provide enough materials for the jacks and jack studs, header, and crooked stud.

As mentioned earlier, there are numerous computer programs available, as well as websites, that can be used to calculate the amount of materials needed to complete a particular construction project. For example, various calculators can be found at: Construction Resource (www.construction-resource.com). In addition, most local building material suppliers will also calculate the amount of materials needed for a construction project.

Interior Doors

Unlike exterior doors that are made of more dense materials, interior doors are typically lightweight, designed to isolate an interior space. They are not ordinarily exposed to weather and, therefore, are not usually subjected to the typical problems associated with an exterior door.

Repairing and Replacing Interior Doors

The most common problems associated with interior doors are doors that bind along their edges, doors that don't latch, and loose hinges.

Repairing Doors That Bind along Their Top Edge

The most common cause for interior doors sticking along or near their top edge is loose hinge plate screws. If the hinge plate screws are loose, try tightening them. This might take care of the problem; however, if the problem persists, then the screw holes might be stripped. If the screw holes are stripped, then replacing the stripped screw with a longer or wider screw usually works. If a longer screw is used, be sure to get a screw long enough to go through the door jamb, through the shim, and into the stud. Typically a 3-inch screw will be sufficient.

Repairing Doors That Bind along Their Bottom Edge

The most common cause for interior doors sticking along or near the bottom is the jamb shifting or pulling away from its fastener (nail). The problem can be corrected by resetting the door jamb. This is accomplished by prying away the bottom section of the casting and renailing the door jamb into place followed by repositioning and tacking the casting back into place.

Repairing Doors That Bind along Their Entire Edge

If a door is not sealed properly (painted or varnished along all edges), moisture can enter the door and cause it to swell, thus preventing the door from not closing properly. Doors that are not sealed properly swell typically during the rainy season of the year. If the door is not closing properly due to an elevation in moisture, trimming the door will work as long as the door's moisture content remains the same. In other words, as soon as the dry season approaches the door will not be properly gapped. Therefore, waiting for the dry season and then trimming (if needed) and sealing the door's edges is the correct way of repairing the door.

Repairing Doors That Don't Latch

Another common problem a facility maintenance technician will have to resolve is a door that won't latch. This is typically caused by the shifting of either the door or the door's frame, resulting in the misalignment of the strike plate or strike. **Refer to Procedure 7-14 Installing Cylindrical Locksets, the section entitled Installing the Striker Plate for step-by-step instructions.** If the misalignment is less than or equal to $\frac{1}{8}$ of an inch, then the simplest approach to resolving the problem is to file the strike plate. If the amount of misalignment is more than 1/8 inch, then the strike plate should be repositioned instead of filing it. **Refer to Procedure 7-14 Installing Cylindrical Locksets for step-by-step instructions.**

Any structure constructed before 1978 might have lead-based paint. Lead-based paint is considered a hazardous material and must be handled and disposed of properly according to the local, state, and federal regulations.

Replacing Interior Doors

Hollow-core interior doors can be easily damaged and may, therefore, require replacing. To replace a door, the following procedure should be followed:

- Remove the old door by removing the hinge pins. When removing the hinge pins, always start from the bottom and work your way to the top. Remove the top hinge pin last.

- Using the old door as a pattern, lay it on top of the new door and mark any excess material (from the new door) that should be removed (trimmed). In addition, the locations of the hinge mortises and the lockset can be transferred.

- Trim the excess material for the new door. *Refer to Procedure 7-5 Hanging Interior Doors for step-by-step instructions*.

- Drill and assemble the lockset. *Refer to Procedure 7-14 Installing Cylindrical Locksets for step-by-step instructions*.

Repairing Interior Door Hardware

Squeaky door hinges, deadbolts that stick, doors that are difficult to open, door knobs that are difficult to turn are often a sign that the hardware needs to be either cleaned or lubricated, or both.

Cleaning Door Hardware

To clean the hardware associated with a door, the following steps should be followed:

- Remove the hardware form the door.

- If necessary, dissemble the hardware and soak it in a cleaning solution.

- After soaking it in a cleaning solution, dry it thoroughly with either compressed air or a clean rag.

- Lubricate the components, using penetrating oil and reassemble (if necessary) and reinstall the hardware on the door.

NOTE: The rest of this chapter outlines various procedures for interior and exterior carpentry maintenance.

REMODELING

Remodeling and Carpentry

Before a structure can be remodeled, there is usually a certain amount of deconstruction or demolition that must first take place. The general procedures for demolishing a kitchen, bathroom, and utility room are outlined below.

Demolishing a Kitchen

One of the areas in a residential structure that is often remodeled is the kitchen. Typically, this is completed when the appliances are either replaced or updated. The general procedure for demolishing a kitchen is:

- Disconnect all utilities servicing the kitchen.

- Install a plastic barrier separating the affected area from the rest of the structure.

CAUTION

Always check with your local and state agencies before starting a remodeling project.

Before beginning the demolition or deconstruction process, the proper permits must be obtained.

- Cover any appliance that is not going to be removed from the area with a protective covering (cardboard).
- Disconnect electrical wiring to any appliance that is to be removed. Any exposed wire should be capped, using wirenuts to prevent exposure to the bare conductors.
- Disconnect any plumbing servicing appliances and/or fixtures to be removed. Any pipes that are exposed should be capped (including vents).
- Remove appliances that are in the affected area. For example, if the technician is demolishing kitchen counter and/or cabinets, then at this stage the vent hood, as well as a drop-in stove and/or cooktop, should be removed. Even if these components are not going to be reused, they can be recycled at an architectural salvage yard.
- Remove plumbing fixtures in the affected area. In the preceding example, the kitchen sink and garbage disposal should be removed. Even if these components are not going to be reused, they can be recycled at an architectural salvage yard.
- For kitchen cabinetry, remove the drawers and hardware. Even if these components are not going to be reused, they can be recycled at an architectural salvage yard.
- Remove the lower kitchen cabinets and countertops.
- Remove the upper kitchen cabinets.
- If the walls are to be taken down, remove the drywall and plaster. Care should be exercised to prevent hitting any utility line that might be located in the walls. Also, if the wall is a load-bearing wall, extreme caution should be taken and an engineer should be consulted before proceeding. These walls have to be replaced with a load-bearing member so that the integrity of the structure is maintained.
- If the ceiling is to be replaced, remove the drywall and/or plaster. If the ceiling is a drop ceiling, try removing the ceiling tiles and bracing carefully so that they can be recycled.
- If the flooring is to be replaced, remove the flooring down to the subfloor.

Demolishing a Bathroom

Another area in a residential structure that is often remodeled is the bathroom. This is typically done when the plumbing fixtures are updated or replaced or when the cabinetry is updated. The general procedure for demolishing a bathroom is:

- Disconnect all utilities servicing the bathroom.
- Install a plastic barrier separating the affected area from the rest of the structure.
- Disconnect electrical wiring to any appliance that is to be removed. Any exposed wire should be capped, using wirenuts to prevent exposure to the bare conductors.
- Disconnect any plumbing servicing appliances and/or fixtures to be removed. Any pipes that are exposed should be capped (including vents).

- Remove plumbing fixtures in the affected area. Even if these components are not going to be reused, they can be recycled at an architectural salvage yard.

- For bathroom cabinetry, remove the drawers and hardware. Even if these components are not going to be reused, they can be recycled at an architectural salvage yard.

- If the walls are to be taken down, remove the drywall and plaster. Care should be exercised to prevent hitting any utility line that might be located in the walls. Also, if the wall is a load-bearing wall, extreme caution should be taken and an engineer should be consulted before proceeding. These walls have to be replaced with a load-bearing member so that the integrity of the structure is maintained.

- If the ceiling is to be replaced, remove the drywall and/or plaster. If the ceiling is a drop ceiling, try removing the ceiling tiles and bracing carefully so that they can be recycled.

- If the flooring is to be replaced, remove the flooring down to the subfloor.

Demolishing a Utility Room

This area in a residential structure is not remodeled as often as a kitchen or a bathroom, but the demolition process is similar. Utility rooms are typically remodeled when the plumbing fixtures are updated or replaced, when the cabinetry is updated, or when appliances are updated and/or replaced. The general procedure for demolishing a utility room is:

- Disconnect all utilities servicing the utility room.

- Install a plastic barrier separating the affected area from the rest of the structure.

- Disconnect electrical wiring to any appliance that is to be removed. Any exposed wire should be capped using wirenuts to prevent exposure to the bare conductors.

- Disconnect any plumbing servicing appliances and/or fixtures to be removed. Any pipes that are exposed should be capped (including vents).

- Remove plumbing fixtures in the affected area. In the preceding example, the utility room sink should be removed. Even if these components are not going to be reused, they can be recycled at an architectural salvage yard.

- For utility room cabinetry, remove the drawers and hardware. Even if these components are not going to be reused, they can be recycled at an architectural salvage yard.

- If the walls are to be taken down, remove the drywall and plaster. Care should be exercised to prevent hitting any utility line that might be located in the walls. Also, if the wall is a load-bearing wall, extreme caution should be taken and an engineer should be consulted before proceeding. These walls have to be replaced with a load-bearing member so that the integrity of the structure is maintained.

- If the ceiling is to be replaced, remove the drywall and plaster. If the ceiling is a drop ceiling, try removing the ceiling tiles and bracing carefully so that they can be recycled.

- If the flooring is to be replaced, remove the flooring down to the subfloor.

SUMMARY

- The two most commonly used types of power fastening tools are pneumatic and power-actuated.

- Wood is classified as hardwood or softwood—hardwood comes from deciduous trees, which shed their leaves each year; softwood comes from coniferous, or cone-bearing, trees, commonly known as "evergreens."

- Lumber, when first cut from the log, is called "green lumber" and is very heavy because most of its weight is water.

- Engineered products are typically stronger than traditional lumber products.

- Drywall is a panel made of gypsum plaster that has been pressed between two thick sheets of paper.

- In the United States and Canada, drywall is available in widths of 48 inches (4 feet) and 54 inches (4 feet $\times$ 6 inches).

- Common lengths of drywall are 72 inches (8 feet), 84 inches (12 feet), and 192 inches (16 feet), and the most common thicknesses of drywall used today is $\frac{1}{2}$ inch and $\frac{5}{8}$ inch.

- Moisture-resistant and fire-rated drywall (Type X fire-code drywall) is also available.

- In most residential applications, $\frac{1}{2}$-inch drywall is used for interior walls.

- Drywall can be installed using adhesive, nails, and/or screws. Drywall screws are much easier to install than drywall nails and also have better holding capability than drywall nails.

- When attaching drywall to existing drywall, panel adhesive is often a better method of attaching.

- A schedule, on a set of construction drawings, is a list added separately to the drawing plans that describes such items as windows, doors, drywall, flooring, fasteners, and the like.

- Before starting the installation process, plan the layout so that it will end up with as few joints as possible.

- If the ceiling height is less than 8 ft 1 inch, then consider installing the drywall in a horizontal fashion.

- When installing drywall, always start with the ceiling, followed by the walls.

- Panels are always applied either parallel or perpendicular to the joist.

- Installing drywall on a curved surface can be easily done by slightly moistening the paper and core of the drywall.

- A multilayer drywall application is one in which one or more layers of drywall are installed over a base layer of drywall, gypsum backing board, or some other gypsum base product.

- When drywall is installed on steel studs, it is attached using $1\frac{1}{4}$-inch self-tapping drywall screws and a cordless screw gun.

- Sound is a type of energy that is created by vibrations.

- Fast vibrations produce a high-pitched sound, whereas slower vibrations produce a lower-pitched sound.

- Controlling sound and/or the transmission of sound in a facility is critical, especially in a commercial environment.
- Acoustical insulation should be provided to isolate the bedroom, living area, and bathrooms. When these surfaces are struck by the sound wave, their surfaces begin to vibrate.
- The sound transmission class (STC) is a rating used to describe the resistance of a section of a building section to the passage of sound.
- The higher the STC number, the better the section of the building is at providing a sound barrier.
- For ceiling and floor construction, insulating between an upper floor and a ceiling below provides an excellent means of increasing the STC of the ceiling and floor.
- Impact noise is classified by its impact noise rating (INR). Like STC, the higher the INR, the more resistant a floor is to conducting impact noise.
- A suspended ceiling is a type of architectural design that, when first introduced, was primarily used in commercial applications.
- Suspended ceilings offer a means of concealing HVAC ductwork, electric cables, and plumbing, while still providing easy access to these utilities.
- Some of the more common types of suspended ceiling are: exposed grid, concealed grid, Bandraster, and suspended drywall ceilings.
- Roof penetrations can lead to problems if they are not properly sealed and, in some cases, if a method for diverting rainwater and/or snow is not provided.
- Sizing and spacing of steel structural members should only be done by trained professionals.
- All steel framing members are coated in zinc to keep them from corroding due to exposure to moisture.
- Steel frame partition wall layout is done basically the same way as wood partition wall layout.
- Wall openings are framed according the type of window or door that is to be installed.
- Metal furring is often used on ceiling and walls by attaching them are right angles to the joist and/or studs. Their spacing should not exceed 24 inches on center.
- A truss is an assembly of structural members joined to form a rigid framework.
- A roof truss consists of upper and lower chords and diagonals called "web members."
- The upper chords act as rafters, and the lower chords serve as ceiling joists.
- Joints are fastened securely with metal or wood gusset plates.
- The most common problems associated with interior doors are binding along the edges, doors that don't latch, and loose hinges.
- The most common cause for interior doors sticking along or near the bottom is the jamb shifting or pulling away from its fastener (nail).
- If a door is not sealed properly (painted or varnished along all edges), moisture can enter the door and cause it to swell, thus preventing the door from not closing properly.
- Before a structure can be remodeled, there is usually a certain amount of deconstruction or demolition that must take place.

PROCEDURE 7-1

Constructing the Grid Ceiling System

 Locate the height of the ceiling, marking elevations of the ceiling at the ends of all wall sections. Snap chalk lines on all walls around the room to the height of the top edge of the wall angle. If a laser is used, the chalk line is not needed since the ceiling is built to the light beam.

- Fasten wall angles around the room with their top edge lined up with the line. Fasten tem into the framing wherever possible, not more than 24 inches apart. If available, power nailers can be used for efficient fastening.

FROM EXPERIENCE

To fasten wall angles to concrete walls, short masonry nails are sometimes used. However, they are difficult to hold and drive. Use a small strip of cardboard to hold the nail while driving it with the hammer.

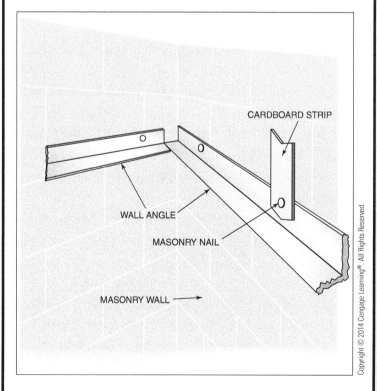

CARDBOARD STRIP

WALL ANGLE

MASONRY NAIL

MASONRY WALL →

B Make miter joints on outside corners and butt joints in interior corners and between straight lengths of wall angle. Use a combination square to lay out and draw the square and angled lines. Cut carefully along the lines with snips.

- From the ceiling sketch, determine the position of the first main runner. Stretch a line at this location across the room from the top edges of the wall angle. The line serves as a guide for installing *hanger lags* or *screw eyes* and *hanger wires* from which main runners are suspended.

- Install the cross tee line by measuring out from the short wall, along the stretched main runner line, a distance equal to the width of the border panel. Mark the line. Stretch the cross tee line through this mark and at right angles to the main runner line.

- Install hanger lags not more than 4 feet apart and directly over the stretched line. Hanger lags should be of the type commonly used for suspended ceilings. They must be long enough to penetrate wood joists a minimum of 1 inch to provide strong support. Hanger wires may also be attached directly around the lower chord of bar joists or trusses.

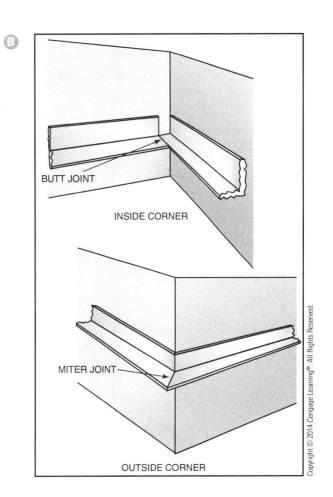

B

BUTT JOINT

INSIDE CORNER

MITER JOINT

OUTSIDE CORNER

CAUTION

Use care in handling the cut ends of the metal grid system. The cut ends are sharp and may have jagged edges that can cause serious injury.

Constructing the Grid Ceiling System (Continued)

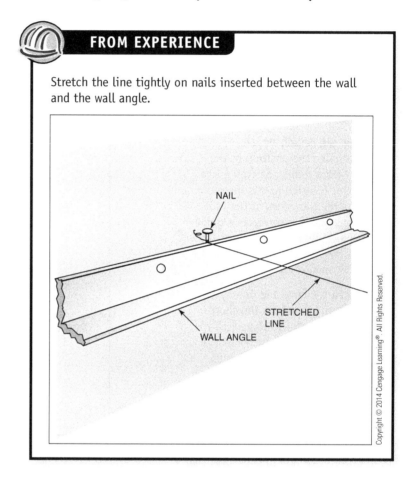

FROM EXPERIENCE

Stretch the line tightly on nails inserted between the wall and the wall angle.

NAIL

STRETCHED LINE

WALL ANGLE

Ⓒ Cut a number of hanger wires, using wire cutters. The wires should be about 12 inches longer than the distance between the overhead construction and the stretched line. Attach the hanger wires to the hanger lags. Insert about 6 inches of the wire through the screw eye. Securely wrap the wire around itself three times. Pull on each wire to remove any kinks. Make a 90° bend where the wire crosses the stretched line. If you are using a laser, make the 90° bend later when the main runner is installed.

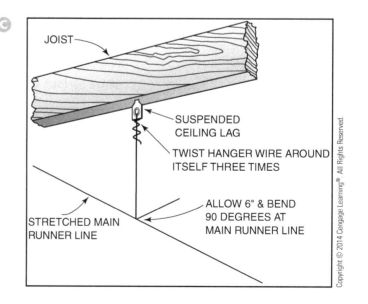

JOIST

SUSPENDED CEILING LAG

TWIST HANGER WIRE AROUND ITSELF THREE TIMES

ALLOW 6" & BEND 90 DEGREES AT MAIN RUNNER LINE

STRETCHED MAIN RUNNER LINE

Ⓓ Stretch lines, install hanger lags, and attach and bend hanger wires in the same manner at each main runner location. Leave the last line stretched tightly in position. It will be used to align the cross tee slots of the main runner.

• At each main runner location, measure from the wall to the cross tee line. Transfer this measurement to the main runner, measuring from the first cross tee slot beyond the measurement, so as to cut as little as possible from the end of the main runner.

• Example: If the first cross tee will be located 23 inches from the wall, then the main runner will be cut.

• Cut the main runners about $\frac{1}{8}$ inch less to allow for the thickness of the wall angle. Backcut the web slightly for easier installation at the wall. Measure and cut main runners individually. Do not use the first one as a pattern to cut the rest. Measure each from the cross tee line.

Ⓔ Hang the main runners by resting the cut ends on the wall angle and inserting suspension wires in the appropriate holes on the top of the main runner. Bring the runners up to the bend in the wires or to the laser light beam. Twist the wires with at least three turns to hold the main runners securely. More than one length of main runner may be needed to reach the opposite wall. Connect the lengths of the main runners together by inserting the tabs into the matching ends. Make sure that the end joints come up tight.

Ⓓ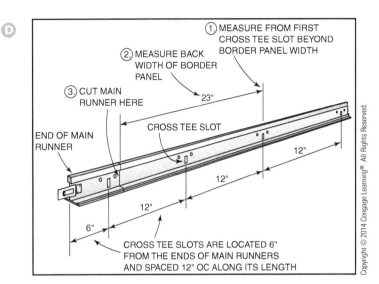

Ⓔ

PROCEDURE 7-1

Constructing the Grid Ceiling System (Continued)

(F) The length of the last section is measured from the end of the last one installed to the opposite wall, allowing about $\frac{1}{8}$ inch less to fit.

- Cross tees are installed by inserting the tabs on the ends into the slots in the main runners. These fit into position easily, although the method of attaching varies from one manufacturer to another. Install all full-length cross tees between main runners first.

- Lay in a few full-sized ceiling panels to stabilize the grid while installing the border cross tees.

- Cut and install cross tees along the border. Insert the connecting tab of one end in the main runner and rest the cut end on the wall angle. It could be crucial to measure and cut cross tees for border panels individually, if walls are not straight or square.

- For 2 × 2 panels, install 2-foot cross tees at the midpoints of the 4-foot cross tees. After the grid is complete, straighten and adjust the grid to level and straight where necessary.

(F)

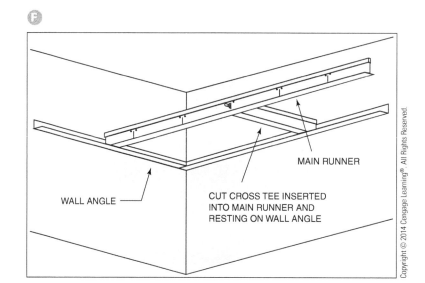

WALL ANGLE

MAIN RUNNER

CUT CROSS TEE INSERTED INTO MAIN RUNNER AND RESTING ON WALL ANGLE

Ⓖ Place Ceiling panels in position by tilting them slightly, lifting them above the grid, and letting them fall into place. Be careful when handling panels to avoid marring the finished surface. Cut and install border panels first and install the full-sized panels last. Measure each border panel individually, if necessary. Cut them $\frac{1}{8}$ inch smaller than measured so that they can drop into place easily. Cut the panels with a sharp utility knife, using a straightedge as a guide. A scrap piece of cross tee material can be used as a straightedge. Always cut with the finished side of the panel up.

Ⓖ

Ⓗ When a column is near the center of a ceiling panel, cut the panel at the midpoint of the column. Cut semicircles from the cut edge to the size required for the panel pieces to fit snugly around the column. After the two pieces are rejoined around the column, glue scrap pieces of panel material to the back of the installed panel. If the column is close to the edge or end of a panel, cut the panel from the nearest edge or end to fit around the column. The small piece is also fitted around the column and joined to the panel by gluing scrap pieces to its back side.

Ⓗ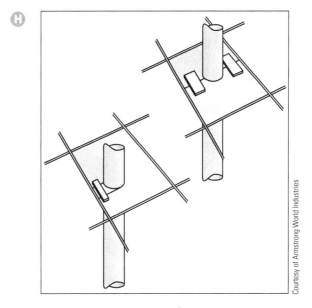

PROCEDURE 7-2

Replacing Broken Ceiling Tiles

- Broken or discolored ceiling tiles can be easily replaced using the following steps:

- After positioning a ladder or scaffolding under the tile or tiles to be replaced, push up on the tile to pop it out of the track.

- Once the tile has been popped out of the track, tilting it an angle will permit you to remove the ceiling tile.

- Use the old ceiling tile as a guide to trim the new ceiling tile.

- The ceiling tile can be trimmed using a straight edge and a utility knife.

- Place the ceiling panels in position by tilting them slightly, lifting them above the grid, and letting them fall into place.

PROCEDURE 7-3

Applying Wall Molding

Ⓐ To snap a line for wall trim, begin by holding a short scrap piece of the molding at the proper angle on the wall. Lightly mark the wall along the bottom edge of the molding. Measure the distance from the ceiling down to the mark.

- Measure and mark this same distance down from the ceiling on each end of each wall to which the molding is to be applied. Snap lines between the marks. Apply the molding so its bottom edge is to the chalk line.

Ⓐ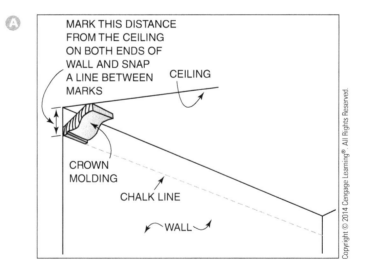

MARK THIS DISTANCE FROM THE CEILING ON BOTH ENDS OF WALL AND SNAP A LINE BETWEEN MARKS

CEILING

CROWN MOLDING

CHALK LINE

WALL

B Apply the molding to the first wall with square ends in both corners. If more than one piece is required to go from corner to corner, the butt joints may be squared or mitered. Position the molding in the miter box the same way each time. Mitering the molding with the same side down each time helps make fitting more accurate, faster, and easier.

B

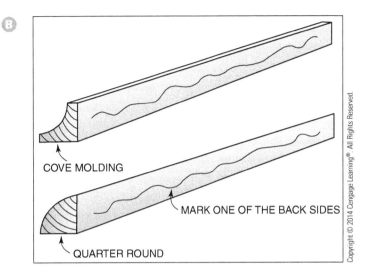

COVE MOLDING

MARK ONE OF THE BACK SIDES

QUARTER ROUND

FROM EXPERIENCE

Since the revealed edges of the molding are often not the same, cut the molding with the same orientation. To do this mark one of the back surfaces with a pencil.

PROCEDURE 7-3

Applying Wall Molding (Continued)

C If a small molding is used, fasten it with finish nails in the center. Use nails of sufficient length to penetrate into solid wood at least 1 inch. If large molding is used, fastening is required along both edges. Nail at about 16-inch intervals and in other locations as necessary to bring the molding tight against the surface. End nails should be placed 2–3 inches from the end to keep the molding from splitting. If it is likely that the molding may split, blunt the pointed end of the nail.

• Cope the starting end of the first piece on each succeeding wall against the face of the last piece installed on the previous wall. Work around the room in one direction. The end of the last piece installed must be coped to fit against the face of the first piece.

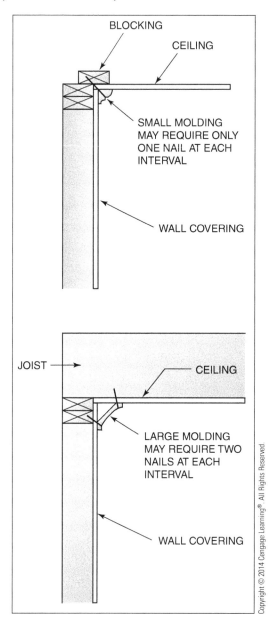

PROCEDURE 7-4

Applying Door Casings

(A) Set the blade of the combination square so that it extends $\frac{5}{16}$ inch beyond the body of the square. Gauge lines at intervals along the side and head jamb edges by riding the square against the inside face of the jamb. Let the lines intersect where side and head jambs meet.

- Cut one miter on the ends of the two side casings. Cut them a little long as they will be cut to fit later. Be sure to cut pairs of right and left miters.

- Miter one end of the head casing. Hold it against the head jamb of the door frame so that the miter is on the intersection of the gauged lines. Mark the length of the head casing at the intersection of the gauged lines on the opposite side of the door frame. Miter the casing to length at the mark.

- Fasten the head casing in position with a few tack nails. Move the ends slightly to fit the mitered joint between head and side casings. Keep the casing inside edge aligned to the gauged lines on the head jamb. The mitered ends should be in line with the gauged lines on the side jambs. Use finish nails along the inside edge of the casing into the header jamb. Straighten the casing as necessary as nailing progresses. Drive nails at the proper angle to keep them from coming through the face or back side of the jamb. Fasten the top edge of the casing into the framing.

(A)

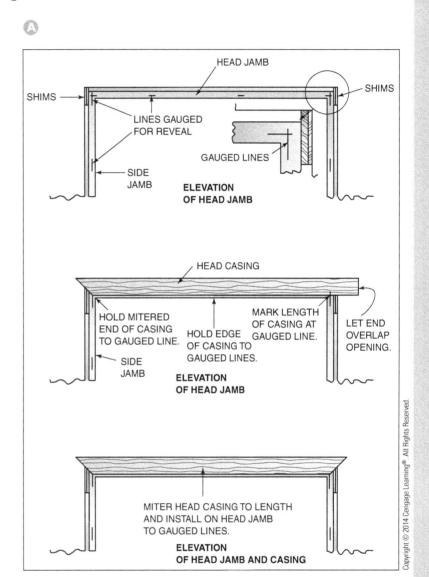

Applying Door Casings (Continued)

- Cut the previously mitered side casing to length. Mark the bottom end by turning it upside down with the point of the miter touching the floor. Mark the side casing in line with the top edge of the head casing. Make a square cut on the casing at that mark. If the finish floor has not been laid, hold the point of the miter on a scrap block of material that is equal in thickness to the finish floor. Replace the side casing in position and try to fit it at the mitered joint. If the joint needs adjusting, trim with a power miter box or use a sharp block plane.

- When the casing is fitted, apply a little glue to the joint. Nail the side casing in the same manner as that of the head casing. Bring the faces flush, if necessary, by shimming between the back of the casing and the wall. Usually, only thin shims are needed. Any small space between the casing and the wall is usually not noticeable or it can be filled later with joint filling compound. Also, the backside of the thicker piece may be planed or chiseled to the desired thinness.

- Drive a 4d finish nail into the edge of the casing and through the mitered joint, and then set all fasteners. Keep nails 2 or 3 inches from the end to avoid splitting the casing.

FROM EXPERIENCE

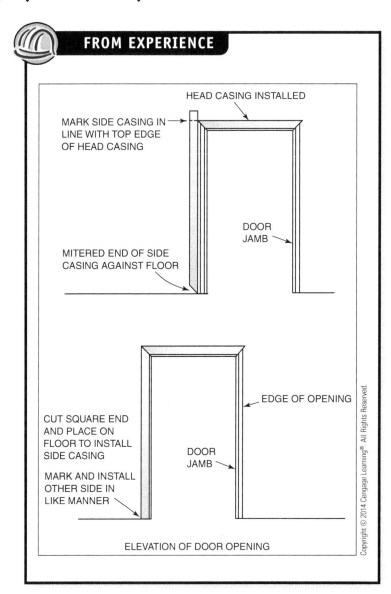

ELEVATION OF DOOR OPENING

Hanging Interior Doors

Setting a Prehung Door Frame

A Remove the protective packing from the unit. Leave the small fiber shims between the door and the jambs to help maintain this space. Cut off the horns if necessary. Remove nail that holds the door closed.

- Center the unit in the opening, so the door will swing in the desired direction. Be sure the door is closed and spacer shims are still in place between the jamb and the door.

B Level the head jamb. Make adjustments by shimming the jamb that is low so that it brings the head jamb level. Adjust a scriber to the thickness of shim and scribe this amount off of the other jamb. Remove frame and cut the jamb. Note that the clearance under the door is being reduced by the amount being cut off.

A

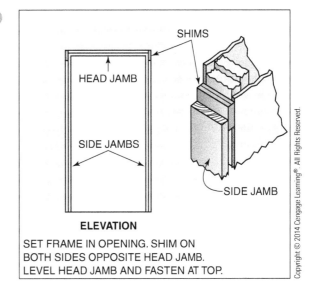

SHIMS

HEAD JAMB

SIDE JAMBS

SIDE JAMB

ELEVATION

SET FRAME IN OPENING. SHIM ON BOTH SIDES OPPOSITE HEAD JAMB. LEVEL HEAD JAMB AND FASTEN AT TOP.

B

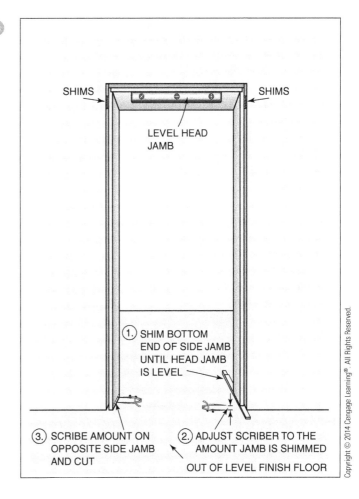

SHIMS   SHIMS

LEVEL HEAD JAMB

1. SHIM BOTTOM END OF SIDE JAMB UNTIL HEAD JAMB IS LEVEL

3. SCRIBE AMOUNT ON OPPOSITE SIDE JAMB AND CUT

2. ADJUST SCRIBER TO THE AMOUNT JAMB IS SHIMMED

OUT OF LEVEL FINISH FLOOR

Hanging Interior Doors (Continued)

Ⓒ Plumb the hinge side jamb of the door unit. A 2-foot carpenter's level may not be accurate when plumbing the sides because of any bow that may be in the jambs. Use a 6-foot level or a plumb bob. Tack the jamb plumb to the wall through the casing with one nail on either side.

Ⓒ

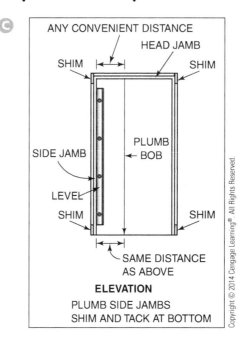

ANY CONVENIENT DISTANCE
HEAD JAMB
SHIM   SHIM
SIDE JAMB
PLUMB BOB
LEVEL
SHIM   SHIM
SAME DISTANCE AS ABOVE
ELEVATION
PLUMB SIDE JAMBS
SHIM AND TACK AT BOTTOM

Ⓓ Open the door and move to the other side. Check that the unit is nearly centered. Install shims between the side jambs and the rough opening at intermediate points, keeping side jambs straight. Shims should be located behind the hinges and lockset **strike plates**. Nail through the side jambs and shims. Remove spacers from door edges.

• Check the operation of the door. Make any necessary adjustments. The space between the door and the jamb should be equal on all jambs. The entire door edge should touch the stop or weather strip.

Ⓓ

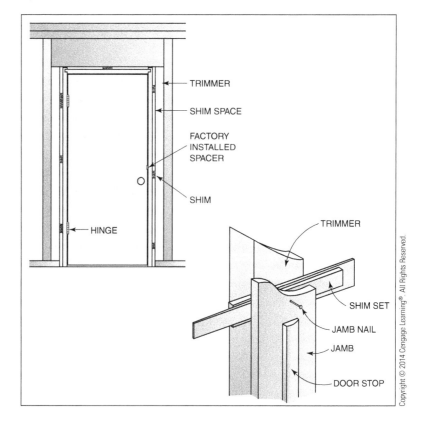

TRIMMER
SHIM SPACE
FACTORY INSTALLED SPACER
SHIM
HINGE
TRIMMER
SHIM SET
JAMB NAIL
JAMB
DOOR STOP

Ⓔ Finish nailing the casing and install it on the other side of the door. Drive and set all nails. Do not make any hammer marks on the finish.

Ⓔ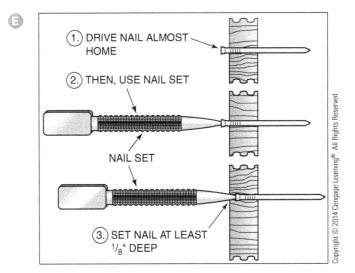

1. DRIVE NAIL ALMOST HOME

2. THEN, USE NAIL SET

NAIL SET

3. SET NAIL AT LEAST ⅛" DEEP

Fitting a Door to a Frame

Ⓐ Begin by checking the door for its beveled edge and the direction of the face of the door. Note the direction of the swing.

• Lightly mark the location of the hinges on the door. On paneled doors, the top hinge is usually placed with its upper end aligned with the bottom edge of the top rail. The bottom hinge is placed with its lower end aligned with the top edge of the bottom rail. The middle hinge is centered between them. On flush doors, the usual placement of the hinge is about 9 inches down from the top and 13 inches up from the bottom, as measured to the center of the hinge. The middle hinge is centered between the two.

• Check the opening frame for level and plumb.

Ⓐ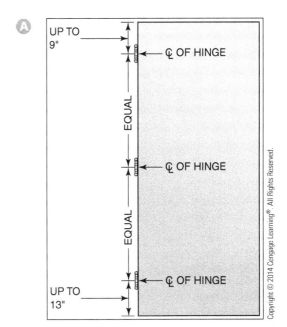

UP TO 9"

ℂ OF HINGE

EQUAL

ℂ OF HINGE

EQUAL

UP TO 13"

ℂ OF HINGE

PROCEDURE 7-5

Hanging Interior Doors (Continued)

B Plane the door edges so that the door fits onto the opening with an even joint of approximately $\frac{3}{32}$-inch between the door and the frame on all sides. A wider joint of approximately $\frac{1}{8}$ inch must be made to allow for the swelling of the door and frame in extremely damp weather. Use a *door jack* to hold the door steady. Do not cut more than $\frac{1}{2}$-inch total from a doors width. Do not remove more than 2 inches from a doors height. Check the fit frequently by placing the door in the opening, even if this takes a little extra time.

B

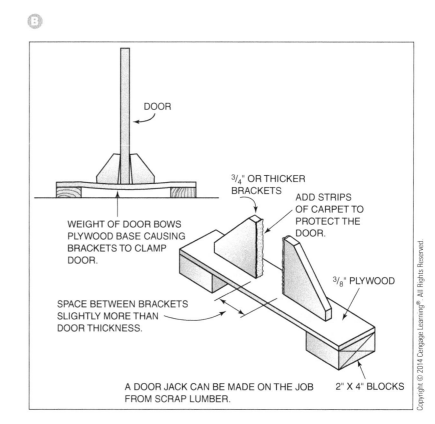

DOOR

$3/4$" OR THICKER BRACKETS

ADD STRIPS OF CARPET TO PROTECT THE DOOR.

WEIGHT OF DOOR BOWS PLYWOOD BASE CAUSING BRACKETS TO CLAMP DOOR.

$3/8$" PLYWOOD

SPACE BETWEEN BRACKETS SLIGHTLY MORE THAN DOOR THICKNESS.

A DOOR JACK CAN BE MADE ON THE JOB FROM SCRAP LUMBER.

2" X 4" BLOCKS

C Place the door in the frame. Shim the door so the proper joint is obtained along all sides. Place shims between the lock edge of the door and the side jamb of the frame. Mark across the door and jamb at the desired location for each hinge. Place a small X on both the door and the jamb, to indicate on which side of the mark to cut the gain.

- Remove the door from the frame. Place a hinge leaf on the door edge with its end on the mark previously made. Score a line along edges of the leaf. Score only partway across the door edge.

C

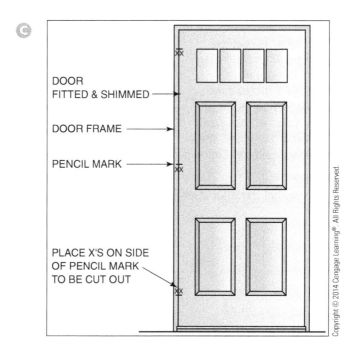

DOOR FITTED & SHIMMED

DOOR FRAME

PENCIL MARK

PLACE X'S ON SIDE OF PENCIL MARK TO BE CUT OUT

D Score the hinge lines, taking care not to split any part of the door. With a chisel, cut small chips from each end of the gain joint. The chips will break off at the scored end marks. With the flat of the chisel down, pare and smooth the excess down to the depth of the gain. Be careful not to slip.

D

- Press the hinge leaf into the gain joint. It should be flush with the door edge; install screws. Center the screws carefully, so the hinge leaf will not shift when the screw head comes in contact with the leaf.

- Place the door in the opening and insert the hinge pins. Check the swing of the door and adjust as needed.

E Apply the *door stops* to the jambs with several tack nails, in case they have to be adjusted when locksets are installed. A **back miter** joint is usually made between molded side and header stops. A butt joint is made between square-edge stops.

E
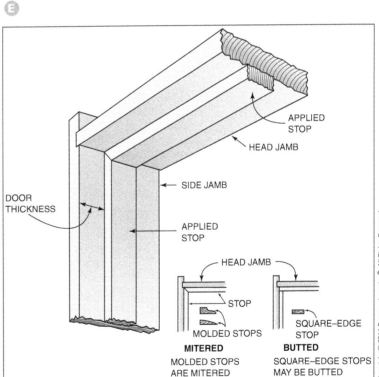

APPLIED STOP

HEAD JAMB

SIDE JAMB

DOOR THICKNESS

APPLIED STOP

HEAD JAMB

STOP

MOLDED STOPS

MITERED

MOLDED STOPS ARE MITERED

SQUARE–EDGE STOP

BUTTED

SQUARE–EDGE STOPS MAY BE BUTTED

PROCEDURE 7-5

Hanging Interior Doors (Continued)

Installing Bypass Doors

Ⓐ Cut the track to length. Install it on the header jamb according to the manufacturer's directions. Bypass doors are installed so that they overlap each other by about 1 inch when closed.

- Install pairs of *roller hangers* on each door. The roller hangers may be offset a different amount for the outside door than the inside door. They are also offset differently for doors of various thicknesses. Make sure that rollers with the same and correct offset are used on each door. The location of the rollers from the edge of the door is usually specified in the manufacturer's instruction sheet.

Ⓐ

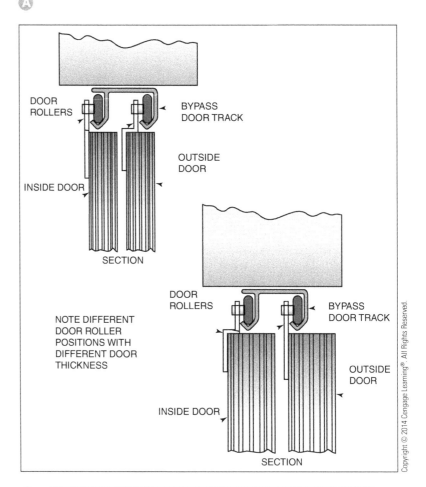

Ⓑ Mark the location and bore holes for *door pulls*. Use flush pulls so that bypassing is not obstructed. The proper size hole is bored partway into the door. The pull is tapped into place with a hammer and wood block. The press fit holds the pull in place. Rectangular flush pulls, also used on bypass doors, are held in place with small recessed screws.

Ⓑ

C Hang the doors by holding the bottom outward. Insert the rollers in the overhead track. Gently let the door come to a vertical position. Install the inside door first, then the outside door.

- Test the door operation and the fit against side jambs. Door edges must fit against side jambs evenly from top to bottom. If the top or bottom portion of the edge strikes the side jamb first, it may cause the door to jump from the track. The door rollers have adjustments for raising and lowering. Adjust one or the other to make the door edges fit against side jambs.

D A *floor guide* is included with bypass door hardware to keep the doors in alignment. The guide is centered on the lap of the two doors to steady them at the bottom. Mark the location and fasten the guide.

C

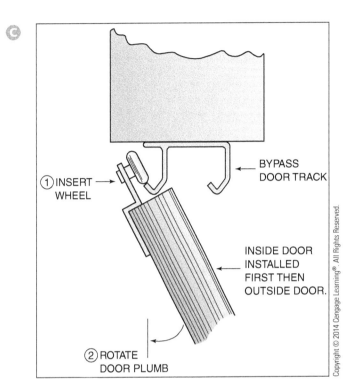

① INSERT WHEEL

BYPASS DOOR TRACK

INSIDE DOOR INSTALLED FIRST THEN OUTSIDE DOOR.

② ROTATE DOOR PLUMB

D

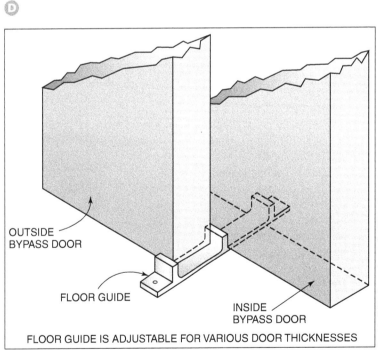

OUTSIDE BYPASS DOOR

FLOOR GUIDE

INSIDE BYPASS DOOR

FLOOR GUIDE IS ADJUSTABLE FOR VARIOUS DOOR THICKNESSES

Hanging Interior Doors (Continued)

Installing Bifold Doors

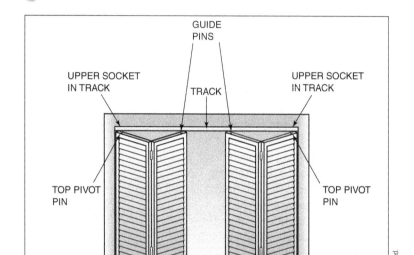

(A) Check that the door and its hardware are all present. The hardware consists of the track, pivot sockets, pivot pins and guides, door aligners, door pulls, and necessary fasteners.

(B) Cut the track to length. Fasten it to the header jamb with screws provided in the kit. The track contains adjustable *sockets* for the door *pivot pins*. Make sure these are inserted before fastening the track in position. The position of the track on the header jamb is not critical. It may be positioned as desired.

• Locate the bottom pivot sockets. Fasten one on each side, at the bottom of the opening. The pivot socket bracket is L-shaped. It rests on the floor against the side jamb. It is centered on a plumb line from the center of the pivot sockets in the track on the header jamb above.

• Install pivot pins at the top and bottom ends of the door in the prebored holes closest to the jamb. Sometimes the top pivot pin is spring loaded. It can then be depressed for easier installation of the door. The bottom pivot pin is threaded and can be adjusted for height. The guide pin rides in the track. It is installed in the hole provided at the top end of the door farthest away from the jamb.

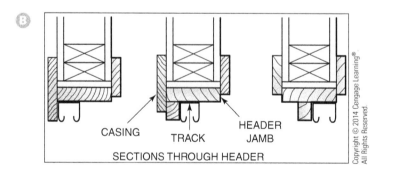

○ Loosen the set screw in the top pivot socket. Slide the socket along the track toward the center of the opening about one foot away from the side jamb. Place the bottom door pivot in position by inserting it into the bottom pivot socket. Tilt the door to an upright position, while at the same time inserting the top pivot pin in the top socket. Slide the top pivot socket back toward the jamb where it started from.

• Adjust top and bottom pivot sockets in or out so that the desired joint is obtained between the door and the jamb. Lock the top and bottom pivot sockets in position. Adjust the bottom pivot pin to raise or lower the doors, if necessary.

• Install second door in the same manner.

• Install pull knobs and door aligners in the manner and location recommended by the manufacturer. The door aligners keep the faces of the center doors lined up when closed.

○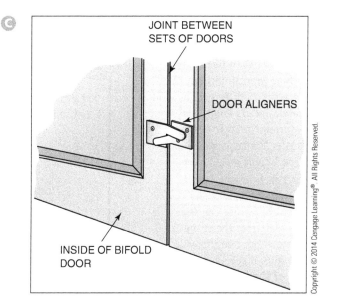

JOINT BETWEEN SETS OF DOORS

DOOR ALIGNERS

INSIDE OF BIFOLD DOOR

PROCEDURE 7-6

Applying Base Moldings

Ⓐ Cut the first piece with squared ends if it fits between two walls. Miter the butt joint if desired. If one piece fits from corner to corner, determine its length by measuring from corner to corner. Then, transfer the measurement to the baseboard. Cut the piece $\frac{1}{2}$ to 1 inch longer. Place the piece in position with one end tight to the corner and the other end away from the corner. Press the piece tight to the wall near the center. Place small marks on the top of the base trim and onto the wall so that they line up with each other. Reposition the piece with the other end in the corner. Press the base against the wall at the mark. The difference between the mark on the wall and the mark on the base is the amount to cut off.

- After cutting, place one end of the piece in the corner and bow out the center. Place the other end in the opposite corner, and press the center against the wall. Fasten in place. Continue in this manner around the room. Make regular miter joints on outside corners.

- If both ends of a single piece are to have regular miters for outside corners, fasten it in the same position as it was marked. Tack the rough length in position with one finish nail in the center. Mark both ends. Remove and cut the miters. Remember that these marks are to the short side of the miter so that the piece will be longer than these marks indicate. Reinstall the piece by first fastening into the original nail hole.

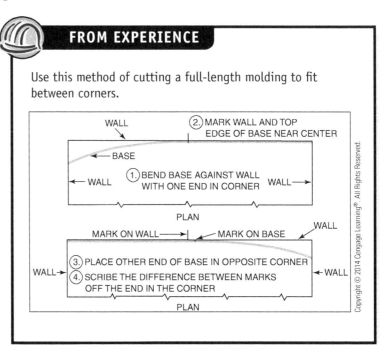

FROM EXPERIENCE

Use this method of cutting a full-length molding to fit between corners.

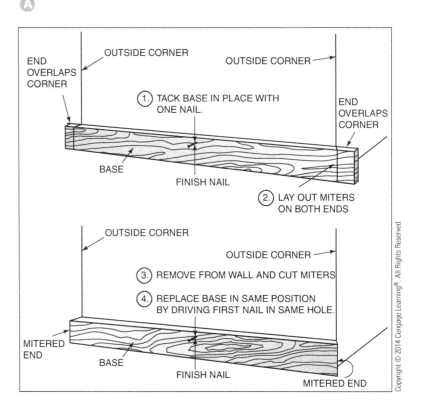

B If a *base cap* is applied, it is done so in the same manner as most wall or ceiling molding. Cope interior corners and miter exterior corners. However, it should be nailed into the floor and not into the baseboard. This prevents the joint under the shoe from opening should shrinkage take place in the baseboard.

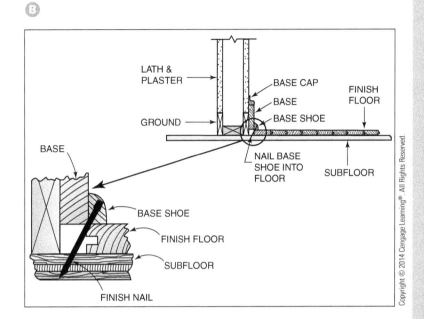

C When the base shoe must be stopped at a door opening or other location, with nothing to butt against, its exposed end is generally *back-mitered* and sanded smooth. Generally, no base shoe is required if carpeting is to be used as a floor finish.

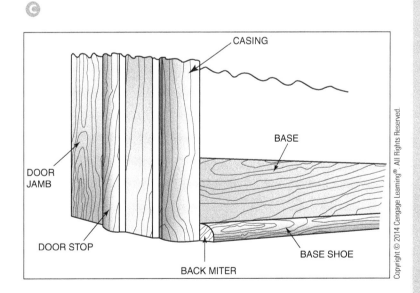

PROCEDURE 7-7

Installing Window Trim

Applying the Stool

Ⓐ Hold a piece of side casing in position at the bottom of the window and draw a light line on the wall along the outside edge of the casing stock. Mark a distance outward from these lines equal to the thickness of the window casing. Cut a piece of stool stock to length equal to the distance between the outermost marks.

• Position the stool with its outside edge against the wall. The ends should be in line with the marks previously made on the wall. Lightly square lines, across the face of the stool, even with the inside face of each side jamb of the window frame.

• Set the pencil dividers or scribers to mark the cutout so that, on both sides, an amount equal to twice the casing thickness will be left on the stool. Scribe the stool by riding the dividers along the wall on both sides. Also scribe along the bottom rail of the window sash.

• Cut to the lines using a handsaw. Smooth the sawed edge that will be nearest to the sash. Shape and smooth both ends of the stool the same as the inside edge.

• Apply a small amount of caulking compound to the bottom of the stool. Fasten the stool in position by driving finish nails along its outside edge into the sill. Set the nails.

FROM EXPERIENCE

Raise the lower sash slightly. Place a short, thin strip of wood under it, on each side, which projects inward to support the stool while it is being laid out. Place the stool on the strips. Raise or lower the sash slightly so the top of the stool is level.

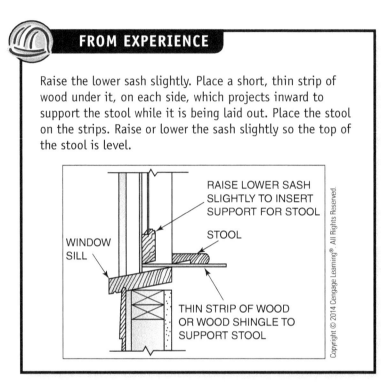

RAISE LOWER SASH SLIGHTLY TO INSERT SUPPORT FOR STOOL

STOOL

WINDOW SILL

THIN STRIP OF WOOD OR WOOD SHINGLE TO SUPPORT STOOL

Ⓐ

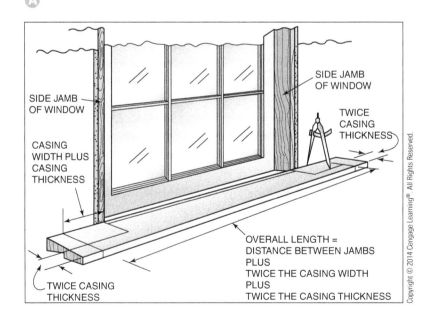

SIDE JAMB OF WINDOW

SIDE JAMB OF WINDOW

CASING WIDTH PLUS CASING THICKNESS

TWICE CASING THICKNESS

TWICE CASING THICKNESS

OVERALL LENGTH = DISTANCE BETWEEN JAMBS PLUS TWICE THE CASING WIDTH PLUS TWICE THE CASING THICKNESS

Applying the Apron

 Cut a length of apron stock equal to the distance between the outer edges of the window casings.

- Each end of the apron is then *returned upon itself*. This means that the ends are shaped the same as its face. To return an end upon itself, hold a scrap piece on the apron. Draw its profile flush with the end. Cut to the line with a coping saw. Sand the cut end smooth. Return the other end upon itself in the same manner.

- Place the apron in position with its upper edge against the bottom of the stool. Be careful not to force the stool upward. Keep the top side of the stool level by holding a square between it and the edge of the side jamb. Fasten the apron along its bottom edge into the wall, then drive nails through the stool into the top edge of the apron.

FROM EXPERIENCE

When nailing through the stool, wedge a short length of 1×4 stock between the apron and the floor at each nail location. This supports the apron while nails are being driven. Failure to support the apron results in an open joint between it and the stool. Take care not to damage the bottom edge of the apron with the supporting piece.

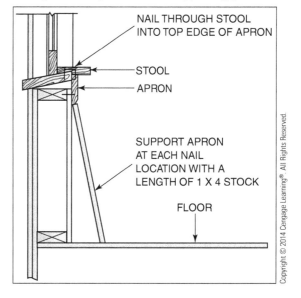

NAIL THROUGH STOOL INTO TOP EDGE OF APRON

STOOL

APRON

SUPPORT APRON AT EACH NAIL LOCATION WITH A LENGTH OF 1 X 4 STOCK

FLOOR

PROCEDURE 7-7

Installing Window Trim (Continued)

Installing Jamb Extensions

A Measure the distance from the jamb to the finished wall. Rip the jamb extensions to this width with a slight back bevel on the side toward the jack stud.

- Cut the pieces to length and apply them to the header, side jambs, and stool. Shim them, if necessary, and nail with finish nails that will penetrate the framing at least an inch.

Applying the Casings

- Cut the number of window casings needed to a rough length with a miter on one end. Cut side casings with left- and right-hand miters.

- Install the header casing first and then the side casings in a similar manner as with door casings. Find the length of side casings by turning them upside down with the point of the miter on the stool in the same manner as door casings.

- Fasten casings with their inside edges flush with the inside face of the jamb or with a reveal. Make neat, tight-fitting joints at the stool and at the head.

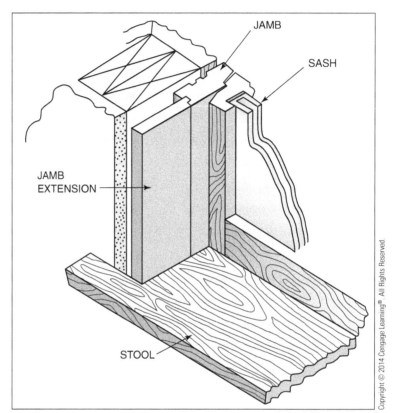

Installing Wood Flooring

Preparation for Installation

- Check the subfloor for any loose areas and add nails where appropriate. Sweep and vacuum the subfloor clean. Scraping may be necessary to remove any unwanted material.

- Cover the subfloor with building paper. Lap it 4 inches at the seams and perpendicular to the direction of the finish floor. This paper prevents squeaks in dry seasons and retards moisture from below that could cause warping of the floor.

- So that the flooring can be nailed into the floor joist, snap chalk lines on the paper showing the centerline of floor. For better holding power, fasten flooring through the subfloor and into the floor joists whenever possible. On $\frac{1}{2}$-inch plywood subfloors, all flooring fasteners must penetrate into the joists.

Starting Strip

- The positioning of the first course is critical. Place a strip of flooring on each end of the room, ¾ inch from the starter wall with the groove side toward the wall. The spacing between the flooring and the wall is used for expansion. Later this spacing will be concealed by the molding.

- Mark along the edge of the flooring tongue. Snap a chalk line connecting the two points. Hold the strip with its tongue edge to the chalk line.

PROCEDURE 7-8

Installing Wood Flooring (Continued)

Ⓐ Face-nail it with 8d finish nails, alternating from one edge to the other, 12–16 inches apart.

- Make sure that end joints between strips are driven up tight.

- Cut the last piece to fit loosely against the wall. Use a piece long enough so that the cut-off piece is 8 inches or longer. This scrap piece is used to start the next course back against the other wall.

Ⓐ

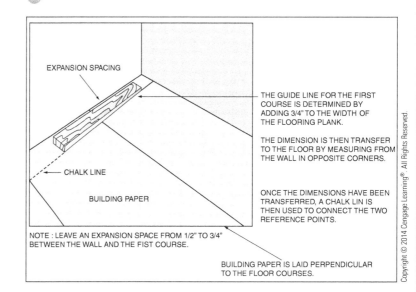

EXPANSION SPACING

THE GUIDE LINE FOR THE FIRST COURSE IS DETERMINED BY ADDING 3/4" TO THE WIDTH OF THE FLOORING PLANK.

THE DIMENSION IS THEN TRANSFER TO THE FLOOR BY MEASURING FROM THE WALL IN OPPOSITE CORNERS.

CHALK LINE

ONCE THE DIMENSIONS HAVE BEEN TRANSFERRED, A CHALK LIN IS THEN USED TO CONNECT THE TWO REFERENCE POINTS.

BUILDING PAPER

NOTE : LEAVE AN EXPANSION SPACE FROM 1/2" TO 3/4" BETWEEN THE WALL AND THE FIST COURSE.

BUILDING PAPER IS LAID PERPENDICULAR TO THE FLOOR COURSES.

Ⓑ After the second course of flooring is fastened, lay out eight or nine loose rows of flooring, end to end. This is called *racking the floor*. Racking is done to save time and material.

- Lay out in a staggered end-joint pattern. End joints should be approximately 6 inches apart. Cut pieces to fit within $\frac{1}{2}$ inch of the end wall. Distribute long and short pieces evenly for the best appearance. Avoid clusters of short strips. Lay out loose flooring.

- Continue across the room. Rack seven or eight courses ahead as work progresses.

Ⓑ

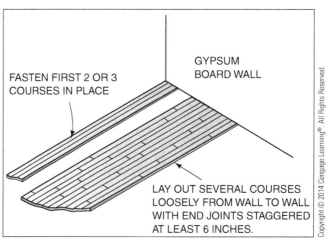

FASTEN FIRST 2 OR 3 COURSES IN PLACE

GYPSUM BOARD WALL

LAY OUT SEVERAL COURSES LOOSELY FROM WALL TO WALL WITH END JOINTS STAGGERED AT LEAST 6 INCHES.

PROCEDURE 7-9

Replacing Broken Tiles

- Before a broken tile can be replaced, the grout must be removed from around the tile you wish to replace. The grout is a bonding agent that binds the tiles together and provides a means of sealing the subfloor.

- To remove soft unsanded grout (typically used for wall tiles), use a sharp object such as a utility knife to scratch the grout out.

- To remove sanded grout (typically used for floor tiles), use a cold chisel, which scores the surface of the grout. Once the surface has been scored, remove it using a utility knife.

- If the grout spacing is sufficient (wide enough), then use a grout saw to remove the grout.

- Once you remove the grout, simply lift the tile away from the remaining tiles if the tile is loose. If the tile is not loose, carefully tap the edges of the tile to be removed using a cold chisel and a hammer until the tile is loosened.

- Once the tile has been removed, using a putty knife, remove any lumps or bumps in the mortar.

Ⓐ Once the hole has been clean of debris, lumps, and bumps, test fit the new tile to ensure that it sets properly in the space. The tile should not be higher than the surrounding tiles or have excessive rocking. If the tile is higher than the surrounding tiles or has excessive rocking associated with it, then remove the tile and continue to scrape more mortar from the surface. If the tile is larger than the hole, then trim the tile to fit. For such small jobs, use a hand tile cutter.

Ⓐ

Replacing Broken Tiles (Continued)

B Fit the tile to a curved hole using a nibbler. Nibblers are used to make small or irregular cuts.

- Once the tile is properly fit, remove the tile and apply a $\frac{1}{8}$-inch layer of adhesive to the back of the tile. The adhesive should not be any closer than $\frac{1}{2}$ inch from the edge of the tile.

- Press the tile using a slight back-and-forth motion to evenly spread the adhesive and to ensure a good bond. Allow the tile to set for 24 hours before applying the grout.

- The grout should be mixed per the manufacturer's instructions.

- Using a damp sponge push the grout into the cracks.

B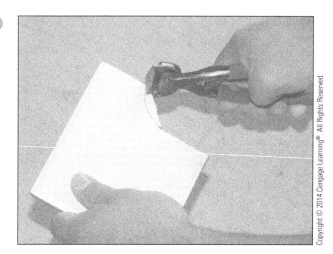

PROCEDURE 7-10

Removing an Existing Carpet

- A carpet can be installed over almost any type of smooth substrate (concrete, plywood, particleboard, etc.). However, before the carpet can be installed, the old carpet (if necessary) must first be removed. This is accomplished by removing all the furniture from the area as well as the molding (around the floor).

- Before removing the old carpet, be sure to wear a respirator to keep from breathing in any dust.

- Using a carpet or a utility knife, cut the old carpet into 2-foot wide strips. This will make handling the old carpet much easier.

- Starting at one end of the carpet, pull it up from the tackless strips and rolling it up as you go.

- If the carpet has an underlayment installed, it should be also removed.

- Remove any existing tack strips and ensure that the area is clean of any debris and dry.

- Inspect and repair any loose floorboards.

 FROM EXPERIENCE

In most cases when a carpet is being replaced, the underlayment will be worn out. Therefore, even if the underlayment appears to be in good shape, it is recommended that you always replace it when installing a new carpet.

Installing or Replacing Carpet

- Before installing a new carpet, be sure to remove the existing one, using the procedure outlined in "Removing an Existing Carpet."

- Install new tackless strips around the perimeter of the room in which the carpet is to be installed. The tackless strips should be installed about $\frac{1}{2}$ inch from the wall with the pins or tacks facing toward the wall. In addition, be sure that tackless strips are butted together. This is especially true in the corners.

Ⓐ When installing the underlayment, place it down so that the edges overlap the tackless strips. The underlayment should be placed so that the edges are butted together but not overlapped.

- Once the underlayment is in position, it must be either glued or stapled along the inside edge of the tackless strip.

- Trim the excess underlayment. Seal all seams using duct tape.

Ⓑ The carpet is installed by laying it over the underlayment.

Ⓒ Bond any seams between sections of carpet using adhesive carpet tape or hot melt tape. This is accomplished by placing the tape between the underlayment and the carpet (adhesive side up).

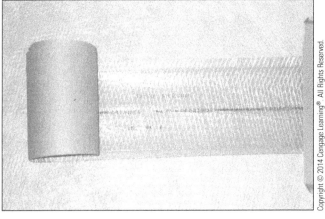

D If hot melt tape is used, then place an electric seaming iron below the carpet and directly onto the adhesive of the tape.

D

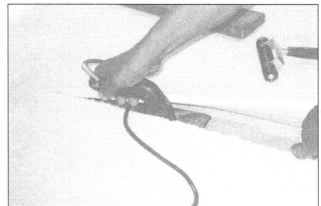

E Press the carpet firmly onto the hot tape, butting the edges together.

E

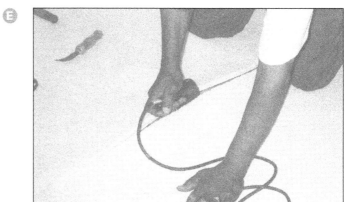

F Roll the seam using a carpet seam roller.

F

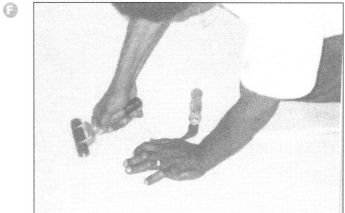

PROCEDURE 7-11

Installing or Replacing Carpet (Continued)

G Once all the seams have been made (as previously described), stretch the carpet into place, using either a knee kicker or a power stretcher.

G

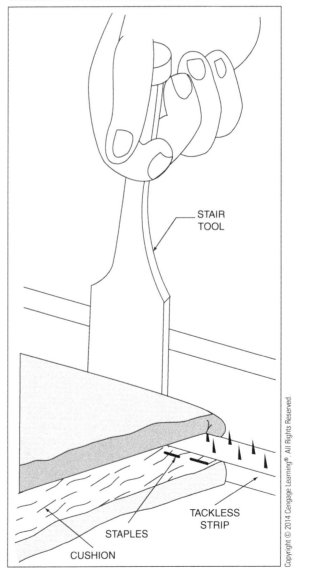

H Press the edge of the carpet into position using a stair tool.

- Trim any excess carpet and replace the base boards.

H

Installing Manufactured Cabinets

Cabinet Layout Lines

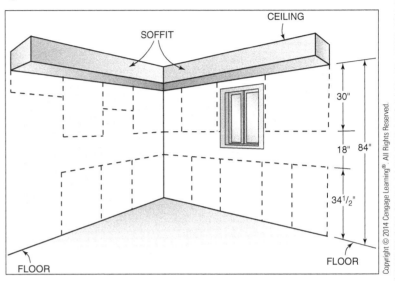

A Measure $34\frac{1}{2}$ inches up the wall. Draw a level line to indicate the tops of the base cabinets. Measure and mark another level line on the wall 54 inches from the floor. The bottom of the wall units are installed to this line.

• Mark the stud locations of the framed wall. Drive cabinet mounting screws into the studs. Lightly tap on and across a short distance of the wall with a hammer. Above the upper line on the wall, drive a finish nail in at the point where a solid sound is heard to accurately locate the stud. Drive nails where the holes will be later covered by a cabinet. Mark the locations of the remaining studs where cabinets will be attached. At each stud location, draw plumb lines on the wall. Mark the outlines of all cabinets on the wall to visualize and check the cabinet locations against the layout.

Installing Wall Units

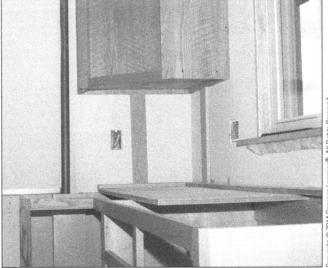

A Use a *cabinet lift* to hold the cabinets in position for fastening to the wall. If a lift is not available, remove the doors and shelves to make the cabinet lighter and easier to clamp together. If possible, screw a strip of lumber so that its top edge is on the level line for the bottom of the wall cabinets or strips of wood cut to the proper length. This is used to support the wall units while the cabinets are being fastened. If it is not possible to screw to the wall, build a stand on which to support the unit near the line of installation.

Installing Manufactured Cabinets (Continued)

B Start the installation of wall cabinets in a corner. On the wall, measure from the line representing the outside of the cabinet to the stud centers. Transfer the measurements to the cabinets. Drill shank holes for mounting screws through mounting rails usually installed at the top and bottom of the cabinet. Place the cabinet on the supporting strip or stand so that its bottom is on the level layout line. Fasten the cabinet in place with mounting screws of sufficient length to hold the cabinet securely. Do not fully tighten the screws. Install the next cabinet in the same manner.

B

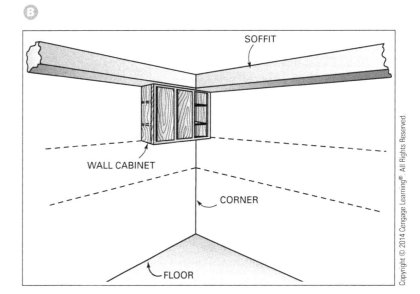

SOFFIT

WALL CABINET

CORNER

FLOOR

C Align the adjoining *stiles* so that their faces are flush with each other. Clamp them together with C-clamps. Screw the stiles tightly together. Continue this procedure around the room. After all the stiles are secured to each other, tighten all mounting screws. If a filler needs to be used, it is better to add it at the end of a run. It may be necessary to scribe the filler to the wall.

C

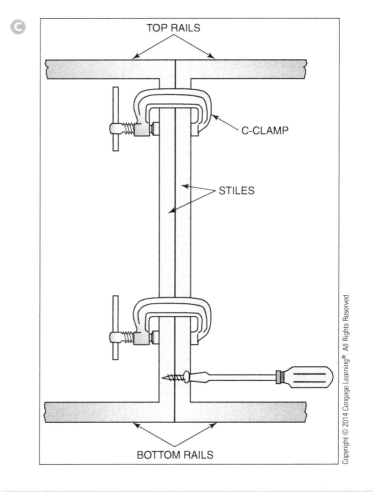

TOP RAILS

C-CLAMP

STILES

BOTTOM RAILS

Ⓓ Procedure for scribing a filler strip at the end of a run of cabinets.

• The space between the top of the wall unit and the ceiling may be finished by installing a soffit.

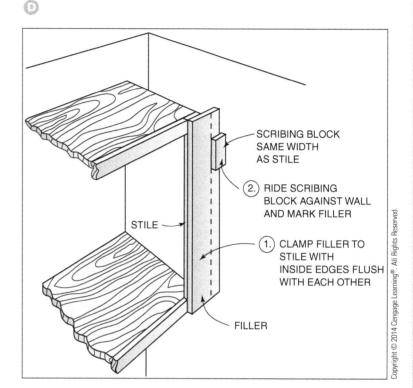

Ⓓ

SCRIBING BLOCK SAME WIDTH AS STILE

2. RIDE SCRIBING BLOCK AGAINST WALL AND MARK FILLER

STILE

1. CLAMP FILLER TO STILE WITH INSIDE EDGES FLUSH WITH EACH OTHER

FILLER

Installing Base Cabinets

Ⓐ Start the installation of base cabinets in a corner. Shim the bottom until the cabinet top is on the layout line. Level and shim the cabinet from back to front. If cabinets are to be fitted to the floor, shim until their tops are level across width and depth. This will bring the tops above the layout line that was measured from the low point of the floor. Adjust the scriber so that the distance between the points is equal to the amount the top of the unit is above the layout line. Scribe this amount on the bottom end of the cabinets by running the dividers along the floor.

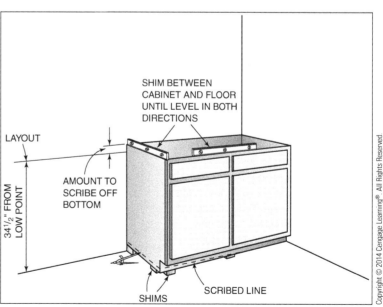

Ⓐ

SHIM BETWEEN CABINET AND FLOOR UNTIL LEVEL IN BOTH DIRECTIONS

LAYOUT

34½" FROM LOW POINT

AMOUNT TO SCRIBE OFF BOTTOM

SHIMS

SCRIBED LINE

PROCEDURE 7-12

Installing Manufactured Cabinets (Continued)

Ⓑ Cut both ends and toeboard to the scribed lines. Replace the cabinet in position. The top ends should be on the layout line. Fasten it loosely to the wall. Install the remaining base cabinets in the same manner. Align and clamp the stiles of adjoining cabinets. Fasten them together. Finally, fasten all units securely to the wall.

Ⓑ

Installing Countertops

- After the base units are fastened in position, cut the countertop to length. Fasten it on top of the base units and against the wall. Scribe the backsplash, limited by the thickness of its scribing strip, to an irregular wall surface. Use pencil dividers to scribe a line on the top edge of the backsplash. Plane or belt sand to the scribed line.

- Fasten the countertop to the base cabinets with screws up through triangular blocks usually installed in the top corners of base units. Take care not to drill through the countertop. Use screws of sufficient length, but not so long that they penetrate the countertop.

- Exposed cut ends of postformed countertops are covered by specially shaped pieces of plastic laminate. Sink cutouts are made by carefully outlining the cutout and cutting with a saber saw or router. The cutout pattern usually comes with the sink. Use a fine-toothed blade to prevent chipping out the face of the laminate beyond the sink. Some duct tape applied to the base of the power tool will prevent scratching of the countertop when making the cutout.

PROCEDURE 7-13

Removing Manufactured Cabinets

- Before attempting to remove and/or replace cabinets, first ensure that the contents of the cabinet have been removed.

- If the cabinets have electrical outlets or plumbing connections in them, turn off the utilities before starting.

- If both base and wall cabinets are to be removed, start by removing the base cabinets. Remove any flooring that may be attached to the toeboard.

- Remove any utilities connected to the cabinets.

- Remove all molding and/or decorative accents attached to the cabinets.

- Remove all nails and/or screws along the inside edge of the cabinets.

- Once all the nails and/or screws have been removed, carefully remove the base cabinets.

- When removing wall cabinets, always have at least one helper.

- Once the cabinet has been unsecured from the wall, remove it from the area.

PROCEDURE 7-14

Installing Cylindrical Locksets

- To install a cylindrical lockset, first check the contents and read the manufacturer's directions carefully. Many kinds of locks are manufactured and the mechanisms vary greatly. Follow the directions included with the lockset carefully.

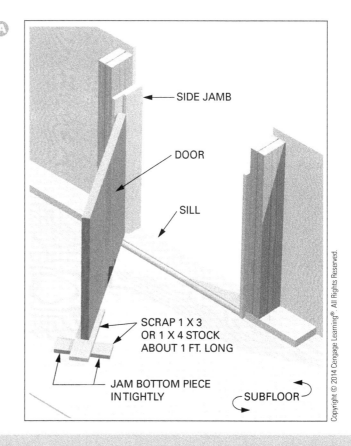

SIDE JAMB

DOOR

SILL

SCRAP 1 X 3 OR 1 X 4 STOCK ABOUT 1 FT. LONG

JAM BOTTOM PIECE IN TIGHTLY

SUBFLOOR

- Ⓐ However, there are certain basic procedures. Open the door to a convenient position. Wedge the bottom to hold it in place. Measure up, from the floor, the recommended distance to the centerline of the lock. This is usually 36 to 40 inches. At this height, square a light line across the edge and stile of the door.

Installing Cylindrical Locksets (Continued)

Marking and Boring Holes

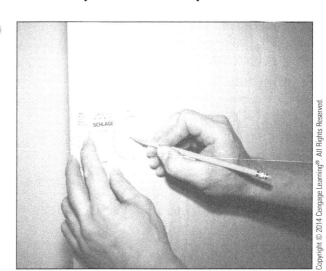

Ⓐ

Ⓐ Position the center of the paper template supplied with the lock on the squared lines. Lay out the centers of the holes that need to be bored. Fold the template over the high corner of the beveled door edge. The distance from the door edge to the center of the hole through the side of the door is called the *backset* of the lock. Usual backsets are $2\frac{3}{8}$ inches for residential and $2\frac{3}{4}$ inches for commercial constructions. Make sure that the backset is marked correctly before boring the hole. One hole must be bored through the side and one into the edge of the door. The manufacturer's directions specify the hole sizes where a 1-inch hole for bolts and $2\frac{1}{8}$-inch hole for locksets are common.

• The hole through the side of the door should be bored first. Stock for the center of the boring bit is lost if the hole in the edge of the door is bored first. It can be bored with hand tools, using an expansion bit in a bit brace. However, it is a difficult job. If you are using hand tools, bore from one side until only the point of the bit comes through, then bore from the other side to avoid splintering the door.

Using a Boring Jig

Ⓐ A **boring jig** is frequently used. It is clamped to the door to guide power-driven **multispur bits**. With a boring jig, holes can be bored completely through the door from one side. The clamping action of the jig prevents splintering.

• After the holes are bored, insert the latchbolt in the hole bored in the door edge. Hold it firmly and score around its faceplate with a sharp knife. Remove the latch unit. Deepen the vertical lines with the knife in the same manner as with hinges. Take great care when using a chisel along these lines. This may split out the edge of the door. Chisel out the recess so that the faceplate of the latch lays flush with the door edge.

Ⓑ Use **faceplate markers**, if available, to lay out the mortise for the latch faceplate. A marker of the appropriate size is held in the bored latch hole and tapped with a hammer. Complete the installation of the lockset by following specific manufacturer's directions.

Installing the Striker Plate

• The **striker plate** is installed on the door jamb, so when the door is closed it latches tightly with no play. If the plate is installed too far out, the door will not close tightly against the stop. It will then rattle. If the plate is installed too far in, the door will not latch.

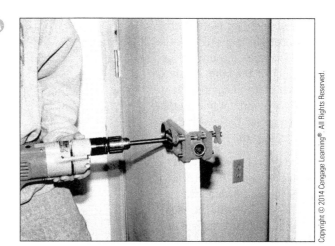

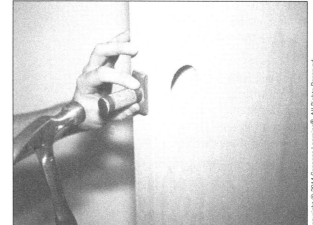

Installing Cylindrical Locksets (Continued)

Ⓐ To locate the striker plate in the correct position, place it over the latch in the door. Close the door snugly against the stops. Push the striker plate in against the latch. Draw a vertical line on the face of the plate flush with the outside face of the door.

● Open the door. Place the striker plate on the jamb. The vertical line, previously drawn on it, should be aligned with the edge of the jamb. Center the plate on the latch. Hold it firmly while scoring a line around the plate with a sharp knife. Chisel out the mortise so that the plate lies flush with the jamb. Screw the plate in place. Chisel out the center to receive the latch.

Rekey Locks

● Take the locking part of the lock out by unscrewing it from the inside. Leave the rest of the lock in place.

● Take the knob with the locking mechanism with its proper key to the hardware store. The hardware store person will replace the pins so that they'll fit another key.

● Put the door lock back together.

Ⓐ

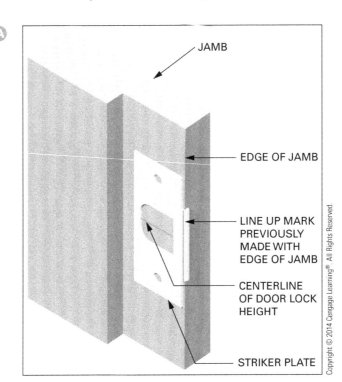

JAMB

EDGE OF JAMB

LINE UP MARK PREVIOUSLY MADE WITH EDGE OF JAMB

CENTERLINE OF DOOR LOCK HEIGHT

STRIKER PLATE

Cutting and Fitting Gypsum Board

A Take measurements accurately to within $\frac{1}{4}$ inch for the ceiling and $\frac{1}{8}$ inch for the walls. Using a utility knife, cut the board by first scoring the face side through the paper to the core. Guide it with a *drywall T-square*, using your toe to hold the bottom. Only the paper facing needs to be cut.

A

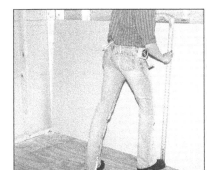

B Bend the board back against the cut. The board will break along the cut face. Score the backside paper.

- Lifting the panel off the floor, snap the cut piece back quickly to the straight position. This will complete the break.

B

FROM EXPERIENCE

Cut only the center section of the backside paper, leaving the bottom and top portions. These will act as hinges for the cut piece when it is snapped back into place.

C To make cuts parallel to the long edges, the board is often gauged with a tape and scored with a utility knife. When making cuts close to long edges, score both sides of the board before the break to obtain a clean cut.

- Ragged edges can be smoothed with a drywall rasp, coarse sanding block, or knife.

C

Installing Sheet Paneling

Starting the Application

A Mark the location of each stud in the wall on the floor and ceiling. Paneling edges must fall on stud centers, even if applied with adhesive over a backer board, in case supplemental nailing of the edges is necessary.

- If the wall is to be wainscoted, snap a horizontal line across the wall to indicate its height.

- Apply narrow strips of paint on the wall from floor to ceiling over the stud where a seam in the paneling will occur. The color should be close to that of the seams of the paneling. This will hide the joints between sheets if they open slightly because of shrinkage.

- Cut the first sheet to a length about $\frac{1}{4}$ inch less than the wall height. Place the sheet in the corner. Plumb the edge and tack it temporarily into position.

B Notice the joint at the corner and the distance the sheet edge overlaps the stud. Set the distance between the points of a scriber to the same as the amount the sheet overlaps the center of the stud. Scribe this amount on the edge of the sheet butting the corner.

- Remove the sheet from the wall and cut close to the scribed line. Plane the edge to the line to complete the cut. Replace the sheet with the cut edge fitting snugly in the corner.

- If a tight fit between the panel and the ceiling is desired, set the dividers and scribe a small amount at the ceiling line. Remove the sheet again and cut to the scribed line. The joint at the ceiling need not be fit tight if a molding is to be used.

A

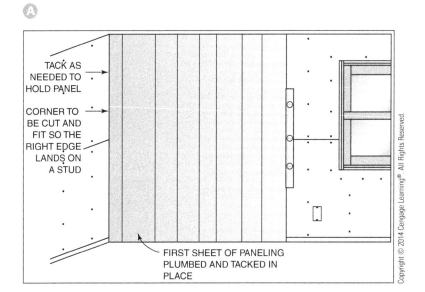

TACK AS NEEDED TO HOLD PANEL

CORNER TO BE CUT AND FIT SO THE RIGHT EDGE LANDS ON A STUD

FIRST SHEET OF PANELING PLUMBED AND TACKED IN PLACE

B

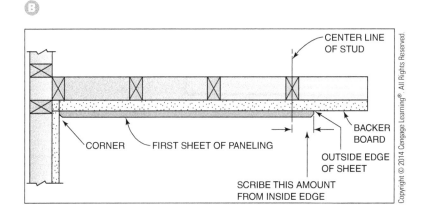

CENTER LINE OF STUD

BACKER BOARD

CORNER   FIRST SHEET OF PANELING

OUTSIDE EDGE OF SHEET

SCRIBE THIS AMOUNT FROM INSIDE EDGE

Wall Outlets

Ⓐ To lay out for wall outlets, plumb and mark both sides of the outlet to the floor or ceiling, whichever is closer. Level the top and bottom of the outlet on the wall out beyond the edge of the sheet to be installed.

⦿ Place the sheet in position and tack. Level and plumb marks from the wall and floor onto the sheet for the location of the opening.

Ⓑ Remove the sheet and cut the opening for the outlet. When using a saber saw, cut from the back of the panel to avoid splintering the face.

Fastening

⦿ Apply adhesive beads 3 inches long and about 6 inches apart on all intermediate studs. Apply a continuous bead along the perimeter of the sheet. Put the panel in place. Tack it at the top when panel is in proper position.

⦿ Press on the panel surface to make contact with the adhesive. Use firm, uniform pressure to spread the adhesive beads evenly between the wall and the panel. Grasp the panel and slowly pull the bottom of the sheet a few inches away from the wall.

⦿ Press the sheet back into position after about 2 minutes. Drive nails as needed and recheck the sheet for a complete bond after about 20 minutes. Apply pressure to ensure thorough adhesion and to smooth the panel surface.

⦿ Apply successive sheets in the same manner. Panels should touch only lightly at joints.

Ⓐ

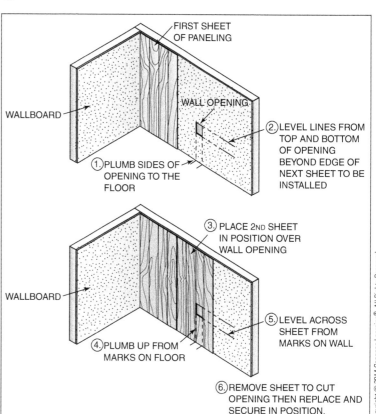

Ⓑ

Installing Sheet Paneling (Continued)

Ending the Application

- Take measurements at the top, center, and bottom. Cut the sheet to width and install. If no corner molding is used, the sheet must be cut to fit snugly in the corner. To mark the sheet accurately, first measure the remaining space at the top, at the bottom, and about the center. Rip the panel about $\frac{1}{2}$ inch wider than the greatest distance.

Ⓐ Place the sheet plumb with the cut edge in the corner and the other edge overlapping the last sheet installed. Tack the sheet in position so that the amount of overlap is exactly the same from top to bottom. Set the scriber for the amount of overlap and scribe this amount on the edge in the corner.

- Cut close to the scribed line and then plane to the line. If the line is followed carefully, the sheet should fit snugly between the last sheet installed and the corner, regardless of any irregularities.

Ⓑ Exterior corners may be finished by capping the joint.

- Use a wood block for more accurate scribing of wide distances.

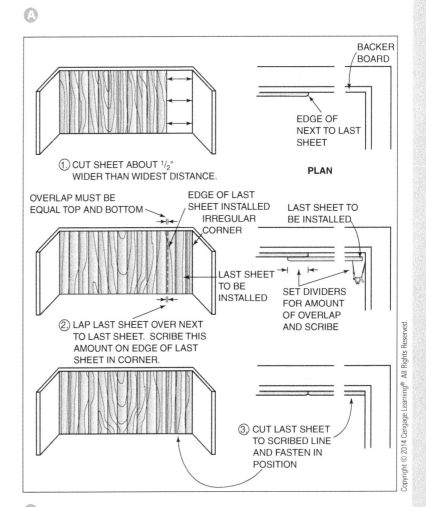

Ⓐ

① CUT SHEET ABOUT ½" WIDER THAN WIDEST DISTANCE.

BACKER BOARD

EDGE OF NEXT TO LAST SHEET

PLAN

OVERLAP MUST BE EQUAL TOP AND BOTTOM

EDGE OF LAST SHEET INSTALLED IRREGULAR CORNER

LAST SHEET TO BE INSTALLED

LAST SHEET TO BE INSTALLED

② LAP LAST SHEET OVER NEXT TO LAST SHEET. SCRIBE THIS AMOUNT ON EDGE OF LAST SHEET IN CORNER.

SET DIVIDERS FOR AMOUNT OF OVERLAP AND SCRIBE

③ CUT LAST SHEET TO SCRIBED LINE AND FASTEN IN POSITION

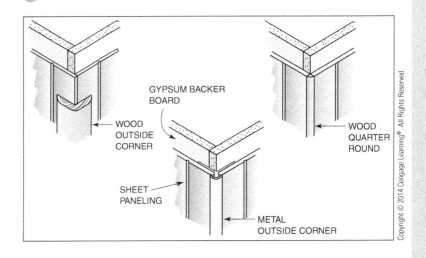

Ⓑ

WOOD OUTSIDE CORNER

GYPSUM BACKER BOARD

WOOD QUARTER ROUND

SHEET PANELING

METAL OUTSIDE CORNER

PROCEDURE 7-17

Installing Solid Wood Paneling

Starting the Application

A Select a straight board with which to start. Cut it to length, about $\frac{1}{4}$ inch less than the height of the wall. If tongue-and-groove stock is used, tack it in a plumb position with the grooved edge in the corner.

- Adjust the scribers to scribe an amount a little more than the depth of the groove. Rip and plane to the scribed line.

- Replace the piece and face-nail along the cut edge into the corner with finish nails about 16 inches apart. Blind-nail the other edge through the tongue.

- Apply succeeding boards by blind-nailing only into the tongue. Make sure that the joints between boards come up tightly. Severely warped boards should not be used.

- As installation progresses, check the paneling for plumb. If it is out of plumb, gradually bring it back by driving one end of several boards a little tighter than the other end. Cut out openings in the same manner as described for sheet paneling.

Installing Solid Wood Paneling (Continued)

Applying the Last Board

(A) Cut and fit the next to the last board and then remove it. Cut, fit, and tack the last board in the place of the next-to-the-last board just removed.

- Cut a scrap block about 6 inches long and equal in width to the finished face of the next-to-the-last board. The tongue should be removed. Use this block to scribe the last board by running one edge along the corner and holding a pencil against the other edge.

- Remove the board from the wall. Cut and plane it to the scribed line. Fasten the next-to-the-last board in position. Fasten the last board in position with the cut edge in the corner.

- Face-nail the edge nearest the corner.

(A)

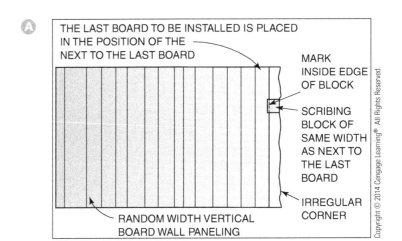

THE LAST BOARD TO BE INSTALLED IS PLACED IN THE POSITION OF THE NEXT TO THE LAST BOARD

MARK INSIDE EDGE OF BLOCK

SCRIBING BLOCK OF SAME WIDTH AS NEXT TO THE LAST BOARD

IRREGULAR CORNER

RANDOM WIDTH VERTICAL BOARD WALL PANELING

Installing Flexible Insulation

Ⓐ Install positive ventilation chutes between the rafters where they meet the wall plate. This will compress the insulation slightly against the top of the wall plate to permit the free flow of air over the top of the insulation.

• Install the air-insulation dam between the rafters in line or on with the exterior sheathing. This will protect the insulation from air movement into the insulation layer from the soffits.

Ⓐ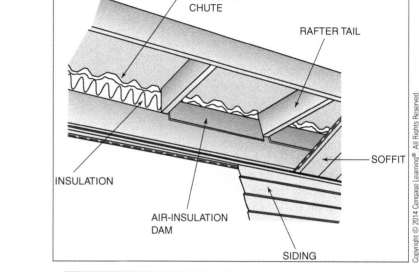

POSITIVE VENTILATION CHUTE

RAFTER TAIL

SOFFIT

INSULATION

AIR-INSULATION DAM

SIDING

Ⓑ To cut the material, place a scrap piece of plywood on the floor to protect the floor while cutting. Roll out the material over the scrap. Using another scrap piece of wood, compress the insulation and cut it with a sharp knife in one pass.

Ⓑ

Ⓒ Place the batts or blankets between the studs. The flanges of the vapor retarder may be stapled either to cover the studs or to cover the inside edges of the studs as well as the top and bottom plates. A better vapor retarder is achieved with fastening to cover the stud, but the studs are less visible for the installation of the gypsum. Use a hand or hammer-tacker stapler to fasten the insulation in place.

Ⓒ

PROCEDURE 7-18

Installing Flexible Insulation (Continued)

Ⓓ Fill any spaces around windows and doors with spray-can foam. Nonexpanding foam will fill the voids with an airtight seal and protect the house from air leakage. After the foam cures, add flexible insulation to fill the remaining space.

• Install ceiling insulation by stapling it to the ceiling joists or by friction-fitting it between them. Push and extend the insulation across the top plate to fit against the air-insulation dam.

Ⓓ

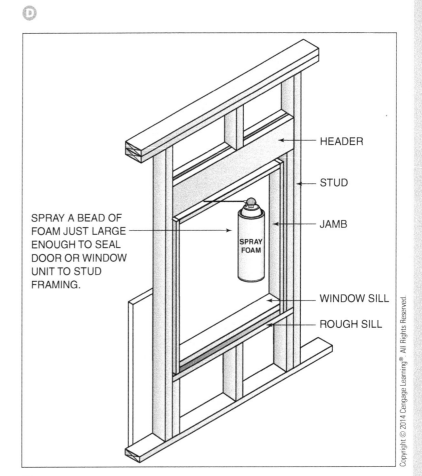

Ⓔ Flexible insulation installed between floor joists over crawl-spaces may be held in place by wire mesh or pieces of heavy-gauge wire wedged between the joists.

Ⓔ

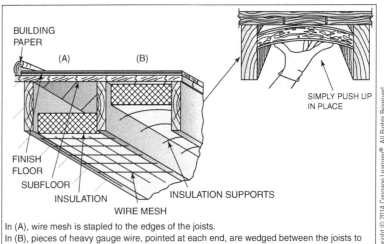

In (A), wire mesh is stapled to the edges of the joists.
In (B), pieces of heavy gauge wire, pointed at each end, are wedged between the joists to support the insulation.

PROCEDURE 7-19

Installing Windows

Ⓐ Place window in the opening after removing all shipping protection from the window unit. Do not remove any diagonal braces applied at the factory. Close and lock the sash. Windows can easily be moved through the openings from the inside and set in place.

Ⓑ Center the unit in the opening on the rough sill with the exterior window casing against the wrapped wall sheathing. Level the window sill with a wood shim tip between the rough sill and the bottom end of the window's side jamb, if necessary. Secure the shim to the rough sill.

• Remove the window unit from the opening and caulk the backside of the casing or nailing flange. This will seal the unit to the building. Replace the unit and nail the lower end of both sides of the window. Next, plumb a side and nail the unit along the sides and top. Check that the sash operates properly. If not, make necessary adjustments.

Ⓒ Flash the head casing by cutting to length of the flashing with tin snips. Its length should be equal to the overall length of the window head casing. If the flashing must be applied in more than one piece, lap the joint about 3 inches. Slice the housewrap just above the head casing, and slip the flashing behind the wrap and on top of the head casing. Secure it with fasteners into the wall sheathing. Refasten the house wrap. Tape all seams in housewrap and over window nailing flanges.

Ⓐ

CAUTION

Have sufficient help when setting large units. Handle them carefully to avoid damaging the unit or breaking the glass. Broken glass can cut through protective clothing and cause serious injury.

Ⓑ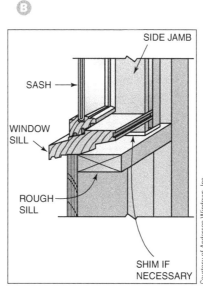

Ⓒ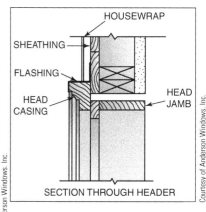

SECTION THROUGH HEADER

PROCEDURE 7-20

Replacing a Damaged Window Screen

- Remove the damaged window screen from the window frame.

- After placing the screen on a sturdy flat surface (large enough to support the damaged screen), remove the rubber spline, thus allowing the old screen to be separated from the frame

- With the old screen removed, measure the length and width of the screen frame. When cutting the new screen, add 2 inches to the measure of the screen frame.

- Lay the new screen onto the frame, using a screen-rolling tool and starting in one corner press the rubber spline and screen firmly into the spline groove.

- Continue this process all the way around the screen frame.

- Using a utility knife carefully trim the excess screen.

Exterior Carpentry Maintenance

PROCEDURE 7-21

Installing Gutters

Ⓐ On both ends of the fascia, mark the location of the bottom side of the gutter. The top-outside edge of the gutter should be in relation to a straight line projected from the top surface of the roof. The height of the gutter depends on the pitch of the roof.

- Stretch a chalk line between the two marks. Move the center of the chalk line up enough to give the gutter the proper pitch from the center to the ends. Snap a line on both sides of the center.

- Fasten the gutter brackets to the chalk line on the fascia with screws. All screws should be made of stainless steel or other corrosion-resistant material. Aluminum brackets may be spaced up to 30 inches on center (OC).

Ⓐ
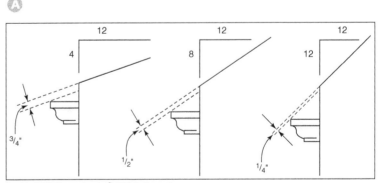

B Locate and install the outlet tubes in the gutter as required, keeping in mind that the downspout should be positioned plumb and square with the building. Add end caps and caulk all seams only on the inside surfaces.

- Hang the gutter sections in the brackets. Use slip-joint connectors to join larger sections. Use either inside or outside corners where gutters make a turn. Caulk all inside seams.

- Fasten downspouts to the wall with appropriate hangers and straps. Downspouts should be fastened at the top and bottom and every 6 feet in between. The connection between the downspout and the gutter is made with elbows and short straight lengths of downspout.

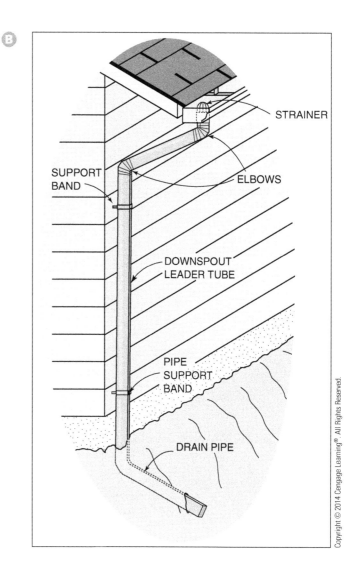

B

STRAINER

SUPPORT BAND

ELBOWS

DOWNSPOUT LEADER TUBE

PIPE SUPPORT BAND

DRAIN PIPE

Installing Gutters (Continued)

C Because water runs downhill, care should be taken when putting the downspout pieces together. The downspout components are assembled where the upper piece is inserted into the lower one. This makes the joint lap such that the water cannot escape until it leaves the bottommost piece.

C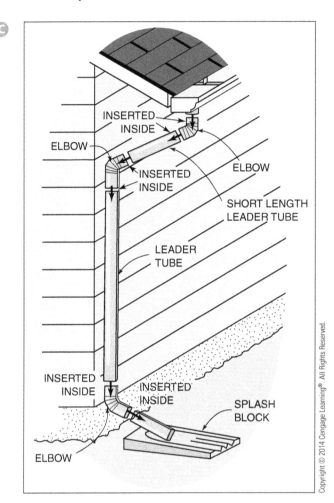

Installing Asphalt Shingles

> **CAUTION**
>
> Installation of roofing systems involves working on ladders and scaffolding as well as on top of the building. Workers should always be aware of the potential for falling. Keep the location of roof perimeter in mind at all times.

(A) Prepare the roof deck by clearing sawdust and debris that will cause a slipping hazard.

- Begin underlayment over the deck at a lower corner. Lap the following courses of felt over the lower course at least 2 inches. Make any end laps at least 4 inches. Lap the felt 6 inches from both sides over all hips.

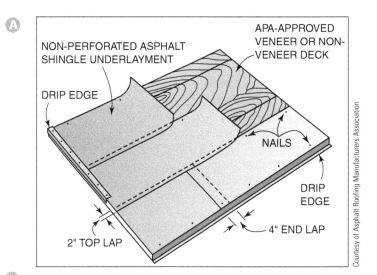

Courtesy of Asphalt Roofing Manufacturers Association

(B) Nail or staple through each lap and through the center of each layer about 16 inches apart. Roofing nails driven through the center of metal disks or specially designed, large-head felt fasteners hold the underlayment securely in strong winds until shingles are applied.

- Install metal drip edge along the perimeter on top of the underlayment. This will help prevent blow-offs.

- Prepare the starter course by cutting off the exposure taps lengthwise through the shingle. Save these tabs, as they may be used as the last course at the ridge. Install the course so that no end joint will fall in line with an end joint or tab cutout of the regular first course of shingles.

Courtesy of APA - The Engineered Wood Association

PROCEDURE 7-22

Installing Asphalt Shingles (Continued)

FROM EXPERIENCE

Use a utility knife to cut shingles from the back side. Cut only halfway through and then fold and break the shingle to complete the cut. When cutting from the granular top surface, use a hook blade.

Ⓒ Determine the starting line, either the rake edge or vertical center-snapped lines. To start from the middle of the roof, mark the center of the roof at the eaves and the ridge. Snap a chalk line between the marks. Snap a series of chalk lines from this one, 4 or 6 inches apart, depending on the desired end tab, on each side of the centerline. When applying the shingles, start the course with the end of the shingle to the vertical chalk line. Start succeeding courses in the same manner. Break the joints as necessary, working both ways toward the rakes.

Ⓓ Starting shingle layout at the rake edge involves placing the first course, with a whole tab at the rake edge. The second course is started with a shingle that is 6 inches shorter; the third course, with a strip that is a full tab shorter; the fourth, with one and one-half tabs removed, and so on. These starting pieces are precut for faster application.

Ⓒ

SNAPPED LINES PERPENDICULAR TO FASCIA OR PARALLEL TO RAKE FASCIA

METAL DRIP EDGE

FIRST SHINGLE OF EACH COURSE STARTS AGAINST CHALK LINE

STARTER STRIP

Ⓓ

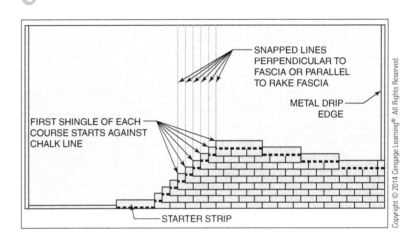

METAL DRIP EDGE APPLIED OVER FELT ALONG RAKE

NAILING

2" HEAD LAP

UNDERLAYMENT

WOOD DECK

EAVES FLASHING STRIP

$5\frac{5}{8}"$

4" END LAP

SELF-SEALING STRIP

METAL DRIP EDGE

① STARTER — BEGIN WITH A FULL STARTER SHINGLE MINUS 3" SO BUTT SEAMS DO NOT ALIGN WITH FIRST COURSE

② START FIRST COURSE WITH FULL STRIP

④ START THIRD COURSE WITH FULL STRIP MINUS FIRST TAB

③ START SECOND COURSE WITH FULL STRIP MINUS 1/2 TAB

E If the cutouts are to break on the thirds, cut the starting strip for the second course by removing 4 inches. Remove 8 inches from the strip for the third course, and so on.

• Fasten each shingle from the end nearest the shingle just laid, which prevents buckling. Drive fasteners straight so that the nail heads will not cut into the shingles. Both ends of the course should overhang the drip edge $\frac{1}{4}$ to $\frac{1}{8}$ inch.

E

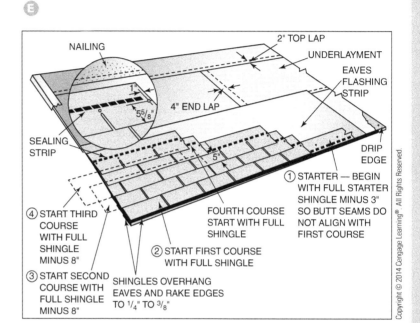

NAILING

2" TOP LAP

UNDERLAYMENT

EAVES FLASHING STRIP

4" END LAP

$5\frac{5}{8}$"

SEALING STRIP

5"

DRIP EDGE

④ START THIRD COURSE WITH FULL SHINGLE MINUS 8"

FOURTH COURSE START WITH FULL SHINGLE

① STARTER — BEGIN WITH FULL STARTER SHINGLE MINUS 3" SO BUTT SEAMS DO NOT ALIGN WITH FIRST COURSE

② START FIRST COURSE WITH FULL SHINGLE

③ START SECOND COURSE WITH FULL SHINGLE MINUS 8"

SHINGLES OVERHANG EAVES AND RAKE EDGES TO $\frac{1}{4}$" TO $\frac{3}{8}$"

F Install vented ridge cap as per manufacturer's instructions. Cut cap shingles and begin installation from one end. Center the cap shingle over the vented ridge cap. Secure each shingle with one fastener on each side.

• Apply the cap across the ridge until 3 or 4 feet from the end, then space the cap to the end in the same manner used to space the shingle course to the ridge. Cut the last ridge shingle to size. Apply it with one fastener on each side of the ridge. Cover the two fasteners with asphalt cement to prevent leakage.

F

VENTED RIDGE CAP

Installing Roll Roofing

Roll Roofing with Concealed Fasteners

A Apply 9-inch-wide strips of the roofing along the eaves and rakes overhanging the drip edge about $\frac{3}{8}$ inch. Fasten with two rows of nails one inch from each edge spaced about 4 inches apart.

- Apply the first course of roofing with its edge and ends flush with the strips. Secure the upper edge with nails staggered about 4 inches apart. Do not fasten within 18 inches of the rake edge.

- Apply cement only to the edge strips covered by the first course. Press the edge and rake edges firmly to the strips. Complete the nails in the upper edge out to the rakes.

- Apply succeeding courses in a similar manner. Make all end laps 6 inches wide. Apply cement the full width of the lap.

- After all courses are in place, lift the lower edge of each course. Apply the cement in a continuous layer over the full width and length of the lap. Press the lower edges of the upper courses firmly into the cement. A small bead should appear along the entire edge of the sheet. Care must be taken to apply the correct amount of cement.

- To cover the hips and ridge, cut strips of 12 inches × 36 inches roofing. Bend the pieces lengthwise through their centers.

- Snap a chalk line on both sides of the hip or ridge down about $5\frac{1}{2}$ inches from the center. Apply cement between the lines. Fit the first strip over the hip or ridge.

A

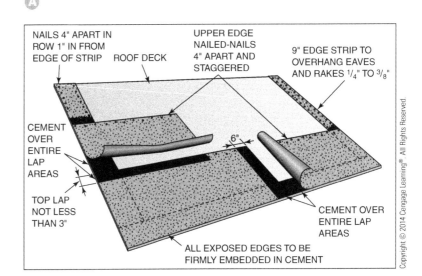

NAILS 4" APART IN ROW 1" IN FROM EDGE OF STRIP

ROOF DECK

UPPER EDGE NAILED-NAILS 4" APART AND STAGGERED

9" EDGE STRIP TO OVERHANG EAVES AND RAKES $\frac{1}{4}$" TO $\frac{3}{8}$"

CEMENT OVER ENTIRE LAP AREAS

6"

TOP LAP NOT LESS THAN 3"

CEMENT OVER ENTIRE LAP AREAS

ALL EXPOSED EDGES TO BE FIRMLY EMBEDDED IN CEMENT

B Press it firmly into place. Start at the lower end of a hip and at either end of a ridge. Lap each strip 6 inches over the preceding one. Nail each strip only on the end that is to be covered by the overlapping piece.

- Spread cement on the end of each strip that is lapped before the next one is applied. Continue in this manner until the end is reached.

Double Coverage Roll Roofing

A Cut the 19-inch strip of *selvage*, nonmineral surface side, from enough double-coverage roll roofing to cover the length of the roof. Save the surfaced portion for the last course at the ridge. Apply the selvage portion parallel to the eaves. It should overhang the drip edge by $\frac{3}{8}$ inch. Secure it to the roof deck with three rows of nails.

B Apply the first course using a full-width strip of roofing. Secure it with two rows of nails in the selvage portion.

B
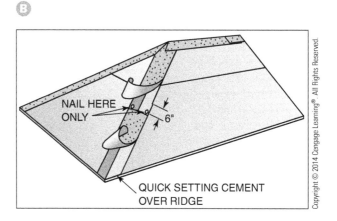

NAIL HERE ONLY   6"

QUICK SETTING CEMENT OVER RIDGE

A
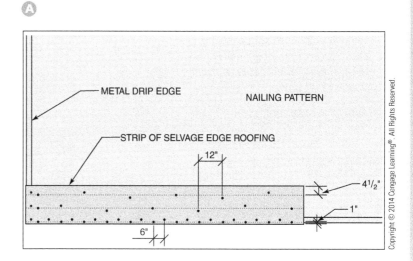

METAL DRIP EDGE   NAILING PATTERN

STRIP OF SELVAGE EDGE ROOFING

12"

4½"

1"

6"

B
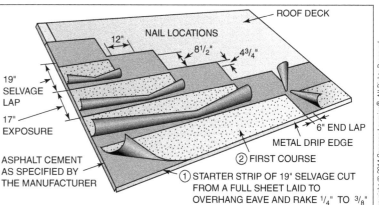

ROOF DECK

NAIL LOCATIONS

12"

8½"

4¾"

19" SELVAGE LAP

17" EXPOSURE

6" END LAP

METAL DRIP EDGE

ASPHALT CEMENT AS SPECIFIED BY THE MANUFACTURER

② FIRST COURSE

① STARTER STRIP OF 19" SELVAGE CUT FROM A FULL SHEET LAID TO OVERHANG EAVE AND RAKE ¼" TO ³⁄₈"

PROCEDURE 7-23

Installing Roll Roofing (Continued)

C Apply succeeding courses in the same manner. Lap the full width of the 19-inch selvage each time. Make all end laps 6 inches wide. End laps are made in the manner shown in the accompanying figure. Stagger end laps in succeeding courses.

- Lift and roll back the surface portion of each course. Starting at the bottom, apply cement to the entire selvage portion of each course. Apply it to within $\frac{1}{4}$ inch of the surfaced portion. Press the overlying sheet firmly into the cement. Apply pressure over the entire area using a light roller to ensure adhesion between the sheets at all points.

D Apply the remaining surfaced portion left from the first course as the last course. Hips and ridges are covered in the same manner shown in the accompanying figure.

- Follow specific application instructions because of differences in the manufacture of roll roofing. Follow specific requirements for quantities and types of adhesive.

C

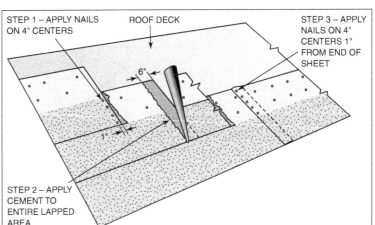

D

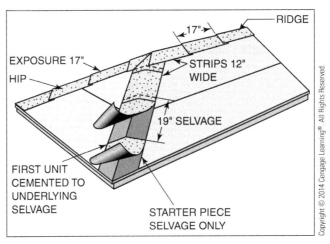

Woven Valley Method

A Install underlayment and starter strip to both roofs.

- Apply first course of one roof, say the left one, into and past the center of the valley. Press the shingle tightly into the valley and nail, keeping the nails at least 6 inches away from the valley centerline. Cut shingles to adjust the butt ends so that there is no butt seam within 12 inches of the valley centerline.

- Apply the first course of the other (right) roof in a similar manner, into and past the valley.

- Apply succeeding courses by repeating this alternating pattern, first from one roof and then on the other.

A

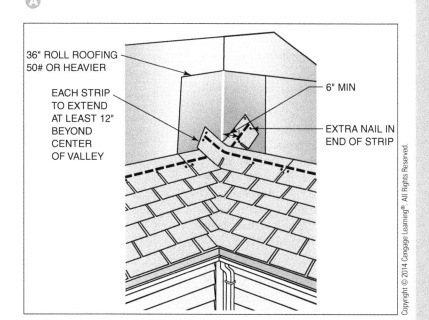

36" ROLL ROOFING 50# OR HEAVIER

EACH STRIP TO EXTEND AT LEAST 12" BEYOND CENTER OF VALLEY

6" MIN

EXTRA NAIL IN END OF STRIP

Closed Cut Valley Method

A Begin by shingling the first roof completely, letting the end shingle of every course overlap the valley by at least 12 inches. Form the end shingle of each course snugly into the valley. Cut shingles to adjust the butt ends so that there is no butt seam within 12 inches of the valley centerline.

- Snap a chalk line along the center of the valley on top of the shingles of the first roof.

- Apply the shingles of the second roof, cutting the end shingle of each course to the chalk line. Place the cut end of each course that lies in the valley in a 3-inch-wide bed of asphalt cement.

A

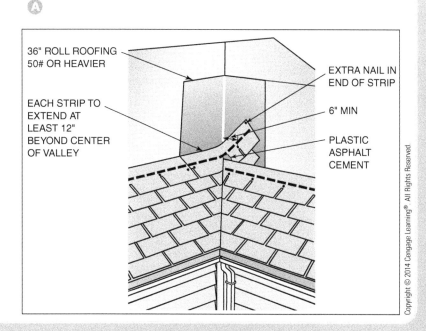

36" ROLL ROOFING 50# OR HEAVIER

EACH STRIP TO EXTEND AT LEAST 12" BEYOND CENTER OF VALLEY

EXTRA NAIL IN END OF STRIP

6" MIN

PLASTIC ASPHALT CEMENT

Step Flashing Method

(A) Snap a chalk line in the center of the valley on the valley underlayment.

- Apply the shingle starter course on both roofs. Trim the ends of each course that meet the chalk line.

- Fit and form the first piece of flashing to the valley on top of the starter strips. Trim the bottom edge flush with the drip edge. Fasten with two nails only in the upper corners of the flashing. Use nails of like material to the flashing to prevent electrolysis.

- Apply the first regular course of shingles to both roofs on each side of the valley, trimming the ends to the chalk line. Bed the ends in plastic asphalt cement. Do not drive nails through the metal flashing. Apply flashing to each succeeding course in this manner.

(A)

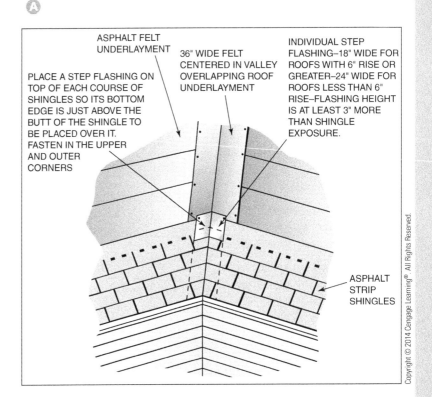

ASPHALT FELT UNDERLAYMENT

36" WIDE FELT CENTERED IN VALLEY OVERLAPPING ROOF UNDERLAYMENT

INDIVIDUAL STEP FLASHING—18" WIDE FOR ROOFS WITH 6" RISE OR GREATER—24" WIDE FOR ROOFS LESS THAN 6" RISE—FLASHING HEIGHT IS AT LEAST 3" MORE THAN SHINGLE EXPOSURE.

PLACE A STEP FLASHING ON TOP OF EACH COURSE OF SHINGLES SO ITS BOTTOM EDGE IS JUST ABOVE THE BUTT OF THE SHINGLE TO BE PLACED OVER IT. FASTEN IN THE UPPER AND OUTER CORNERS

ASPHALT STRIP SHINGLES

Installing Horizontal Siding

Ⓐ The siding should be positioned so that it is about equal both above and below the window sill. This is done by first determining the siding exposure. Divide the overall height of each wall section by the maximum allowable exposure. Round up this number to get the number of courses in that section, then divide the height again by the number of courses to find the exposure. These slight adjustments in exposure will not be noticeable to the eye.

Ⓐ

EXAMPLE: Consider the overall dimensions in the accompanying figure. Divide the heights by the maximum allowable exposure, 7 inches in this example. Round up to the nearest number of courses that will cover that section. Divide the section height by the number of courses to find the exposure.

$40\frac{1}{2} \div 7 = 5.8 \Rightarrow 6$ courses $\qquad 40\frac{1}{2} \div 6 = 6.75$ or $6\frac{3}{4}$"-inch exposure

$45\frac{1}{5} \div 7 = 6.5 \Rightarrow 7$ courses $\qquad 45\frac{1}{2} \div 7 = 6.5$ or $6\frac{1}{2}$"-inch exposure

$12\frac{1}{2} \div 7 = 1.8 \Rightarrow 2$ courses $\qquad 12\frac{1}{2} \div 2 = 6.25$ or $6\frac{1}{4}$"-inch exposure

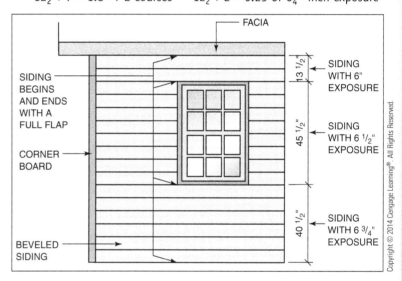

FACIA

SIDING BEGINS AND ENDS WITH A FULL FLAP

CORNER BOARD

BEVELED SIDING

13 1/2" SIDING WITH 6" EXPOSURE

45 1/2" SIDING WITH 6 1/2" EXPOSURE

40 1/2" SIDING WITH 6 3/4" EXPOSURE

Installing Horizontal Siding (Continued)

B Install a starter strip of the same thickness and width of the siding at the headlap fastened along the bottom edge of the sheathing. For the first course, a line is snapped on the wall at a height of the top edge of the first course of siding.

• From this first chalk line, lay out the desired exposures on each corner board and each side of all openings. Snap lines at these layout marks. These lines represent the top edges of all siding pieces.

• Install the siding per manufacturer's recommendations, staggering the butt joints in adjacent courses as far apart as possible. Use a small piece of felt paper behind the butt seams to ensure the weathertightness of the siding.

C When applying a course of siding, start from one end and work toward the other end. With this procedure, only the last piece will need to be fitted. Tight-fitting butt joints must be made between pieces. Measure carefully and cut the piece slightly longer. Place one end in position. Bow the piece outward, position the other end, and snap it into place. Take care not to move the corner board with this technique. Do not use this technique on cementitious siding.

D Siding is fastened to each bearing stud or about every 16 inches. On bevel siding, fasten through the butt edge just above the top edge of the course below. Do not fasten through the lap. This allows the siding to swell and shrink with seasonal changes without splitting of the siding. Blind-nail cementitious siding by fastening only along the top edge. Blind-nailing is not recommended in high-wind areas.

B

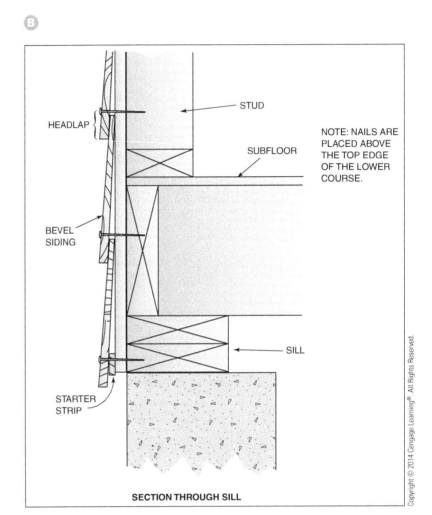

NOTE: NAILS ARE PLACED ABOVE THE TOP EDGE OF THE LOWER COURSE.

SECTION THROUGH SILL

C

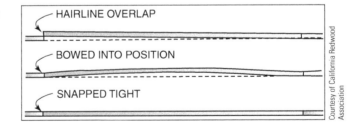

HAIRLINE OVERLAP

BOWED INTO POSITION

SNAPPED TIGHT

 **FROM EXPERIENCE**

Setting up a comfortable work station for cutting will allow the carpenter to work more efficiently and safely. This will also reduce waste and improve workmanship.

6" & NARROWER	8" & WIDER
PLAIN	**PLAIN**
USE ONE CASING NAIL PER BEARING TO BLIND NAIL.	USE TWO SIDING OR BOX NAILS, 3-4" APART TO FACE NAIL.
USE ONE SIDING OR BOX NAIL TO FACE NAIL ONCE PER BEARING. 1" UP FROM BOTTOM.	APPROXIMATE 1/8" GAP FOR DRY MATERIAL 8" AND WIDER. 1/2" = FULL DEPTH OF RABBET. USE TWO SIDING OR BOX NAILS, 3-4" APART, PER BEARING.
BOARD AND BATTEN 1/2"	**BOARD AND BATTEN** / **BOARD ON BOARD**
RECOMMEND 1/2" OVERLAP. ONE SIDING OR BOX NAIL PER BEARING.	INCREASE OVERLAP PROPORTIONATELY. USE TWO SIDING OR BOX NAILS, 3-4" APART.
SIDING USUALLY APPLIED HORIZONTALLY	

6" & NARROWER	8" & WIDER
PLAIN	**PLAIN**
RECOMMEND 1" OVERLAP. ONE SIDING OR BOX NAIL PER BEARING, JUST ABOVE THE 1" OVERLAP.	RECOMMEND 1" OVERLAP. ONE SIDING OR BOX NAIL PER BEARING, JUST ABOVE THE 1" OVERLAP.
RABBETED EDGE	**RABBETED EDGE** APPROXIMATE 1/8" GAP FOR DRY MATERIAL 8" AND WIDER. 1/2" = FULL DEPTH OF RABBET
ALLOWS FOR 1/2" OVERLAP. ONE SIDING OR BOX NAIL PER BEARING. 1" UP FROM BOTTOM EDGE.	ALLOWS FOR 1/2" OVERLAP. ONE SIDING OR BOX NAIL PER BEARING. 1" UP FROM BOTTOM EDGE.
USE SIDING OR BOX NAIL TO FACE NAIL ONE PER BEARING, 1 1/2" UP FROM BOTTOM EDGE.	APPROXIMATE 1/8" GAP FOR DRY MATERIAL 8" AND WIDER. 1/2" = FULL DEPTH OF RABBET. USE TWO SIDING OR BOX NAILS, 3-4" APART, PER BEARING, TO FACE NAIL.
T&G PATTERN / **SHIPLAP PATTERN**	**T&G PATTERN** / **SHIPLAP PATTERN** APPROXIMATE 1/8" GAP FOR DRY MATERIAL 8" AND WIDER. 1/2" = FULL DEPTH OF RABBET
USE CASING NAILS TO BLIND NAIL T&G PATTERNS, ONE NAIL PER BEARING. USE SIDING OR BOX NAILS TO FACE NAIL SHIPLAP PATTERNS, 1" UP FROM BOTTOM EDGE.	USE TWO SIDING OR BOX NAILS, 3-4" APART, TO FACE NAIL, 1" UP FROM BOTTOM EDGE.
SIDING USUALLY APPLIED HORIZONTALLY	

Courtesy of Western Wood Products Association

PROCEDURE 7-28

Installing Vertical Tongue-and-Groove Siding

A Slightly back-bevel the ripped edge. Place it vertically on the wall with the beveled edge flush with the corner similar to making a corner board. Face-nail the edge nearest the corner.

• Fasten a temporary piece on the other end of the wall, projecting below the sheathing by the same amount. Stretch a line to keep the bottom ends of other pieces in a straight line.

B Apply succeeding pieces by toenailing into the tongue edge of each piece. Make sure that the edges between boards come up tight. Drive the nail home until it forces the board up tight. Make sure to keep the bottom ends in a straight line. If butt joints are necessary, use a mitered or rabbeted joint for weathertightness.

C To cut the piece to fit around an opening, first fit and tack a siding strip in place where the last full strip will be located. Level from the top and bottom of the window casing to this piece of siding and mark the piece.

A

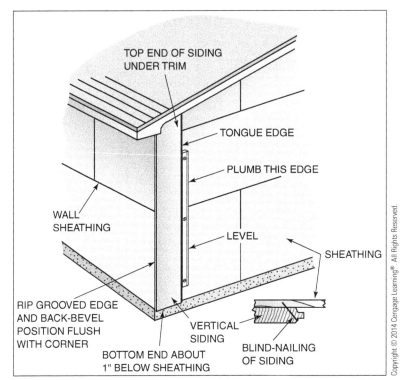

TOP END OF SIDING
UNDER TRIM

TONGUE EDGE

PLUMB THIS EDGE

WALL SHEATHING

LEVEL

SHEATHING

RIP GROOVED EDGE AND BACK-BEVEL POSITION FLUSH WITH CORNER

VERTICAL SIDING

BLIND-NAILING OF SIDING

BOTTOM END ABOUT 1" BELOW SHEATHING

B

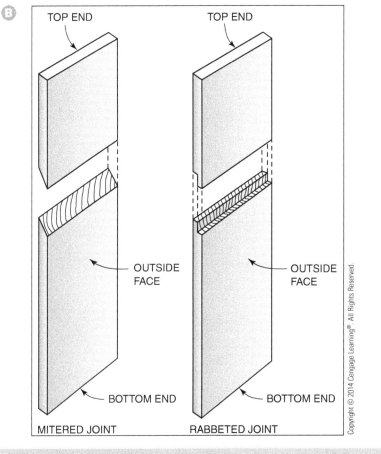

TOP END

TOP END

OUTSIDE FACE

OUTSIDE FACE

BOTTOM END

BOTTOM END

MITERED JOINT

RABBETED JOINT

- Next, use a scrap block of the siding material, about 6 inches long, with the tongue removed. Be careful to remove all of the tongue, but no more. Hold the block so that its grooved edge is against the side casing and the other edge is on top of the tacked piece of siding. Mark the vertical line on the siding by holding a pencil against the outer edge of the block while moving the block along the length of the side casing. Remove and cut the piece, following the layout lines carefully. Set this piece aside for the time being. Cut and fit another full strip of siding in the same place as the previous piece. Fasten both pieces in position.

- Continue the siding by applying the short lengths across the top and bottom of the opening as needed.

D Fit the next full-length siding piece to complete the siding around the opening. First tack a short length of siding scrap above and below the window and against the last pieces of siding installed. Tack the length of siding to be fitted against these blocks in the grooves. Level and mark from the top and bottom of the window to the full piece. Lay out the vertical cut by using the same block with the tongue removed, as used previously. Hold the grooved edge against the side casing. With a pencil against the other edge, ride the block along the side casing while marking the piece to be fitted.

- Remove the piece and the scrap blocks from the wall. Carefully cut the piece to the layout lines. Then fasten in position. Continue applying the rest of the siding.

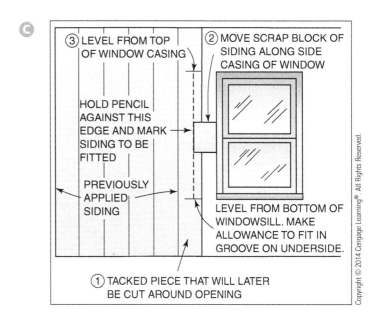

C
③ LEVEL FROM TOP OF WINDOW CASING

② MOVE SCRAP BLOCK OF SIDING ALONG SIDE CASING OF WINDOW

HOLD PENCIL AGAINST THIS EDGE AND MARK SIDING TO BE FITTED

PREVIOUSLY APPLIED SIDING

LEVEL FROM BOTTOM OF WINDOWSILL. MAKE ALLOWANCE TO FIT IN GROOVE ON UNDERSIDE.

① TACKED PIECE THAT WILL LATER BE CUT AROUND OPENING

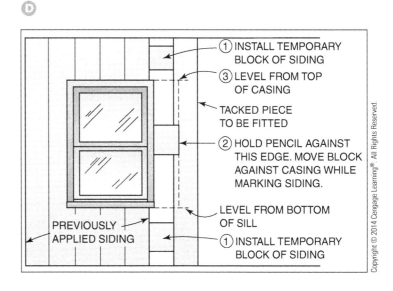

D
① INSTALL TEMPORARY BLOCK OF SIDING

③ LEVEL FROM TOP OF CASING

TACKED PIECE TO BE FITTED

② HOLD PENCIL AGAINST THIS EDGE. MOVE BLOCK AGAINST CASING WHILE MARKING SIDING.

LEVEL FROM BOTTOM OF SILL

PREVIOUSLY APPLIED SIDING

① INSTALL TEMPORARY BLOCK OF SIDING

PROCEDURE 7-29

Installing Panel Siding

Ⓐ Install the first piece with the vertical edge plumb. Rip the sheet to size, putting the cut edge at the corner. The factory edge should fall on the center of a stud. Panels must also be installed with their bottom ends in a straight line. It is important that horizontal butt joints be offset and lapped, rabbeted, or flashed. Vertical joints are either shiplapped or covered with **battens**.

• Apply the remaining sheets in the first course in like manner. Cut around openings in a similar manner as with vertical tongue-and-groove siding. Carefully fit and caulk around doors and windows. Trim the end of the last sheet flush with the corner.

Ⓐ
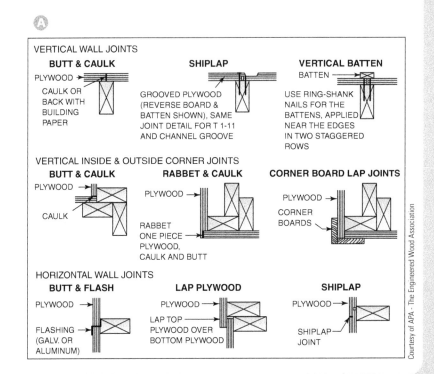

VERTICAL WALL JOINTS

BUTT & CAULK   **SHIPLAP**   **VERTICAL BATTEN**

PLYWOOD → ▢   BATTEN →

CAULK OR BACK WITH BUILDING PAPER

GROOVED PLYWOOD (REVERSE BOARD & BATTEN SHOWN), SAME JOINT DETAIL FOR T 1-11 AND CHANNEL GROOVE

USE RING-SHANK NAILS FOR THE BATTENS, APPLIED NEAR THE EDGES IN TWO STAGGERED ROWS

VERTICAL INSIDE & OUTSIDE CORNER JOINTS

BUTT & CAULK   **RABBET & CAULK**   **CORNER BOARD LAP JOINTS**

PLYWOOD →

CAULK

PLYWOOD →

RABBET ONE PIECE PLYWOOD, CAULK AND BUTT

PLYWOOD →

CORNER BOARDS →

HORIZONTAL WALL JOINTS

BUTT & FLASH   **LAP PLYWOOD**   **SHIPLAP**

PLYWOOD →

FLASHING (GALV. OR ALUMINUM)

PLYWOOD →

LAP TOP PLYWOOD OVER BOTTOM PLYWOOD

PLYWOOD →

SHIPLAP JOINT

Courtesy of APA - The Engineered Wood Association

PROCEDURE 7-30

Installing Wood Shingles and Shakes

Ⓐ Fasten a shingle on both ends of the wall with its butt about 1 inch below the top of the foundation. Stretch a line between them from the bottom ends. Fasten an intermediate shingle to the line to take any sag out of the line. Even a tightly stretched line will sag in the center over a long distance.

• Fill in the remaining shingles to complete the undercourse. Take care to install the butts as close to the line as possible without touching it and remove the line.

Ⓐ
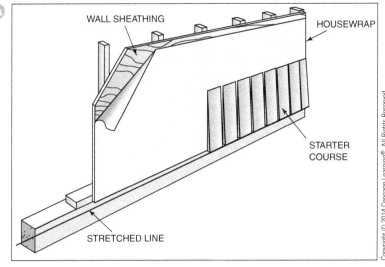

WALL SHEATHING

HOUSEWRAP

STARTER COURSE

STRETCHED LINE

B Apply another course on top of the first course. Offset the joints in the outer layer at least $1\frac{1}{2}$ inches from those in the bottom layer. Shingles should be spaced $\frac{1}{8}$ to $\frac{1}{4}$ inch apart to allow for swelling and prevent buckling. Shingles can be applied close together if factory-primed or if treated soon after application.

C To apply the second course, snap a chalk line across the wall at the shingle butt line. Using only as many finish nails as necessary, tack 1 × 3 straightedges to the wall with their top edges to the line. Lay individual shingles with their butts on the straightedge.

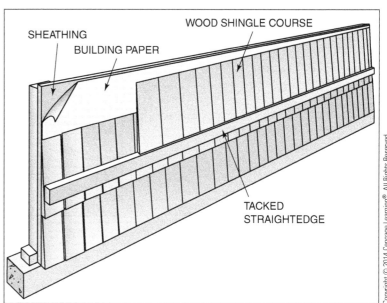

SHEATHING
BUILDING PAPER
WOOD SHINGLE COURSE
TACKED STRAIGHTEDGE

Applying Horizontal Vinyl Siding

Ⓐ Snap a level line to the height of the starter strip all around the bottom of the building. Fasten the strips to the wall with their edges to the chalk line. Leave a $\frac{1}{4}$-inch space between them and other accessories to allow for expansion. Make sure that the starter strip is applied as straight as possible. It controls the straightness of entire installation.

Ⓐ

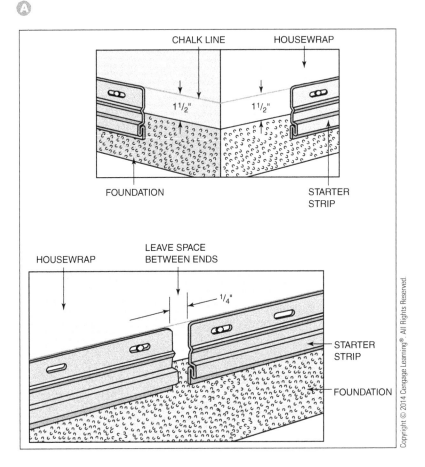

B Cut the corner posts so that they extend $\frac{1}{4}$ inch below the starting strip. Attach the posts by fastening at the top of the upper slot on each side. The posts will hang on these fasteners. The remaining fasteners should be centered in the nailing slots. Make sure that the posts are straight, plumb, and true from top to bottom.

B

TOP END

STARTING NAIL

CORNER POST

CHALK LINE

OUTSIDE CORNER POST

STARTER STRIP

FOUNDATION

$1/_4$" LOWER THAN STARTER STRIPS

CHALK LINE

INSIDE CORNER POST

HOUSEWRAP

FOUNDATION

STARTER STRIP

Applying Horizontal Vinyl Siding (Continued)

Ⓒ Cut each J-channel piece to extend, on both ends, beyond the casings and sills a distance equal to the width of the channel face. Install the side pieces first by cutting a $\frac{3}{4}$-inch notch, at each end, out of the side of the J-channel that touches the casing. Fasten in place.

• On both ends of the top and bottom channels, make $\frac{3}{4}$-inch cuts at the bends leaving the tab attached. Bend down the tabs and miter the faces. Install them so the mitered faces are in front of the faces of the side channels.

Ⓒ

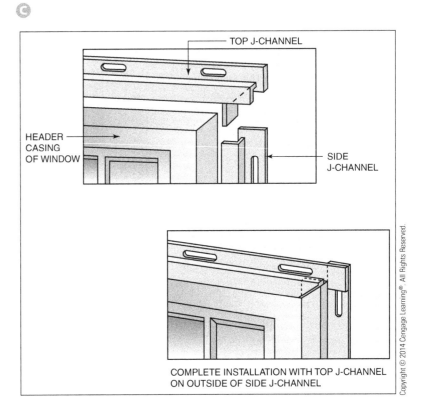

TOP J-CHANNEL

HEADER CASING OF WINDOW

SIDE J-CHANNEL

COMPLETE INSTALLATION WITH TOP J-CHANNEL ON OUTSIDE OF SIDE J-CHANNEL

D Snap the bottom of the first panel into the starter strip. Fasten it to the wall. Start from a back corner, leaving a $\frac{1}{4}$-inch space in the corner post channel. Work toward the front with other panels. Overlap each panel about 1 inch. The exposed ends should face the direction from which they are least viewed.

• Install successive courses by interlocking them with the course below and staggering the joints between courses.

D

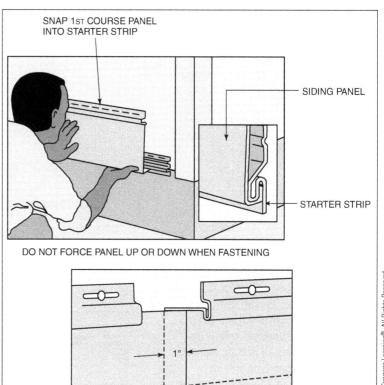

SNAP 1ST COURSE PANEL INTO STARTER STRIP

SIDING PANEL

STARTER STRIP

DO NOT FORCE PANEL UP OR DOWN WHEN FASTENING

1"

LAP PANELS AT LEAST 1"

E To fit around a window, mark the width of the cutout, allowing $\frac{1}{4}$-inch clearance on each side. Mark the height of the cutout, allowing $\frac{1}{4}$-inch clearance below the sill. Using a special *snaplock punch,* punch the panel along the cut edge at 6-inch intervals to produce raised lugs facing outward. Install the panel under the window and up in the undersill trim. The raised lugs cause the panel to fit snugly in the trim.

E

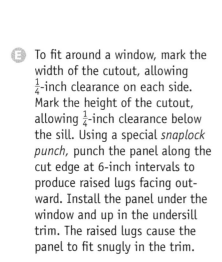

UNDERSILL TRIM

SIDING PANEL

RAISED LUGS   SNAPLOCK PUNCH

Applying Horizontal Vinyl Siding (Continued)

F Panels are cut and fit over the windows in the same manner as under them. However, the lower portion is cut instead of the top. Install the panel by placing it into the J-channel that runs across the top of the window.

F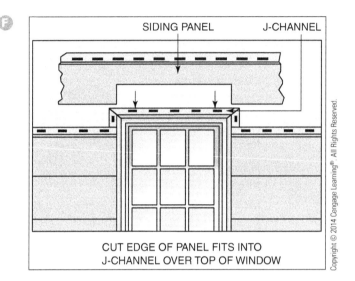

SIDING PANEL J-CHANNEL

CUT EDGE OF PANEL FITS INTO
J-CHANNEL OVER TOP OF WINDOW

G Install the last course of siding panels under the soffit in a manner similar to fitting under a window. An *undersill trim* is applied on the wall and up against the soffit. Panels in the last course are cut to width. Lugs are punched along the cut edges. The panels are then snapped firmly in place into the undersill trim.

G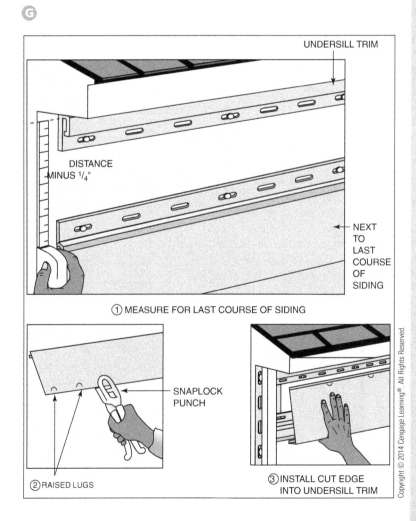

UNDERSILL TRIM

DISTANCE MINUS ¹/₄"

NEXT TO LAST COURSE OF SIDING

① MEASURE FOR LAST COURSE OF SIDING

SNAPLOCK PUNCH

② RAISED LUGS

③ INSTALL CUT EDGE INTO UNDERSILL TRIM

PROCEDURE 7-32

Applying Vertical Vinyl Siding

 Measure and lay out the width of the wall section for the siding pieces. Determine the width of the first and last pieces.

- Cut the edge of the first panel nearest to the corner. Install an undersill trim in the corner board or J-channel with a strip of furring or backing. This will keep the edge in line with the wall surface. Punch lugs along the cut edge of the panel at 6-inch intervals. Snap the panel into the undersill trim. Place the top nail at the top of the nail slot. Fasten the remaining nails in the center of the nail slots.

- Install the remaining full strips making sure that there is $\frac{1}{4}$-inch gap at the top and bottom. Fit around openings in the same manner as with fitting vertical siding. Install the last piece into undersill trim in the same manner as for the first piece.

Example: What is the starting and finishing widths for a wall section that measures 18 feet to 9 inches for siding that is 12 inches wide?

Convert this measurement to a decimal by first dividing the inches portion by 12 and then adding it to the feet to get 18.75 feet.

Divide this by the siding exposure, in feet: $18.75 \div 1$ foot $= 18.75$ pieces.

Subtract the decimal portion along with one full piece giving 1.75 pieces. Next $1.75 \div 2 = 0.875$, multiplied by 12 gives $10\frac{1}{2}$ inches.

This is the size of the starting and finishing pieces. Thus there are 17 full-width pieces and two 10 ½-inch wide pieces.

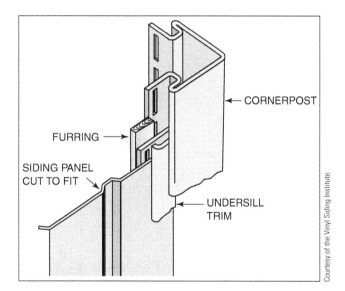

CORNERPOST

FURRING

SIDING PANEL
CUT TO FIT

UNDERSILL
TRIM

Courtesy of the Vinyl Siding Institute

REVIEW QUESTIONS

Select the most appropriate answer.

1 True or false? The most common method of classifying wood is by its source.

2 List three common hardwoods.

3 List three common softwoods.

4 Define the term "engineer panels."

5 List the steps for estimating the amount of drywall needed for a particular application.

6 What is the purpose of a roofing cricket?

7 What is sound? How is it transmitted from one room to another?

9 What is the fire rating of the following drywall:

Standard drywall

$\frac{5}{8}$" type X

$\frac{1}{2}$" type X

$\frac{3}{4}$" type X

8 List the methods for estimating the number of wall studs necessary for a wall.

10 List the types of suspended ceilings commonly used today

Name: _____

Date: _____

Carpentry

IDENTIFY CHARACTERISTICS OF WOOD

Upon completion of this job sheet, you should be able to understand how wood is classified.

1. Go to your local lumber yard or building supply center where there are various types of wood. Make a list of the different types of wood available and classify it as either hardwood or softwood.

Name of the Wood	Classification

2. Are the woods available native to your area? (Y/N) If not, where does the wood come from?

3. At your local lumber yard or building supply center, list the available types of engineered panels they make.

Type of Panel	How They Are Made

INSTRUCTOR'S RESPONSE:

_____

_____

_____

Name: _____

Date: _____

Carpentry

CALCULATING DRYWALL MATERIALS

Upon completion of this job sheet, you should be able to calculate the
amount of drywall and materials needed for particular application.

1. For the room below, calculate the amount of drywall necessary, the ceiling has
an 8-8-foot ceiling height.

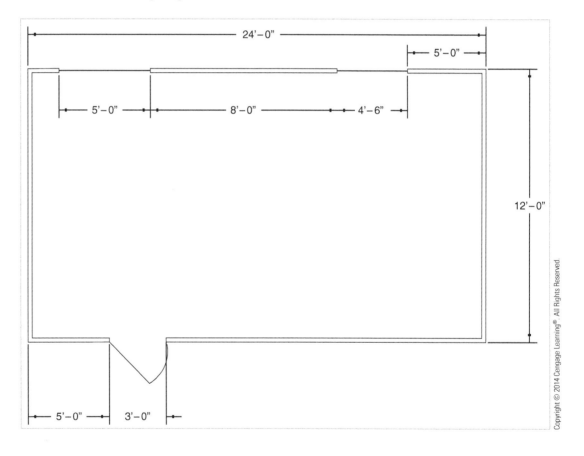

INSTRUCTOR'S RESPONSE:

_____

_____

_____

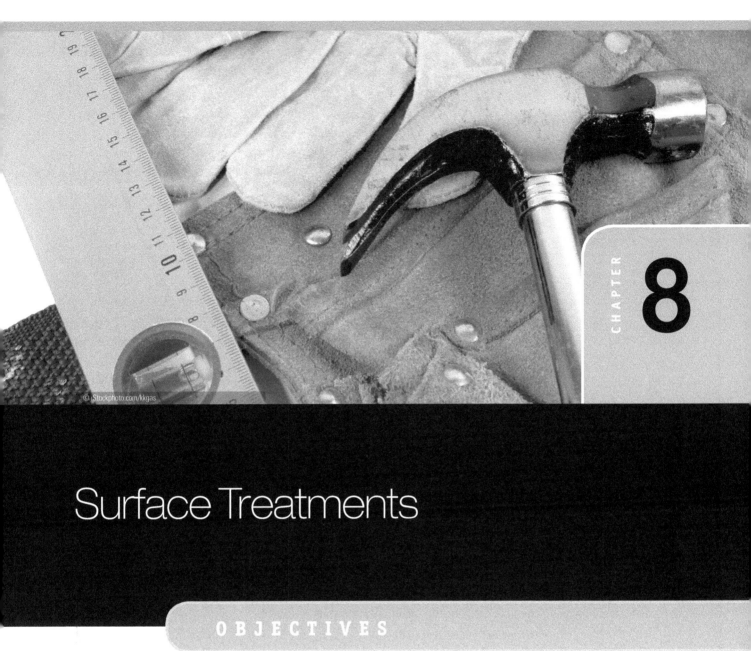

© iStockphoto.com/kkgas

Surface Treatments

OBJECTIVES

By the end of this chapter, you will be able to:

KNOWLEDGE-BASED

- Identify and select proper surface finishes.
- Identify and select proper finishing tools for different types of finishes.
- Understand the necessity of checking the local building code in regard to lead paint before starting a painting project in an area that might contain lead paint.

SKILL-BASED

- Prepare the surface and site properly for finishing, including sanding, caulking, and covering exposed surfaces.

- Apply paint using roller and brush according to manufacturer and job specifications.
- Apply paint, using a paint sprayer, according to manufacturer and job specifications.
- Clean and store paint materials, including brushes, rollers, thinners, and spray guns, according to manufacturer's specifications and OSHA regulations.

Mask a faux finish used to apply a new color over a dry base coat to create an image or shape

Mural a faux finish used to give the illusion of scenery or architectural elements

Kill spot an inconspicuous area used to start and end wallpaper in a room

Booking is the process of keeping the wallpaper paste from drying out once the water has activated the wallpaper paste

Introduction

Interior surfaces are used primarily to define spaces and hide structural, electrical, and plumbing elements. If these surfaces are constructed using drywall/gypsum board, then they will need a surface treatment to protect the construction. There are an unlimited number of ways to treat or finish drywall/gypsum board surfaces. These finishes can include paint, wallpaper, texturing, faux finish, or a combination of these applications. The function of the space, along with cost, durability, acoustics, and building codes, is a key factor in the type of surface finish chosen. Whether treating the surface of a new drywall or replacing an existing treatment, achieving a professional result requires proper surface preparation, tool and material selection, and application.

Painting Safety

Before attempting any painting and finishing project, always follow these general safety rules:

- Before performing any painting-related work in areas that are heavily populated, be sure that all the occupants have been notified.
- Never use painting products near an open flame and/or heat source.
- Never smoke while handling painting supplies and/or painting.
- Always use an approved respirator mask when painting or using thinners.
- For indoor projects, always provide adequate ventilation before starting the painting project.
- To prevent skin irritation, always use vinyl gloves when handling paints and/or thinners.
- Always use painter's goggles to prevent splatters from entering the eyes.
- If you should get paint in your eye(s):
    1. Flush your eye(s) for 10 minutes using cool water.
    2. Seek medical attention.
- Always wear a hat to prevent paint from getting into your hair.
- Always read and follow the manufacturer's instructions and specifications for handling all epoxy materials, thinners, catalysts, paint removers, and so on.
- When using a ladder for a painting project, always make sure that the ladder's safety devices are properly locked into place before using it.

- Check the ladder's placement and ensure that it is properly leveled before attempting to use it.
- Use ladders only for their intended purpose, and always follow the manufacturer's recommendations.

Fixing Drywall Problems

Address all drywall problems before attempting to add a treatment to a wall. Drywall can have nail pops, which are visible dimples in the drywall caused by underlying nails. Drywall can also have minor defects caused by small nails used for hanging pictures or thumbtacks, or small dents caused by furniture. In some cases, you may see large holes in the drywall that must be patched properly to ensure a smooth finish after the new wall treatment is complete (Figure 8-1). The steps for fixing a number of drywall problems are given next.

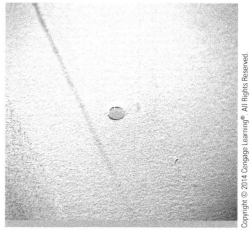

FIGURE 8-1: Wall with nail holes.

Steps for Fixing Nail Pops

1. If you can remove the nail without damage, do so. Refasten the drywall with drywall screws. The screw should be driven until it is recessed. The screws should not be permitted to break or puncture the paper covering on the drywall.

2. If the nail cannot be removed, drive it back in place. Fill the dent and screw holes with joint compound (Figure 8-2). Allow the compound to dry, then sand it smooth.

Steps for Repairing Holes or Scrapes

1. Remove any loose drywall plaster, as well as any torn paper, using a utility knife (Figure 8-3).

2. Using sandpaper, sand the edges of the hole. This will roughen the edges of the hole, thus providing a better surface for the joint compound to bond to. Before starting to repair hole with joint compound, wipe dust away from the hole (Figure 8-4).

FIGURE 8-2: Repair flaws in the wall.

FIGURE 8-3: Utility knife.

FIGURE 8-4: Rough up hole with sandpaper.

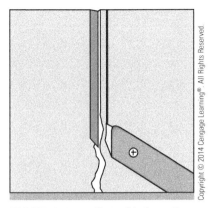

FIGURE 8-5: Cut a V along the length of the crack.

3. Using a putty knife, fill the area with joint compound. Allow the joint compound to dry, then sand it smooth.

Steps for Repairing Cracks

1. Use a utility knife to cut a V along the length of the crack. Starting on one side of the crack, cut at an angle from just outside the crack to its center. Repeat this process on the opposite side of the crack to complete the V (Figure 8-5).

2. Using a putty knife, fill the area with joint compound. Allow the joint compound to dry, then sand it smooth.

Steps for Patching a Large Drywall Hole

1. Using a carpenter's rule or a straight edge to mark out a rectangle around the hole. (Figure 8-6).

2. Use a keyhole saw to cut out the rectangle or score along the lines using a utility knife. Use several more passes to finalize the rectangular cutout with the utility knife (Figure 8-7).

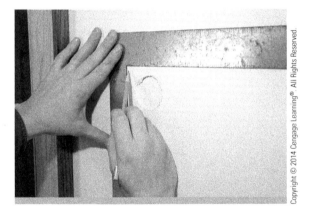

FIGURE 8-6: Draw a rectangle around the hole.

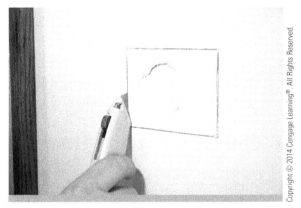

FIGURE 8-7: Cut a rectangular hole in the wall.

3. Cut a drywall patch 2 inches larger than the hole. Next, remove the 2-inch perimeter, but leave the facing paper (Figure 8-8).

4. Apply joint compound around the outside perimeter of the hole and along its inside edges (Figure 8-9).

FIGURE 8-8: Cut a drywall patch.

FIGURE 8-9: Joint compound on patch.

5. Place the patch into position, and hold it in place for several minutes while it begins to adhere. With a drywall knife, spread more joint compound as needed. Allow the compound to dry completely (Figure 8-10).

6. Sand, prime, and paint (Figure 8-11).

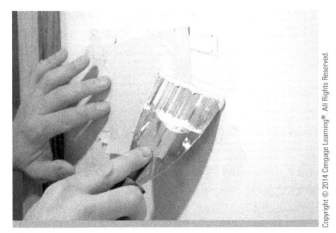

FIGURE 8-10: Position patch on the hole be repaired.

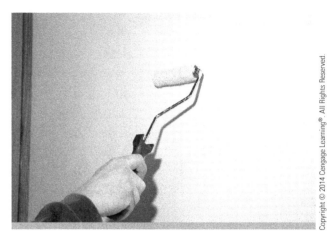

FIGURE 8-11: Paint patch.

REMODELING

Painting and Remodeling

Although painting is not often thought of as a form of remodeling, it can add value to an existing structure. Before starting any painting project, it is important to determine whether the structure contains lead-based paint. Prior to its use being banned in 1977, lead was used in the production of paint, and it is considered a hazardous material. If it is determined that lead paint is present and will need to be removed, then proper removal and disposal techniques must be performed by licensed technicians. If it is determined that lead paint is present and is going to be painted over, before beginning any painting be sure to check your local building codes on painting over existing lead-based-paint covered surfaces.

Always check with your local and state agencies before starting a remodeling project.

Before beginning the demolition or deconstruction process, the proper permits must be obtained.

Painting as a Surface Treatment

A common way of finishing and protecting drywall/gypsum is by painting the surface. The proper sequence for painting a room is from the ceiling to the floor (also known as from the top down). Begin with the ceiling, followed by the walls, followed by the windows and/or doors and finally finishing with the baseboards.

Surface Preparation

Before starting the actual painting, prepare the surfaces properly to get a professional-looking paint job. To start the paint process, select the proper paint for the type and location of the surface and use of the area.

Types of Paints

There are several types of paint available on the market today. Each of these types of paints is specifically designed with a particular purpose and surface as well as a specific location in a facility.

- **Latex paints**—These are long-lasting, water-soluble paint that produces less fumes and odors than oil-based products. In addition, latex paints can be easily cleaned using soap and water. It is for these reasons that latex has become the preferred paint for most interior surfaces.

- **Oil paints**—These are durable and resist scraping and wear and tear. Most contain alkyd, a soya-based resin that dries harder than latex. Paint thinner must be employed when cleaning up painting equipment used with oil paints, as well as spills.

- **No- or low-VOC (volatile organic compound) paints**—These include acrylic and oil-based paints. VOCs are volatile organic compounds that are outgassed during application and for years after the paint has dried. Paints and primers with no VOCs and low VOCs have fewer toxic chemicals that are released during outgassing.

Types of Finishes

The more resin a paint contains, the greater the sheen (or shine) it has after drying. The resin adds durability and makes maintenance/cleaning much easier. However, the shinier the surface, the more pronounced the imperfections of the wall surface and/or the application process. Also, prices increase along with the level of sheen.

- **Gloss paints**—These are typically oil-based and include resin to give them a hard-wearing quality. The higher the gloss level, the higher the shine and easier they are to maintain.

- **Flat paints**—These paints are perfect for low-traffic areas because of their lack of sheen and relatively low cost. Low-traffic areas typically include: the formal dining area and master bedroom. These paints provide a beautiful matte coating that hides minor surface imperfections. However, this finish stains easily and is difficult to clean. Flat paints are typically used for ceiling paint.

- **Eggshell paints**—These paints offer a smooth finish and a subtle sheen that is slightly more glossy than flat paint. They are washable; durable; and ideal for bedrooms, hallways, home offices, and family rooms.

- **Satin paints**—These are a step above eggshell in their ability to be cleaned, providing a nice balance between being washable and having a subtle gloss. They have excellent wear characteristics and aesthetics, making them an excellent choice for any room.

- **Semigloss paints**—These guarantee maximum permanence. They are commonly employed in children's rooms as well as high-moisture areas and for trim.

- **High-gloss paints**—Because of their ability to reflect light, these paints work well for highlighting details, such as trim and decorative molding. They are also the best choice for doors, cabinets, or any area that sees a high level of abuse.

- **Ceiling flats**—These are designed specifically for ceilings and are usually extra-spatter-resistant.

Identify and Select Proper Tools

Selecting the proper tools will save you a lot of time and additional effort. Always choose top-quality tools.

Types of Finishing Tools

- **Pressure washers**—These are used to remove dirt and grime in a short time and with little effort.
- **Paint scrapers**—These are used to remove old paint from a surface to be painted. Paint scrappers are most effective for removing loose or peeling paint. Currently, there are a number of different styles of paint scrappers available. Examples include the classic putty knife (Figure 8-12), the 11-in-1 multipurpose tool, and the double-edge wood scraper.
- **Power sanders**—These help with the removal of old peeling paint.

FIGURE 8-12: Putty knife.

Types of Brushes

Selecting the proper brush for a particular painting task will make the job much easier by providing the technician with the best results possible. Selecting the correct paint brush for a particular task is done according to the type of paint or finish, the surface to be painted, and personal preference. Two basic categories of brushes are water-based and oil-based paint brushes. Choose the appropriate type of brush for the type of paint, finish, and surface.

Water-Based Paint Brushes

There are three main types of water-based paint brushes (see Figure 8-13).

1. **Polyester**—These bristle brushes hold and release more paint, which provides a smoother finish. This type of brush is typically more thorough and faster to clean up than other synthetic brushes.

2. **Nylon**—These brushes typically have stiffer bristles and wear longer than those with any other filament. They are well suited for rough surfaces.

3. **Poly-nylon**—These brushes are blends that provide longer wear, maximum resiliency, and easy clean up.

Oil-Based Paint Brushes

There are two basic types of oil-based paint brushes (see Figure 8-14).

1. **White China bristles**—These are the best choice for varnishes, polyurethane, and stains. They are finer than black China bristles and usually produce a finer finish.

2. **Black China bristles**—These are best used with oil-based paint, primer, and enamels.

FIGURE 8-13: Polyester, nylon, and poly-nylon brushes.

FIGURE 8-14: White china and black china bristle brushes.

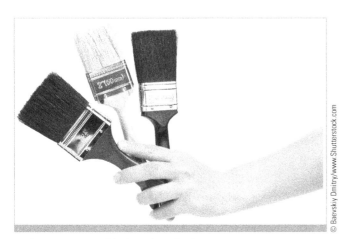

FIGURE 8-15: Various brush sizes.

Brush Sizes and Shapes

Paint brushes are available in various shapes and sizes. Sizes range from 1 inch to 4 inches in width (see Figure 8-15).

- **Angular brushes**—Also referred to as "sash brushes," these are great for angular or narrow surfaces. They are also good to use when the painting surface is hard to reach. These are perfect for general-purpose brushes.

- **Flat and oval brushes**—These are used on all surfaces but are best used on flat surfaces such as wide trim, doors, cutting-in walls, and ceilings.

Types of Rollers

When selecting a paint roller, the most important factor to consider is the surface you are going to paint. A general rule of thumb for using a roller cover is: the smoother the surface being painted, the shorter the nap; the rougher the surface, the longer the nap. A high-quality roller cover should have a phenolic core, which will not soften in water but will withstand every type of paint solvent.

Roller Covers

Rollers are available in both natural (mohair or lamb's wool) and synthetic materials (nylon, polyester, or a combination of the two). Natural materials are best with oil-based paints, whereas synthetic materials are best with water-based paints. For latex paint, use only synthetic rollers. Natural materials are too absorbent (see Figure 8-16).

- **Smooth surface**—Select a short nap ranging from $\frac{1}{8}$ inch to $\frac{1}{4}$ inch. When a roller with a longer nap is used, it has a tendency to leave a pronounced "orange peel" effect on the surface being painted. Use this cover on smooth plaster, sheet rock, wallboard, smooth wood, Masonite, and Celotex.

- **Slightly rough surface**—Select a medium nap, ranging from $\frac{3}{8}$-inch to $\frac{1}{2}$ inch. Longer fibers have a tendency to push the paint into rough surfaces without producing the orange peel effect. This type of roller cover is best suited for sand finish plaster, textured plaster, acoustical tile, poured concrete, rough wood, and shakes.

- **Rough surface**—Select a long-nap ($\frac{3}{4}$-inch to $1\frac{1}{4}$-inch) cover. Longer fibers push the paint into the deep valleys of rough surfaces This type of roller cover is best suited for concrete block, stucco, brick, Spanish plaster, cinder block, corrugated metal, and asphalt or wood shingles.

FIGURE 8-16: Paint roller and pan.

Roller Frames

Roller cage frames are available in various styles (see Figure 8-17). U-shaped frames are generally sturdier. When selecting a roller frame, it is important to choose one that is sealed on the ends to help keep the paint on the roller, where it belongs.

Prepare the Surface for Painting

Proper surface preparation is critical to professional, long-lasting results.

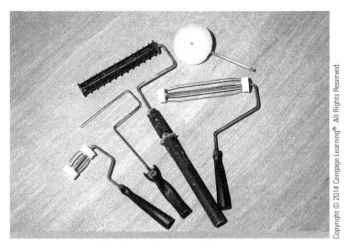

FIGURE 8-17: Various roller cage frame types.

Steps for Preparing the Surface for Painting

1. Remove as much furniture as possible from the room. Heavier items and items that cannot be removed should be grouped in the center of the room and covered with drop cloths (Figure 8-18).

2. Mask the baseboards with 2-inch masking tape and old newspaper or with 12-inch baseboard masking. Use a drop cloth to protect the floor (a canvas drop cloth is less slippery and creates a safer work area).

3. Repair drywall. Drywall issues are typically easy to locate. Nails from time to time will pop out from the drywall; corners become dented, scraped, scared or otherwise damaged. Tape can split. Dents, gouges, and holes appear.

4. Disconnect the power, at the service panel, to the affected switches/outlets before removing the switch plates and receptacle plates. Never try to tape or paint over switch plates or receptacle plates. Always remove them. Be sure to tape the screws to the plate so that they are not lost (Figure 8-19).

5. Examine the area around the windowpanes foe sign of loose or missing putty. If any damaged putty is located, replace it with new putty and allow it too dry before painting.

Painting the Ceiling

If the walls are going to be painted, there is no need to be concerned with protecting them. However, if you are painting only the ceiling, you should protect the walls by taping plastic drop cloths to the top of the walls.

FIGURE 8-18: Furniture covered with drop cloth.

FIGURE 8-19: Switch plate has been removed.

When moving furniture, be extremely careful to protect yourself and the furniture. Consider the following:

- If you know you must move large furniture, get a helper.

- Use a back support to protect your back.

- Walk large, heavy pieces by moving them back and forth on their legs to the left and right in small increments.

- Turn a piece of scrap carpet upside down and position it under the object; this will make it easier to slide across the floor.

- Use a dolly to move heavy pieces.

Steps for Painting the Ceiling

1. Start by painting a 2-inch-wide strip around the edges of the ceiling.

2. Once the edge of the ceiling has been painted, using a roller brush and an extension pole, start in the corner and paint a 3-square-foot area using a zigzag pattern. Doing so will disperse the paint evenly on the roller.

3. Completely fill in this 3-foot section without reloading the roller.

4. Continue to covering the ceiling by painting 3-foot-square sections while moving across the ceiling's shortest dimension. Make sure to overlap your strokes while the paint is wet to minimize lap marks (Figure 8-20).

FIGURE 8-20: Paint the ceiling in small sections.

Painting the Walls

Once the ceiling has been completed, move to the walls. Be sure that the walls are fully prepared to be painted.

Steps for Painting with a Brush and Roller

1. Prime the walls. Proper priming is key to a long-lasting job (see Table 8-1).

2. Using a brush start by painting a 2-inch strip at the ceiling.

3. Using a brush to paint 2-inch strips in corners, around windows, doors, cabinets, and baseboards.

4. Switch to a roller. Using a zigzag pattern paint wall in a vertical direction. On the first stroke push the roller upward then forming an "M" pattern to evenly distribute the paint on the roller. Work in 3-square-foot area sections. Fill in the "M" pattern without reloading the roller until you have completely covered the area. Continue in 3-foot sections until the wall is completely covered. Touch up spots that you might have missed while the paint is still wet to help reduce sheen differences in the paint (Figure 8-21).

FIGURE 8-21: Painting a wall with a roller.

Surface and Paint to Be Used	Recommended Preparation
Painting oil or acrylic on new drywall	Use acrylic primer.
Painting oil on new plaster	After appropriate curing time (approximately 4 weeks), use an oil-based primer.
Painting acrylic on new plaster	Use a diluted first coat of the final paint.
Drywall previously painted with latex or oil	Scrape any loose or flaking paint and sand smooth. Lightly sand entire wall to remove gloss. Use acrylic primer or prep coat.
Drywall previously painted with an intense pigment (very bright or very dark color)	Scrape any loose or flaking paint and sand smooth. Lightly sand entire wall to remove gloss. Use stain-killing primer.
Drywall previously painted with builder's flat	Scrape any loose or flaking paint and sand smooth. Use acrylic primer.
Plaster wall previously painted with latex or oil	Scrape any loose or flaking paint. Lightly sand entire wall to remove gloss. Use acrylic primer or primer/sealer.
Any wall with stains	Use stain-killing primer.
Any wall with mold	Spray affected area thoroughly with distilled white vinegar and allow to dry, then use stain-killing primer. Note: This method will kill 82% of the mold. Bleach will not kill mold.

TABLE 8-1: Wall Preparation for Painting.

Types of Paint Primers

- **Primers/sealers**—Also known as drywall repair clears (DRCs). These oil- or acrylic-based coatings are designed to penetrate, seal, and protect the wall while priming the surface, to provide a firm bond between the substrate and the previous paint or new paint layer.

- **Stain-killing primers**—These oil- or acrylic-based primers do not actually remove stains from a surface. They use resin to provide an impenetrable seal between the surface and the final coat of paint. Without a stain-killing primer, many stains and intense colors will continue to bleed through all subsequent coats of paint.

Steps for Painting with a Paint Sprayer

1. Using a bin or a bucket, pour the paint through a strainer. Be sure to avoid lumps or odd bits of nonpaint material (Figure 8-22).

2. Thin the paint. Do not thin paint more than what is recommended by the manufacturer or it will not cover properly.

3. Be sure that you are covered from head to toe. Wear a long-sleeved shirt, gloves, and dust mask or respirator (Figure 8-23).

4. Start from the top and paint downward in smooth, steady strokes. Keeping your strokes steady and smooth, start at a corner and work from the top down. A lot of paint is going on the surface in a short period of time. It is better to paint using several light coats than a single coat (Figure 8-24).

Acrylic- and oil-based primers are available in no- and low-VOC versions. Paints and primers with no or low VOC have fewer toxic chemicals that are released during outgassing. This creates a safer environment during the painting process and for those who will live or work within the building after the installation is complete.

FIGURE 8-22: Paint strainer.

FIGURE 8-23: Protected painter.

Faux Finishes

Once the base coat has been applied, the painting process can be taken to the next level by creating a faux finish. Faux (pronounced "foe") is a French word meaning "false." This technique is also called Trompe l'Oeil (pronounced "trome play"). Again, this is a French phrase meaning "fools the eye." Both terms refer to the technique of using paint to create the illusion of other, more expensive, material.

Tools for Creating Decorative Faux Finishes

An endless variety of tools can be used to create decorative finishes, from specialty tools purchased at home improvement stores to improvised tools that can be found around the home. Some of the most common tools are shown in Figure 8-25(A) through (H). Their uses will be discussed in the following section.

General Categories of Faux Finishes

Five general categories of faux finishes are antiquing or distressing, combing, ragging or sponging, masking, and murals. A faux finish can consist of a single category or a combination of two or more categories.

Antiquing or Distressing

Antiquing or distressing gives the appearance of age or wear. This finish is created by adding color washes

FIGURE 8-24: Paint Sprayer.

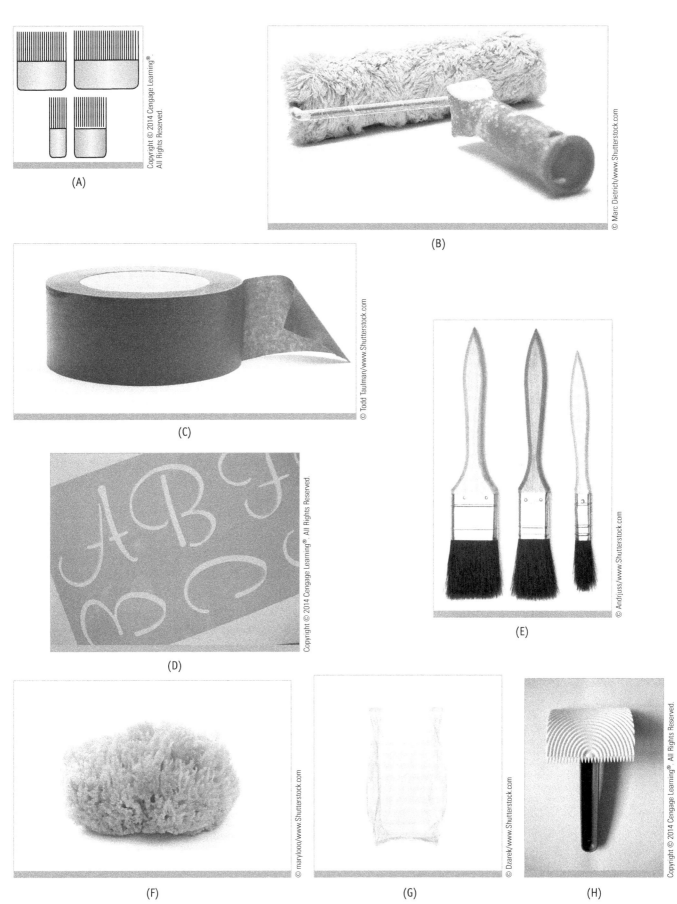

FIGURE 8-25: Tools for decorative faux finishes include (A) comb, (B) rag roller, (C) painter's tape, (D) stencil, (E) stipple brush, (F) sponge, (G) bag, and (H) wood grainer.

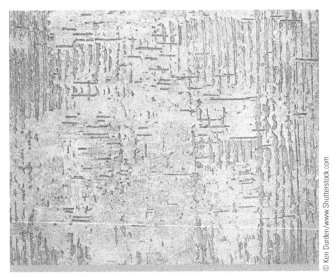

FIGURE 8-26: Distressed faux finish.

FIGURE 8-27: Combed faux finish.

FIGURE 8-28: Ragged faux finish.

or glazes to a dry base coat to create shade variations that mimic aged surfaces. Crackling can also be used to give the appearance of age. It intentionally creates cracks in the top coat to allow an undercoat to show through. It is done by applying a special clear coat that dries very quickly, causing the paint layer just below it to crack (see Figure 8-26).

Combing

Combing can mimic the look of fabric. It is created by adding a top coat of paint to a dry base coat, then dragging a brush or comb through the wet top coat. The comb can be dragged in a hatch or cross-hatch pattern to give the illusion of woven threads in fabric. The two coats of paint usually have a similar hue but different shades (for example, a medium blue base coat with a light blue combed coat on top). This technique can also be used to mimic wood grain. Instead of a comb or brush, a special wood-graining tool is simultaneously rocked and dragged through the wet paint of the top coat (see Figure 8-27).

Ragging or Sponging

Ragging or sponging creates the illusion of marble or stone. The texture is created by adding or subtracting color, using a sponge, crumpled rag, plastic bag, or stipple brush. The additive method adds one or more colors to a dry base coat using one of these tools. In the subtractive method, new layers of paint are rolled onto the surface of the dry base coat, then subtracted using one of the tools listed (see Figure 8-28).

Masking

Masking mimics wallpaper or border. A **mask** is placed on a dry base coat and a new color is applied over the mask to create an image or shape. Masking requires either

FIGURE 8-29: Stenciled faux finish.

painter's tape or a stencil. Painter's tape can be used to create stripes by masking off a dry base coat at regular intervals, adding a second color, then removing the painter's tape. Squares or diamonds can also be created by varying the placement of the painter's tape.

Murals

Murals give the illusion of scenery or architectural elements. This type of finish is done by an artist and depicts an image or scene painted on a wall or ceiling. A mural can be very simple, such as white clouds painted on a blue base coat to create a skyscape, or it can be very complex, such as a detailed depiction of a landscape or fantasy scene (for example, cherubs). Finally, a mural can give the illusion of architectural elements, such as columns, intricate molding, and windows. Murals can be created using regular paint or fresco. Fresco uses wet pigments as paint on uncured plaster (see Figure 8-30).

To keep the top coat from bleeding under the painter's tape, a clear coat can be applied over the tape before adding the top coat. A stencil is a Mylar sheet with an image cut out at the center. When the Mylar is placed against a surface, it creates a mask around the cutout image. Paint is stippled, or gently "pounced," over the Mylar with a sponge or stipple brush to reproduce the image on the surface. Stencils are available in a wide range of shapes, patterns, and images. They can be used to generate a continuous border or an overall pattern (see Figure 8-29).

FIGURE 8-30: Mural.

Cleaning and Storing Equipment and Supplies

It is important to clean all paint equipment thoroughly, store equipment and supplies properly, and discard empty containers appropriately.

FIGURE 8-31: Rinse paint roller under tap.

FIGURE 8-32: Removing paint from brush.

FIGURE 8-33: Clean brush with water.

Steps for Cleaning Paint Rollers

1. Disassemble paint roller.
2. If you are using oil-based paint, submerge cover in solvent.
3. Wash and rinse cover. Using a mild detergent, rinse the roller in clean clear water (see Figure 8-31).
4. If you are using oil-based paint, remove paint from frame and hardware, using mineral spirits or a natural alternative such as Citrus Solvent; otherwise, use water.
5. Hang roller to dry.

Steps for Cleaning Paint Brushes

1. Remove all excess paint by either wiping it onto old newsprint of using the edge of a paint scrapper (Figure 8-32).
2. Use water to clean water soluble paints. Water soluble paints can merely be rinsed out under a running tap. However, to save water and energy, first work use a bucket of water to remove much of the paint build-up. This makes cleaning them under the tap faster and easier. Before finishing, make sure that the water running through the brush is clear (Figure 8-33).
3. Use mineral spirits or a natural alternative such as Citrus Solvent to clean oil-based paints. Brushes used in oil-based paints should first be cleaned in a bucket of white spirit. Once this has been accomplished, all excess paint is removed by shaking it out of the brush as well as working the brush in brush cleaner. Brush cleaner is water-soluble solution that allows the brush to be rinsed under a running tap. Always wash the brushes using soap and water. Rinse the brush under running water.
4. Remove all excess water from the brush by shaking it or pressing it between your palms.
5. To retain the shape of the brush, it should be wrapped in paper for storage.

Steps for Cleaning a Spray Gun

1. Remove any unused paint from the container.

2. Use water to clean water-based paints. Fill container with water and spray until it emerges clear.

3. Use mineral spirits or a natural alternative such as Citrus Solvent to clean oil-based paints. Fill container with white spirit instead of water. When it is clear of paint, clean the container with a mix of hot water and detergent.

4. Dismantle the machine and clean all parts with a damp cloth.

Proper Paint and Trash Disposal

Water-thinned (latex) and oil-based paints should be used completely. A small amount of these paints should be kept for touch-ups. Dry, empty paint containers may be recycled in a recycling program. In most states, latex paint can be dried out in the can and disposed of in your domestic garbage as long as the lid is off revealing that the contents of the can are dried.

Finishing Up

Let the paint dry thoroughly, then remove the masking tape carefully. Use a snap-off razor if necessary to separate the tape from the paint. Pull tape slowly back over itself at a steep angle. No job is complete until the entire area is cleaned up completely and checked for quality. Review all painted areas. Screw all plates back on the wall and check for straightness. Check floors, walls, nonpainted areas, and room entrances for paint spatters or drips and clean as needed. If a client is involved, ask the client to conduct a final assessment of the area with you. Walk through the newly painted areas and point out your work. Note any dissatisfaction and remedy the problem immediately and eagerly. Consider providing the client with a small bottle or can of the original paints, clearly labeled, so that they can do minor touch-ups if needed.

Brushes that permitted paint to dry on them can be cleaned by soaking the brush in brush restorer. Afterward, you can use a wire brush to remove any dried paint still on the brush. Because the brush can be easily damaged, do not use the wire brush too harshly. Finally, if you are using a wire brush, always brush in the direction of the bristles, away from the handle.

Wall Coverings as a Surface Treatment

Another technique for protecting and decorating drywall/gypsum board is by covering the surface with wallpaper. After fixing any drywall problems as discussed at the beginning of the chapter, additional surface preparation is required for a wallpaper application.

Surface Preparation for Wall Coverings

Proper preparation will help the wallpaper adhere to the wall correctly. If the wallpaper does not bond solidly to the wall, seams will be exposed during the drying process and, over time, the paper may begin to peel away from the surface. If, however, the wallpaper bonds too tightly to the surface, it will be extremely difficult to remove and may even damage the surface of the wall.

Primers are used to create a more uniform substrate for improved bonding with the wallpaper. Primers and sealers also increase the slide of the wallpaper. This allows the wallpaper to slide over the surface of the wall for easier placement during the installation process. However, not all primers permit the wallpaper to slide easily during installation. Always consult manufacturer guidelines for the chosen wallpaper and adhesive. If no guidelines are recommended, use the following information to prepare the wall(s) based on the type of surface being prepared.

Steps for Wall Preparation

1. Remove as much furniture as possible from the room. Heavier items and items that cannot be removed should be grouped in the center of the room and covered with drop cloths.

2. Protect the floor with a drop cloth (a canvas drop cloth is less slippery and creates a safer work area).

3. Refer to the "Fixing Drywall Problems" section earlier in this chapter. Make all necessary repairs.

4. For existing installations, use a sponge to wash all installation surfaces with a trisodium phosphate (TSP) solution and rinse with water. Allow walls to dry for 24 hours. This will help remove excess oil, grease, and dirt from the surface. Do not wash new, unpainted drywall or plaster surfaces.

5. Disconnect the power at the service panel to the affected switches/outlets before removing the switch and receptacle plates. Be sure to tape the screws to the plate so that they are not lost.

6. Paint any surfaces such as ceilings or trim before priming or wallpapering the walls (see Table 8-2).

7. Use the manufacturer's guidelines or the following chart to determine and apply the proper primer, sealer, or sizing.

Types of Wallpaper Primers

- **Primer-sealers**—Also referred to as drywall repair clears, or DRCs. These oil- or acrylic-based coatings are designed to penetrate and seal the wall while priming the surface to provide a firm bond between the substrate and the wallpaper. This type of primer protects the drywall and allows the wallpaper to slide more easily during the installation process.

- **Primers-sizes**—These acrylic coats provide a tacky surface for increased bonding with wallpaper.

- **Stain-killing primers**—These oil- or acrylic-based primers do not actually remove stains from a surface. They use resin to provide an impenetrable seal between the surface and the wallpaper. Without a stain-killing primer, many stains and intense colors will bleed or show through wallpaper.

- **Size or sizing**—Some wallpaper manufacturers plainly specify the use of sizing for the installation of their products. Size is commonly sold as a powder to be mixed with water according to the manufacturer's specifications. Sizing is used on plaster walls to prevent excessive amounts of wallpaper paste from being absorbed into the plaster during installation. Sizing is occasionally used on drywall for increased bonding between the wall and the covering.

Surface	Recommended Preparation
New drywall	Use primer/sealer.
New plaster	After appropriate curing time (approximately 4 weeks), use size or a diluted solution of wallpaper adhesive.
Drywall painted with latex or oil	Scrape any loose or flaking paint. Lightly sand entire wall to remove gloss. Use acrylic primer or prep coat.
Drywall painted with an intense pigment (very bright or very dark color)	Scrape any loose or flaking paint. Lightly sand entire wall to remove gloss. Use stain-killing primer.
Drywall painted with builder's flat	Use primer/sealer.
Plaster wall painted with latex or oil	Scrape any loose or flaking paint. Lightly sand entire wall to remove gloss. Use acrylic primer or primer/sealer.
Any wall with stains	Use stain-killer primer.
Any wall with mold	Spray affected area thoroughly with distilled white vinegar and allow to dry; then use stain-killer primer. Note: This method will kill 82% of the mold. Bleach will not kill mold.
Wallpaper	Every effort should be made to remove old wallpaper. Puncture or score the surface with a puncture roller or heavy grit sand paper to allow moisture to penetrate. Use a steamer or a damp sponge to loosen the adhesive; then scrape old wallpaper away using a broad knife. If removal is not an option, check all corners and edges to confirm good contact with the wall. Reglue or remove loose wallpaper, spackle and sand where necessary. Use an oil-based primer.
True paper wall covering	Check for proper bond on all old corner and edges; then use size.

TABLE 8-2: Wall Preparation for Wallpapering.

Basic Categories of Wall Coverings

The basic categories of wallpaper are listed here. Generally, wallpaper contains either a paper or a cloth substrate laminated to a paper, vinyl, or fabric decorative face.

True paper has a paper substrate, and the decorative layer is printed directly on this substrate. Most papers are uncoated, but some seal in the decorative inks with a slight top coating. This type of wallpaper is very delicate and difficult to install, and it is more suited to areas that will see minimal abuse.

Vinyl-coated paper has a underling layer made of paper. The decorative surface is coated with vinyl or polyvinyl chloride (PVC). This type of wallpaper is durable and is easy to clean and remove. It is resistant to moisture and grease, making it a good option for bathrooms, kitchens, and basements.

Coated fabric has a fabric substrate that has been covered with liquid vinyl or acrylic. The decorative face is printed directly on this vinyl or acrylic layer. Because this type of wallpaper allows more moisture to "breathe" through the surface, it is best for use in low-moisture rooms that will not compromise the integrity of the adhesive, such as living areas.

Paper-backed vinyl/solid sheet vinyl has a paper underling layer laminated to the decorative surface that consists of a solid sheet of vinyl. This type of wallpaper is extremely durable and is easy to clean and remove. Solid sheet vinyl can be used in most areas of the facility because of its ability to resist grease, stains, mold and mildew. This makes it a good option for bathrooms, kitchens, and basements. However, it is not a good candidate for areas that receive a lot of physical wear (for example, mudrooms or storage rooms).

Fabric-backed vinyl has a woven or nonwoven cloth ground (mesh-like textile backing) underlining layer that is laminated to a solid vinyl decorative face. This type of wall covering is extremely durable and is used almost exclusively for commercial and institutional applications.

Specialty may use either a paper or a fabric substrate. The decorative surface may include foils, synthetic fiber flocking, natural elements such as cork or sisal, or other exotic elements. This type of wallpaper is usually expensive and difficult to install. It is usually long-lived, but not easily cleaned and does not withstand abuse. It is used mainly in highly decorative, low-traffic installations.

Anaglypta and *supaglypta* have a paper substrate. Anaglypta has a pulp surface and supaglypta has a cotton fiber surface. The surface is deeply embossed and left uncoated so that it can be painted after installation. Because this is extremely thick wallpaper, it is an excellent choice for covering up wall imperfections. The look can be quickly changed and adapted with a fresh coat of paint. The embossing is easily damaged with sharp objects or furniture. Depending on the surface paint used, this wallpaper can be used in any room that does not see a high level of abuse.

Environmentally friendly paper has a paper substrate often from recycled material. The surface, or face of the wallpaper, is generally an organic or recycled material (rice paper, grass, wood, or newspaper). It improves interior air quality because it has no or low VOC and no vinyl. It often comes pre-pasted with toxin-free adhesives. Because this type of covering is breathable, air bubbles and mildew are minimized. Check with the manufacturer to determine if the paper you choose will work well in high-humidity locations such as bathrooms and kitchens. Continual improvements have made ecofriendly wallpaper competitive with traditional wallpaper choices, but this option remains somewhat more expensive and may not be suitable for high-abuse areas such as mudrooms and storage rooms.

> *Remove masking tape before beginning the wallpapering process.*

Determining the Type and Amount of Wall Coverings

It can be tricky to estimate the amount of wallpaper needed for a project. Extra time spent measuring and planning will greatly reduce the risk of coming up short at the end.

1. Calculate the square footage of the space. Multiply the height by the width of each wall to find the overall square footage. Multiply the height by the width of each door, window, fireplace, and so forth, then subtract this total from the overall square footage to find the final coverage area. Add 10 percent to the total for waste. Round your total up to the nearest foot (see Figure 8-34).

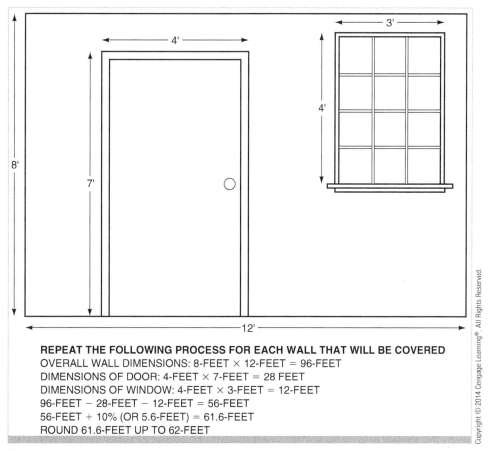

REPEAT THE FOLLOWING PROCESS FOR EACH WALL THAT WILL BE COVERED
OVERALL WALL DIMENSIONS: 8-FEET × 12-FEET = 96-FEET
DIMENSIONS OF DOOR: 4-FEET × 7-FEET = 28 FEET
DIMENSIONS OF WINDOW: 4-FEET × 3-FEET = 12-FEET
96-FEET − 28-FEET − 12-FEET = 56-FEET
56-FEET + 10% (OR 5.6-FEET) = 61.6-FEET
ROUND 61.6-FEET UP TO 62-FEET

FIGURE 8-34: Determining the amount of wall covering.

2. Determine the starting place where you will begin (and end) hanging the wallpaper. Since it is virtually impossible to wallpaper an entire room and get the final strip to fit perfectly, plan to start (and end) the wallpaper in an inconspicuous area, for example, in the corner behind the entrance door, above a door, or window that is not directly across from the entry. This area is known as the **kill spot**.

3. Determine the number of strips of wallpaper to cover the space. This can be accomplished by using the width of the wallpaper to mark off the number of strips it will take to cover the entire area. For example, if the wallpaper is 27-inch wide, begin at the kill spot and measure out horizontally 27 inches. Place a small mark on the wall, and measure another 27 inches from this mark. Number each strip marked on the wall consecutively, for example, 1, 2, 3. . . . Continue around the room until you come back to the kill spot (see Figure 8-35).

4. Determine the repeat or drop in the pattern. This information is usually determined by the manufacturer and included on the label of the wallpaper itself. It refers to the linear space on the roll between identical pattern elements. The amount of vertical space between the pattern repeat will be important when aligning the strips of wallpaper

FIGURE 8-35: Measuring width of strips on a wall.

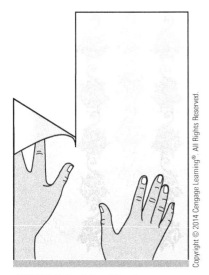

FIGURE 8-36: Aligning the patter on two strips of wallpaper.

horizontally. Multiply the amount of pattern drop by the number of strips used to cover the space (see Figure 8-36). For example, if you determine in Step 3 that it will take 22 strips to complete the room, and the wallpaper has a 6-inch drop; multiply 22 × 6. Divide this total by 12 to convert the results to feet (22 inches × 6 inches = 132 inches) (132/12 = 11 feet). This will give you the total amount of wallpaper that will be lost when aligning each strip to match the pattern. Round your total up to the nearest foot (see Figure 8-36).

5. Consult the dealer or refer to the wallpaper label to find the total square footage contained on each roll. Most American double rolls of wallpaper cover about 70 square feet. Combine the totals from the final coverage area (Step 1) with the pattern drop loss (Step 4) and divide the square footage contained in one roll of wallpaper. For example, if the final coverage area including 10 percent waste is 220 square feet, and the pattern drop loss is 11 feet (220 feet + 11 feet = 231 feet) (231 feet ÷ 70-foot roll coverage = 3.3 rolls). Round up to the next whole roll. So the room in the example would require four rolls.

6. Check the label or with the distributor to find out if the wallpaper chosen is prepasted. If it is not, use the wallpaper manufacturer's guidelines to determine the type and amount of paste you will need.

Table 8.3 provides a checklist of tools that are necessary when wallpapering (see also Figure 8-37).

> *If the trial layout created in Step 2 leaves awkward areas such as long narrow strips (less than 3 inches wide) next to a door or corner, adjust your starting point or the width of your starting strip to eliminate the awkward areas. Be sure to mark the new strip areas on the wall. When adjusting the width of the starting strip, always trim from the kill spot side of the strip.*

Vinyl smoother (recommended) or wallpaper smoothing brush
❏ Metal square or yard stick
❏ Vinyl-to-vinyl wallpaper paste
❏ Ladder
❏ Thirty-inch water tray
❏ Bucket
❏ Seam roller
❏ Sponge
❏ Eight-inch broad knife
❏ Carpenter pencil
❏ Scissors
❏ Painter's masking tape
❏ Snap-off razor and extra pack of blades
❏ Level or plumb bob
❏ Paint roller with $\frac{3}{8}$-inch nap cover (if the paper is not prepasted)
❏ Metal square or yard stick
❏ Basswood table or a sheet of plywood and sawhorses to create a cutting surface

TABLE 8-3: Tools Used for Wallpaper Installation.

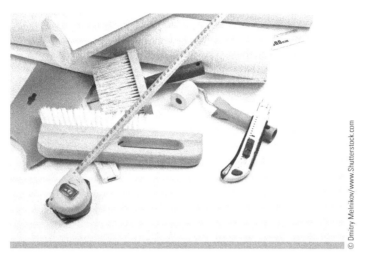

FIGURE 8-37: Wallpaper tools.

Applying Wallpaper

The installation of wallpaper is not a difficult task to complete as long as the facility maintenance technician follows a few basic steps, which are listed next. However, it should be noted that all manufacturers' instructions and recommendations should be followed.

1. Read and follow all instructions included with the wallpaper. Failure to follow these guidelines could void any warranty and cause the wallpaper to fail.

2. Using the markings created during the trial layout, start in the most visible area (usually directly across from the kill spot). With a pencil, extend the vertical mark of first strip from floor to ceiling using a level or plumb bob. A proper vertical line will make hanging the wallpaper much easier and greatly enhance the final appearance of the job (see Figure 8-38).

3. Measure the wall height followed by adding 4 inches to the measurement for trim allowance. Unroll the wallpaper face down on the cutting surface. Mark your measurements using a pencil.

4. Use a metal square or straight edge to make a straight cut across the paper. If the wallpaper is a straight match, you can continue cutting out the remainder of the strips. Keep the trial layout in mind for areas that do not require a full strip (below a window or above a fireplace; see Figure 8-39).

5. If the wallpaper has a drop match pattern:
   a. Turn the strip face up and unroll the next strip face up to the left of the first.
   b. Extend the top of the second strip until the pattern aligns across the two strips.
   c. Use the bottom of the first strip as the measuring line to cut the second strip.
   d. Label the back of this strip "left," and save it for later use. This strip will be longer than the first to allow for the drop match.
   e. Unroll the next strip face up to the right of the first.
   f. Extend the top of the second strip until the pattern aligns across the two strips.
   g. Use the bottom of the first strip as the measuring line to cut the second strip.
   h. Label the back of this strip "right." This strip will be longer than the first to allow for the drop match.

FIGURE 8-38: Creating an initial plumb line.

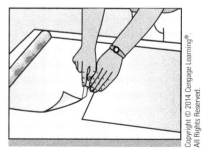

FIGURE 8-39: Using a straight edge to cat the wallpaper.

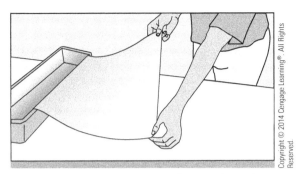

FIGURE 8-40: Removing wallpaper from the tray.

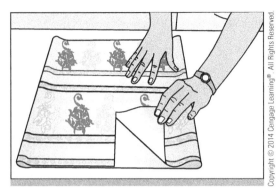

FIGURE 8-41: Booking wallpaper.

    i. Continue using Steps e through g to cut out the remaining strips to the kill spot. Use Steps A through C to finish the other half of the area around to the kill spot. Before installing each strip, be sure to use it as a guide for cutting the next strip.

6. If the paper is not prepasted, use a paint roller to apply the paste to the back of the wallpaper according the wallpaper and paste instructions. Roll the paste across the narrow expanse of the paper, then along the length of the strip to the top and bottom edges to evenly distribute the paste. Follow points D and E in Step 7 to allow the paper to relax before installation.

7. If the wallpaper is prepasted, consult the wallpaper's guidelines for booking. **Booking** is the process used to activate the paste. Manufacturer's guidelines vary for soaking and relaxing times, but the general process is as follows:

    a. Roll the measured and cut strip face inward.

    b. Place the rolled strip in the water tray (filled with warm water unless otherwise specified) and gently swish it side to side and allow it to soak for the prescribed amount of time.

    c. Grasp the end of the strip and pull slowly out of the tray, unwinding the strip face down (see Figure 8-40).

    d. Once it is completely out of the water, fold half of the wallpaper back over itself to the approximate center of the strip so that the pasted sides are together. Fold the other half back over itself to meet and overlap the opposite edge by a few inches. *Do not* crease the paper at the "folds;" it should create a small loop (see Figure 8-41).

    e. Starting with the folded ends, roll the booked wallpaper loosely together to form a scroll and allow it to relax for the time indicated in the guidelines.

8. Unroll the wallpaper. Make sure the pattern is facing up. Separate the edges of the pasted side only on the top half. While allowing a few inches to extend past the ceiling/trim line, align the wallpaper with the vertical line marked on the wall and firmly press the exposed portion of the paste to the wall. Although it is tempting, do not try to align the edge of the wallpaper with the ceiling/trim. Allow for overhang to be trimmed later for a better fit and more professional appearance (see Figure 8-42).

9. Once the wallpaper is in position, use a vinyl smoother or wallpaper brush to gently work out wrinkles and air bubbles. Be careful not to squeeze excessive amounts of paste out during this process. It could leave the wallpaper with too little adhesive for a proper bond. If a wrinkle is not coming out, pull the far edge (the edge farthest away from the plumb line or previous strip) of the wallpaper away from the wall and smooth it back down with the vinyl smoother (see Figure 8-43).

FIGURE 8-42: Align the first strip with the plumb line.

FIGURE 8-43: Using the wallpaper brush to smooth out air bubbles.

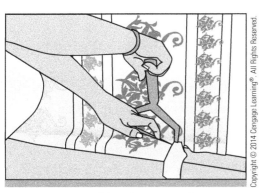

FIGURE 8-44: Trimming seam allowances.

FIGURE 8-45: Aligning the pattern on two strips of wallpaper.

FIGURE 8-46: Rolling the edges with a seam roller.

10. Place the 8-inch broad blade firmly along the trim line at the ceiling/ trim (see Figure 8-44). Use a snap-off razor to trim the excess by cutting along the edge of the broad blade. The broad blade should always be placed between the installed wallpaper and the blade. This will hold the paper in place and protect the wallpaper from damage if the knife slips. Snap off old blades after a few cuts. A dull blade will tear the wallpaper.

11. Once the top half is smoothed and trimmed, unbook the bottom half, smooth, and trim.

12. Book the next strip and install next to the first being careful to align the pattern and butting the edges snugly together (minute shrinkage will occur during drying, so any gaps at this point will increase once the project is complete) (see Figure 8-45).

13. Use the seam roller on the abutting seams being careful not to force too much adhesive out of the joint, which could cause poor adhesion. Note: Do not use a seam roller on anaglypta or supaglypta wallpaper (see Figure 8-46).

14. Use a bucket of warm water and a sponge to clean paste off of face of the wallpaper before it dries. Replace the warm water in the water tray when it becomes cold and/or slimy.

15. Work in one direction around to the kill spot. Then, starting from the first strip installed, work in that direction to the kill spot. At the kill spot, place the broad knife vertically in the center of the overlapping area. Use the broad knife as a guide to make a double-cut through the top and bottom strips simultaneously. Remove the waste from the top cut, then peel back the newly installed paper to reveal the waste on the bottom. Remove the waste and smooth the wallpaper back into place. Because of the double-cut, the two edges should line up perfectly (see Figure 8-47).

Wallpaper rarely matches perfectly throughout the entire length of the seam due to minor paper stretching and wall imperfections. Use eye level to align the pattern since mismatches at the ceiling or floor will be less noticeable.

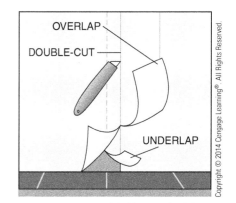

OVERLAP

DOUBLE-CUT

UNDERLAP

FIGURE 8-47: Making a double-cut through two strips of wallpaper.

Wallpapering Corners and around Obstacles

Corners (inside or outside) are never straight and plumb. They can be crooked from floor to ceiling and bulge or sink in the middle. Also, walls can shift slightly over

FIGURE 8-48: Using a pencil and broad knife as a guide for cutting.

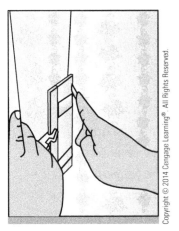

FIGURE 8-49: Finish cutting the paper at the top and bottom of overlap areas.

time. To keep the wallpaper from twisting or showing seam gaps at corners and to ensure that the paper continues to be plumb and straight after turning a corner, use the following steps.

Wallpapering an Inside Corner

1. Wrap and smooth the paper folding it around the corner.
2. Using painter's masking tape, tape a carpenter pencil along the length of the broad knife's edge. This will create a guide approximately $\frac{1}{4}$-inch wide (see Figure 8-48).
3. Place the broad knife vertically in the corner with the pencil against the wall that has just been papered. Use a snap-off razor to cut the wallpaper as you slide the broad knife down the length of the corner. Use scissors to finish cutting the paper at the top and bottom overlap areas (see Figure 8-49).
4. Peel the remainder of the strip from the wall. If the strip is less than 4 inches, discard the strip and begin with a new strip. If the strip is more than 4-inch wide, rebook it to keep the adhesive wet during the next step.
5. Measure the width of the next piece (either the piece just cut or a full strip of wallpaper). Use this measurement to create a plumb line on the wall after the corner.
6. Hang the cut piece or new strip on the wall using the plumb line as the placement guide. Smooth the wallpaper over the $\frac{1}{4}$-inch overlap into the corner.
7. Peel the top piece back slightly in the corner. Apply vinyl-to-vinyl wallpaper paste to the overlap area, and smooth the paper back into place.
8. Trim the excess from the top and bottom as usual. You may need to peel the corner out slightly and use scissors to make the final cut in the corner; then smooth the corner back down.
9. This will create a $\frac{1}{4}$-inch overlap at each corner to help mask wall movement and to ensure that the paper will come out of the corner plumb. This will create a slight pattern mismatch in the corner; this is normal and will be less obvious once the job is complete (see Figure 8-50).

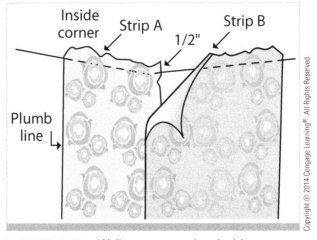

FIGURE 8-50: Wallpaper around an inside corner.

Wallpapering an Outside Corner

1. Hang and smooth paper around an outside corner as usual. Cut a vertical slit in the top and bottom trim allowances at the corner to allow the wallpaper to wrap around the corner. Carefully remove all bubbles that tend to form at the peak of the corner.

2. Check the edge of the paper coming out of the corner for plumb. If it is plumb, continue hanging wallpaper from there (see Figure 8-51).

3. If the edge of the wallpaper coming out of the corner is not plumb, use the following steps:

   a. Measure the distance from the last strip to the corner and add 1 inch. Cut the width of the strip that will go around the corner to this length. Install and smooth this strip. Cut a slit in the top and bottom trim allowances to allow the wallpaper to wrap around the corner smoothly.

   b. Measure the width of the cut strip (or a new strip if the remainder is less than 4 inches). Add $\frac{1}{4}$ inch to this measurement, and create a new plumb line on the wall after the corner.

   c. Install the cut strip (or new strip) starting with the plumb line and working backward toward the corner. The edge should fall $\frac{1}{4}$ inch short of the corner. If the edge were installed too close to the corner, it might fray with the abuse that outside corners usually endure (see Figure 8-52).

   d. Peel back the top layer at the corner and use vinyl-to-vinyl wallpaper adhesive where the paper overlaps.

   e. Trim the allowances from the top and bottom of the wall as usual.

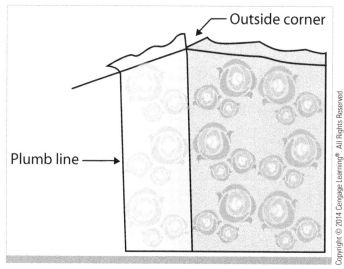

FIGURE 8-51: Wallpapering around an outside corner.

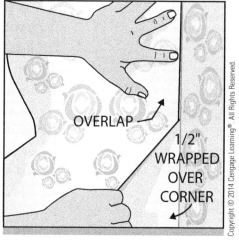

FIGURE 8-52: Wallpapering around an outside corner that is plumb.

Wallpapering around Obstacles

Obstacles might include anything from windows, doors, and built-in shelves to fireplaces. This section will specifically discuss windows, but the same general principles will apply to most obstacles you will encounter.

1. Hang the wallpaper in the normal way, aligning the edge with the previous strip. Use the vinyl smoother to smooth the paper to the vertical edge of the window.

2. In the area that overlaps the window, use scissors to cut a diagonal slit from the approximate center to the upper corner of the window trim face. Use a snap-off razor to carefully continue this diagonal slit from the face to the trim to the wall (along the depth of the trim; see Figures 8-53 and 8-54).

3. Smooth the top of wallpaper up to the ceiling and down to the top of the window.

FIGURE 8-53: Using scissors to cut diagonally to the edge of Window trim.

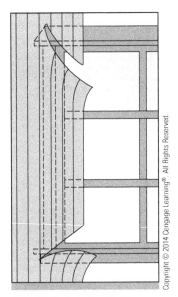

FIGURE 8-54: Cut lines for wallpapering over window.

FIGURE 8-55: X-cut and trim lines for and outlet.

4. Again, in the area that overlaps the window, use scissors to cut a diagonal slit from the approximate center to the lower corner of the window trim face. Use a snap-off razor to carefully continue this diagonal slit from the face to the trim to the wall (along the depth of the trim).

5. This area may be more complicated because of the many angles created by the window sill. Take your time and use minidiagonal slits for each corner you encounter.

6. Smooth the lower portion of wallpaper up to the bottom of the window and down baseboard.

7. Trim around the window by placing the broad knife against the trim as a guide. Trim into the wall, not the window facing.

8. Use a bucket of warm water and a sponge to clean paste off windows and trim before it dries.

Wallpapering around Outlets and Switches

Although only outlets are referred to in this section, use the same technique for both outlets and switches.

1. Hang the wallpaper in the normal was, aligning the edge with the previous strip. Use the vinyl smoother to smooth the paper to the vertical edge of the outlet.

2. Use a snap-off razor to make an X-cut over the outlet. The size of the X should be just enough to allow the paper continue smoothly over the outlet. Do not extend the edges of the X to the corners of the outlet box yet.

3. Continue smoothing the entire strip onto the wall. Trim the allowances from the top and bottom of the wall.

4. Use scissors to extend each line of the X into the corners of the outlet box. Trim the excess either by running a snap-off razor along the inside edge of the outlet box or by allowing a $\frac{1}{2}$-inch overhang and wrapping the edges into the outlet box (see Figure 8-55).

5. Use a bucket of warm water and a sponge to clean paste off the face of outlet before it dries.

Finishing Up

Drain water tray and bucket, rinse adhesive from tools with warm water, and dry. Review the area for problems or edges that might need more paste. Screw all plates back on the wall and check for straightness. Check floors, walls, and room entrances for paste spatters or drips and clean as needed. Remove drop cloths and replace furniture. If a client is involved, ask the client to conduct a final assessment of the area with you after the wallpaper has completely dried.

Texture as a Surface Treatment

The final technique we will discuss for protecting and decorating drywall/gypsum board is to create a surface texture. After fixing major drywall problems as discussed at the beginning of the chapter, walls should be primed before being textured. Primer is used to create a more uniform substrate for improved bonding with the texture. Use the following information to prepare the wall(s) based on the type of surface being prepared.

Steps for Wall Preparation

1. Remove as much furniture as possible from the room. Heavier items and items that cannot be removed should be grouped in the center of the room and covered with drop cloths.

2. Protect the floor with a drop cloth (a canvas drop cloth is less slippery and creates a safer work area).

3. Refer to the "Fixing Drywall Problems" section earlier in this chapter. Repair major defects such as jutting nails; minor defects will be concealed by the texture.

4. For existing installations, use a sponge to wash all installation surfaces with a trisodium phosphate (TSP) solution and rinse with water. Allow walls to dry for 24 hours. This will help remove excess oil, grease, and dirt from the surface. Do not wash new, unpainted drywall or plaster surfaces.

5. Disconnect the power at the service panel to the affected switches/outlets before removing the switch and receptacle plates. Be sure to tape the screws to the plate so that they are not lost.

6. Paint any surfaces such as ceilings or trim before priming the walls (see Table 8-4).

Surface and Location	Recommended Preparation
Bathrooms and basements	Use oil-based primer in high-humidity areas.
Kitchens	Use oil-based primer in areas that may accumulate excessive grease.
New drywall	Use acrylic primer.
New plaster	Use oil-based primer.
Drywall painted with latex or oil	Scrape any loose or flaking paint. Lightly sand entire wall to remove gloss. Use acrylic primer.
Plaster wall painted with latex or oil	Scrape any loose or flaking paint. Lightly sand entire wall to remove gloss. Use acrylic primer.
Any wall with stains or an intense pigment	Scrape any loose or flaking paint. Lightly sand entire wall to remove gloss. Use stain-killer primer.
Any wall with mold	Spray affected area thoroughly with distilled white vinegar and allow to dry, then use stain-killer primer. Note: This method will kill 82% of the mold. Bleach will not kill mold.
Wallpaper	Every effort should be made to remove old wallpaper. Puncture or score the surface with a puncture roller or heavy grit sand paper to allow moisture to penetrate. Use a steamer or a damp sponge to loosen the adhesive, then scrape old wallpaper away using a broad knife. The moisture in drywall texture will loosen wallpaper adhesive. Applying texture over wallpaper is not recommended.

TABLE 8-4: Wall Preparation for Texturing.

Texture Mediums

- **All-purpose drywall joint compound**—Also called drywall mud, this comes premixed or in powder form. It is generally considered the easiest texture for beginners.
- **Gypsum plaster**—Gypsum, lime, and sand. It is more expensive than joint compound, but it creates a tougher surface that is more resistant to abuse.
- **Stucco**—Cement, lime, and sand
- **Texture paint**—About as thick as pancake batter or wet plaster
- **Environmentally friendly plaster**—This is made from natural clays and recycled aggregates. Installation is similar to other textures, but it may require several applications to build up the surface slowly to prevent cracks.
- **Sand**—A 30- or 70-mesh (the larger the mesh, the larger the grain) white quartz sand can be added to any texturing medium for a more rustic, old-world appearance.

Types of Textures

Application techniques and the resulting variety of textures created are limited only by the imagination. However, texture is most often applied by using a blower, trowel, or paint roller. The resulting textures generally fall within the following categories:

- **Popcorn texture**—This can be sprayed or can be rolled on with a roller. The roller method creates peaks or stalactites to give this texture its signature cottage cheese appearance. It was commonly used for ceilings during the 1960s and 1970s (see Figure 8-56).
- **Orange peel texture**—This must be sprayed on the wall as a thin mixture. The resulting spatters resemble the peel of an orange (see Figure 8-57).

FIGURE 8-56: Popcorn texture on drywall.

FIGURE 8-57: Orange peel texture on drywall.

- **Knockdown texture**—This begins with the application of the orange peel texture. The peaks that are formed during the application are then "knocked down" using a trowel (see Figure 8-58).

- **Skip trowel**—The material is rolled on with a paint roller. The surface is then troweled as with the knockdown texture. Depending on the thickness of the texture and the amount of pressure used during troweling, this surface can be very subtle or very rough (see Figures 8-59 and 8-60).

- **Spanish texture**—Also called a "knife texture," it is applied randomly with a trowel. The trowel marks are left in the texture to give it an old-world appearance (see Figure 8-61).

FIGURE 8-58: Knockdown texture on drywall.

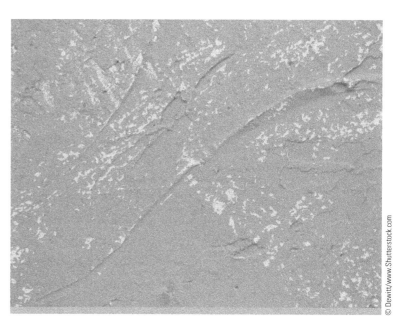

FIGURE 8-59: Subtle trowel texture.

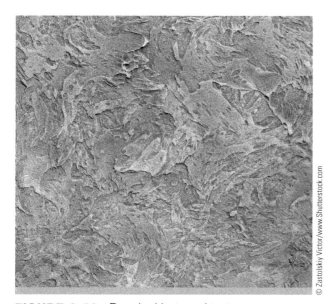

FIGURE 8-60: Rough skip trowel texture.

FIGURE 8-61: Spanish texture on drywall.

- **Sand texture**—This can be applied with a sprayer, roller, or brush. Sand is added to texture paint and applied to the wall. This is an easy texture to apply, but distributing the sand evenly across the surface can be tricky (see Figure 8-62).
- **Stamp texture**—This is applied using a paint roller; then it is stamped with an object to give the desired results. The object used for stamping can be anything that transfers a pattern into the texture medium, ranging from a foam image stamp to a stipple brush to a human hand (see Figure 8-63).
- **Drag texture**—This is applied using a paint roller, then an object such as a comb or wallpaper brush is dragged across the surface to create the desired effect (see Figure 8-64).
- **Stencil texture**—The base texture is applied using a paint roller or trowel and allowed to dry. A mud stencil is then placed on the wall, and texture is applied over the stencil. When the stencil is removed, the texture leaves a relief of the stencil design on the wall (see Figure 8-65).

Table 8-5 provides a list of tools that are necessary when texturing drywall.

Spraying or Blowing Texture on Drywall

1. Spraying or blowing texture can be extremely messy. Carefully mask all surfaces that will not be textured by using painter's masking tape and plastic drop cloths. Using a plastic drop cloth cover the floor. Tape the edges of the drop clothes together to create a secure seal.

2. Cover clothing and shoes; always use goggles and a mask.

3. Turn off the electricity at the circuit panel, then remove all outlet and switch covers. Protect the switches and outlets by tucking newspaper around them.

4. If you are using all-purpose drywall joint compound (premixed or dry), add water according to manufacturer's instructions and mix thoroughly to achieve the proper consistency for spraying.

5. Attach the spray gun to the compressor. Adjust the spray nozzle and the air pressure to achieve the desired results. Generally, a $\frac{3}{8}$-inch opening will work well at 60 psi. As you decrease the air pressure, you will need to increase the nozzle opening.

FIGURE 8-62: Sand texture on drywall.

FIGURE 8-63: Stamp texture on drywall.

FIGURE 8-64: Drag texture on drywall.

FIGURE 8-65: Texture relief using stencil and plaster.

❏ Texturing medium
❏ Drop cloths
❏ Painter's masking tape
❏ Ladder
❏ Electric drill with ribbon attachment
❏ Texture blower or spray gun and compressor
❏ Paint roller, cover, and tray
❏ Trowel

TABLE 8-5: Tools for Texturing Drywall.

6. If you are using a knockdown technique:
    a. Wait 10–20 minutes before trying to trowel the surface, or you will smear the texture.
    b. Drag a trowel lightly across the surface at a low angle to flatten the peaks created by the sprayer. Scrape excess texture back into the container each time you raise the trowel from the surface.

7. Allow the texture to dry for at least 24 hours.

8. Paint the surface with a semigloss paint (the higher gloss reduces the amount of paint that is absorbed by the texture). Take special care to coat all surfaces in deep texture applications.

9. Remember, the key to a professional texture is consistency.

Rolling or Troweling Texture on Drywall

1. Mask all surfaces that will not be textured by using masking tape and plastic drop cloths. Cover the floor with plastic drop cloths. If you are texturing only one wall, mask off 12 inches of the adjacent walls using masking tape and newspaper.

2. Turn off electricity at the circuit panel, then remove all outlet and switch covers. Protect the switches and outlets by tucking newspaper around them.

3. When texturing an entire room, it is difficult to texture around inside corners without damaging the texture on the previous wall. Texture facing walls and allow them to dry. Mask off 12 inches of these textured walls at the corners, using masking tape and newspaper, before texturing the adjoining walls. This will protect the texture on each wall at the corner.

4. Mix powdered all-purpose drywall compound, using manufacturer's guidelines. If you are using ready-mixed all-purpose drywall compound:
   a. Remove half of the mud and transfer it to an airtight container.
   b. Thin the remaining mud with one cup of water, and mix it thoroughly with the electric drill, using the ribbon attachment. Add more water in small amounts until the mud has the consistency of pancake batter or a milkshake (somewhat thicker for a knockdown texture).

5. If you are using a trowel, start at one edge of the wall and begin applying texture.
   a. Use one edge of the trowel to scoop up about a cup of mud.
   b. Set the mud edge of the trowel against the wall. The clean edge of the trowel should be raised away from the wall several inches.
   c. Sweep the trowel across the wall in the desired pattern. Play with the pressure and motion of the trowel in a small area until get the look you want. If you make a mistake while the mud is still wet, simply scrape the mud off the wall with the trowel and try again.
   d. Work in small areas from one side of the wall to the other. Always keep a wet edge to the mud.
   e. Use less texture at corners and trim.

6. If you are using a roller, start at one edge of the wall and begin applying texture.
   a. Fill the paint tray with mud and place the lid on the remainder.
   b. Roll texture onto the roller brush and roll the texture onto the wall. Two factors will determine the look of the texture: the nap or type of roller brush used and the speed of the application. The slower the brush is rolled across the surface, the higher the peaks in the texture.
   c. If you are using a knockdown technique, drag a trowel lightly across the surface at a low angle to flatten the peaks created by the roller. Scrape excess texture back into the container each time you raise the trowel from the surface. You may need to apply texture to a small area then trowel before moving on, or have one person apply texture while a second person trowels.

    d. Work in a small area, and play with application techniques to achieve the results you want.

    e. Always keep a wet edge on the mud.

    f. Use less texture at corners and trim.

7. Allow the texture to dry for at least 24 hours.

8. Paint the surface with a semigloss paint (the higher gloss reduces the amount of paint that is absorbed by the texture). Take special care to coat all surfaces in deep texture applications.

9. Remember, the key to a professional texture is consistency.

Finishing Up

Clean the paint sprayer according to manufacturer's instructions; rinse tools with warm water and dry them. Review the area for inconsistencies. Allow the texture to dry completely. Screw all plates back on the wall and check for straightness. Check floors, walls, and room entrances for spatters or drips and clean as needed with warm water. Remove drop cloths and replace the furniture. If a client is involved, ask the client to conduct a final assessment of the area.

Other Specialty Finishes

This chapter has discussed three common types of finishing techniques for drywall. However, there are many more ways to treat and protect a surface. These specialty finishes are typically more labor intensive and costly. It is worth noting, however, that almost anything relatively flat, from tile, to leather to precious metals, can be applied to a surface for protection and decoration.

SUMMARY

- Never use painting products near an open flame and/or heat source or smoke while handling painting supplies and/or painting.

- Use ladders only for their intended purpose, and always follow the manufacturer's recommendations.

- Address all drywall problems before attempting to add a treatment to a wall.

- The correct order for painting a room is always start at the top and paint to the bottom.

- Before starting any painting project, it is important to determine whether the structure contains lead-based paint.

- Before starting the actual painting, prepare the surfaces properly to get a professional-looking paint job.

- Latex paint is a water soluble, durable paint that emits a lower odor than that of oil based paints.

- Paint thinner must be employed when cleaning up painting equipment used with oil paints as well as spills.

- VOCs are the volatile organic compounds that are outgassed during paint application and for years after the paint has dried.
- The greater the gloss level, the higher the shine and easier the paint is to maintain.
- Eggshell paints provide a smooth finish with a subtle sheen that is slightly glossier than flat paint.
- Satin paints are a step above eggshell in their ease of cleaning and provide a balance between being washable and having a subtle gloss.
- Semigloss paints ensure maximum durability.
- Ceiling flats are designed specifically for ceilings.
- Generally, wallpaper contains either a paper or a cloth underlining layer that is laminated to a paper, vinyl, or fabric decorative face.
- Popcorn texture can be sprayed or can be rolled on with a roller.
- Orange peel texture must be sprayed on the wall as a thin mixture.
- Knockdown texture begins with the application of the orange peel texture.
- Skip trowel texture is rolled on using a paint roller.
- Spanish texture also called a "knife texture."
- Sand texture can be applied with a sprayer, roller, or brush.
- Stamp texture is applied using a paint roller, then it is stamped with an object to give the desired results.
- Drag texture is applied using a paint roller, and then an object such as a comb or wallpaper brush is dragged across the surface to create the desired effect.
- Stencil texture the base texture is applied using a paint roller or trowel and allowed to dry.

REVIEW QUESTIONS

1   List three types of paints commonly used in residential and light commercial applications.

4   List the steps for preparing a surface for painting.

2   List the three main types of brushes for water-based paint.

5   List the steps for fixing a nail pop.

3   List the two basic types of brushes for oil-based paint.

6   List the steps for repairing a small hole in drywall.

Name: _____

Date: _____

Painting

IDENTIFY AND SELECT PROPER SURFACE FINISH

Upon completion of this job sheet, you should be able to demonstrate your ability to identify and select the proper surface finish and explain the selection process. Describe the surface being painted:

What needs to be painted:	Bedroom	Bathroom	Living room	Dining room
	Kitchen	Hall	Doors	Trim

1. Are surfaces exposed to continual washing or scrubbing? Y/N
   If yes, what is the best type of paint to use?

   _____

   _____

2. Does the surface need to stand up to washing? Y/N
   If yes, what is the best type of paint to use?

   _____

   _____

3. If the room to be painted is a bathroom or kitchen, what is the best type of paint to use?

   _____

   _____

4. If you are painting a high-traffic area, such as a door or trim, what is the best type of paint to use?

   _____

   _____

5. If you need to hide minor imperfections, what is the best type of paint to use?

   _____

   _____

6. If you are painting a ceiling, why is it important to use ceiling paint?

   _____

   _____

INSTRUCTOR'S RESPONSE:

_____

_____

_____

Name: _____

Date: _____

Painting

SELECT PROPER FINISHING TOOL

Upon completion of this job sheet, you should be able to demonstrate your ability to select the proper finishing tools and explain the selection process.

1. Why is it important to use high-quality tools?

   _____

   _____

2. If it is necessary to remove loose paint from the surface to be painted, what tool(s) should be used to perform this task?

   _____

   _____

3. Paint brushes are selected based on the type of finish being used and the surface being painted. List the two types of brushes and the characteristics of each type.

   _____

   _____

4. Rollers are an alternative to using brushes. Explain the different types of rollers and the types of surfaces they would be used on.

   _____

   _____

   _____

INSTRUCTOR'S RESPONSE:

_____

_____

_____

Name: _____

Date: _____

Painting

PROPERLY PREPARING SURFACE FOR PAINTING

Upon completion of this job sheet, you should be able to inspect the painting surface and determine what preparations need to be done to the surface before painting. Inspect the following, indicating whether inspected surfaces need to be repaired prior to painting.

Surfaces	Repair	Replace
Walls		N/A
Baseboard		
Crown molding		
Window trim		
Door		
Door jambs		

If the surface can be repaired, explain what process will be done to repair it. If the surface needs to be replaced, fill out a work order for the work to be done. Preparation checklist:

TASKS COMPLETED

1. Surface repaired or replaced
2. Furniture moved from room or covered
3. Wall plates and receptacles removed
4. Exposed surfaces covered with caulking
5. Drop cloth or other protective material in place

INSTRUCTOR'S RESPONSE:

_____
_____
_____

Name: _____

Date: _____

Painting

APPLYING PAINT BY USING A BRUSH, ROLLER, OR SPRAYER

Upon completion of this job sheet, you should be able to paint the surface using a brush, roller, or sprayer.

1. List the steps for repairing a crack in drywall.

   _____

   _____

2. Is the ceiling being painted? If yes, describe the process that you will use to paint the ceiling.

   _____

   _____

3. Explain the process for painting the walls.

   _____

   _____

4. Why is it suggested that you paint in a zigzag or "M" pattern?

   _____

   _____

5. If you are using a sprayer, why is it important to pour the paint into a bucket through a strainer?

   _____

   _____

6. When you are using a sprayer, why do paint manufacturers recommend thinning the paint first?

   _____

   _____

INSTRUCTOR'S RESPONSE:

   _____

   _____

   _____

Name: _____

Date: _____

Painting

CLEANUP AND STORING EQUIPMENT

Upon completion of this job sheet, you should be able to demonstrate the ability
to clean up and store all equipment.

TASKS COMPLETED

1. Disassemble paint roller.

2. Clean roller and frame in a cleaning solvent.

3. Hang roller to dry.

4. Clean paint brushes in a cleaning solvent.

5. Wrap brushes in paper.

6. If you are using sprayer, remove any unused paint from sprayer.

7. Fill sprayer with appropriate solvent and spray until what emerges is clear.

INSTRUCTOR'S RESPONSE:

_____
_____
_____

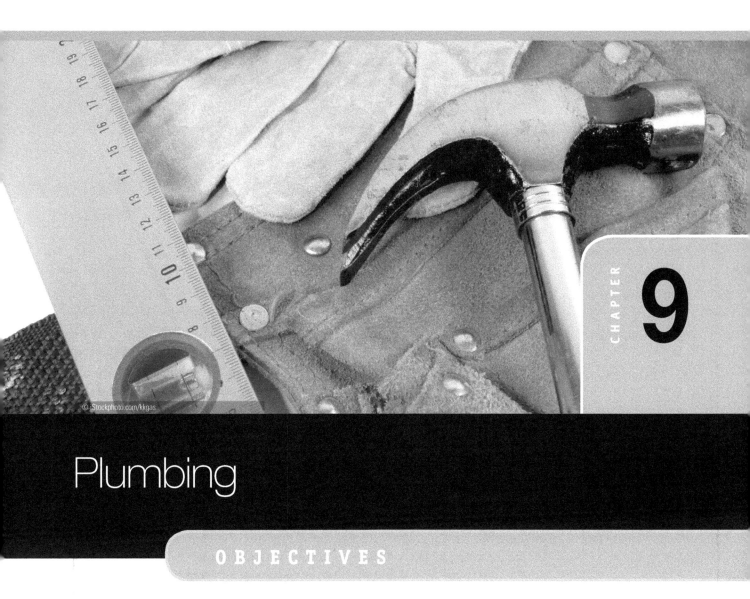
© iStockphoto.com/kkgas

CHAPTER 9

Plumbing

OBJECTIVES

By the end of this chapter, you will be able to:

KNOWLEDGE-BASED

- Identify and select basic plumbing tools for a specific application.
- Identify and select the proper type of plastic piping.
- Identify and select the proper type of copper tubing.
- Identify and select the proper type of metallic pipe.
- Identify and select the proper type of pipe fitting.
- Identify and select pipe hangers and supports.
- Identify, select, and apply caulking.
- Identify and apply caulking.
- Describe the basic steps for performing demolition on a plumbing system in a kitchen, bathroom, and utility room.

SKILL-BASED

- Correctly measure and cut copper tubing.

- Fabricate plastic pipe with correct fittings to correct dimensions as required for job without any leaks.
- Assemble compression fittings without any leaks.
- Clean and replace traps, drains, and vents including the use of sink snake or rod to clean drain lines.
- Caulk and seal fixtures according to manufacturer's specifications.
- Fabricate and solder copper pipe with correct fittings as required for job without any leaks.
- Test and set hot water temperature according to manufacturer's specifications.
- Follow and apply all national and local building codes.
- Locate and repair leaks in pipes and fixtures.
- Install shower seals.
- Repair, replace, and/or rebuild plumbing fixtures and connections to job specifications without any leaks.

Thermostat a device used to control the temperature of water by controlling the heat source

Potable water water that is safe for human consumption

Fittings devices used to connect pipes together in a plumbing system

Valves devices used to control the flow of water, waste, gas, and so on

Introduction

Today, society relies extensively on the use of plumbing systems to maintain its current standard of living. In a residential setting, they are commonly used to transport drinking water to a home, while conveying the waste products produced away from the home. Therefore, it is essential that the plumbing systems be properly maintained to ensure the health and welfare of the people the system services.

Plumbing Safety

It is the responsibility of everyone to ensure that a work site is safe. Although there are general safety procedures to be followed when working on a jobsite, each trade will have its own set of safety rules to be followed to ensure the safety of the workers and help prevent property damage. If anyone is creating a safety issue, that person should be stopped from doing that and immediately corrected. The following are a few of the safety rules that should be followed as well as some of the dangers of working with or around plumbing.

- Installing an incorrect piping product in a potable water system could contaminate the drinking water supply.
- Installing thin-wall pipe not designed for pressure in a pressure system can cause injury and property damage.
- SV gasket lubricant can irritate the eyes; a material safety data sheet (MSDS) should be reviewed and kept in the work location.
- Solvent welding requires the use of glue and, in most instances, a solvent primer and cleaner.
- These chemicals should be included in a company's Hazardous Communication Program. Refer to an MSDS for information about proper use, dangers, and medical treatment.
- PVC can shatter if over pressurized or damaged.
- Not all fittings are safe for potable drinking water systems. Incorrect fitting selection for drinking water systems can cause serious illness to a consumer.
- Because copper tubing can have burrs and sharp edges, globes should be worn to prevent cuts. If a cut should occur as a result of handling copper tubing, clean and bandage immediately to avoid infection.

- Heating polyethylene (PE) piping with a torch creates dangerous fumes and, although this practice may be common, it should be avoided.

- When plastic tubing is burnt, it produces toxic fumes. Because all plastic tubing is flammable never use an open flame near it. Also never install plastic piping in areas designed for metal pipe applications.

- Because of the weight associated with cast iron pipe, proper techniques must be used to eliminate back injury.

- Long sections of pipe should only be carried using two people. Both people should be on the same side of the pipe when carrying.

- Before using joint compound or cutting oil read the MSDS. Read and follow all manufacturers' instructions as well as safety information. Also understand and know the medical treatments in case of accidental consumption of oil or compound.

- When handling and storing flexible gas tubing, always protect the outer sheathing from being nicked by sharp objects. Keep ends of tubing covered so debris cannot enter the tubing.

Plumbing Tools

Before any repairs can be made to a plumbing system, some basic tools will be required. These tools can be purchased in most home improvement stores, and they range in price from a few to a couple hundred dollars. However, as in most trades and jobs, high-quality tools do have a tendency to perform better and last longer.

FIGURE 9-2: Angled-jaw pliers.
Copyright © 2014 Cengage Learning®. All Rights Reserved.

- **Locking tape measure**—Used to make measurements for plumbing fixtures (e.g., sinks; Figure 9-1).

- **Angled-jaw pliers**—Used to make adjustments for valves and fixtures (Figure 9-2).

- **Fourteen-inch pipe wrench**—Used to loosen and tighten metal pipe and fittings (Figure 9-3).

Note: Typically when tightening fittings on a steel pipe, a second pipe wrench is required: one to hold the pipe and one to turn the fitting.

- **Hacksaw**—Used for cutting metal and plastic pipes (Figure 9-4).

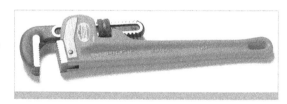

FIGURE 9-3: Pipe wrench.
Copyright © 2014 Cengage Learning®. All Rights Reserved.

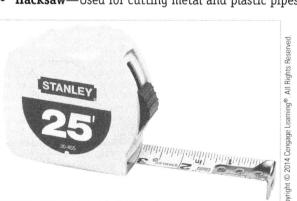

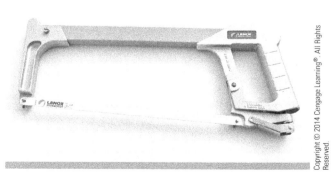

FIGURE 9-1: Tape measure.

FIGURE 9-4: Hacksaw.

- **Plunger**—Used to unclog (unstop) a fixture (sink, toilet, etc.; Figure 9-5).
- **Six- and/or eight-inch adjustable wrench**—Used to loosen and tighten valves and fittings (Figure 9-6).
- **Closet auger**—Used to unstop toilet fixture by removing the foreign object that is causing the problem (Figure 9-7).

FIGURE 9-5: Plunger.

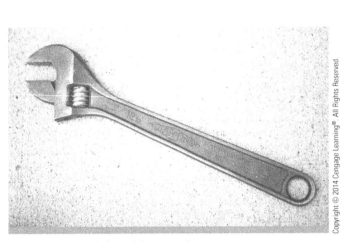

FIGURE 9-6: Adjustable wrench.

FIGURE 9-7: Closet auger.

FIGURE 9-8: Electric spin drain cleaner.

- **Electric spin drain cleaner**—Used to unstop kitchen sinks and lavatory. Typically, a $\frac{1}{4}$-inch cable is used for kitchen sinks (Figure 9-8).
- **Torpedo level**—Used to ensure that fixtures are installed correctly (level) and waste piping systems have the necessary slope (Figure 9-9).
- **Midget copper cutter**—Used for cutting copper tubing (Figure 9-10a).
- **Plastic pipe cutter**—Used for cutting plastic tubing (Figure 9-10b).
- **Toolbox**—Used to store all the plumbing tools (Figure 9-11).
- **Miter box**—Used to cut various materials at either straight (square) or various precision angles (Figure 9-12).
- **Reamer**—Used to remove burrs from piping (Figure 9-13).

FIGURE 9-9: Topedo level.

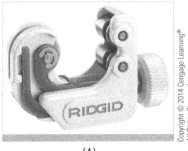

(A)

(B)

FIGURE 9-10: (A) Copper tubing cutter. (B) Plastic pipe cutter.

FIGURE 9-11: Plumber's toolbox.

FIGURE 9-12: Miter box.

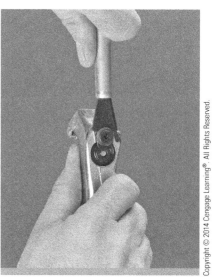

FIGURE 9-13: Reamer.

Piping

Pipe is used to bring **potable water** (water that is safe for human consumption) and gas into a structure and to allow sewage and wastewater to drain from it.

Types of plastic pipes include:

- **Polyvinyl chloride (PVC) pipe**—Available in various sizes and thicknesses (schedules), it is most commonly used in wastewater applications (see Figure 9-14).

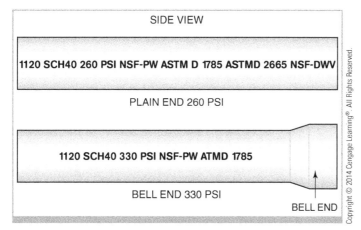

SIDE VIEW

1120 SCH40 260 PSI NSF-PW ASTM D 1785 ASTMD 2665 NSF-DWV

PLAIN END 260 PSI

1120 SCH40 330 PSI NSF-PW ATMD 1785

BELL END 330 PSI

BELL END

FIGURE 9-14: PVC pipe.

- **Chlorinated polyvinyl chloride (CPVC) pipe**—This is a yellowish-white flexible pipe used in water distribution systems. In residential facilities, this pipe is identified as SDR 11 (Figure 9-15).

- **Acrylonitrile butadiene styrene (ABS) pipe**—This is a black plastic pipe typically used in wastewater venting systems (Figure 9-16).

- **Polyethylene (PE) pipe**—This is used only for exterior water services. This type of pipe is not well suited for use in places where it receives direct sunlight (Figure 9-17).

SIDE VIEW

SDR 11 CPVC 4120   400 PSI@ 73°F   ASTM D-2846 NSF-PW DRINKING WATER
100 PSI@ 180°F

FIGURE 9-15: CPVC pipe.
Copyright © 2014 Cengage Learning®. All Rights Reserved.

FIGURE 9-16: ABS pipe.
Copyright © 2014 Cengage Learning®. All Rights Reserved.

SIDE VIEW

ASTM F-628 ABS DWV NOT FOR PRESSURE

FIGURE 9-17: PE pipe.
Copyright © 2014 Cengage Learning®. All Rights Reserved.

- **Cross-linked polyethylene (PEX) pipe**—This is typically a whitish-colored pipe used for water distribution systems (Figure 9-18).

  Types of metal pipes include:

- **Cast iron pipe**—Often used in residential wastewater systems because it is a much quieter pipe for draining applications than plastic (see Figure 9-19).

- **Galvanized and black steel pipe**—Used for residential gas supply piping (Figure 9-20).

- **Brass pipe**—Used widely in the early twentieth century for water distribution systems; today, it is used in the manufacture of faucets, valves, and other products used for potable water. Brass pipe is more corrosive resistant than galvanized steel. The threads commonly used comply with NTP standard. Brass pipe is cut, threaded and assembled the same way as galvanized and black steel pipe. Because it is chrome plated, care must be used when working with chrome to protect the finish. Normally strap wrenches are used to keep from scarring the finish.

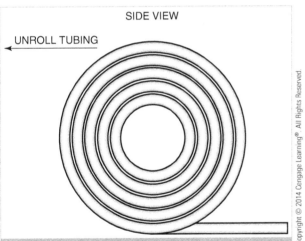

SIDE VIEW

UNROLL TUBING

Copyright © 2014 Cengage Learning®. All Rights Reserved.

FIGURE 9-18: PEX pipe.

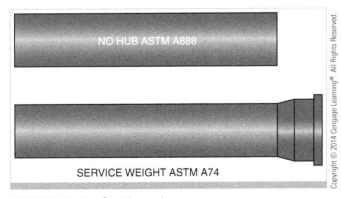

NO HUB ASTM A888

SERVICE WEIGHT ASTM A74

Copyright © 2014 Cengage Learning®. All Rights Reserved.

FIGURE 9-19: Cast Iron pipe.

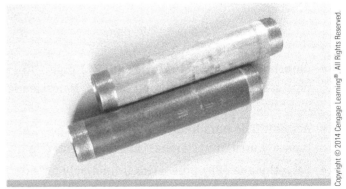

Copyright © 2014 Cengage Learning®. All Rights Reserved.

FIGURE 9-20: Black pipe.

Currently, three types of copper tubings—types K, L, and M—are used for domestic water (see Figure 9-21).

- **Type K copper tubing**—A thick-walled tubing used primarily for underground water service; it can be purchased either in soft rolls or in 20-foot stock lengths.

- **Type L copper tubing**—A thin-walled tubing used in above-ground installations; it can be purchased either in soft rolls or in 20-foot stock lengths.

- **Type M copper tubing**—A hard thin-walled tubing that is sold only in 20-foot stock lengths; it should be used only for indoor applications in which the tubing can be easily accessed (Figures 9-22 and 9-23).

Copper Type	Potable Water	DWV	Underground	Available in Roll
DWV		✓		
Type M	✓	✓		
Type L	✓	✓	✓	✓
Type K	✓	✓	✓	✓

FIGURE 9-21: Types and basic uses of copper tubing.

FIGURE 9-22: Copper tubing.

FIGURE 9-23: Unrolling copper tubing.

Pipe Fittings

Regardless of the type of piping materials used in a plumbing system, the basic components (fittings) used to connect the piping together into a usable system are the same (just made from different materials). **Fittings** are named based on their unique characteristics and material type. Fittings are classified according to their design. Some of the more common fittings used are (Figure 9-24):

- Offset—Used to change the direction of a piping system.
  - Pressure systems use shorter (tighter) radius changes, whereas DWV systems use larger radius.

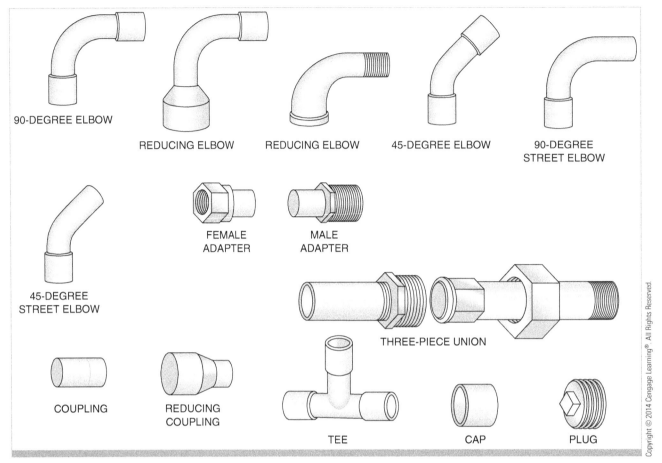

FIGURE 9-24: Commonly used pipe fittings.

- Two 45° fittings can be used to create a 90° offset.
- In a DWV system, two $22\frac{1}{2}$° fittings can be used to create a 45° offset.
- When an offset is created using two fittings, this is referred to as a "swing joint."
- Twenty-two-degree elbow, used in drainage waste and vent (DWV) systems.
- Ninety-degree elbow, used to change the direction of a pipe by 90°.
- Forty-five-degree elbow, used to change the direction of a pipe by 45°. It is typically used in pairs to offset a section of pipe around an obstacle.
- A cast iron 90° fitting is ordered as a quarter bend.
- A cast iron 45° fitting is ordered as an eighth bend.
- A cast iron $22\frac{1}{2}$° fitting is ordered as a sixteenth bend.

- **Tee**—Used to create a branch line.
  - A fitting with three connections, used for pressure systems, is known as a tee.
  - For sizing and ordering purposes, the opening of a tee is identified as either run or branch.
  - Run openings are those in the direction of flow through a tee.
  - The branch opening is perpendicular to the run opening.

- **Cap**—Used to terminate a section of pipe; a cap is a fitting that fits around the outside of a pipe.

- **Coupling**—Used to connect two pieces of pipe having the same diameter.

- **Plug**—Used to terminate a section of pipe, a plug is a fitting that screws into the inside of a section of pipe.
- **Union**—Used to connect two pieces of pipe, while allowing for them to be disconnected without being cut.
- **Reducer**—Used to reduce a section of pipe to a smaller diameter.
- **Adapter**—Used to connect pipes of different sizes or materials together.
- **Bushing**—Used to reduce a section of pipe like a reducer, but threads inside a fitting to create the reduction.

  There are two basic connection types for fittings:
- Fittings that can receive a pipe (hub) and fittings that can be inserted into a hub (socket) of a pipe end.
- No-hub cast iron (NHCI) fittings that are connected using a specifically designed clamp.

Valves

Regardless of the application, all plumbing systems use devices and valves. **Valves** are used to control the flow of water, waste, gas, and so on. They can be operated manually or automatically by using motors and a computer interface to control their actions. When installing a valves or device onto a potable water system, all plumbing codes must be observed. Threaded valves and devices typically have female threads, while a lot of the valves and devices used with copper tubing are soldered connections Plastic valves and devices are typically welded using solvents. A single fixture can be isolated by using a special valve known as a "stop." A valve designed specifically to isolate the gas supply to a fixture is known as a "gas cock." To protect a portable water supply and system, a device known as a "backflow device" is used. A pressure-reducing valve reduces the pressure in a piping system.

Isolation Valves

The code dictates that every residence have a minimum of at least one isolation valve installed in a location in which the water can be cut off in the event of an emergency or a repair (see Table 9-1). In most locations, the codes dictate that residential water service cannot be any smaller than $\frac{3}{4}$ inch. Many plumbing codes require a full-port design for the main isolation valve for a residential application. A full-port valve has the same inside diameter as the pipe it is connected to.

Type	Residential Usage
Ball valve	Water and gas
Gate valve	Water
Stop valve	Water
Stop-and-waste valve	Water
Gas cock	Water

TABLE 9-1: Common Isolation Valves Used in Residential.

Piping Support and Hangers

In order for a plumbing system to work properly, it must be correctly sized and then properly supported. Typically, the type of support used for a particular section of piping depends on the kind of pipe used, the position of the pipe, and the purpose of that pipe. In general, five common types of plumbing supports are typically used for residential settings (see Figures 9-25 and 9-26).

- **Clamp**—Used to anchor a section of pipe to an architectural support.
- **Hanger strap**—Made from a flexible material containing predrilled holes that are equally spaced throughout the length of the hanger.
- **U-bolts**—Typically used to secure a section of pipe to a structural steel member.
- **Wire hanger**—Similar to a U-bolt, this type of pipe hanger is typically used on smaller, lighter pipe.
- **Pipe staples**—Typically installed using a staple gun; this type of hanger is used to secure piping to an architectural support.

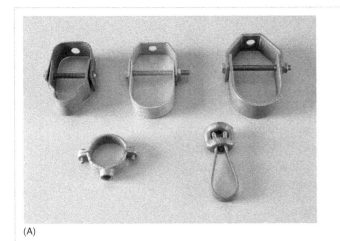

(A)

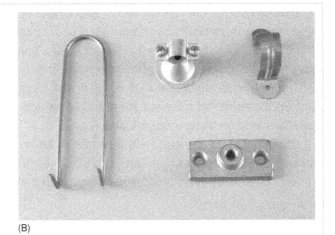

(B)

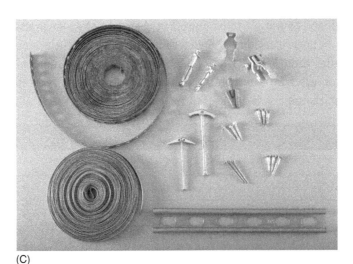

(C)

FIGURE 9-25: Various types of support and hangers.

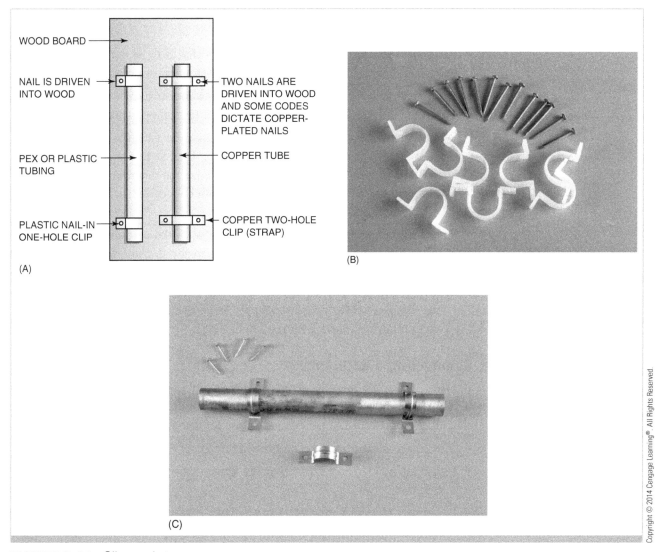

FIGURE 9-26: Clips and straps.

REMODELING

Remodeling and Plumbing Systems

When a structure is remodeled that contains plumbing fixtures, the fixtures are most often removed and either replaced and reused. In either case, the general procedures for remodeling a structure that contains plumbing fixtures are as follows.

Remodeling a Kitchen

One of the areas in a residential structure that is often remodeled is the kitchen. Typically, this is completed when the appliances are either replaced or updated. The general procedure for demolishing a kitchen is:

- Disconnect all utilities servicing the kitchen.

- Install a plastic barrier separating the affected area from the rest of the structure.

Always check with your local and state agencies before starting a remodeling project. Before beginning the demolition or deconstruction process, the proper permits must be obtained.

- Cover any appliance that is not going to be removed from the area with a protective covering (cardboard).
- Disconnect electrical wiring to any appliance that is to be removed. Any exposed wire should be capped using wirenuts to prevent exposure to the bare conductors.
- Disconnect any plumbing servicing appliances and/or fixtures to be removed. Any pipes that are exposed should be capped (including vents).
- Remove appliances that are in the affected area. For example, if the technician is demolishing kitchen counter and/or cabinets, then at this stage the vent hood, as well as a drop in stove and/or cook top should be removed. Even if these components are not going to be reused they could be recycled at an architectural salvage yard.
- Remove plumbing fixtures in the affected area. In the preceding example, the kitchen sink and garbage disposal should be removed. Even if these components are not going to be reused, they can be recycled at an architectural salvage yard.
  - If the kitchen sink is equipped with a disposal, be sure that you disconnect the disposal's electrical power supply and wirenut all exposed wires, before attempting to remove from the kitchen sink.

Remodeling a Bathroom

Another area in a residential structure that is often remodeled is the bathroom. This is typically done when the plumbing fixtures are updated or replaced or when the cabinetry is updated. The general procedure for demolishing a bathroom is:

- Disconnect all utilities servicing the bathroom.
- Install a plastic barrier separating the affected area from the rest of the structure.
- Disconnect electrical wiring to any appliance that is to be removed. Any exposed wire should be capped using wirenuts to prevent exposure to the bare conductors.
- Disconnect any plumbing servicing appliances and/or fixtures to be removed. Any pipes that are exposed should be capped (including vents).
- Remove plumbing fixtures in the affected area. Even if these components are not going to be reused, they can be recycled at an architectural salvage yard.
  - Many municipalities have recycling facilities that will take old toilets. Before disposing of an old toilet, always check with your local municipality.

Remodeling a Utility Room

This area in a residential structure is not remodeled as often as a kitchen or a bathroom, but the demolition process is similar. Utility rooms are typically remodeled when the plumbing fixtures are updated or replaced or when the cabinetry is updated, or when appliances are updated and/or replaced. The general procedure for demolishing a utility room is:

- Disconnect all utilities servicing the utility room.
- Install a plastic barrier separating the affected area from the rest of the structure.
- Disconnect electrical wiring to any appliance that is to be removed. Any exposed wires should be capped using wirenuts to prevent exposure to the bare conductors.

- Disconnect any plumbing servicing appliances and/or fixtures to be removed. Any pipes that are exposed should be capped (including vents).

- Remove plumbing fixtures in the affected area. In the preceding example, the utility room sink should be removed. Even if these components are not going to be reused, they can be recycled at an architectural salvage yard.

- If the ceiling is to be replaced, remove the drywall and plaster. If the ceiling is a drop ceiling, try removing the ceiling tiles and bracing carefully so that they can be recycled.

Measuring and Cutting Pipes

To correctly measure copper tubing, always measure it from the centers of two fittings or from the end of the pipe run to the center of a fitting (Figures 9-27 and 9-28).

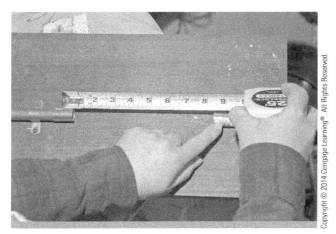

FIGURE 9-27: Center-to-Center Measurement of Copper Tubing.

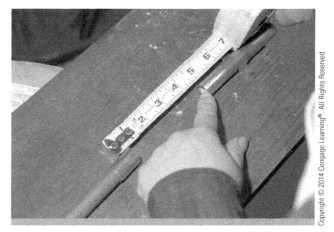

FIGURE 9-28: Center-to-end measurement of copper tubing.

Although copper tubing can be cut with either a hacksaw or a tubing cutter, it is recommended that, when possible, a tubing cutter should be used. A tubing cutter will ensure that the end of the copper tubing is square. However, if copper tubing is cut with a hacksaw, it is recommended that a miter box or fixture be used (Figure 9-29).

Using a Tubing Cutter

To cut copper tubing using a tubing cutter, do the following:

1. Use a locking tape measure to measure and mark the copper tubing to the desired length (Figure 9-30).

2. Place copper tubing into fixture (Figure 9-31).

3. Unwind the pipe cutter so that the pipe can be inserted.

4. Once the cutter has been positioned so that the wheel sits on the pipe where it needs to be cut, tighten the pipe cutter.

FIGURE 9-29: Miter box.

5. Tighten the cutter $\frac{1}{4}$ turn and then rotate it around the pipe before tightening it again (Figure 9-32).

6. After tubing has been completely cut, remove any burrs inside the tubing with a round file (Figure 9-33).

Using a Hacksaw

To cut copper tubing with a hacksaw do the following:

1. Use a locking tape measure to measure and mark the copper tubing to the desired length (Figure 9-30).

2. Place copper tubing into fixture (Figure 9-31).

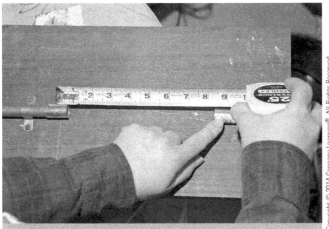

FIGURE 9-30: Measuring copper tubing.

FIGURE 9-31: Copper tubing in a fixture.

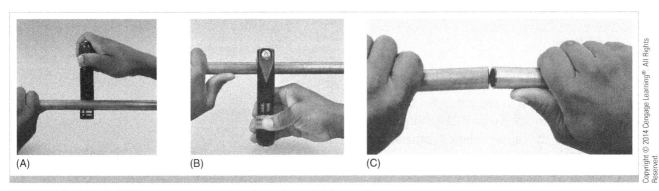

(A)　　　　(B)　　　　(C)

FIGURE 9-32: A–C The proper procedure for using a tubing cutter.

FIGURE 9-33: Cutting copper tubing using a hacksaw.

3. Start the cutting process by pulling and pushing the hacksaw back and forth (Figure 9-33).

4. After tubing has been completely cut, remove any burrs inside the tubing with a round file (Figure 9-34).

Unclogging Pipes

One of the most frequent types of calls that a facilities maintenance technician will receive is one regarding either a clogged lavatory or toilet. The following situations can usually be resolved without the assistance of a professional by following a few basic steps.

Using a Plunger to Unclog a Kitchen Sink or Lavatory

A kitchen sink or lavatory can be unclogged with a plunger. Care should be taken when unclogging a sink that has a disposal attached using a plunger. The pressure generated by the plunger can damage the flange gasket or tailpiece of the disposal. The following steps should be used when unclogging a kitchen sink or lavatory using a plunger:

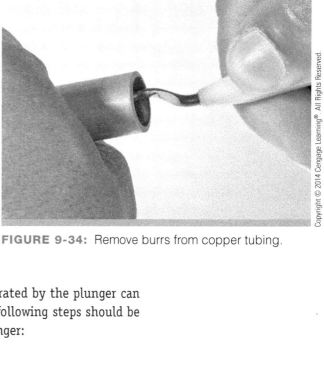

FIGURE 9-34: Remove burrs from copper tubing.

1. Remove the sink's strainer or plug (Figure 9-35).

2. If the sink does not already have water in it, fill the sink to the halfway mark.

3. Place the plunger's tuber globe over the drain carefully, making sure that the entire drain opening is covered by the plunger (Figure 9-36).

4. Using forceful strokes plunge the sink drain at least 15 times before removing the plunger to see if the sink will drain.

5. If the sink is still clogged, repeat Steps 3 and 4.

6. Once the sink has been unclogged, run hot water down the drain for several minutes. This will help clean out anything remaining in the system.

FIGURE 9-35: Sink strainer.

FIGURE 9-36: Positioning a plunger over a drain opening.

Using a Drain Cleaner to Unclog a Kitchen Sink or Lavatory

When using liquid drain cleaner to unclog a kitchen sink or lavatory, perform the following steps:

1. Before using a liquid drain cleaner, carefully read and follow all instructions on the package.
2. When handling liquid drain cleaner, take care *not* to spill it on any of the surrounds or on your skin. If you do come in contact with the drain cleaner, refer to the packaging for the steps to take or consult a doctor immediately.

A Green Approach to Unclogging a Kitchen Sink or Lavatory

In some cases, a clog can be removed using a more natural, environmentally friendly approach. This approach uses baking soda and vinegar. When baking soda is mixed with vinegar, a chemical reaction is produced in which carbon dioxide is released.

1. Remove any contents from the sink.
2. Pour approximately $\frac{1}{4}$ cup of baking soda into the drain opening.
3. Pour approximately 1 cup of vinegar into the drain opening.
4. Cover the drain opening with a lid for 15 minutes.
5. Uncover the drain opening and test the drain.
6. If the drain is still clogged, repeat Steps 2–5.
7. Rinse the drain with hot water.

   Note: A plunger may be required to help loosen the clog.

Care should be taken when using a plunger; there is the possibility of damaging the wax ring.

Using a Plunger to Unclog a Toilet

Do the following when using a plunger to unclog a toilet:

1. Insert the plunger into the toilet bowl, fully covering the drain opening (Figure 9-37).
2. Pushing down and pulling up the handle of the plunger, vigorously plunge the toilet 15–20 times.
3. Lift the plunger from the drain opening to see if the toilet drains.
4. If the toilet is still clogged, repeat Steps 1–3.

Using a Drain Cleaner to Unclog a Toilet

A liquid drain cleaner should be used on a toilet only as a last resort because of its possible impact on the environment. However, if you use a drain cleaner to unclog a toilet, follow these steps:

1. Before using a liquid drain cleaner, carefully read and follow all instructions on the package. Also, before using a liquid drain cleaner on a toilet, make sure that the cleaner is safe to use with porcelain.

FIGURE 9-37: Plunger covering the drain opening.

2. When handling liquid drain cleaner, take care *not* to spill it on any of the surrounds or on your skin. If you do come in contact with the drain cleaner, refer to the packaging for the steps to take or consult a doctor immediately.

3. After the toilet is unclogged, flush the toilet several times to check the water flow and remove the drain cleaner.

> *If the clog is not removed using a screwdriver, then a sink snake can be used to unclog the drain.*

Cleaning a Slow-Draining Lavatory

If a lavatory is slow draining, try running hot water into the sink for about 10 minutes. Running hot water can sometimes loosen contaminates that might be causing the sink to drain slowly. If using hot water does not completely open the drain, try using an environmentally safe cleaner (baking soda and vinegar—see the previous section on using a more natural approach).

While unclogging a lavatory with a cleanout, do the following:

1. Place a bucket under the sink to catch wastewater.

2. Using a wrench, open the cleanout (Figure 9-38).

3. Use a screwdriver to probe around and pull out the clog (Figure 9-39).

4. Replace the cleanout cover and gasket.

5. Run hot water into the sink for 10 minutes.

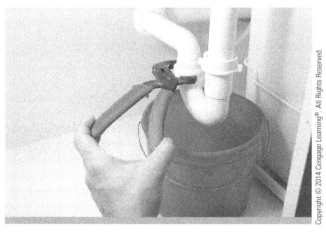

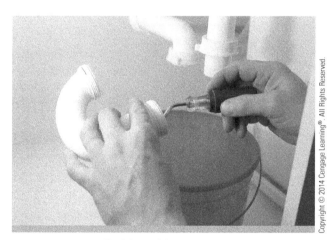

FIGURE 9-38: Removing a sink trap using angled-jaw pliers.

FIGURE 9-39: Probing with a screwdriver.

Using a Sink Snake to Unclog a Lavatory

Another way to try unclogging a lavatory is with a cleanout and a sink snake. This can be accomplished as follows:

1. Place a bucket under the sink to catch wastewater (Figure 9-40).

2. Using a wrench, open the cleanout (Figure 9-41).

3. Carefully push the snake into the cleanout opening, moving the tape back and forth to help it navigate the drain pipe (Figure 9-42).

4. When the snake reaches the clog, twist and push it until the clog is removed.

5. Replace the cleanout cover and gasket (Figure 9-43).

6. Run hot water into the sink for 10 minutes.

FIGURE 9-41: Remove a sink trap using angled-jaw pliers to unclog a drain with a sink snake.

FIGURE 9-42: Using a sink snake.

FIGURE 9-40: Placing a bucket under a sink.

FIGURE 9-43: Replacing a sink trap using angled-law pliers.

FIGURE 9-44: Using a plunger in a bathtub to unclog a drain.

Using a Plunger to Unclog a Bathtub

A plunger can also be used to unclog a bathtub drain:

1. Place the plunger's tuber globe over the drain, carefully making sure that the entire drain opening is covered by the plunger (Figure 9-44).

2. Using forceful strokes, plunge the tub drain at least 15 times before removing the plunger to see if the tub will drain.

3. If the tub is still clogged, repeat Steps 3 and 4.

4. Once the tub has been unclogged, run hot water down the drain for several minutes. This will help clean out anything remaining in the system.

Using a Drain Cleaner to Unclog a Bathtub

When using drain cleaners to unclog a bathtub drain, perform the following steps:

1. Before using a liquid drain cleaner, carefully read and follow all instructions located on the package.

2. When handling liquid drain cleaner, care should be taken *not* to spill it on any of the surrounds or on your skin. If you do come in contact with the drain cleaner, refer to the packaging or consult a doctor immediately.

A Green Approach to Unclogging a Bathtub

Using environmentally friendly cleaners to unclog a bathtub drain is another option:

1. Pour approximately $\frac{1}{4}$ cup of baking soda into the drain opening.

2. Pour approximately one cup of vinegar into the drain opening.

3. Cover the drain opening for 15 minutes with a lid.

4. Uncover the drain opening and test the drain.

5. If the drain is still clogged, repeat Steps 2–5.

6. Rinse the drain with hot water.

Using a Toilet Auger to Unclog a Toilet

1. Loosen the setscrews on the auger and push the cable into the drain, moving it back and forth until the clog is reached (Figures 9-45 and 9-46).

2. Tighten the set screws on the toilet auger.

3. While pushing on the toilet auger, crank on the auger clockwise until the obstruction is cleaner.

4. Remove the toilet auger from toilet.

5. Test the flow of the toilet by flushing it several times.

FIGURE 9-45: Setting the set screw on a toilet auger.   **FIGURE 9-46:** Pushing auger cable into drain opening.

Caulking

Caulk is primarily used to seal cracks caused by mating parts (recall the discussion in Chapter 4). However, in plumbing, caulk is used mostly for waterproofing, for example, for the crack produced where a sink meets the mating wall.

Applying Caulk

1. Remove all dust, dirt, and any other grime from around area to apply caulk.

2. If water or a solvent is used to clean the area to be caulked, make sure the area is dry.

3. Insert caulk tube into caulk gun (Figure 9-47).

4. Using a knife, remove the tip of the caulk, cutting away as little as possible. The amount of tip removed will affect the size of the bead of caulk that the caulk gun applies. Make sure the tube of caulk does not contain a second seal. To check, stick a nail into the hole made and cut off the tip (Figure 9-48).

5. Hold the caulk at about a 5–10° angle, in the direction of travel. The direction of travel is the direction in which you will be pulling the caulk gun (Figure 9-49).

6. After applying the caulk to the crack, use your finger to lightly work the caulk into the crack. (Note: Keeping your finger wet will help ensure a smoother caulk bead (see Figure 9-50).)

7. Use a wet (moist) towel to wipe away all excess caulk.

FIGURE 9-47: Caulk tube and gun.

FIGURE 9-48: Caulk tube with cut angle on end.

FIGURE 9-49: Applying caulking.

FIGURE 9-50: Smoothing caulking with a finger.

Selecting a Caulk

The four most commonly used caulk today are:

1. **Acrylic latex caulk**—Fast-drying all-purpose caulk designed for dry applications. It is well suited for painting and is easily cleaned up using soap and water

2. **Vinyl latex caulk**—Designed for wet applications and a good choice for showers and tubs

3. **Silicone caulk**—Long-lasting, mildew-resistant, watertight adhesive caulk that does not discolor; typically, silicone caulk cannot be painted or cleaned using soap and water

4. **Butyl rubber caulk**—Designed for use in outdoor applications and can be used to fill large joints or cracks

Plumber's Putty

Plumber's putty is a pliable sealant, used to seal pipe joints and fittings, that can be used for quick emergency repairs (Figure 9-51). To apply plumber's putty, follow these steps:

1. Clean the surface of the fitting or fixture where the putty is to be applied.

2. Apply a bead of plumber's putty to the mating part. In the case of a sink drain, apply the putty under the rim of the flange and place the drain into the drain outlet (see Figure 9-52).

3. Tighten the fitting, causing the plumber's putty to spread.

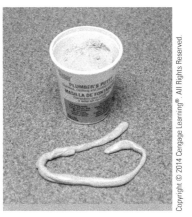

FIGURE 9-51: Plumber putty.

FIGURE 9-52: Applying caulk on drain outlet.

Assembling Pipes Using PVC Cement

Refer to Procedure 9-2 Cutting PVC with a PVC Tubing Cutter, Procedure 9-3 Cutting PVC with a Hacksaw and Procedure 9-4 Joining PVC Pipe for step-by-step instructions.

Soldering a Copper Fitting onto a Piece of Copper Pipe

Refer to Procedure 9-5 Soldering Using a Propane Torch for step-by-step instructions.

National and Local Plumbing Codes

Because local plumbing codes vary from location to location in the United States, it is recommended that you review all local plumbing codes before attempting any plumbing installation and/or repair. The national plumbing code can be accessed

When using a propane torch, always keep a fire extinguisher handy. Always read and carefully follow all directions on the propane torch. Always carefully read the instructions on the fire extinguisher before using the propane torch.

at http://emarketing.delmarlearning.com/downloads/StayCurrent_02_05.pdf. Local plumbing codes can be obtained by contacting the state code and administration department.

Adjusting the Temperature of a Water Heater

The U.S. Department of Energy states that a temperature at the tap of 90°F is adequate for most household applications. However, to obtain this temperature the water leaving the tank should be no less than 130°F. This temperature prevents dangerous bacterial growth.

Testing the Water Temperature

1. At one of the sinks in the structure serviced by the hot water tank, turn on the hot water. Be sure to leave the water on for a few moments, making sure the tap water has reached its full temperature.
2. Fill a glass with the hot water.
3. Place a thermometer in the glass of hot water.
4. Remove the thermometer and read the recorded temperature.

Adjusting the Temperature of an Electric Water Heater

The water temperature is usually controlled by a **thermostat** positioned on the side of the unit. To reset or set the temperature of an electric water heater, simply set the thermostat to the desired temperature.

In some cases, the thermostat may be concealed; if this is the case, do the following:

1. Turn off power to the water heater.
2. Using a screwdriver, remove the covering, revealing the thermostat.
3. Adjust the thermostat to the desired temperature.
4. Replace the cover.
5. Turn on electricity.

Adjusting the Temperature of a Gas Water Heater

Most gas water heaters have the thermostat located on the side of the water heater; it can be adjusted by simply setting the thermostat to the desired temperature (Figure 9-54). In a few cases, the thermostat may be concealed. If this is the case, then the following steps should be followed:

1. Using a screwdriver, remove the thermostat cover.
2. Adjust the thermostat to the desired temperature.
3. Replace the thermostat cover.

Some electric water heaters have multiple thermostats. If your water heater has multiple thermostats, both thermostats must be set on the same temperature (Figure 9-53).

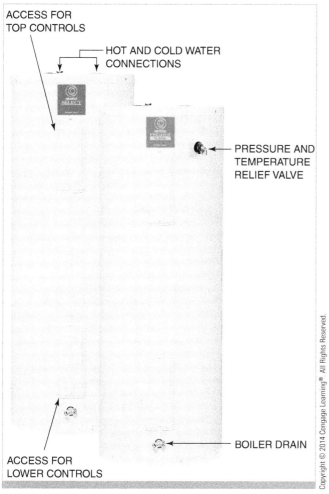

FIGURE 9-53: Electric water heater.

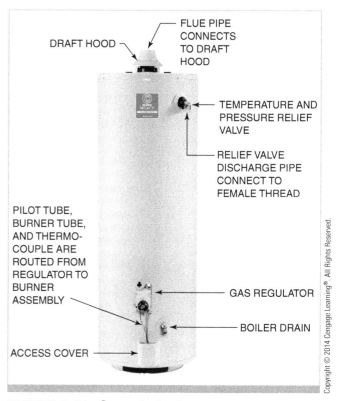

FIGURE 9-54: Gas water heater.

Basic Water Heater Replacement

When you are replacing a water heater, stick with what is already there. If you have an electric water heater, replace it with an electric water heater unless you are willing to run gas lines and exhaust vents. If you are exchanging a gas water heater, replace it with a gas water heater unless you are willing and able to install new electrical service.

1. Turn utilities (gas, electricity and water) to the water heater.
2. Drain the water heater while opening a hot water faucet to allow air to enter into the system (Figure 9-55).
3. Disconnect the water lines (Figure 9-56).
4. Move the new water heater to its position by using an appliance cart, dolly, or hand truck. Position the new water heater so the piping and the gas vent are aligned or will most easily reach.
5. If you removed the shutoff valve, replace it (Figure 9-57).
6. Install the water lines and pressure-relief line (Figure 9-58).

Follow manufacturer's instructions for a specific step-by-step list on how to install specific water heaters such as gas or electric heater.

FIGURE 9-55: Draining the water heater.

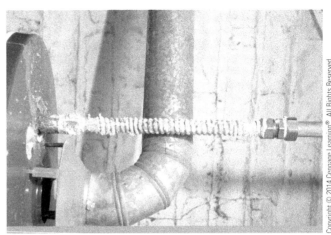

FIGURE 9-56: Disconnect water lines.

FIGURE 9-57: Water heater shutoff valve.

FIGURE 9-58: Install water lines.

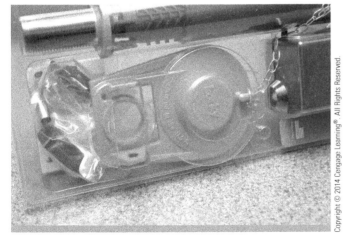

FIGURE 9-59: Flapper assembly package.

Plumbing Leaks

Plumbing leaks can cause thousands of dollars-worth of damage if left unchecked. To check a plumbing system for leaks, the facilities maintenance technician should periodically check under sinks, around toilets, and in crawlspaces. Once a leak has been detected, correct the problem or consult a plumber.

One common source of plumbing leaks is the flapper assembly in a toilet. If the flapper is leaking, then the defective flapper should be replaced. A new flapper can be obtained at your local plumbing supply. To replace the flapper in a toilet, consult the instructions on the flapper package (Figure 9-59).

Shower Seals

If the bathroom floor is getting wet after a shower, then possibly the shower door seal should be replaced. Shower seals can be obtained at your local plumbing supply. To replace the shower seal, consult the instructions located on the packaging of the shower seal.

Repairing a Faucet

Several types, makes, and models of sink faucets are available on the market today. The most common types of sink faucets available are single and double handled. Faucets manufactured today are either a compression or a washerless type.

Compression faucets regulate the flow of water by applying pressure onto a rubber washer located within the faucet, while washerless faucets use a cartridge, ball, or disk to control the flow of water (Figure 9-60).

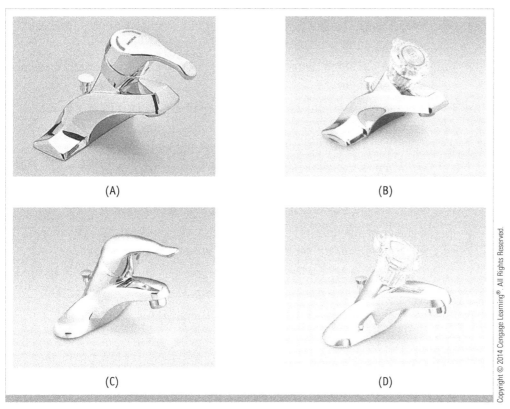

(A)

(B)

(C)

(D)

FIGURE 9-60: Various faucet styles.

Before repairs are made to a faucet, always close the sink drain to prevent parts of the faucet from going down the drain. In addition, never use a pipe wrench on a polished fixture without first applying tape to the pipe wrench jaws. Applying tape to the pipe wrench jaws will prevent the wrench jaws from damaging the fixture's finish. Finally, shut off all water supplies to the faucet.

Because a number of types and styles of faucets are available on the market today, facilities maintenance technicians should consult the manufacturer's documentation at the local plumbing supply company when repairing a sink faucet. Only manufacturer-approved parts should be used. In addition, always read and follow the faucet manufacturer's instructions.

SUMMARY

- Before any repairs can be made to a plumbing system, some basic tools are required.

- Types of plastic pipes include: polyvinyl chloride (PVC) pipe, chlorinated polyvinyl chloride (CPVC) pipe, acrylonitrile butadiene styrene (ABS) pipe, PE pipe, cross-linked polyethylene (PEX) pipe, cast iron pipe, galvanized, and black steel pipe

- Currently, three types of copper tubing—types K, L, and M—are used for domestic water.

- Regardless of the type of piping materials used in a plumbing system, the basic components (fittings) used to connect the piping together into a usable system are the same (just made from different materials).

- Some of the more common fittings used are: 90° elbow, 45° elbow, tee, cap, coupling, plug, union, reducer, and adapter.

- In order for a plumbing system to work properly, it must be correctly sized and then properly supported.

- Five common types of plumbing supports are typically used for residential settings are: clamp, hanger strap, U-bolts, wire hanger, and pipe staples.

- When a structure is remodeled that contains plumbing fixtures, the fixtures are most often removed and either replaced and reused.

- Although copper tubing can be cut using either a hacksaw or a tubing cutter, it is recommended that when possible a tubing cutter be used.

- One of the most frequent types of calls that a facilities maintenance technician will receive is one regarding either a clogged lavatory or toilet.

- In plumbing, caulk is used mostly for waterproofing, for example, for the crack produced where a sink meets the mating wall.

- The U.S. Department of Energy states that a temperature at the tap of 90°F is adequate for most household applications.

- Most gas water heaters have the thermostat positioned on the side of the unit; it can be adjusted by simply setting the thermostat to the desired temperature.

Flaring Copper Tubing

- Using the tubing cutter and a tape measure, cut a piece of copper tubing to length. When cutting the copper tubing place the copper tubing between the cutting head and the rollers.

Ⓐ Tighten the cutting wheel until it is against the tubing.

Ⓑ Tighten the cutting wheel $\frac{1}{4}$ and $\frac{1}{2}$ revolution.

Ⓒ Rotate the tubing cutter two or three revolutions.

Ⓓ Repeat Steps C and D until the piece has been removed.

- Using the reamer, remove any burrs from the cut end of the copper tubing. Using a piece of emery cloth or steel wool, remove any burrs from the outside of the tubing.

- Using the flaring block, place the copper tubing into the flaring block. Use the portion of the flaring tool below the slot as a height gage.

- Place flaring tool on the flaring block.

- Tighten the cone into the tubing a few turns, and then back it out.

- Continue this process until the flare is made.

Cutting PVC with a PVC Tubing Cutter

- When cutting PVC place the pipe between the cutting head and lower part of the cutter.

Ⓐ Squeeze the blade down onto the pipe until the tubing is cut.

Ⓑ Using the reamer, remove any burrs from the cut end of the PVC. Using a piece of emery or steel wool to cloth, remove any burrs from the outside of the tubing.

Ⓐ

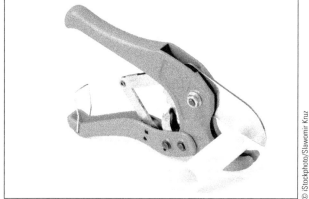

Cutting PVC.

Cutting PVC with a Hacksaw

- When cutting PVC with a hacksaw, whenever possible use a jig or miter box to securely hold the PVC in place.

Ⓐ Mark the piece to be cut using a carpenter's pencil.

Ⓑ When starting the cut, use the saws center section of teeth on the cut line.

Ⓒ Push the saw using short slow strokes until the cut get started.

Ⓓ Using the reamer, remove any burrs from the cut end of the PVC. Using a piece of emery cloth or steel wool, remove any burrs from the outside of the tubing.

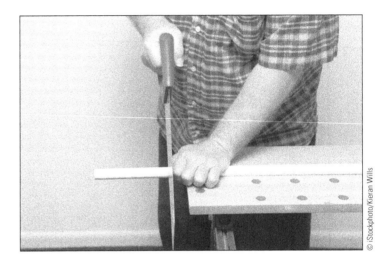

Ⓑ

Cutting PVC with a Hacksaw.

© iStockphoto/Kieran Wills

PROCEDURE 9-4

Joining PVC Pipe

- Mark the piece to be cut using a carpenter's pencil.

Ⓐ Cut the pipe using either a hacksaw or a tubing cutter (See "Cutting PVC with a PVC Tubing Cutter or Cutting PVC with a Hacksaw").

- Using the reamer, remove any burrs from the cut end of the PVC. Using a piece of emery cloth remove, any burrs from the outside of the tubing.

- Before applying cement, always try to dry-fit the piping arrangement to ensure that the connections are correct length and orientation. If necessary, make an adjustment to the length. Once the cement has set, they cannot be separated.

Ⓐ Apply primer, to both the male and female portions of the joint.

Ⓑ Apply cement to both the male and female portions of the joint

Ⓒ Insert the male end of the fitting into the female end and rotate the pipe $\frac{1}{4}$ turn.

- Hold the pipe and fitting together for approximately 1 minute to prevent the pipe from pulling out of the fitting.

Ⓑ

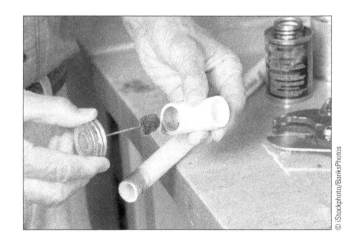

Applying PVC Cement.

⟨ **CAUTION** ⟩

Always follow the manufacturer's instructions and safety guidelines when working with PVC primer and PVC cements. Only use these chemical for their intended use. When possible use the primer and cement in well-ventilated areas. If this is not possible, limit your exposure time to the fumes from these chemicals.

PROCEDURE 9-5

Soldering Using a Propane Torch

- Before soldering copper tubing, always check the local building codes for solder requirements (lead-free or near lead free solder). Never use acid core solder when soldering copper tubing.

A When soldering a fitting such as a coupling, en elbow, or a tee always solder all connections at one time.

- If soldering a fitting to an existing section of pipe, start by using a piece of steel wool or emery cloth to clean the outside of the copper tubing until it is bright all around. Also, using a reamer, remove any burrs from the inside of the copper tubing.

A Using a pipe tubing cleaning brush, clean the inside of the fitting.

A

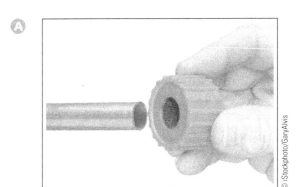

Cleaning Copper Tubing.

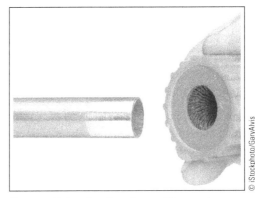

Copper Tube that has been Cleaned.

- If soldering a new section of tubing to an existing fitting, start by using a tape measure, measure and mark the section of pipe to be joined.

Ⓐ Cut the section of pipe using either a hacksaw or a tubing cutter.

- After cleaning the inside of the fitting and the outside of the tubing, apply flux, using a flux brush to the inside of the fitting as well as the outside of the pipe.

Ⓐ

Cleaning Copper Fitting.

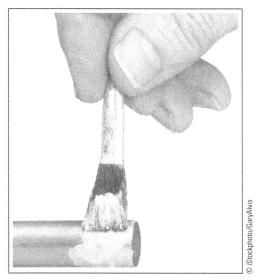

Applying Flux.

Soldering Using a Propane Torch (Continued)

Using a Propane Torch

- Before igniting a propane torch, always check the torch for leaks. If the torch has a leak, do not use it.

- Open the valve on the tank. Open the valve enough to hear gas escaping from the torch.

- Using a striker ignite the torch

- Adjust the torch valve to set the flame to the desired height and heat.

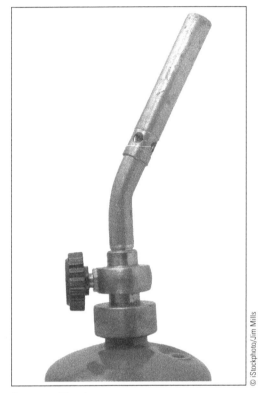

Propane Torch.

© iStockphoto/Jim Mills

Applying Heat

- Regardless of the type of heat source used to solder copper tubing, the following guidelines should be followed.

Ⓐ Start by applying heat to the edge of the fitting and tubing and then work the heat to the center of the fitting.

Ⓑ Never apply heat directly to the solder or an area that has been already soldered.

Ⓒ When heating copper tubing for soldering, start by holding the flame perpendicular to the copper tubing. This action is known as preheating and will evenly heat the inside of the socket of the fitting and the tubing. Do not move the flame onto the tubing, but instead move the flame onto the fitting away from the tubing a distance equal to the length of the socket of the fitting. To test the temperature of the tubing and the fitting, touch the solder to the joint; if it melts then the joint is ready to be sweated (soldered). If the solder does not melt, then reapply the heat.

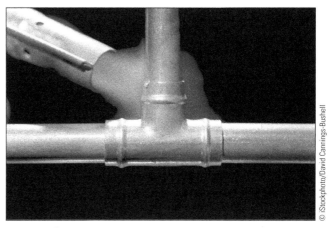

Heating Copper Tubing.

Shutting off the Torch

Ⓐ To extinguish the torch when finished, turn the valve to the off position.

CAUTION

- **Always wear safety glasses when soldering or brazing to avoid personal injury.**
- **Never point the torch toward another person while igniting or using the torch.**
- **Never use a match or cigarette lighter to light the torch.**
- **No not over heat the joint because this will cause the solder to run off the joint instead of being drawn into the joint.**

REVIEW QUESTIONS

1   List three types of plastic pipes commonly used in plumbing.

2   List the two main types of metal pipes used in residential and commercial plumbing.

3   List the three types of copper tubing used in residential and commercial plumbing.

4   List the steps for cutting copper tubing with a hacksaw.

5   List the steps for using a plunger to unclog a kitchen sink.

6   List the steps for unclogging a kitchen sink using a natural approach.

7   List the steps for unclogging a lavatory with a cleanout.

8   List the steps for using a drain cleaner to unclog a bathtub.

© iStockphoto.com/icphoto

Name: _____

Date: _____

Plumbing

TOOL IDENTIFICATION

Upon completion of this job sheet, you should be able to identify and know the use of any tools used for plumbing tasks at your facilities.

1. Inspect your facility. Identify and list the use of any tools used for plumbing tasks.

Tool	Use

2. Where are the tools stored?
3. Are instructions available for using the tools correctly? If not, do you think they should be?

INSTRUCTOR'S RESPONSE:

_____

_____

_____

Name: _____

Date: _____

Plumbing

PIPE IDENTIFICATION

Upon completion of this job sheet, you should be able to identify and know the use of pipes used at your facility.

Pipe Type	Use

Inspect your facility. Identify and list the uses of various types of pipes.

INSTRUCTOR'S RESPONSE:

_____

_____

_____

Name: _____

Date: _____

Plumbing

PIPE IDENTIFICATION

Upon completion of this job sheet, you should be able to identify the correct piping material for a particular task.

Match the pipe material to the application.

A. Copper   _____ 1 Gas pipe connected to a central heating unit.

B. Cast iron   _____ 2 Piping connecting a submergible potable pump to a storage tank.

C. Black steel   _____ 3 Primarily used for drain, waste vent application (black in color).

D. Galvanized steel _____ 4 Yellowish pipe used for potable hot water.

E. PVC   _____ 5 Sold in hard and soft versions and used in most plumbing applications and used by the HVAC industry.

F. ABS   _____ 6 Used primarily for commercial DWV and storm drainage systems.

G. Polyethylene   _____ 7 White flexible plastic-type tubing used for potable water and heating systems.

H. PEX   _____ 8 Black steel pipe lined with a protective coating that allows it to be used for potable water applications.

I. CPVC   _____ 9 Used for cold water applications and DWV.

INSTRUCTOR'S RESPONSE:

_____

_____

_____

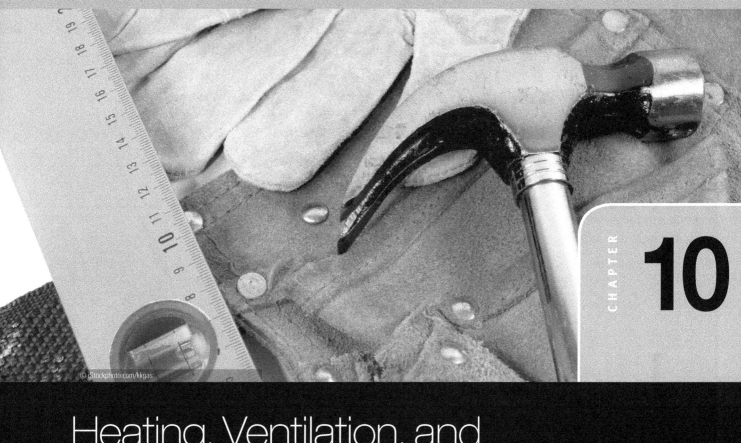

© iStockphoto.com/kkgas

CHAPTER 10

Heating, Ventilation, and Air-Conditioning Systems

OBJECTIVES

By the end of this chapter, you will be able to:

KNOWLEDGE-BASED

- Explain the importance of properly installing air filters.
- List the three common types of furnaces used.

SKILL-BASED

- Perform general maintenance procedures including the following.
- Perform general maintenance on a furnace.
- Tighten and/or replacing belts.
- Adjust and/or replacing pulleys.
- Replace filters on HVAC units.
- Maintain the heat source on gas-fired furnaces.

- Perform general maintenance of hot water or steam boilers.
- Perform general maintenance of an oil burner and boiler.
- Repair and replace electrical devices, zone valves, and circulator pumps.
- Light a standing pilot.
- Perform general maintenance of a chilled water system.
- Clean coils.
- Lubricate motors.
- Follow systematic diagnostic and troubleshooting practices.
- Maintain and service condensate systems.
- Replace through-the-wall air conditioners.

Evaporator a device used to transfer heat from the inside of a structure to the outside by passing air over its coils

Compressor the system component that raises the pressure and temperature of the refrigerant in the system

Condenser the system component that is used to eject heat from the refrigerant to the surrounding outdoor environment

Metering device the system component that removes from the system the remaining heat and pressure previously introduced by the compressor

Carbon monoxide the product of incomplete combustion

Carbon dioxide a gas created by burning natural gas

Introduction

Most people think that an HVAC system is strictly responsible for controlling the temperature of a structure. However, in reality HVAC systems are responsible for much more. They control not only the indoor air temperature but also the quality of the indoor air itself, that is, the purity and humidity levels are also affected by HVAC systems. This chapter on HVAC will provide a practical description and overview of the various HVAC equipment and systems used in both residential and commercial buildings.

HVAC Safety

As stated earlier, it is the responsibility of everyone to ensure that a work site is safe. When working on a jobsite, one should behave in a safe and professional manner. Accidents on the job often result from carelessness. It is very important for workers to be aware of their surroundings at all times and to evaluate the immediate area for possible safety hazards. The following are a few of the safety rules that should be followed when working with or around HVAC:

- Always deenergize electric circuits before working on them.
- Never work on electric circuits while standing on a wet floor or when not wearing rubber-soled boots. Shocks received when standing in a wet location are quite often deadly as the current passes through the heart, causing it to stop. Water and electricity do not mix! Stay dry and stay safe.
- Should a wire come loose from inside an air-conditioning system and come in contact with the casing of the equipment, electric shock can result by simply touching the surface of the unit.
- Never leave tools or other materials on the top platform of the ladder. If the ladder is moved by another individual, the object can fall, causing injury.
- Never engage in horseplay while at work.
- Always be aware of your surroundings and potential hazards.
- Dress properly for work, wearing long pants, long-sleeved shirts, and work boots.

- Remove metallic jewelry, as it is a good conductor of heat and electricity.
- Use safety glasses, ear plugs, and gloves for additional protection from dangerous conditions on the jobsite.
- Power tools and equipment should be grounded to protect against electric shock.
- Electric shock occurs when the body becomes part of an electric circuit.
- Ground wires and prongs should never be cut or disconnected.
- The ground fault circuit interrupter (GFCI) deenergizes a circuit when a current leak to ground is sensed.
- Be familiar with fire extinguishers, which are classified by the types of fires they are designed to be used on.
- Understand fire extinguisher use: pull, aim, squeeze, sweep (PASS).
- Always use tools for the tasks they are intended to perform.
- Handle and use chemicals according to the manufacturer's directions.
- Be prepared for injuries on the job and have a first aid kit handy.
- Be familiar with OSHA, NFPA, and ANSI guidelines, which help ensure safety in the work place.

The Refrigeration Process

The refrigeration process can be defined as the process of transferring heat from one location to another location. Typically, this consists of removing heat from within the structure and transferring it outdoors. The basic air-conditioning system is made up of four major components that enable the heat transfer process. These components are:

- Evaporator
- Compressor
- Condenser
- Metering device

The Evaporator

The **evaporator** is the system component that is directly responsible for performing the actual heat transfer in the occupied space (for more information regarding heat transfer see Chapter 18). In other words, it is the evaporator that is responsible for actual cooling of the structure. Heat is transferred to the refrigerant flowing through the evaporator from the air passing over the evaporator coils.

The Compressor

The HVAC **compressor** is the system component that raises the pressure and the temperature of the refrigerant leaving the evaporator. It is the increase in temperature that allows the heat transferred into the system to be released from the refrigerant.

The Condenser and the Metering Device

Once the temperature in the refrigerant has been raised by the compressor, it is the job of the **condenser** to eject the heat from the refrigerant into the surrounding outdoor environment. Once the refrigerant leaves the condenser, the remaining heat and high pressure introduced into the system by the compressor are reduced by the **metering device**.

Perform General Furnace Maintenance

There are three common types of furnaces: gas, electric, and oil. Because each furnace manufacturer has a different set of specifications, the following are general guidelines for performing furnace maintenance. For more detailed instructions on maintaining a furnace, read the manufacturer's specifications or call a qualified repair person. Some common problems associated with gas furnaces are outlined in Table 10-1.

Problem	Possible cause	Solution	Comments
Furnace won't run.			
	No Power	Check for: blown fuses tripped circuit breakers.	
	Switch off	Turn on power switch near the furnace.	
	Motor overload	Allow motor to cool and press the reset button.	
	Pilot light out	Relight pilot.	
	No gas	Confirm the gas valve is fully open.	
Not enough heat.			
	Thermostat set too low	Readjust the thermostat settings.	
	Filter dirty	Replace filter.	
	Blower clogged	Clean blower assembly.	
	Registers closed or blocked	Confirm all registers are open and not blocked.	If necessary consult an HVAC professional.
	System out of balance	Consult an HVAC professional.	
	Blower belt loose or broken	Adjust fan belt. If necessary replace it.	

TABLE 10-1: Common Problems with Gas Furnaces.

Problem	Possible cause	Solution	Comments
Pilot won't light.			
	Pilot opening blocked	Clear pilot opening.	
	No gas	Confirm that the pilot light is depressed. Confirm that the gas valve is fully open.	
Pilot won't stay lit.			
	Loose or faulty thermocouple	Tighten connections on thermostat. If necessary replace them.	
	Pilot flame set too low	Consult an HVAC professional.	
Furnace turns on and off repeatedly.			
	Filter dirty	Replace filter	
Blower won't stop running.			
	Blower control set wrong	Confirm that thermostat is set to "AUTO."	
	Limit switch set wrong	Consult an HVAC professional.	
Furnace noisy.			
	Access panels loose	Tighten screws to access panel. If necessary consult an HVAC professional.	
	Belts sticking, worn, or damaged	Replace belts.	
	Blower belts too loose or too tight	Adjust belts.	
	Motor and/or blower needs lubrication	Lubricate motor.	If necessary consult an HVAC professional.

TABLE 10-1: (*Continued*)

Tightening Belts

Loose belts on an air distribution system can cause the following:

- Insufficient airflow
- Evaporator coil freezing
- Inadequate cooling

If a belt is slipping, the inside surfaces of the pulleys will become polished to a near-mirror finish. If such is the case, the pulleys must be replaced. See the "Replacing Pulleys" section on page 390 for more on this. If the pulleys are not polished, proceed to adjust the belts:

1. Make certain that the power to the blower is off and that the blower itself has come to a complete stop.

Never attempt to stop rotating equipment or machinery with your hands. Severe personal injury can result.

Do not use a screwdriver or other similar object to pry the belt off the pulleys. Doing so can result in the slippage of the tool, causing severe personal injury.

2. Open the blower access panel/service door on the furnace. Check manufacturer's specifications to locate the blower access panel/service door.

3. With a pencil, mark the position of the motor mounts on the furnace and the bolt positions on the motor base (Figure 10-1).

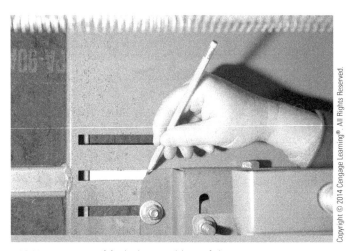

FIGURE 10-1: Mark the position of the motor mounts.

FIGURE 10-2: Loosen the motor mounts.

4. Using the proper size wrench, loosen the motor mount bolts so that the motor and the motor mount can move freely. Do not completely remove these bolts. Make certain to leave the motor secured to the motor mount itself (Figure 10-2).

5. Gently move the motor closer to the blower shaft and pulley to further loosen the belt (Figure 10-3).

6. Slide the belt off the pulleys.

7. Turn the belt inside out and inspect the underside for any cracks, missing pieces of the belt material, or other signs of excessive belt wear. (See Figures 10-4 and 10-5A and B.)

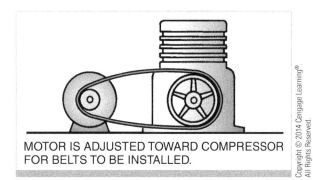

MOTOR IS ADJUSTED TOWARD COMPRESSOR FOR BELTS TO BE INSTALLED.

FIGURE 10-3: Adjust the motor to remove the belt.

FIGURE 10-4: Technician inspecting the belt.

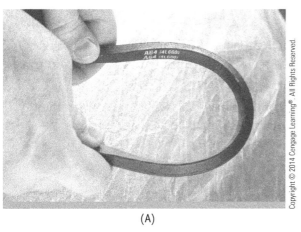
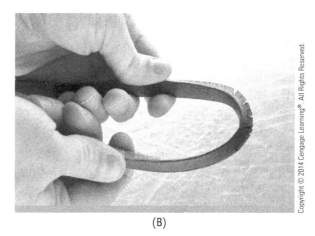

(A)   (B)

FIGURE 10-5: (A) A good belt. (B) This belt is cracked and needs to be replaced.

8. If the belt is worn, it needs to be replaced. Refer to the "Replacing Belts" section for more information on this. If the belt is in good condition, go to Step 9.

9. With the belt removed, inspect the interior surfaces of the pulleys. The interior surfaces of the pulleys should look rough and should not be shiny. Shiny, polished surfaces on the pulleys are an indication that the belts have been slipping. Belt slippage will cause premature wear. Polished pulleys should be replaced. See the "Replacing Pulleys" section for more on this (see Figure 10-6).

10. If both the belt and pulleys are in good condition, position the motor close to the blower pulley and replace the belt on the pulleys.

11. Gently push the motor away from the blower pulley to increase the belt tension.

12. When the motor mount is slightly past the pencil markings that were made earlier, begin to tighten the motor mounts to the chassis. Make certain that the new position of the motor mount on the chassis is parallel to the original markings to help ensure proper pulley alignment.

13. Check the belt tension by placing your thumb and fingers on the opposite sides of the belt and gently squeezing them together. Basically, you are trying to squeeze the two opposite sides of the belts together. There should be some play in the belts, but no more than one inch of deflection (see Figure 10-7).

14. If the sides of the belts can be pushed in more than 1 inch, loosen the motor mounting bolts again and repeat Steps 11 through 13.

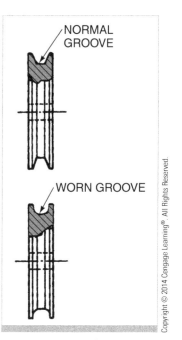

NORMAL GROOVE

WORN GROOVE

FIGURE 10-6: Comparison between normal and worn pulleys.

 CAUTION

Do not use a screwdriver or other similar object to replace the belt on the pulleys. Doing so can result in the slippage of the tool, causing severe personal injury.

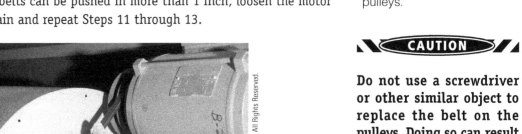

FIGURE 10-7: Check the belt tension.

Replacing Belts

If the belt is broken, damaged, or worn, do not reinstall that belt on the system. Damaged and worn belts should be replaced immediately to help ensure the satisfactory, continued operation of the system.

1. Obtain the belt information from the old belt. If the information on the old belt cannot be read, refer to the "Estimating Belt Sizes" section (Figure 10-9).

2. Obtain a new, exact replacement for the old belt.

FIGURE 10-8: Tension gauge.

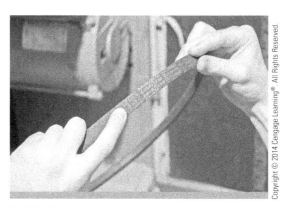

FIGURE 10-9: Belt information.

3. If the belt has been removed from the pulleys and is damaged, proceed to Step 6. If the belt simply broke, continue with Step 4.

4. With a pencil, mark the position of the motor mounts on the furnace and the bolt positions on the motor base.

5. Using the proper size wrench, loosen the motor mount bolts so that the motor and the motor mount can move freely. Do not completely remove these bolts, and make certain to leave the motor secured to the motor mount itself.

6. Position the motor close to the blower pulley, and replace the belt on the pulleys.

7. Gently push the motor away from the blower pulley to increase the belt tension.

8. When the motor mount is slightly past the pencil markings that were made earlier, begin to tighten the motor mounts to the chassis. Make certain that the new position of the motor mount on the chassis is parallel to the original markings to help ensure proper pulley alignment.

9. Check the belt tension by placing your thumb and fingers on the opposite sides of the belt and gently squeezing them together. Basically, you are trying to squeeze the two opposite sides of the belts together. There should be some play in the belts, but no more than 1 inch of deflection.

10. If the sides of the belts can be pushed in more than 1 inch, loosen the motor mounting bolts again and repeat Steps 7 through 9.

Estimating Belt Sizes

There will be times when you will not be able to read the information on an old belt. This can be the result of excessive amounts of dirt, age, or simply the destruction of the belt itself. In order to determine important belt information, follow these steps.

1. Measure the center-to-center distance (in inches) between the motor shaft and the blower shaft (Figure 10-10).
2. Multiply the measurement in Step 1 by 2.
3. Measure the diameters of the drive pulley and the driven pulley (Figure 10-11).
4. Add the two diameters in Step 3 together.
5. Divide the previous answer obtained in Step 4 in half.
6. Multiply the result from Step 5 by 3.14.
7. Add the result from Step 6 to the result from Step 2. This gives you the approximate length of the required belt.
8. Measure the width of the old belt. An "A" belt has a width of $\frac{17}{32}$ inch, while a "B" belt has a width of $\frac{21}{32}$ inch (Figure 10-12).
9. The results from Steps 7 and 8 provide the type of belt and the approximate length of the required belt.

FIGURE 10-10: Measure center to center.

FIGURE 10-11: Measure pulley diameters.

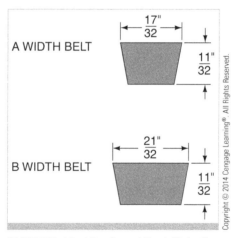

FIGURE 10-12: A and B width belts.

Sample calculation: Estimate the length of a belt that is used to connect an 8-inch pulley to a 10-inch pulley that is installed on motor and blower shafts that are 16 inches apart.

Here is the step-by-step solution, which corresponds to the steps in the original procedure:

1. The center-to-center distance between the motor shaft and the blower shaft is 16 inches.
2. $16 \times 2 = 32$ inches
3. Pulley diameters = 8 inches and 10 inches
4. 8 inches + 10 inches = 18 inches
5. 18 inches ÷ 2 = 9 inches

CAUTION

If the belt sits deep into one or both pulleys, determine the "effective" diameter of the pulley, which is the diameter of an equivalent pulley if the belt was resting at the outer edge of the pulley.

6. 9 inches $\times$ 3.14 = 28.26 inches

7. 28.26 inches + 32 inches = 60.26 inches = 60 inches

Adjusting Pulleys

Quite often, the cause of belt slippage, breakage, and premature wear is misaligned pulleys. To check pulley alignment:

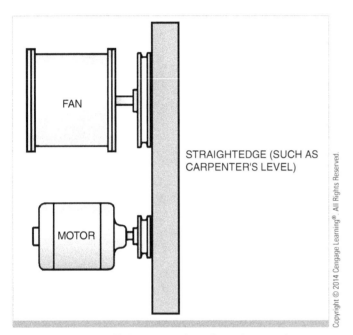

FIGURE 10-13: Pulleys must be aligned properly.

1. Make certain that the system is off and all rotating equipment has come to a complete stop. Never attempt to stop rotating equipment by hand. Severe personal injury can result.

2. Remove the access panel on the blower compartment. Check the manufacturer's specifications to locate the blower access panel/service door.

3. Place a straight edge such as a wooden ruler against the faces of both the drive and the driven pulleys (Figure 10-13).

4. The straight edge should touch all four sides of the pulleys:
   a. Outside edge of the drive pulley
   b. Outside edge of the driven pulley
   c. Inside edge of the drive pulley
   d. Inside edge of the driven pulley

5. This will determine not only if the pulleys are lined up, but also if they are parallel to each other.

6. If the pulleys are not properly aligned but are parallel to each other, either the drive pulley or the driven pulley will have to be repositioned on the respective shaft. See the section on "Repositioning Pulleys."

7. If the pulleys are not parallel to each other, the motor mount or the blower mount must be adjusted to correct this situation.

Replacing Pulleys

If you have decided to replace one or more pulleys, the original pulley must be removed from the shaft. The shaft may be dirty and/or rusty, making this project potentially very time-consuming. Here are some tips to accomplish this:

1. Clean the shaft completely.

2. Spray the shaft with a loosening agent (rust remover). See Figure 10-14.

3. Allow the loosening agent to seep into the joint between the shaft and the hub of the pulley.

4. Completely remove the set screw that holds the pulley to the shaft. This may be a square set screw or an Allen wrench (Figure 10-15).

FIGURE 10-14: Spray the pulley hub and shaft.

FIGURE 10-15: Use an Allen wrench to loosen set screw.

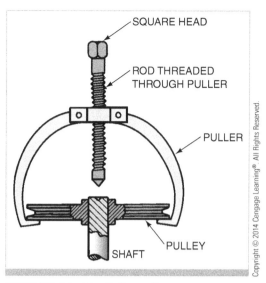

FIGURE 10-16: Pullet puller.

5. Be sure to place the set screw in a place where it will not be lost.

6. Spray loosening agent in the set screw hole, and allow it to seep into the space between the pulley and the shaft.

7. If the preceding tips do not help in the pulley removal process, a pulley puller should be used (Figure 10-16).

8. Once the pulley has been removed, clean the shaft completely.

9. Obtain the new pulley and position the new pulley on the shaft so that it is perfectly aligned with the other pulley. Refer to the previous section on "Adjusting Pulleys" for more on this.

10. Once the pulley has been properly positioned, tighten the pulley securely to the shaft.

Repositioning Pulleys

There are times when the pulley is in good shape but needs to be repositioned on the shaft. Use the procedures, steps, and tips in the previous two sections to loosen, reposition, align, and retighten the pulley on the shaft.

Replacing Filters on HVAC Units

The efficiency and overall performance of an air-conditioning system can be affected by many different factors, such as overcharging or undercharging the system; however, for a facility maintenance technician ensuring that the air filter is replaced on a regular schedule will go a long way in helping maintain the systems efficiency and performance. When air filters are not replaced on a regular schedule, contaminated air can be pushed across the evaporator coil and into the conditioned space. It is recommended that the facility maintenance technician change air-conditioning filters monthly.

1. Make certain that the system is off and all rotating equipment has come to a complete stop. Never attempt to stop rotating equipment by hand. Severe personal injury can result.

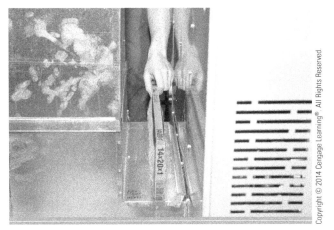

FIGURE 10-17: Remove the filter.

FIGURE 10-18: Measure the filter rack.

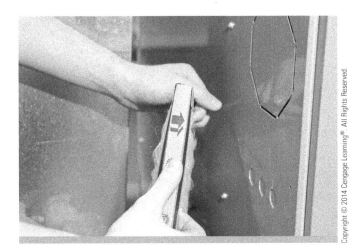

FIGURE 10-19: Directional arrow on the filter.

2. Remove existing filter(s) from the system (Figure 10-17).

3. Inspect the channel that holds the filters to be sure that the channels are in good shape and that the filter is supported on at least two sides.

4. Measure the filter channel (Figure 10-18).

5. Make certain that the replacement filter is the same size as the filter channel, not necessarily the size of the filter that came out of the unit. (Someone may have put the wrong size filter in the unit.)

6. Obtain the correct size filter.

7. Locate the arrow on the edge of the filter (Figure 10-19).

8. Install the filter in the channel with the arrow on the filter pointing in the direction of airflow, which is toward the blower.

9. Mark the edge of the filter with the date and your initials (Figure 10-20).

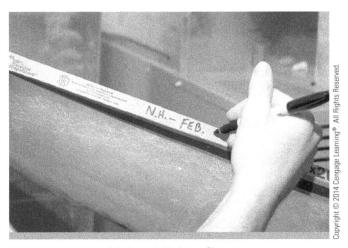

FIGURE 10-20: Mark or initial the filter.

10. Once the filter has been installed, inspect the filter and the channel to be certain that no air is able to bypass the filter.

11. Seal any and all air leaks to prevent/eliminate air bypass.

12. Make certain that any filter channel covers are replaced and secured.

Maintaining the Heat Source on Gas-Fired Furnaces

Servicing fossil fuel systems should be done by trained professionals, but there are a number of things that the maintenance technician can do to help ensure that the equipment remains in good working order.

1. Typically, the fuel-burning portion of the system does not need to be adjusted each year. So, for best results, do not change any of the current settings on the system.

2. Perform a visual inspection of the burners, pipes, and manifold arrangement. Make certain that these components are free from dirt, dust, and rust. If you find excessive rust and rust-related damage, call for professional service immediately.

3. If excessive dirt or rust is present, you can carefully remove the gas manifold for cleaning. Follow these steps:
   a. Close the manual gas valve that feeds gas to the appliance.
   b. Disconnect the manifold, making certain to carefully disconnect any components that are attached to the piping (Figure 10-21).
   c. Blow out the manifold with pressurized air, making certain to wear the appropriate personal protection equipment to protect yourself from airborne particles (Figure 10-22).
   d. After the cleaning is complete, reassemble the manifold.
   e. Make certain that all piping connections are tight.
   f. Open the manual gas valve.
   g. Restart the system.

Notes to keep in mind about filter replacement:

- Air that bypasses the filter will cause dirt and dust to accumulate on the coils, blowers, air distribution system components, supply registers, and ultimately end up back in the occupied space.

- Be sure to have an ample supply of filters on hand.

- As an absolute minimum, change filters at the beginning of the heating and cooling seasons. It is recommended, however, that you change them every month.

- If the filters are metal permanent filters, they can be cleaned.

- If it has been determined that air filters are missing or too small, be sure to visually inspect the return side of the evaporator coil and make certain it is free from dirt and dust.

FIGURE 10-21: Disconnect the gas manifold.

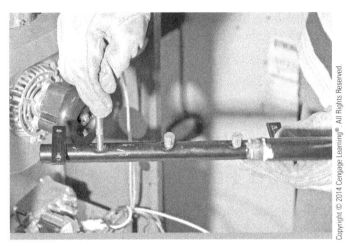

FIGURE 10-22: Clean the manifold.

4. Observe the burners when they are lit. The flames should be bright blue with slightly orange tips. If the flames are yellow or are blue with yellow tips, **carbon monoxide** (the product of incomplete combustion) is present. If yellow flames and/or tips are noticed, seek the assistance of a professional immediately (Figure 10-23A).

5. When burning, the flames should rest just above the burners. Flames that are too high above the burner indicate that there is too much air being introduced. Call for help (Figure 10-23B).

6. Flames should be uniform. Erratic flames may be an indication of a system in need of professional adjustment (Figure 10-23C).

7. Schedule a combustion test/analysis on an annual basis, before the beginning of the heating season.

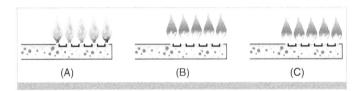

FIGURE 10-23: Proper and Improper flames.

Perform General Maintenance of a Hot Water or Steam Boiler

Only experienced contractors should work on boilers. If any of the following conditions are present, a professional should be called in to examine, troubleshoot, and remedy the situation:

1. Water accumulates on the floor around the boiler.

2. Water drips from the pressure relief valve.

3. Heat source fails to energize after one attempt to reset/restart the system has failed.

4. Boiler fails to operate after water has been added to the system.

5. Individual zones fail to heat after attempts to bleed air from the system have failed.

6. Individual zones fail to heat after attempts to troubleshoot zone valves have failed.

Perform General Maintenance of an Oil Burner and Boiler

Oil burners typically require regular service to ensure continued satisfactory system operation. Here is a list of items that must be addressed as well as some suggestions for keeping oil-fired heating systems in tip-top shape.

1. Schedule a combustion analysis before the start of the heating season. If the oil-fired equipment is used to supply domestic hot water year round, this should be done more frequently. As shown in Figure 10-24, the combustion analysis test should include:
   a. Smoke test
   b. Carbon monoxide level
   c. Carbon dioxide level (is created by burning natural gas)
   d. Stack temperature
   e. Draft test

2. Keep the oil tank as full as possible. The more oil there is in the tank, the less likely that condensation will form and accumulate in the oil. Water mixed with the oil can result in combustion and operational problems.

3. Clean the heat exchanger on the equipment. To do this, follow the manufacturer's recommendations. The steps are:
   a. Disconnect the flue pipe connection.
   b. Use a boiler brush and vacuum to clean the spaces between the boiler sections.
   c. Wear a protective dust mask (Figure 10-25).

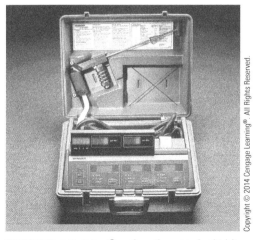

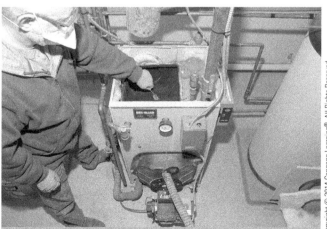

FIGURE 10-24: Combustion analysis kit.   **FIGURE 10-25:** Boiler being cleaned.

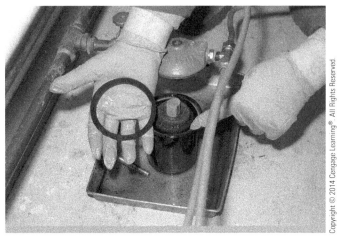

FIGURE 10-26: A gasket.

4. Replace the oil filter.
   a. Make certain that the oil valve line is in the closed position.
   b. Unscrew the existing oil filter.
   c. Replace the filter, making certain that the filter canister gasket is in place (Figure 10-26).
   d. Make certain that the new filter is tight.
   e. Dispose of the oil filter as you would any other hazardous material (Figure 10-27).
5. Check and clean the flue pipe.
   a. Be sure to wear a protective dust mask.
   b. Make certain that a high-quality (filtering) vacuum is used (Figure 10-28).

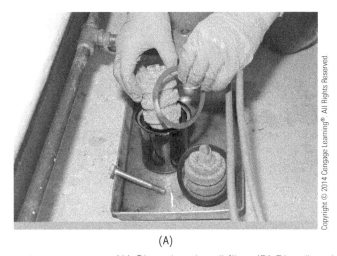

(A) (B)

FIGURE 10-27: (A) Changing the oil filter. (B) Bleeding the line.

FIGURE 10-28: Cleaning the flue.

   c. Be sure to reassemble the flue pipe when finished.
   d. Make certain that the flue pipe is sloped upward toward the chimney.
   e. Inspect the chimney if possible and remove any obstructions or debris from the chimney.
6. Check the oil tank for water accumulation.
7. Inspect the area around the unit for traces of oil.
8. Visually inspect the oil lines for damage.
9. Make certain that fill pipe and vent caps are in place.
10. Check for oil leaks in the tank area.
11. Check for unusual oil odors.
12. Inspect the sight glass (steam boilers only).
    a. If the water level is low, add water to the system via the feed valve.
    b. If the system is losing water at a very fast rate, call for service (Figure 10-29).

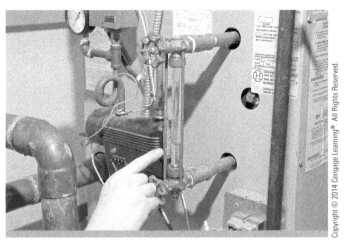

FIGURE 10-29: Sight glass on a steam boiler.

FIGURE 10-30: Disconnect the oil line from the burner.

13. Remove the burner from the unit.
    a. Place a drop cloth or other barrier between the boiler and the floor.
    b. Make certain that the main oil line is closed.
    c. Disconnect the oil line to the burner, making certain that you have rags on hand for any oil droplets (Figure 10-30).
    d. Disconnect power to the boiler.
    e. Disconnect the wiring connections to the burner, making certain to discharge any capacitors (Figure 10-31).
    f. Place the burner on the floor or work bench.

14. Inspect the combustion chamber.
    a. Look for cracks in the refractory.
    b. Look for and clean up any soot build-up (Figure 10-32).

FIGURE 10-31: Disconnect the power supply.

FIGURE 10-32: Inside a combustion chamber.

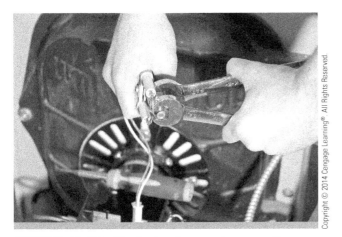

FIGURE 10-33: Replacing the nozzle on an oil heating system.

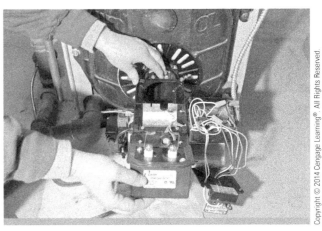

FIGURE 10-34: Opening the oil burner.

15. Replace the nozzle (Figure 10-33) and check the firing assembly. You will need to access the manufacturer's guidelines for this, as each manufacturer and each oil burner model has different procedures for accessing/removing the firing assembly.
    a. Make certain that the new nozzle is an exact replacement for the existing part. Be sure to have plenty of spares on hand.
    b. Be careful to not damage the electrode porcelains.
    c. Inspect porcelains for damage.

16. Clean the cad cell (on systems that are equipped with them).
    a. Open the top of the oil burner (Figure 10-34).
    b. Locate the cad cell.
    c. Wipe the cell down to remove accumulated dirt (Figure 10-35).

17. Clean the transformer springs.
    a. Double check to make certain there is no power being supplied to the unit.
    b. Open the top of the oil burner.
    c. Clean the transformer springs (Figure 10-36).

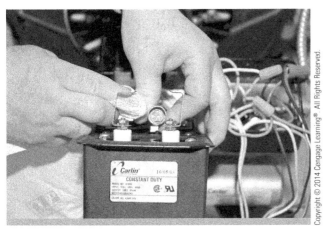

FIGURE 10-35: Cleaning the cad cell.

FIGURE 10-36: Cleaning the springs.

Repair and Replace Electrical Devices, Zone Valves, and Circulator Pumps

Here are some general suggestions and tips for replacing electrical devices on heating and air-conditioning equipment.

1. Make certain that *all* electrical power sources are deenergized. Keep in mind that some heating and air-conditioning systems are powered by more than one power source, so even if one source is off, there may still be power to some system components.

2. Make certain that all system capacitors are discharged to avoid receiving an unexpected electric shock.

3. Make certain that an exact replacement, whenever possible, for the component being replaced has been obtained.

4. In the event an exact replacement component is not available, make certain that the replacement component ratings match those of the original as closely as possible.

5. When replacing components that are directly connected to a water-carrying piping arrangement, make certain that the water from the system has been completely drained to avoid an unexpected flood.

   a. If a zone valve motor is being replaced, there is no need to drain the water system, as the valve mechanism will remain intact.

   b. If the entire zone valve is being replaced, however, the water system must be drained, as the water circuit will be accessed.

   c. If the motor on a circulator pump is being replaced and the linkage/impeller assembly is remaining in the system, the water does not need to be drained.

   d. If the entire circulator is being replaced, the system must be drained.

6. Disconnect the electrical splices one at a time, making certain to label or tag the wires so that you will be able to identify the wires when it comes to reconnecting them (Figure 10-37).

7. Carefully remove the defective component, making certain to place any screws and other small parts in a container such as a cup to prevent them from getting lost (Figure 10-38).

8. Mount the new component in place before making the electrical connections.

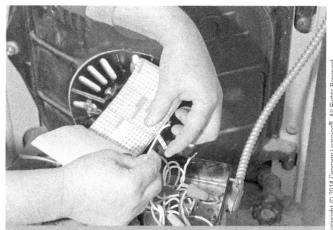

FIGURE 10-37: Tagging the wires.

FIGURE 10-38: Place screws and small parts in a plastic bag.

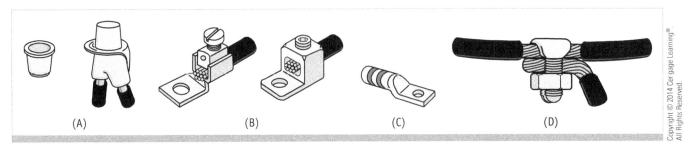

FIGURE 10-39: Types of wire connectors.

9. Once the device has been securely mounted, begin connecting the wires one at a time, making certain that the new wiring corresponds to the wiring of the original device. Be sure to use wire nuts or other mechanical connectors (Figure 10-39).

10. Once completed, make certain that the electrical connections are tight and that no bare wire is extending from the underside of the wire nuts (Figure 10-40).

FIGURE 10-40: Side-by-side not connections done correctly and incorrectly.

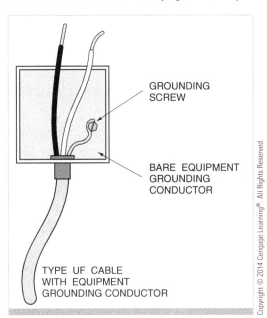

GROUNDING SCREW

BARE EQUIPMENT GROUNDING CONDUCTOR

TYPE UF CABLE WITH EQUIPMENT GROUNDING CONDUCTOR

FIGURE 10-41: Ground wire connected to a metal box.

11. Make certain that no bare current-carrying conductors are making contact with the frame or casing of the device.

12. Make certain that all ground wires are properly connected (Figure 10-41).

Lighting a Standing Pilot

When you perform any procedure on a piece of equipment, it is always recommended that you use the manufacturer's recommendations and procedures before using these general procedures and suggestions. Here are the steps used to light or relight a standing pilot:

1. If the gas valve is in the ON position, close the valve by turning the knob to the OFF position and allow 15 minutes for any unburned fuel to rise through the appliance.

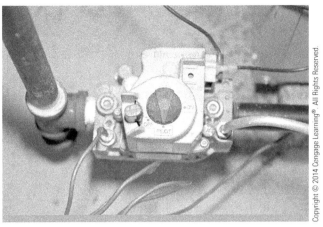

FIGURE 10-42: Standing gas valve.

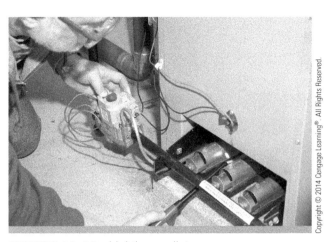

FIGURE 10-43: Lighting a pilot.

2. Set the gas valve knob to the PILOT position and push the knob into the valve (Figure 10-42).

3. At the same time, light the pilot by using a large barbeque-type match to avoid coming in close contact with the pilot light (Figure 10-43).

4. After the pilot light is lit, continue pressing the knob on the gas valve for approximately 1 minute, after which release the knob.

5. If the pilot light fails to remain lit repeat Steps 2, 3, and 4.

6. Position the knob on the gas valve to the ON position (Figure 10-44).

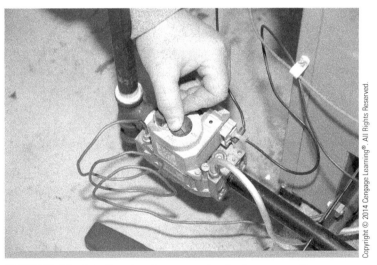

FIGURE 10-44: Turning the gas valve to the on position.

Perform General Maintenance of a Chilled Water System

Since chilled water systems contain refrigerants, only qualified air-conditioning technicians should access the refrigeration circuits. The EPA requires that all technicians who work on the refrigeration circuits of air-conditioning and refrigeration system be certified under Section 608 of the Clean Air Act. However, a number of items can be checked by uncertified maintenance personnel.

1. Inspect the piping circuits for signs of leakage, oil, and damage.
2. Inspect the water pumps.
   a. Measure the amperage of the pumps and verify that the amperage is within an acceptable range.
   b. Listen to the pumps and make note of any unusual or abnormal noises and vibrations.
3. Check the system thermometers.
   a. Water being supplied to the chilled water coil should be in the range of 45°F.
   b. Water returning from the chilled water coil should be in the range of 55°F.
4. Measure the temperature difference between the return air temperature and the supply air temperature. This difference should be between 16°F and 20°F.
5. Make certain that all air filters on the air distribution system are clean.
6. Make certain that the blower/motor assembly is operational.
   a. Check pulley alignment.
   b. Check belts for cracks and damage.
   c. Check motor amperage.

Clean Coils

If air filters are properly installed and air is not permitted to bypass the filters, there should be very little, if any, dirt accumulation on the return air side of the evaporator coil. On occasion, though, system air filters are removed or not properly installed and air is permitted to bypass. To determine that the evaporator coil is actually clean, visually inspect the coil. This may be a difficult task given the configuration of the system and the installation practices employed. Examining the coil may be as easy as removing the access panel on the air handler or may involve having to cut an access door into the duct system if the coil is mounted on top of a furnace. In any event, once the coil has been inspected and it is determined that the coil is indeed in need of a cleaning, here are some tips and suggestions for doing so:

1. Make certain that the system has been turned off and that all rotating machinery has stopped.
2. Make certain that the area is well ventilated as some cleaning agents give off fumes that may irritate the skin or eyes.
3. Wear proper personal protection equipment such as safety glasses and gloves.
4. Using a brush, remove as much of the dirt as possible (Figure 10-45).
5. Avoid using a rigid wire brush, as the bristles may cause damage to the coil.
6. Avoid flattening the fins on the coil, as this will have a negative effect on the airflow through the coil.

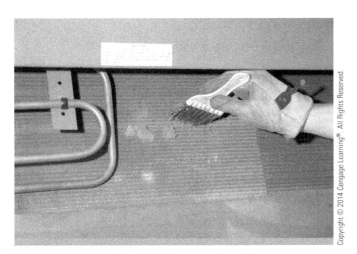

FIGURE 10-45: Brush the evaporator coil.

FIGURE 10-46: Directions on the coil cleaner.

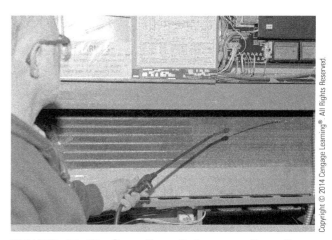

FIGURE 10-47: Spray the coil cleaner on the coil.

7. Avoid getting cut on the evaporator coil fins. They are sharp and cuts received from them are very painful.

8. Mix the coil cleaner as directed on the product label (Figure 10-46).

9. Apply the water/cleaner mixture to the coil with a high-pressure sprayer and allow it to sit, allowing the chemicals to break up the accumulated dirt on the coil surface. Make certain that the strength of the high-pressure sprayer is weak enough to prevent the bending of the coil fins (Figure 10-47).

10. After the manufacturer's suggested time period, rinse the coil with high-pressure water.

11. Do not use a water hose connected to the building's water supply, as this can damage the coil fins and saturate the duct system.

12. Depending on the amount of dirt on the coil, it may be necessary to repeat Steps 9 and 10 to ensure that the coil is as clean as possible.

13. Once the coil has been cleaned, reseal the duct if it was necessary to cut in an access door. If such is the case, be sure to avoid damaging or piercing the refrigerant lines with screws (Figure 10-48).

FIGURE 10-48: Comparison of clean and dirty coil.

Lubricate Motors

Periodic motor lubrication is often required to keep the motors in good working order. Permanently lubricated motors do not need to be lubricated, but all others do. Unless otherwise specified, use a medium (20-weight) oil.

1. Remove the oil port plugs from the oil ports on the motor. Typically, there are two oil ports on a motor, so be sure to remove them both (Figure 10-49).

2. Place the oil plugs in a safe place to avoid losing them. Not replacing the oil plugs can allow dirt and dust into the motor, affecting the operation of the component.

3. Insert the oil port on the oil container into the oiling tubes (Figure 10-50).

4. Depending on the motor, you should add between three and six drops of motor oil into each port.

5. Make certain to replace the oil port plugs.

6. Do not over-lubricate motors.

7. If the motor is equipped with grease fittings, the lubrication process is different:
   a. Loosen and remove the relief screw on the motor. The relief screw is located on the opposite side of the motor as the grease fitting (Figure 10-51).
   b. Using a grease gun, pump grease into the motor until grease leaves through the relief screw.
   c. Replace the relief screw.
   d. Repeat this for both sides of the motor.

FIGURE 10-49: Remove oil plug from motor.

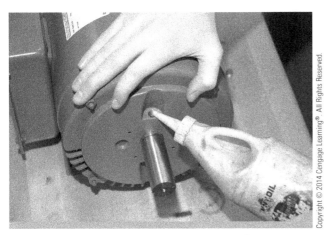

FIGURE 10-50: Insert oil spout into the oil port.

Follow Systematic Diagnostic and Troubleshooting Practices

When you attempt to diagnose a system problem, using a systematic approach is best. Consider these tips when you encounter a problem:

1. If the system appears to be OFF when it is turned on and operation is desired, check the main power supply to the piece of equipment. Very often, a circuit breaker or switch has been inadvertently turned off.

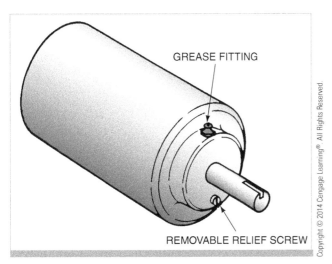

GREASE FITTING

REMOVABLE RELIEF SCREW

FIGURE 10-51: Remove relief screw.

2. Check for proper airflow through the air distribution system (furnace and air-conditioning applications).
   a. Reduced airflow can be responsible for a multitude of problems.
   b. Check air filters.
   c. Check belts and pulleys.

3. Check the operational controls on the system.
   a. If a zone valve is not opening, make certain that the thermostat for that zone is calling for heat.
   b. If a boiler is not operating and the water pipes are hot, the water temperature may have reached the desired temperature.
   c. If an air-conditioning system is not operating, make certain that the thermostat is set for cooling before calling for service.
   d. Check for low voltage by switching the fan switch to the ON position (furnace and cooling applications).
   e. If a boiler fails to operate, check safety devices such as high-limit controls, pressure controls, and/or low water cutoff switches.

4. Check for safety or trouble lights.
   a. Many newer controls have trouble and/or diagnostic lights.
   b. Keep all manufacturers' literature on hand. This paperwork contains valuable information regarding steps you can take to keep equipment up and running.
   c. Manufacturers' literature contains the trouble codes that often appear on the control's display.

5. Check system operating temperatures and/or pressures.
   a. Determine which parameters are within acceptable ranges.
   b. Determine which parameters are not acceptable.

6. Write down your findings.
   a. Keep logs of your findings.
   b. This may help evaluate future system problems.
   c. Keep records of what was done and when.
   d. This will also help service technicians who do the job.

7. Narrow your search.
   a. Eliminate items that are definitely operating properly.
   b. Make a list of possible system problems and examine/eliminate them as needed.

8. Ask yourself *why*?
   a. Be sure to fix the cause, not the effect.
   b. Fixing the effect does not fix the underlying problem.

Maintain and Service Condensate Systems

An integral part of the operation of an air-conditioning system is the ability to remove condensate from the structure. Quite often, condensate pumps are used to accomplish this. Depending on the location of the system, condensate pump failure can result in water damage to the structure. Even if there is no condensate pump

FIGURE 10-52: Damaged condensate pan.

FIGURE 10-53: Pour water into condensate drain pan.

being used, a gravity-type condensate removal system can cause damage in the event that that line becomes clogged. Here are some tips and suggestions for maintaining condensate removal systems.

1. Inspect the condensate pan under the evaporator for:
   a. Signs of rust
   b. Signs of damage
   c. Dirt, dust, and debris accumulation
   Repair and clean as needed (Figure 10-52).

2. Test condensate lines by pouring a significant amount of water into the drain pan located under the evaporator coil.
   a. Observe the rate of water drainage.
   b. Stop pouring water into the line if water does not drain.
   c. Inspect the area around the condensate drain pan for signs of water (Figure 10-53).

3. Inspect the termination point of the line.
   a. If the line terminates outdoors, observe the end of the line before introducing water to the line and again afterward.
   b. Make certain that the water is actually leaving the structure.
   c. If the line terminates in a condensate pump, make certain that the water is indeed ending up in the pump.

4. If the line is not draining, use pressurized air to blow out the line. Repeat Steps 2 and 3 to ensure that the line is now draining.

5. On systems with condensate pumps, make certain to check the operation of the pump.

6. Make certain that the pump remains plugged in or, better yet, have the pump hard-wired to ensure that there is constant power to the pump (Figure 10-54).

7. Test the pump operation by adding water to the pump and inspecting the end of the discharge pipe connected to the outlet of the pump, as in Step 3.

8. Inspect the area around the condensate pump for signs of water.

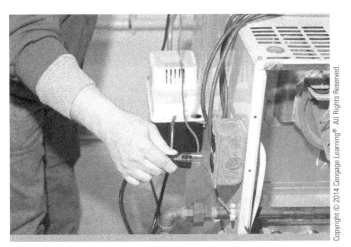

FIGURE 10-54: Condensate pump.

Replace Through-the-Wall Air Conditioners

If you have decided to replace a through-the-wall air conditioner, perform the following steps to replace it (Figure 10-55):

1. Remove the existing unit from the sleeve (Figure 10-56).

Note: Be sure to place a drop cloth or other protective barrier on the floor below the unit to protect the floor from any sharp edges on the unit.

2. Obtain all information from the unit, including the make, model, serial number, voltage rating, amperage rating, and plug type from the unit.

3. Take all unit measurements as well as the internal (daylight opening) measurements of the existing sleeve.

4. Slide the existing unit back into the sleeve.

5. With the acquired information, obtain a replacement unit, making certain that *all* measurements and specifications match those of the existing unit.

6. Uncrate and inspect the new unit.

7. Check to make certain that the sizes are correct, the voltage and amperage ratings are the same as the old unit, and the plug is the same.

8. Remove the old unit from the sleeve.

9. Clean and vacuum out the existing sleeve.

10. Slide the new unit into the existing sleeve and secure it according to the manufacturer's installation literature.

11. Make certain that all shipping materials have been removed from the unit and that the air filter is in place before putting the unit into operation.

FIGURE 10-55: Through-the-wall air conditioner.

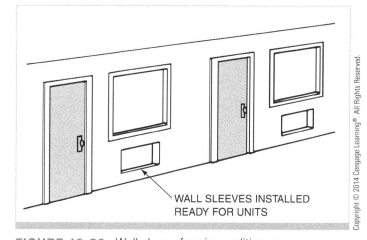

FIGURE 10-56: Wall sleeve for air conditioner.

SUMMARY

- There are three common types of furnaces in use today: gas, electric, and oil.

- Loose belts on an air distribution system can cause insufficient airflow, evaporator coil freezing, and inadequate cooling.

- Misaligned pulleys are often the cause of belt slippage, breakage, and premature wear.

- When removing a pulley from a dirty shaft, use steel wool or fine grit sandpaper to remove the dirt.
- For the efficiency and overall performance of an air conditioning system, a facility maintenance technician should be sure the air filter is replaced on a regular schedule.
- Only experienced contractors should work on boilers.
- Oil burners typically require regular service to ensure continued satisfactory system operation.
- When you perform any procedure on a piece of equipment, it is always recommended that you use the manufacturer's recommendations and procedures before using these general procedures and suggestions.
- If air filters are properly installed and air is not permitted to bypass the filters, there should be very little, if any, dirt accumulation on the return air side of the evaporator coil.
- Periodic motor lubrication is often required to keep the motors in good working order. Permanently lubricated motors do not need to be lubricated, but all others do.

REVIEW QUESTIONS

1 List three common types of furnaces used in residential and commercial environments.

2 List three symptoms of a loose belt on an air distribution system.

3 List the steps for tightening the belt on an air distribution system.

4 List the steps for replacing a belt on an air distribution system.

REVIEW QUESTIONS

5 Estimate the length of a belt used to connect a 6-inch pulley to a 12-inch pulley that is installed on motor and blower shafts that are 14 inches apart. Show your work.

7 List the steps for lubricating a motor.

6 How often should an air filter be replaced on a HVAC system?

8 List the steps for replacing a through-the-wall air conditioner.

Name: _____

Date: _____

HVAC

PERFORM GENERAL INSPECTION ON A FURNACE

Upon completion of this job sheet, you should be able to perform a general inspection on a furnace system.

Type of furnace: _____

Model: _____

Last maintenance date: _____

Maintained by: _____

1. Describe the general running condition of the furnace.

   _____

   _____

2. List four possible fan motor mechanical problems.

   _____

   _____

3. Arrange the following troubleshooting steps in the correct order.
   - Test the system operation.
   - Gather information.
   - Verify the complaint.
   - Complete the service call.
   - Perform the visual inspection.
   - Isolate the problem.
   - Correct the problem.

4. List the three key indicators that pulleys are aligned properly on the blowers.

   _____

   _____

   _____

INSTRUCTOR'S RESPONSE:

   _____

   _____

   _____

HVAC

REPLACING BELTS

Upon completion of this job sheet, you should be able to replace belts on a furnace motor.

Type of furnace: _____

Model: _____

Last maintenance date: _____

Maintained by: _____

Check off the following tasks as they are completed:

Task Completed		Task:
____	1	Obtain belt information from the old belt.
____	2	Obtain a new, exact replacement for the old belt.
____	3	With a pencil, mark the position of the motor mounts on the furnace and the bolt position on the motor base.
____	4	Loosen the motor mounts so the motor and motor mounts can move freely.
____	5	Replace the belt on the pulleys.
____	6	Position the motor, based on the marks in Step 3, to increase belt tension and tighten the motor mounts.
____	7	Check the belt tension.

INSTRUCTOR'S RESPONSE:

_____

_____

_____

Name: _____

Date: _____

HVAC

LIGHTING A STANDING PILOT

Upon completion of this job sheet, you should be able to light a pilot light on a gas furnace.

Type of furnace: _____

Model: _____

Last maintenance date: _____

Maintained by: _____

Check off the following tasks as they are completed:

Task Completed		Task:
____	1	Take the access cover off the furnace and look for the gas control knob.
____	2	Turn the knob to OFF and allow 15 minutes to allow unburned fuel to rise through the furnace.
____	3	Turn the knob until the arrow is pointing to the word "Pilot."
____	4	Push the knob to start the flow of gas.
____	5	Hold a long match or large barbeque-type match up to the pilot to light the pilot.
____	6	After the pilot light is lit, continuing pressing the knob on the gas valve for approximately 1 minute after which release the knob.
____	7	Turn the knob to the ON position.

INSTRUCTOR'S RESPONSE:

_____

_____

_____

CHAPTER 11

Appliance Repair and Replacement

OBJECTIVES

By the end of this chapter, you will be able to:

SKILL-BASED

- Replace a gas stove.

- Replace an electric stove.

- Replace a heating element on an electric stove.

- Replace an oven heating element.

- Repair a range hood.

- Replace a range hood.

Spark ignition a device used to ignite a gas stove

Heating Element a device used to transform electricity into heat through electrical resistance

Bounty program a program offered by most utilities and municipalities whereby they purchase (in the form of a rebate) an old appliance when a person purchases any appliance that qualifies as an energy star appliance

Introduction

One of the most common duties of a facilities maintenance technician is the replacement and repair of gas and electric appliances. Although some appliances require specialty equipment and training to repair, in most cases the facilities maintenance technician can still perform basic maintenance on these appliances.

Appliance Safety

When facility maintenance technicians work on an appliance, they are exposed to numerous dangers that include, but are not limited to, electrical, gas, and heavy lifting. Therefore, you should following these safety guidelines to reduce the likelihood of injury, death, and property damage and loss.

- Never wear loose clothing and jewelry that could get caught in an appliance.
- Always wear gloves to minimize the possibility of cuts.
- Wear safety shoes to reduce the possibility of injury from appliances that are dropped.
- Always wear safety glasses.
- Use the proper lifting techniques, which are outlined in Chapter 3, when lifting heavy appliances.
- Never try to move a heavy appliance by yourself.
- Use the correct tools that are clean and free from defects.
- Keep your work area clean and in good condition.
- Always follow the manufacturer's instructions and recommendations.
- When installing a new appliance, always check the local building codes before beginning.
- When disposing of an appliance, always remove the doors to help prevent accidental entrapment and suffocation.

REMODELING

Demolishing a Kitchen

One of the areas in a residential structure as that is often remodeled is the kitchen area. Typically this is done when the appliances are either replaced or updated. Before beginning the demolition or deconstruction process, the proper permits must be obtained. Once the proper permits have been obtained, certain steps should be taken, depending on the scope of the demolition project.
The general procedure for demolishing a kitchen is:

- Disconnect all utilities servicing the kitchen.

- Install a plastic barrier separating the affected area from the rest of the structure.

- Cover any appliance that is not going to be removed from the area with a protective covering (cardboard).

- Disconnect electrical wiring to any appliance that is to be removed. Any exposed wire should be capped using wirenuts to prevent exposure to the bare conductors.

- Disconnect any plumbing servicing appliances and/or fixtures to be removed. Any pipes that are exposed should be capped (including vents).

- Remove appliances that are in the affected area. For example, if the technician is demolishing kitchen counter and/or cabinets, then at this stage the vent hood, as well as a drop in stove and/or cook top should be removed. Even if these components are not going to be reused, they could be recycled at an architectural salvage yard.

- Remove plumbing fixtures in the affected area. In the preceding example, the kitchen sink and garbage disposal should be removed. Even if these components are not going to be reused they could be recycled at an architectural salvage yard.

CAUTION

Always check with your local and state agencies before starting a remodeling project.
    Before beginning the demolition or deconstruction process, the proper permits must be obtained.

Repairing or Replacing a Gas Stove

When repairing or replacing a gas stove, always follow the manufacturer's instruction as well as any local and state building codes. In addition, always ensure that the area is properly ventilated and free from open flame. Some of the more common problems associated with gas stoves are listed next.

Gas Burner Will Not Light

One of the most common problems related to a gas is a burner that will not light. In this case, follow these steps:

1. Before attempting to performance maintenance on a gas stove always start by turning off the stoves gas supply. Next, lift the top of the stove.

2. The burner unit can be removed by lifting the back of the unit and sliding the front off the gas-supply lines (Figure 11-1).

3. Use a needle or other sharp object to poke into the pilot hole and clean out any debris. Brush the remaining debris away from the tip with a toothbrush. Hold

FIGURE 11-1: Burner unit.

a lit match to the opening to relight the pilot. Lower the lid and turn on your burners to test them (Figure 11-2).

4. Identify a **spark ignition** (a device used to ignite a gas stove) range by a little ceramic nub located between two burners. Look for wires running to it (Figure 11-3).

5. Brush away gunk around and on the igniter with an old toothbrush. Clean the metal "ground" above the igniter wire as well. It must be clean to conduct a spark. Close the lid and turn the burner knob to "light" to test the burner.

FIGURE 11-2: Cleaning pilot holes on a gas burner.

FIGURE 11-3: Spark ignition range on a gas burner.

Gas Stove Will Not Heat Properly

In the case of a gas stove that will light but does not heat properly, follow this procedure:

FIGURE 11-4: Removing an oven door.

1. Turn off the gas to the stove.

2. Remove the oven door (Figure 11-4).

3. To catch any soap and/or water that might spill during the cleaning process, spread a protective covering over the bottom pan.

4. If the stove contains an upper burner, using a screwdriver remove the upper burner cover, typically located on the roof of the oven. (Figure 11-5).

5. Remove any debris from the burner flame opening with a scrub brush and warm soapy water. Welding tip cleaners can be used to clean out any dirt that might have accumulated in the openings (Figure 11-6).

6. The lower burner can be exposed by removing the bottom pan. This is typically accomplished by first taking the protective covering out of the stove and then lifting the pan out of the stove. Once the burner has been exposed, the bottom burner can be cleaned with a scrub brush and soap and water. Welding tip cleaners can be used to clean out any dirt that might have accumulated in the openings.

7. Wipe up any water or dirt that collects under the burner. Let all the parts dry thoroughly; then reassemble the oven.

FIGURE 11-5: Upper burner in the oven of a gas stove.

FIGURE 11-6: Cleaning burner flame opening.

Replacing a Gas Stove

When a gas stove can no longer be repaired or it is no longer feasible to repair, then the gas stove must be replaced. Although replacing a gas stove is typically a simple procedure, the same safety procedure outlined in repairing a stove should be followed.

1. Shut off the gas line and check for gas leaks (Figure 11-7).

2. Lay down a protective floor covering to avoid damaging the floor while moving the stove away from the wall. Disconnect the stove from the gas lines.

3. Remove the cover plate at the base of the stove with a screwdriver.

4. If the structure is equipped with old copper tubing, then you must first disconnect the line at the fitting to the stove. Loosen the nut and slide this back. This is a flared copper-type fitting (Figure 11-8).

5. Move the unit onto the protective floor covering while avoiding damage to the gas pipe. Being careful not to crimp the copper line, disconnect the gas line at the coupling. Be careful not to crimp the copper piping (Figure 11-9).

6. If you have natural gas, skip to Step 11. If you have LP gas, you will need to replace the range orifices or spuds (Figure 11-10A and B).

FIGURE 11-7: Turn off gas line.

FIGURE 11-8: Disconnect the copper tubing.

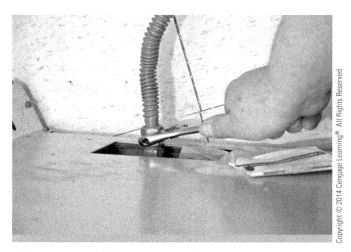

FIGURE 11-9: Disconnect gas line.

7. Orifices and spuds are individually sized with marks and colors. Loosen the old spud and replace with the new one. The oven orifices will then need to be adjusted.

8. Using an appropriate-sized wrench, tighten down the brass fitting.

9. There may be two orifices, one for the broiler and one for the main oven.

10. Underneath the unit, you will need to reverse the plastic pin in the regulator. Remove the hex nut, and flip the plastic pin and reassemble the nut (Figure 11-11).

11. To connect the gas line, use a new connector line with the threads on the pipe identical to the old connector (Figure 11-12).

12. To make the connection, coat the threads with pipe compound.

13. Tighten the fitting, and test it by applying soapy water. If it bubbles in a few seconds, you have a leak and will need to fix the leak (Figure 11-13).

(A)

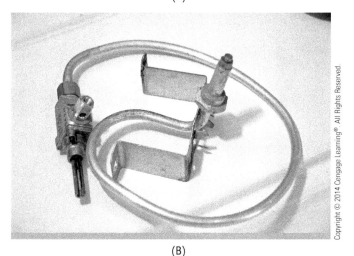

(B)

FIGURE 11-10: Range orifices and spuds (A,B).

FIGURE 11-11: Regulator underneath the range unit.

FIGURE 11-12: Connector line connecting gas lines.

14. The anti-tip bracket is mounted to the wall behind the stove per the manufacturer's instructions (Figure 11-14).

15. Move the new stove into place.

16. Level the new stove by adjusting the legs (Figure 11-15).

17. Open the gas line. If there is any smell of gas, shut off the supply and call a qualified service technician.

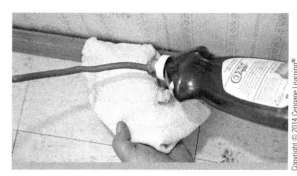

FIGURE 11-13: Testing gas line.

FIGURE 11-14: Antitip bracket.

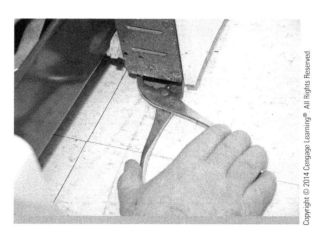

FIGURE 11-15: Level stove.

Repairing and Replacing an Electric Stove

Like gas stoves, an electric stove will malfunction from time to time and, therefore, will require maintenance. Typically with an electric stove, the **heating element** (a device used to transform electricity to heat through resistance) will be the source of trouble. In other words, it eventually burns out. Occasionally, a burned out element can be identified by the blister or bubbles that sometime form on the element. Also, from time to time, an element will burn completely in two, thus making identifying it much easier. Because these elements cannot be repaired, when a heating element burns out, it must be replaced.

Burned-out heating elements are one cause of a burner not working correctly, but this is not the only one. Before you replace the elements, troubleshoot and identify the problem.

Troubleshooting the Problem

Before attempting to repair an electric stove, always read and follow the manufacturer's instructions. Also, a technician should have a good understanding of electricity; therefore, if necessary, review Chapter 5 on electrical theory before attempting to troubleshoot or repair an electric stove. In addition, be sure that any replacement parts have the same electrical rating as replacement parts. Some common problems associated with electric stoves are given in Table 11-1 as well as being outlined in the following steps:

1. Turn off power to the appliance before working on it.

2. Ensure that the element is not wired directly into the appliance but can be easily unplugged from a receptacle. If it plugs in, move on to Step 3. If the element is direct-wired (Figure 11-16), move on to Step 5.

3. After removing the element from its receptacle, inspect the element's prongs. If the prongs are damaged in any way, then the element and the receptacle both must be replaced. To remove the element, simply lift up on the front of the element while pulling on the element itself. (Figure 11-17).

4. If the element's prongs are clean, then the element should be tested before replacing it. Sometimes reseating the element is enough to get the element to start working. Reseating the element consists of reinstalling the element in the

Range will not heat.	
No voltage at the element (outlet)	Correct voltage.
Blown fuse or tripped breaker	Replace fuse or reset breaker. If problem persists, contact an electrician.
Broken or burnt wire in power cord	Check the continuity of the cord and replace it if necessary. If the problem persists, contact an electrician.
Faulty range plug receptacle	Replace the range plug receptacle.
Burnt or oxidized prongs on range plug	Replace the range plug.
Surface burner does not heat.	
Loose connections at the element	Tighten the connections.
Burnt, corroded, or oxidized control switch	Clean the contacts using sandpaper. If necessary, replace the control switch.
Burned-out element	Replace the elements.
Surface burner too hot.	
Switch connection reversed or incorrect	Consult the manufacturer's instructions.
Oven does not heat.	
Faulty oven control	Adjust and/or replace the control.
Incorrect voltage at the element	Contact an electrician.
Loose connections at the element	Tighten the connections.
Burnt, corroded, or oxidized control switch	Clean contacts using sandpaper. If necessary, replace control switch.
Burned-out element	Replace elements.
Faulty thermostat	Replace thermostat.
Oven overheats.	
Faulty oven control	Adjust and/or replace control.
Incorrect element	Install correct element.
Stove heats unevenly.	
Electric stove tilted	Level electric stove.

TABLE 11-1: Common Problems Associated with an Electric Stove.

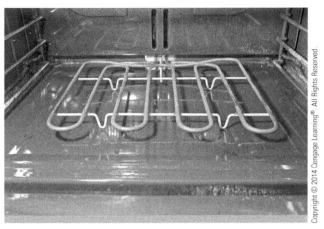

FIGURE 11-16: Electric stove heating elements.

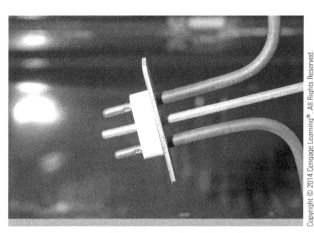

FIGURE 11-17: Prongs on a heating element.

receptacle and turning the stove on. If the element starts heating, then the problem has been addressed. If the element does not start heating, then it should be replaced with another element of the same size and tested again. This time, if the element starts heating, then the original element was bad and must be replaced. If the element does not start heating, then the receptacle should be tested and, if necessary, replaced.

5. For an element that is wired directly into the unit, start by lifting the front of the element, exposing the porcelain insulators with clips on each side (Figure 11-18).

6. The insulator can be separated into it two halves by using a straight slot screwdriver to open the insulator and then removing the clips.

FIGURE 11-18: Porcelain insulators.

7. Once the screws that hold the element to its wiring have been removed, the element can be exchanged for a new element of the same size and shape. Next, reassemble the element so that no bare wires are exposed, and then turn on the burner. If the new element works, the original element must to be replaced.

Replacing an Element

As mentioned earlier, one of the most common problems associated with an electric stove is the failure of the electric heating elements. To replace a heating element, do the following:

1. Turn off power to the appliance before working on it.

2. Get a new heating element identical to the one you are replacing (Figure 11-19).

3. If the stove uses plug in type elements, then installing the new element is as simple as plugging the new element into the receptacle. If the stove's elements are wired directly to

FIGURE 11-19: Removing a heating element on an electric stove.

FIGURE 11-20: Installing a new heating element on an electric stove.

FIGURE 11-21: Removing the old receptacle.

the unit, then installing the new element would consist of screwing the new element to its wiring, reassembling the two halves of the porcelain insulator, and snapping the clips in place (Figure 11-20).

4. Test the element to make sure that it's operating correctly.

Replacing a Receptacle

Over time, the contacts in a receptacle will oxidize and/or corrode and, therefore, will need to be replaced. When this happen, follow this procedure:

1. Turn off power to the appliance before working on it.

2. Depending upon how the receptacle is fastened to the top of the cooktop, removing the old receptacle can be as easy as using a screwdriver to unscrew it from the cooktop. If the receptacle is held into place by steel spring clips, then it can be removed by spreading the clamps and pull out the receptacle (Figure 11-21).

3. Access to the receptacle's wiring can be gained by simply lifting the cooktop (Figure 11-22).

4. Before removing the wires connected to the receptacle, mark the conductor so that they can be correctly installed on the new receptacle. Once the wires have been properly marked, they can be cut, allowing the receptacle to be removed (Figure 11-23).

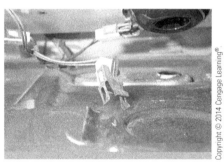

FIGURE 11-22: Cooktop lifted to access receptacle wiring.

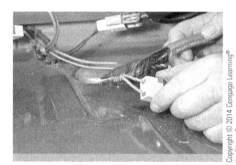

FIGURE 11-23: Cutting the receptacle wiring.

5. To install the new receptacle, start by stripping the insulation from the existing wires, using wire strippers. This should be done in accordance with the manufacturer's specifications and instructions. Next, twist the appropriate wires together and install wirenuts onto the connections. Reinstall the receptacle in its proper location, lower the cooktop, and replace the heating element (Figure 11-24A and B).

Oven Does Not Heat Properly

As mentioned earlier besides a burnt-out heating element, a fault thermostat will cause an oven not to heat properly. The purpose of the thermostat is to regulate the temperature in the oven by permitting current to flow to the element.

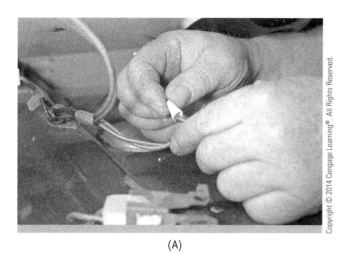

(A)   (B)

FIGURE 11-24: (A) and (B) Installing the new receptacle.

Testing the Thermostat

Although some technicians believe that when in doubt about a piece of equipment or a device it is better to replace it, this is not always the right approach. Before attempting to replace a thermostat, always verify that the thermostat should be replaced. This is accomplished by doing the following:

1. Place an oven thermometer inside the oven, and shut the door (Figure 11-25).

2. Turn on the oven, set it for 350°F, and let it heat for 30 minutes.

3. Check the thermometer. Most thermostats are accurate to within 25°F. If the thermostat is off by more than 50°F, the thermostat is bad and you will need to have a professional replace it. If the thermostat is off by less than 50°F, adjust the thermostat.

4. Locate the adjustment screw. On some thermostats, the adjustment screw is on the back of the thermostat knob; on others, it is inside the thermostat shaft (Figure 11-26).

5. To make a temperature adjustment on the back of a knob, remove the knob and loosen the retaining screws on the back. Turn the center disk toward "hotter" or "raise" to increase the temperature or toward "cooler" or "lower" to decrease the temperature. Tighten the screws, reinstall the knob, and test the oven. Readjust the knob if necessary (Figure 11-27).

6. To make a temperature adjustment inside the shaft, remove the knob and slip a thin flat-head screwdriver into the knob until it engages the adjustment screw in

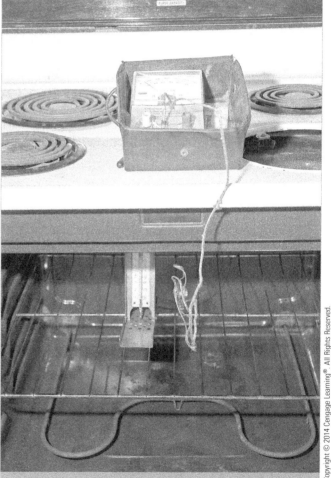

FIGURE 11-25: Oven thermometer.

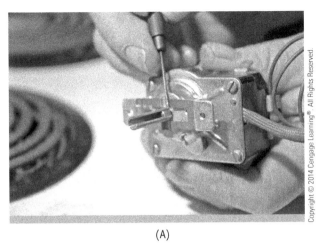
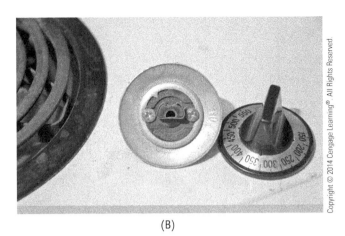

(A)

(B)

FIGURE 11-26: (A) Thermostat and screw. (B) Thermostat knob.

the bottom. Turn the screwdriver clockwise to raise the temperature and counterclockwise to lower it. Each quarter-turn will change the temperature about 25°F.

7. Reinstall the knob and test the oven. Readjust the temperature if necessary.

Replacing the Oven Heating Element

As stated earlier, one of the leading causes of an oven not heating properly is a burnt-out heating element. To replace a heating element follow this procedure:

1. Remove the oven racks so that you have access to the element (Figure 11-27A and B).

2. Remove the two screws from the element mounting plate, which sits flush against the back wall of the oven.

3. Pull the element gently out as far as the wire will allow.

4. Remove the supply wires from the element terminals (Figure 11-28).

5. Replace the element with a new identical one.

6. Put the new element in place and reconnect the leads. Usually only two wires go to the element; it doesn't matter which wire attaches to which terminal, as long as they're screwed on tight.

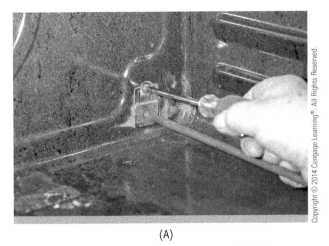

(A)

(B)

FIGURE 11-27: Oven heating element (A,B).

FIGURE 11-28: Element terminals.

FIGURE 11-29: Mounting brackets.

7. Push excess wire back behind the insulation.

8. Line up the holes, and reinstall the mounting bracket using the same screws you removed earlier (Figure 11-29).

9. Replace the oven racks.

Replacing an Electric Stove

As with a gas stove, when the repair of an electric stove becomes too extensive or the replacement parts are no longer available, it might be necessary to replace it. When replacing an electric stove, always follow the manufacturer's instructions and recommendations. Also, it might be necessary to update the wiring and/or breaker in the fuse box. Never purchase an oversized breaker or allow the stove to be installed with a breaker that is too small. This can cause serious electrical problems that can lead to equipment damage and/or malfunction.

1. Lay down a protective covering over the floor to prevent damaging the floor, and move the stove away from the wall.

2. Unplug the stove from the electrical outlet.

3. Remove the old stove and replace it with the new one.

4. Plug in the new stove.

5. Move the new stove into place.

6. Level the new stove by adjusting the legs.

Troubleshooting and Repairing an Ice Maker in a Refrigerator

If the ice maker does not make ice but you can see the arm swing into motion and you hear a buzz for about 10 seconds after it is finished, this normally means that there is a problem with the water supply line.

1. Ensure that the supply line servicing the ice maker is not kinked behind or beneath the refrigerator. If the ice maker has frozen up, it will need to be thawed.

2. Unplug the refrigerator.

3. Remove the ice bin and loose ice from the ice maker.

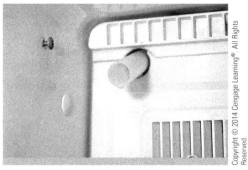

FIGURE 11-30: Fill tube.

Be careful not to melt the plastic parts.

4. Find the fill tube, the white hose that delivers water into the ice maker, and pull the small metal cap off of the housing that holds the fill tube down (Figure 11-30).

5. Warm the hose and surrounding mechanism to melt any ice blocking the mechanism. This can be done by using a hair dryer or soaking the supply tubing in hot water.

Replacing an Ice Maker

Always follow the manufacturer's instructions on replacing the ice maker. Also, whenever possible use only genuine manufacturer's replacement parts. This will ensure that the parts fit and function properly.

Repairing a Refrigerator

Major problems will require a trained refrigeration technician. However, many times the problem is simple and can be corrected. The potential problems that can easily be fixed by a facilities maintenance technician are given next.

Adjusting Controls

Because models will vary, refer to the specific refrigerator manufacturer's instructions.

Testing and Replacing Door Gaskets

1. Test the door seal in several places by closing a piece of paper in the door and then pulling it out. There should be some resistance, indicating that the door is sealed.

2. Remove the old gasket one section at a time. Some gaskets are held on by retaining strips, others by screws or even adhesive.

3. Install an identical gasket, using the retaining strips, screws, or new adhesive.

Always disconnect the dishwasher from it electrical source before attempting to perform any maintenance.

Troubleshooting Dishwasher Problems

As with any appliance, troubleshooting is necessary to determine why a dishwasher is not operating correctly. Table 11-2 outlines some common problems associated with dishwashers.

Water Is on the Floor around the Dishwasher

Water around the dishwasher could mean that either the gasket is damaged or the sprayer is clogged. In the case of leaking or pooling water around the dishwasher, take the following steps to rectify the problem.

Damaged Gasket

1. Check your gasket for cracks or deterioration (Figure 11-31).

Dishwasher doesn't work (no power).	
Blown fuse or tripped breaker	Replace the fuse or reset the breaker. If the problem persists, contact an electrician.
Dishwasher is unplugged	Reconnect the dishwasher.
Broken or burnt wire in power cord	Check the continuity of the cord, and replace it if necessary. If the problem persists, contact an electrician.
Faulty door latch	Replace the door latch.
Faulty door switch	Replace the door switch.
Faulty timer motor	Replace the timer motor.
Faulty selector switch	Replace the selector switch.
Faulty motor	Test and replace the motor relay if necessary. Test and replace the motor if necessary.
Dishwasher motor does not start but the motor hums.	
Check water supply	If the water supply is off, turn on.
Faulty door latch	Replace the door latch.
Faulty door switch	Replace the door switch.
Check inlet valve filter screens	Clean the inlet valve filter screens.
Check fill tube for kinks	Straighten fill line.
Dishwasher does not drain.	
Faulty drain valve	Replace the drain valve
Sink drain and/or drain hoses are restricted	Clear the restriction
Dishwasher leaks water and/or there is soap around the door.	
Incorrect detergent	Make sure that detergent is intended for a dishwasher.
Faulty door latch	Replace the door latch.
Faulty door switch	Replace the door switch.
Damaged door seal	Replace the door seal.
Damaged spray arm	Replace the spray arm.
Dishwasher does not dry properly.	
Burned-out element	Replace the elements.
Faulty thermostat	Replace the thermostat.

TABLE 11-2: Common Problems Associated with a Dishwasher.

2. If the gasket is damaged, remove it from the dishwasher by unscrewing it or prying it out with a screwdriver. Replace it with the same type of gasket. Soak the new gasket in hot water before attempting to install it. Soaking the new gasket will make it more flexible.

3. Check water supply connection under dishwasher for leaks.

Clogged Sprayer

1. To clean the sprayer, first remove the sprayer and soak it in a solution of warm white vinegar for a few hours to loosen mineral deposits. Then, clean out each spray hole with a pointed device such as a needle, awl, or pipe cleaner (Figure 11-32).

A dishwasher can also leak because it is not level.

FIGURE 11-31: Gasket around dishwasher.

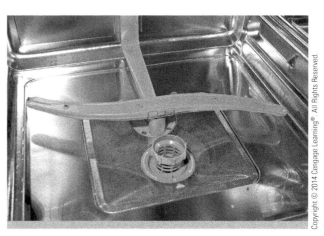

FIGURE 11-32: Dishwasher sprayer.

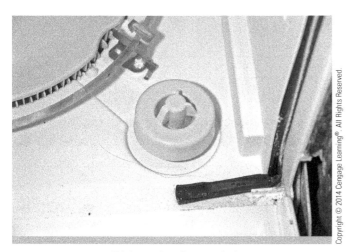

FIGURE 11-33: Dishwasher floor switch.

Dishwasher Overflows

If the dishwasher is overflowing and the drain and drain valve are not obstructed, check the float and float switch. This is accomplished by doing the following:

1. Locate the float switch by opening the dishwasher. It should be a cylindrical piece of plastic and may be set to one side along the front of the cabinet or near the sprayer head in the middle of the machine (Figure 11-33).

2. Check the float to make sure that it moves freely up and down on its shaft. (You may have to unscrew and remove a protective cap to get to the float.) If the float sticks, you'll need to clean away any debris or mineral deposits that are causing it to jam.

3. Pull the float off the shaft, and then clean the inside of the float with a bottle brush. Clean the shaft with a scrub brush.

4. Reinstall the float and check that it moves smoothly.

5. Set the dishwasher to fill, and check to see if it overflows.

Replacing a Dishwasher

When it becomes inefficient to repair a dishwasher or the replacement parts are no longer available, then it is necessary to replace the dishwasher. When replacing a dishwasher, always follow all local and state building codes as well as all manufacturer's instructions and recommendations.

1. At the main service panel, turn the power off to the dishwasher circuit.

2. Locate the shutoff valve for the hot water supply to the dishwasher. Normally, the hot water shutoff valve will be located under the kitchen sink. However, if the hot water is supplied to the unit through the floor or through the wall, then the shutoff valve could be located behind the dishwasher or even in a basement area.

FIGURE 11-34: Accessing lower panels.

FIGURE 11-35: Electrical box on dishwasher.

3. Remove the access panel and lower panels located at the base of the dishwashers (Figure 11-34).

4. Remove the electrical box from the old dishwasher (Figure 11-35).

5. Remove the wirenuts by unscrewing them, exposing the bare connections. Once the bare connections have been exposed, pull the wires apart. The order in which the wires are removed is not critical.

6. Depending upon how the drain hose is connected, disconnect the drain hose from the waste tee on the drain line or the inlet on a disposer. Typically, this is accomplished by using pliers to force open the spring clamp or a screwdriver to open a screw-type clamp. Once it is free, remove the hose from the dishwasher as well. For ease of access, the hose can be removed from the dishwasher at a later time, once the unit has been completely removed (Figure 11-36).

7. The water supply to the unit should be disconnected. Before disconnecting the supply line, have a bucket handy to catch any water that might drain from the supply line (Figure 11-37).

8. Lay down a piece of plywood to drag the old dishwasher onto.

9. Removing the dishwasher from under the counter can be accomplished by first removing the screws attach the unit it the underside of the countertop. Once

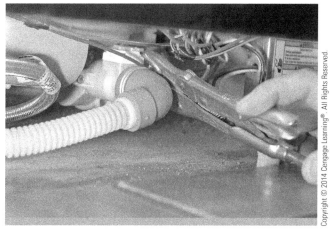

FIGURE 11-36: Drain Hose.

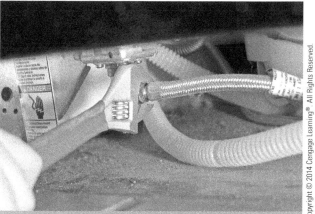

FIGURE 11-37: Water supply line.

FIGURE 11-38: Removing the screws in counter.

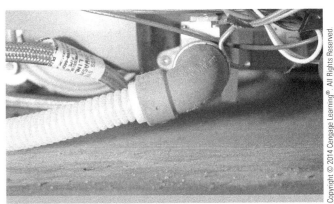

FIGURE 11-39: Drain line connection.

FIGURE 11-40: Leveling the legs of the dishwasher.

these screws have been removed, readjust the leveling screws to their lowest settings, thus allowing the unit to be pulled out of the opening (Figure 11-38).

10. Take the new dishwasher out of the box, and check the back to verify that all of the connections are in place.

11. Take the cap off the drain line connection at the dishwasher (Figure 11-39).

12. Attach the drain line to the dishwasher.

13. Using pliers, crimp the clamp around the hose to secure it.

14. Once the door to the dishwasher has been close and locked, the dishwasher can be slid into place.

15. Following the manufacture's recommendations, adjust the leveling screws on the front of the dishwasher (Figure 11-40). Use a level to verify that the unit is level. Install the mounting screws into the underside of the counter.

16. Reconnect the water supply to the dishwasher as well as the drain simply by reversing the removal procedure outlined previously. Also connect the dishwasher to the electrical supply. Cut off the exposed ends of electrical wires, and use wire strippers to strip about $\frac{1}{2}$ inch of insulation from the ends. Twist the wires together (white-to-white, black-to-black) and put on new wire connectors. Secure the ground (green) wire. Tighten the strain-relief connector.

17. Install the decorative panels.

18. To ensure that the spacing on both sides of the dishwasher is even, read and follow the manufacturer's instructions to adjust the door.

19. Slowly open the supply valve while checking for leak in the water lines as well as the connections. Restore power and operate the machine to check for drain leaks.

20. Reinstall the lower and front access panels.

Keep an old towel or tee shirt and bucket handy to clean up spills.

Repairing a Range Hood

A range hood not adequately removing smoke and smells from your kitchen is usually caused by one of the following:

- The grease filter or some part of the exhaust ductwork may be clogged.
- The fan may be bad.

Unclogging the Exhaust Fan

1. Remove the filter and soak it until the grease has been dissolved in a degreasing solution (Figure 11-41).

2. All traces of the degreaser should be removed by washing with warm, soapy water. Also, a filter may be put in the upper rack of the dishwasher and run through a normal cycle.

3. To remove the exhaust fan. Unplug the fan and remove it from the hood (Figure 11-42).

4. Using an old toothbrush, clean the fan blades while dipping them into a cleaning solution.

5. Using a heavy rag and a plumber's snake, the inside of the exhaust ductwork can be cleaned. Soak the rag in cleaning solution, and use the snake to push the rag through the duct. Rinse out the rag and repeat the operation until the duct appears to be clean (Figure 11-43).

6. Also clean the exhaust hood that is attached to the outside of the structure (Figure 11-44).

7. Reinstall the grease filter.

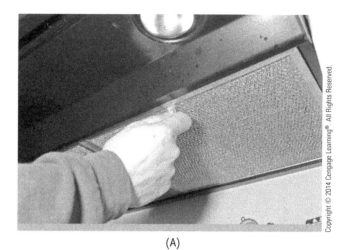

(A)

(B)

FIGURE 11-41: Exhaust fan filter (A,B).

FIGURE 11-42: Remove exhaust fan.

FIGURE 11-43: Clean exhaust ductwork.

FIGURE 11-44: Exhaust hood on the outside of structure.

FIGURE 11-45: Range hood.

Replacing a Range Hood

1. Before attempting to perform maintenance on a range hood, always turn off the electricity. Next, remove the old range hood (Figure 11-45).

2. On the new range hood, remove the filter, fan, and electrical housing cover. Remove the knockouts for the electrical cable and the duct (Figure 11-46).

3. Using heavy cardboard to protect the cooktop's surface, set the range hood on top of the stove, then connect the structures wiring to the hood. Start by connecting the two black wires together, followed by the two white wires. When connecting the ground wires, connect the ground wire under the ground screw and tighten the cable clamp onto the structure's wiring.

4. Install the range hood with the mounting screws. With the hood in position, slide the hood toward the wall until the hood's mounting screws have been engaged. Once they are engaged, tighten the mounting screws using a screwdriver. Once the hood has been secured, replace the bottom of the hood.

5. Using duct tape to secure the joints, fasten the ductwork to the range hood. All joint should be made airtight (Figure 11-47).

6. Install the light bulbs and replace the filters. Turn on the power at the service panel, and check for proper operation.

FIGURE 11-46: Range hood with knockouts.

FIGURE 11-47: Hood ductwork.

Repairing a Microwave Oven

Because a leaking microwave can not only be extremely hazardous but also present a risk of electrocution because of the high wattage that is present, only a qualified service technician should attempt to repair a damaged unit. However, minor repair, such as changing a light bulb, can be performed by a facility maintenance technician, if the light bulb is easily accessible. Another task that can be performed by a facility maintenance technician is ensuring that the unit is getting power.

Troubleshooting a Washer

If the washer is not getting either hot or cold water, refer to your owner's manual to ensure that the washer is operating as it should. If the washer isn't operating as it should, there may be a problem with the water inlet valve.

Checking the Water Inlet Valve

1. Disconnect the appliance's power supply.

2. Locate the washer's water inlet valve. It will be at the back of the washer, and it will have water hoses hooked up to its back (Figure 11-48).

3. Shut off the supply of water to the washer (Figure 11-49).

4. Disconnect both hoses at the back of the washer. Point the hoses into a bucket or a sink, and then turn on the water supply again. Do this to confirm that you are receiving adequate water pressure and that there is not some sort of blockage in the line (Figure 11-50).

5. Inspect the screens found inside the valve. Clean out any debris you find. You should be able to pop them out with a flat-head screwdriver. Use caution when handling the screens, as they are irreplaceable.

FIGURE 11-48: Inlet valve.

FIGURE 11-49: Shutting off water supply.

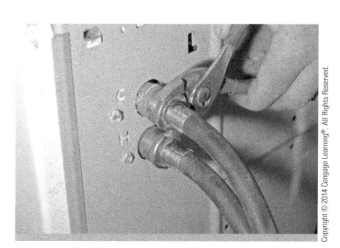

FIGURE 11-50: Disconnect hoses.

Replacing Washer Inlet Valves

1. Disconnect the appliance's power supply.

2. Shut off the water supply to your washer (Figure 11-51).

3. Locate the washer's water inlet valve. It will be at the back of the washer, and it will have water hoses hooked up to its back (Figure 11-52).

4. Disconnect both hoses at the back of the washer. Point the hoses into a bucket or a sink; then turn on the water supply again. Do this to confirm that you are receiving adequate water pressure and that there is not some sort of blockage in the line.

5. Remove the screws that hold the inlet valve in place (Figure 11-52).

6. Remove the hose connecting the valve to the fill spout.

7. Connect your new water inlet valve to the water line.

8. Attach both water supply hoses. Turn the water on and check for leaks.

9. Reconnect the washer to the power supply.

FIGURE 11-51: Water supply.

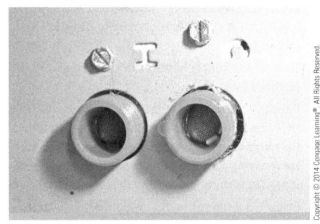

FIGURE 11-52: Inlet valves.

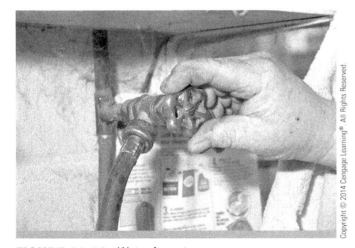

FIGURE 11-53: Water faucets.

Washing Machine Fills Slowly

When a washing machine fills slowly, the problem is usually a clogged intake screen.

1. Turn off the water faucets that feed the machine (Figure 11-53).

2. Unplug the washer, and pull it far enough away from the wall so that you can get behind it to work.

3. Remove each water-supply hose (Figure 11-54).

4. Locate the screens and gently pry out the screens with a small flat-head screwdriver (Figure 11-55).

5. Clean the screens with an old toothbrush.

6. Reinstall the screens.

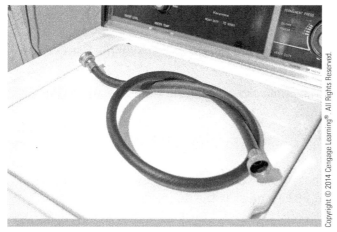

FIGURE 11-54: Water supply hose.

FIGURE 11-55: Intake screen.

7. Reinstall the hoses and tighten the couplings securely.

8. Turn on the water to check for leaks, then plug in the machine and push it back into position.

Installing a New Washer

The installation of a new washing machine is a simple process; however, moving the washing machine into position and out of the facility can be tricky as well as dangerous. Whenever possible, it is strongly recommended that you have help when moving the new machine into position and the old machine out of position.

1. Turn off power to the old washer.

2. Remove the old washer.

3. Clean and dry the floor, and move the new washer into place to be connected.

4. Fasten the drain hose to the washer with a hose clamp. Be sure not to tighten it too much or you might strip the screw (Figure 11-56).

5. Attach the water hoses to the washer. Hot and cold on the taps and on the washer are usually clearly marked. Red indicates hot; blue indicates cold (Figure 11-57).

FIGURE 11-56: Fasten drain hose.

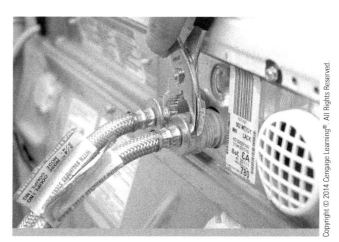

FIGURE 11-57: Attach hose to the washer.

6. Plug the washing machine in and move it into place, placing the drain hose in the drainpipe when you can reach it.

7. Push the washer the rest of the way into the space, being careful not to crimp the hoses.

8. Leave about an inch and a half of space around the washer to allow room for it to vibrate.

9. Turn the water faucets on.

10. Turn the power back on.

11. Run a cycle without clothes or detergent before you use the machine to clear the water pipes and make sure that the drainage is adequate.

Troubleshooting a Dryer

Dryers typically consist of very few parts and, therefore, are usually simple to troubleshoot and maintain. Most dryer problems can be attributed to either a burnt-out heating element or a clogged vent.

Dryer Takes a Long Time to Dry Clothes

If a dryer is taking too long to dry clothing, it is likely that the heating element is partially or completely burned out. Follow the manufacturer's instructions for the specific dryer brand to test and replace the heating element if necessary.

The Vent Is Clogged

If the dryer feels really hot, but the clothes take long time to dry, a clogged vent could be the problem.

1. Check the vent flap or hood on the outside of the structure. Ensure that the airflow through the vent is strong when the dryer is running. If necessary use a clothes hanger to clean out the vent. (Figure 11-58).

2. If the vent is not clogged, then check the dryer duct to ensure that it is not kinked or sagging, thus impeding airflow. If necessary, straighten the dryer duct.

3. If duct is not kinked, then disconnect the check for a blockage inside the dryer by disconnecting the dryer duct and looking inside the dryer, as well as the duct. To remove the blockage from the duct start by shaking duct to remove the blockage or running a wire through the duct. If the duct is damaged, then replace it (Figure 11-59).

Appliance Disposal

Before disposing of any appliance, always check with your local utility to see if it has a bounty program. A **bounty program** is a program offered by

FIGURE 11-58: Vent flap.

FIGURE 11-59: Dryer duct.

most utilities and municipalities in which they purchase in the form of a rebate the old appliance for any appliance that qualifies as an energy star appliance. Check the rules and regulations associated with the bounty program to ensure that your appliance will be accepted. Often these regulations include the appliance being in working condition, of a minimum vintage, and/or of a minimum dimension. If a bounty program is not available or your appliance does not qualify for a bounty program, then you should contact your municipal department or public works for information regarding the procedure for disposal of appliances.

Some municipalities require that all refrigerated appliances have their refrigerant recovered by an EPA certified technician before the appliance can be disposed of. If the company that recovers the refrigerant is not also the final disposer of the appliance, then the EPA requires [40 CFR 82.156(f)(2)] a signed statement containing the name and address of the person who recovered the refrigerant, and the date that the refrigerant was recovered.

CAUTION

Handling refrigerants improperly may result in physical injury and possible death. Only properly trained and certified technicians, using EPA-approved refrigerant recovery equipment, can legally remove refrigerant from an appliances.

SUMMARY

- One of the most common problems related to gas stoves is that a burner will not light.
- When a gas stove can no longer be repaired or it is no longer feasible to repair it, the gas stove must be replaced.
- Typically, with an electric stove, the heating element (a device used to transform electricity to heat through resistance) will be the source of trouble.
- Over time, the contacts in a receptacle will oxidize and/or corrode and, therefore, will need to be replaced.
- The purpose of the thermostat is to regulate the temperature in the oven by permitting current to flow to the element.

- If the ice maker does not make ice but you can see the arm swing into motion and you hear a buzz for about 10 seconds after it is finished, this normally means that there is a problem with the water supply line.
- The potential problems related to a refrigerator that can easily be fixed by a facility maintenance technician are: adjusting controls and testing and replacing door gaskets.
- Water around the dishwasher could mean that either the gasket is damaged or the sprayer is clogged.
- If the dishwasher is overflowing but the drain and drain valve are not obstructed, check the float and float switch.
- A range hood not sufficiently eliminating smoke and odors from a kitchen is usually an indication that the range hood has a bad fan or the grease filter is clogged.
- When a washing machine fills slowly, the problem is usually a clogged intake screen.
- Most dryer problems can be attributed to either a burnt-out heating element or a clogged vent.
- If the drying time for cloths in a dryer increases and the dryer feels hot, check for a clogged vent.
- If the company that recovers the refrigerant is not also the final disposer of the appliance, then the EPA requires (40 CFR 82.156(f)(2)) a signed statement containing the name and address of the person who recovered the refrigerant and the date that the refrigerant was recovered.

REVIEW QUESTIONS

1   List the steps for unclogging a vent.

2   What should the technician check if a dryer does not dry clothes?

3   List the steps for installing a washer.

4   List the steps for replacing a washer inlet valve.

5   List the steps for replacing a range hood.

6   List the steps for unclogging an exhaust fan.

REVIEW QUESTIONS

7   List the steps for replacing a dishwasher.

9   What should the technician check if an electric oven does not heat properly?

8   List the steps for replacing a gas stove.

10 What should the technician check if a dishwasher does not dry properly?

Name: _____

Date: _____

Replacing an Electric Stove

Upon completion of this job sheet, you should be able to replace
an electric stove. In the space provided below, list the steps for replacing
an electric stove.

INSTRUCTOR'S RESPONSE:

_____

_____

_____

Name: _____

Date: _____

Replacing a Dishwasher

Upon completion of this job sheet, you should be able to replace a dishwasher. In the space provided below, list the steps for replacing a dishwasher.

INSTRUCTOR'S RESPONSE:

_____

_____

_____

© iStockphoto.com/kkgas

CHAPTER

12

Trash Compactors

OBJECTIVES

By the end of this chapter, you will be able to:

KNOWLEDGE-BASED

- Explain the purpose of a safety interlock safety device.

SKILL-BASED

- Perform general maintenance procedures.
- Perform general maintenance of hydraulic devices.

- Perform a test of the interlock safety device.
- Check the general condition of a dumpster.

Garbage waste or food that has spoiled as well as other household refuse

Safety interlock a device used to help prevent a machine from

causing damage to itself or injury to the operator by disengaging the device when the interlock is tripped

Dumpster a large waste receptacle

Litter waste that is unlawfully disposed of outdoors

Introduction

In an office or a residential environment, trash is typically stored in a can until it is picked up as part of garbage collection. From there the **garbage** (waste or food that has spoiled as well as other household refuse) is either sent to a processing center (typically an outside firm such a waste management company), it is burned, or it is compacted and then hauled away. When the trash is compacted, this is typically done with a trash compactor. A trash compactor reduces the amount of space necessary for a particular volume of trash to approximately $\frac{1}{12}$ of its original volume.

General Maintenance

In reality, most appliances are rather simple in nature, that is, they consists of a handful of mechanical parts with the rest being electrical control boards, switches, and relays. A trash compactor is no different. Mechanically a trash compactor consists of a motor, the compression ram, and a drive screw. Electrically, a trash compactor is made up of door, limit switches, and exterior controls (see Figure 12-1).

Common Trash Compactor Problems and Solutions

As mentioned earlier, mechanically a trash compactor consists of very few moving components; however, as with all equipment, malfunctions and equipment breakage can happen.

If the compactor will not start:

- Ensure that the compactor door is closed completely.

- Ensure that the unit is plugged in and the circuit breaker or fuse that is servicing the unit is not blown or tripped.

- Inspect the unit's electrical cord to be sure it is not damaged.

FIGURE 12-1: Trash compactor.

If the motor runs, but trash is not compacted:

- As a general rule of thumb, a trash compactor will not compact trash until its drawer is at least one-third full.
- Using a voltage ensure that the outlet that is servicing the trash compactor is supplying the correct voltage (see Figure 12-2).
- Inspect the unit's power nuts for wear. Always follow the manufacturer's specifications.
- Check the power screws for wear or obstructions. See manufacturer's specifications.
- Lubricate the power screws. See manufacturer's specifications.
- Check the drive belt, chain, and/or gears. See manufacturer's specifications.

If the compactor starts but does not complete its cycle (the ram is stuck):

- Inspect the unit for an object that is triggering the compactor door tilt switch.
- Ensure that the unit does not have a loose connection.

If the drawer is stiff or difficult to open:

- Check the drawer tracks for obstructions. If the tracks have obstructions or contain grime, clean them.
- Ensure that the unit's rollers are not damaged.

If the door will not open:

- Make sure that the ram is in the top position.
- If the dense pack switch is on, turn it off.
- While restarting the trash compactor, push the door closed.
- Ensure that the power screw is free of obstructions. See manufacturer's specifications.
- Check the power nuts for obstructions. See manufacturer's specifications.
- Check the ram for obstructions.

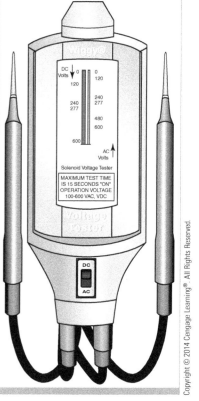

FIGURE 12-2: Voltage tester.

CAUTION

Unless it is absolutely necessary, do not service a trash compactor without first disconnecting the power to the device and locking out and tagging its circuit.

Cleaning and Deodorizing

The interior of a trash compactor should be cleaned on a regular basis. In a trash compactor, the ram is the part that applies pressure to the garbage. Because it comes in contact with the platform, which compacts the trash, it is recommended that the ram be regularly cleaned with a degreasing agent as well as an antibacterial agent. This holds true for any other aspect of the compactor that comes in contact with garbage.

For routine cleaning, use the following steps:

1. Unless absolutely necessary, never work on a trash compactor or any other appliance without first disconnecting it from the power source.
2. Always use gloves and safety glasses (see Figure 12-3).
3. Start by removing the compactor bag and caddy.

FIGURE 12-3: Rubber gloves.

4. Consult and follow the manufacturer's cleaning instructions.

5. Vacuum the interior.

6. To clean the inside and the outside of the unit use an environmentally friendly soap and water solution.

7. Insert a new bag into the caddy and close the drawer.

8. Establish a routine maintenance schedule in which the air freshener and/or charcoal filtration medium is replaced.

9. Reconnect the electrical power to the unit.

Because you are putting food waste products into the trash compactor, and the unit is typically located in a warm location, it is common for it to contain odor-causing bacteria. Setting up a routine cleaning and maintenance schedule will help eliminate the bacteria cause odors and possibly diseases.

General Maintenance of Hydraulic Devices

Check the compactor's preventive maintenance schedule for its most recent maintenance; also check to see if hydraulic fluid lines are adequate and intact. See the manufacturer's specifications for details.

FIGURE 12-4: After disconnecting the power to the compactor, locking out the distribution panel prevents someone from turning on the power.

Performing a Test of the Interlock Safety Device

The **safety interlock** is a device used to help prevent a machine from causing damage to itself or injury to the operator, by disengaging the device when the interlock is tripped. For example, a washing machine will disengage the drum when the lid is opened during operation. It is the safety interlock that prevents operation when the door is open. Test the trash compactor safety device by opening the trash compactor door and pressing the buttons on the trash compactor to make sure it does not start up. If the compactor runs with the door open, follow the manufacturer's instructions to correct the problem (Figure 12-4).

Checking the General Condition of a Dumpster and Dumpster Area

In addition to maintaining the trash compactors, the facilities maintenance technician is responsible for maintaining the **dumpsters** (a large waste receptacle) and the dumpster area. This includes the following:

- Control **litter** (waste that is unlawfully disposed of outdoors).

- Ensure that the company you lease your dumpsters from cleans and maintains them on a regular basis.

- If the dumpster starts leaking, call the leasing company to arrange for the repair of the dumpster.

- If for some reason you have to clean the dumpster, always properly dispose of the contents first. Then, use a dry cleanup method followed by rinsing. Always properly dispose of the water used in the rinsing.

CAUTION

When working around the dumpsters, wear leather gloves to prevent accidental cuts from sharp object that might be in the trash area (metal, glass, etc.)

SUMMARY

- Mechanically, a trash compactor consists of a motor, the compression ram, and a drive screw. Electrically, a trash compactor is made up of door and limit switches and exterior controls.

- In a trash compactor, the ram is the part that applies pressure to the garbage. Because it come in contact with the platform, which compacts the trash, it is recommended that the ram be regularly cleaned with a degreasing agent as well as an antibacterial agent.

- On at trash compactor, the safety interlock prevents operation when the door is open.

- In addition to maintaining the trash compactors, the facilities maintenance technician is responsible for maintaining the dumpsters.

REVIEW QUESTIONS

1 When troubleshooting a trash compactor, what should a contractor check first?

2 When a trash compactor starts but does not complete its cycle, what should the contractor check?

3 List the steps for cleaning a trash compactor.

Name _____

Date: _____

Trash Compactor

CLEANING AND DEODORIZING A TRASH COMPACTOR

Upon completion of this job sheet, you should be able to clean and deodorize a trash compactor.

Task Completed	Task:
____	1 Disconnect the trash compactor from the power source. Unless absolutely necessary, never work on a trash compactor or any other appliance, without first disconnecting it from the power source.
____	2 Always use gloves and safety glasses
____	3 Start by removing the compactor bag and caddy
____	4 Consult and follow the manufacturer's cleaning instructions.
____	5 Vacuum the interior.
____	6 To clean the inside and the outside of the unit, use an environmentally friendly soap and water solution.
	7 Insert a new bag into the caddy and close the drawer.
	8 Establish a routine maintenance schedule in which the air freshener and/or charcoal filtration medium is replaced.
____	9 Reconnect the electrical power to the unit.

INSTRUCTOR'S RESPONSE:

_____

_____

_____

13

Elevators

OBJECTIVES

By the end of this chapter, you will be able to:

SKILL-BASED

- Check and inspect floor leveling.
- Check operation of elevators.
- Perform a test on elevator doors.

Occupants people riding on the elevator

Elevator Platform the floor of an elevator

Americans with Disabilities Act (ADA) a law that prohibits discrimination based on disability

Illumination lighting

Introduction

An elevator is defined as a device consisting of a platform used to raise and lower occupants and/or equipment from one floor to another. It is the goal of elevator maintenance to make sure that the elevator operates as designed and in a safe and efficient manner. Failure to keep an elevator properly maintained will result in decreased reliability and increased response time. In addition, the safety of the elevator's **occupants** (the elevator's riders) will be compromised. A full maintenance contract can and should be purchased from the manufacturer of the equipment. Purchasing a maintenance contract will ensure that the manufacturer takes full responsibility for that equipment. However, even with a maintenance contract in place, the elevator should be checked on a routine basis by the facilities maintenance technician to ensure that the elevator is operating properly.

Checking and Inspecting Elevators

As mentioned, although elevator maintenance and repair should only be performed by certified elevator mechanics, the facilities maintenance technician should still perform routine inspection of the elevator to ensure that the elevator continues to operate efficiently and safely. If any of the following situations should arise, then it is the responsibility of the facilities maintenance technician to report the problem to his or her supervisor and/or the elevator maintenance company.

Unless you are a certified elevator mechanic never attempt to make any adjustments, corrections, or repairs to the elevator.

- Increase in response time (waiting for the elevator).
- The platform of the elevator being not level with the floor when the door opens.
- The elevator doors not opening fully at the destination floor for any reason.
- Unusual noises during the operation of the elevator.
- Overheating.

Long Response Time (Waiting for the Elevator)

In addition to inconveniencing the occupants of a facility, a long response time can be an indicator that the elevator control system has or is developing a problem. On a regular basis, the time spent waiting for an elevator, as well as its speed, should be logged and compared to the manufacturer's specification. This is done by checking the time required for the elevator to travel from the bottom floor to the top floor during peak and nonpeak times.

The Platform of the Elevator Being Not Level

When the **elevator platform** (the floor of the elevator) does not meet the floor of the building, it creates a tripping hazard. This can be as a result of the elevator platform sticking either above or below the structure's floor (Figures 13-1 and 13-2). According to accessibility codes, elevators must automatically level within $\frac{1}{2}$ inch.

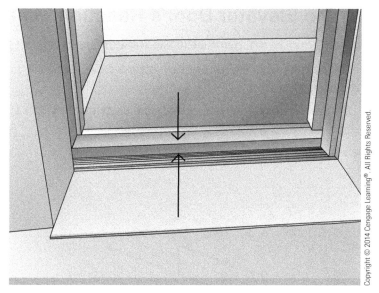

FIGURE 13-1: The elevator that is not leveling properly—the elevator is above the floor.

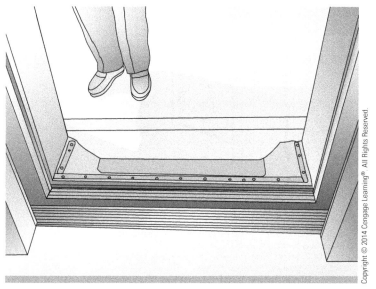

FIGURE 13-2: The elevator that is not leveling properly—the elevator is below the floor.

Overheating

Overheating is a common problem associated with electrical traction elevators because the equipment is usually located above the facility roof in a penthouse. Because the mechanical room is seldom climate controlled (heated and/or cooled; instead, they are ventilated—typically by louvers), they experience temperature swings according to the outside air temperature. If the mechanical room's ventilation is located near the floor, then the outside air temperature can exceed 100°F. It should be noted that, as the temperature in the mechanical room increases, so does the possibility of breakdowns.

Checking the Elevator Door's Reaction Time

With a stopwatch, check the reaction time of the elevator doors (Figure 13-3).

1. Record a baseline time. This will be the time recorded the first time you measure the time the doors take to close.

2. On a regular basis, measure the time, record, and compare it against the baseline time (see Table 13-1).

3. Report the results to your supervisor.

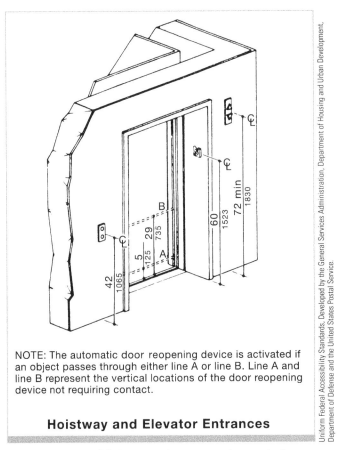

NOTE: The automatic door reopening device is activated if an object passes through either line A or line B. Line A and line B represent the vertical locations of the door reopening device not requiring contact.

Hoistway and Elevator Entrances

Uniform Federal Accessibility Standards, Developed by the General Services Administration, Department of Housing and Urban Development, Department of Defense and the United States Postal Service.

FIGURE 13-3: Minimum entrance requirements for an elevator.

Elevator Door Reaction Time				
Baseline Reaction Time				
Date	Time	Open	Close	Comments
Actual Reaction Time				
Date	Time	Open	Close	Comments

TABLE 13-1: Elevator Door Reaction Time

A typical reaction time chart for an elevator located in an apartment complex lobby might look something like the following example:

Elevator Door Reaction Time				
Baseline Reaction Time				
Date	Time	Open	Close	Comments
12/5/2012	2:00 pm	5 seconds	5 seconds	Times were taken after the elevator was inspected and serviced by a rained elevator technician.
Actual Reaction Time				
Date	Time	Open	Close	Comments
1/6/2013	8:30 am	5 seconds	5 seconds	Regular inspection by FM technician
2/4/2013	9:00 am	5 seconds	5 seconds	Regular inspection by FM technician
3/7/2013	10:00 am	5 seconds	5 seconds	Regular inspection by FM technician
4/4/2013	9:00 am	5 seconds	5 seconds	Regular inspection by FM technician

Elevators and ADA Requirements

An elevator is usually the easiest way to provide access to upper floors when access is required above the ground floor. If an elevator is installed, it must meet the **Americans with Disabilities Act (ADA)** requirements whether the elevator was required. Every floor that the elevator serves will have to meet ADA access requirements regardless. The elevator must be automatic and have proper call buttons,

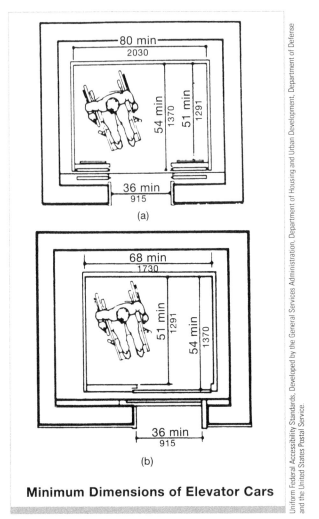

Minimum Dimensions of Elevator Cars

FIGURE 13-4: Minimum dimensions of elevator cars must be considered in the initial building design, as well as the reaction time of elevator doors.

car control, **illumination** (lighting), and door timing and reopening, as shown in Figure 13-3. The elevator floor is part of the access route, so it must be similar to the floor described in walks and hallways. The shape and size of the elevator must be such that it is usable by a person in a wheelchair, as shown in Figure 13-4.

SUMMARY

- An elevator is defined as a device consisting of a platform used to raise and lower occupants and/or equipment from one floor to another.
- Elevator maintenance and repair should only be performed by certified elevator mechanics. The facilities maintenance technician should report the following problems to his or her supervisor and/or the elevator maintenance company:
  - Increase in response time (waiting for the elevator).
  - The platform of the elevator being not level with the floor when the door opens.

- The elevator doors not opening fully at the destination floor for any reason.
- Unusual noises during the operation of the elevator.
- Overheating.

• Elevators must meet the ADA requirements.

• The elevator must be automatic and have proper call buttons, car control, illumination, and door timing and reopening.

REVIEW QUESTIONS

1   True or false? If the platform of the elevator is not level with the floor when the door opens, the technician can make the required adjustments without consulting an elevator service technician.

2   True or false? If the elevator door does not open completely for any reason, the facilities maintenance technician should report it as soon as possible to a qualified elevator technician or a supervisor.

3   What should a facilities maintenance technician look for when inspecting an elevator?

4   What does a long response time for an elevator typically indicate?

5   Define the term "elevator."

6   True or false? Only elevators that are accessible by the public are required to meet ADA requirements.

Name: _____

Date: _____

Elevators

Upon completion of this job sheet, you should be able to identify any safety requirements not being met by the elevator.

	Elevator Checklist		Comments
Regular Inspection	Yes ❑	No ❑	
Regular Service	Yes ❑	No ❑	
Elevator Capacity Displayed	Yes ❑	No ❑	
Emergency Phone	Yes ❑	No ❑	
ADA Compliant	Yes ❑	No ❑	

INSTRUCTOR'S RESPONSE:

_____

_____

_____

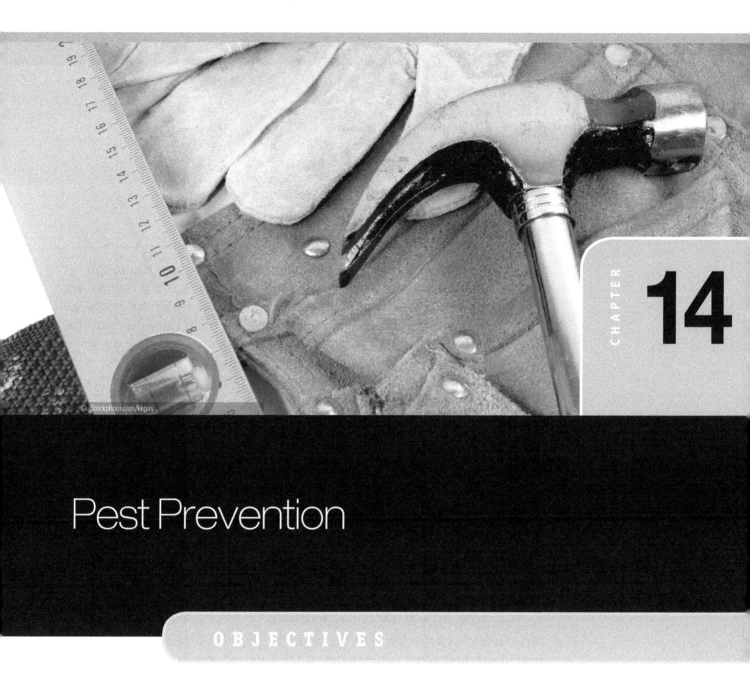

© iStockphoto.com/kkgas

Pest Prevention

OBJECTIVES

By the end of this chapter, you will be able to:

KNOWLEDGE-BASED

- Follow applicable safety procedures.

SKILL-BASED

- Recognize the sources of damage caused by pests.

- Select and apply proper techniques, chemicals, and/or materials to eradicate and/or prevent pest infiltration.

Insecticide chemicals used to kill insects

Pesticide chemicals used to kill pests (rodents or insects)

Introduction

Pest prevention is normally not a high-priority task, until a problem occurs. It is not until a problem is reported that steps are taken to eliminate pests and then prevent the pest problems in the future. Pests can be a noteworthy problem for people and property. The pesticides that are commonly used in pest control may pose potential health hazards to the occupants as well as the environment. Therefore, whenever possible, it is highly recommended that alternative methods be employed. For facilities in which the pest control is contracted out, the facility maintenance technician is usually the primary contact person for the facility.

Types of Pests and Pest Controls

The first step to controlling pests is to determine the type of pest that is infesting a facility. The following are the most common pests a facilities maintenance technician will have to deal with, as well as the methods used to control them.

Cockroaches

Cockroaches are not easy to control because, in order to eliminate them, the **insecticide** must be delivered to the insect. The key to controlling cockroaches is to kill baby cockroaches as they hatch. If this fails, call your Environmental Health Department or pest control contractor.

Ants

Before applying powdered insecticide to the nest, pour boiling water over it. Wherever ants enter into the structure an insecticide lacquer can be used. Typical areas include around door thresholds and/or wall–floor junctions. Ant bait works by letting the ant carry the bait back to the nest, killing the whole colony after a few days. Place it along where ants run.

Spiders

Spiders eat other insects and, therefore, it is not necessary to exterminate them; instead, simply place a carton over a spider, and then slip a piece of thin cardboard under the carton. Next, take the spider outdoors and let it go.

Rodents

Seal off entry points into the facilities. Ensure areas around the facilities, including the trash container areas, are clean. Poison is available as proprietary, ready-mixed bait. If the infestation is serious or persistent, then you should consult a professional a pest control contractor.

Flies (House and Fruit)

Controlling flies requires the following:

1. **Inspection**—Locate their breeding and larval developmental sites.
2. **Sanitation**—Eliminate their larval breeding and developmental sites. This step should reduce the majority of the flies so that other means of extermination will be more effective.
3. **Mechanical controls**—These typically consist of removing garbage receptacles or replacing them with containers that have tight-fitting lids, having tight-fitting windows and doors, making sure windows are securely screened if they can be opened, using self-closing doors, sealing all penetrations through exterior walls for utilities, securely screening all vents, and using of air curtains, insect light traps, sticky-surfaced traps, and the like.
4. **Insecticide application**—Use only appropriately labeled pesticides in accordance to the manufacturer's instructions.
   - **Outdoors**—Boric acid can be placed around the bottom of dumpsters.
   - **Indoors**—Typically this requires a room-by-room determination of the best course of action. This can include the use of automatic/metered dispensers and/or ultra-low volume (ULV) applications, with the low-oil formulations being more desirable.

Stinging/Biting Insects (Bees, Mosquitoes, Wasps, Hornets, and Ticks)

Remove the pest's food, water, and shelter by keeping the facilities clean and sanitizing outdoor areas. Keep tight-fitting lids on garbage cans and empty them regularly.

Termites

Like ants, termites are social insects that mostly feed on dead plant material (wood) and animal waste. Currently, there are approximately 4,000 species of termites in the world. Termites, when left alone, can cause serious structural damage to structures and plantation forest. Detecting and controlling termites is a job for the professionals. Although termite control should be left to the professionals, a facilities maintenance technician can do a few things to help eliminate and prevent termites. These are:

- Eliminate moisture problems resulting from leaky pipes, AC units, storm, and drainage.
- Remove food sources such as dead trees, stumps, fire wood, and damaged wood structures.

Pest Control

Before facilities maintenance technicians can effectively apply pest control to a structure, they must identify the type of pest(s) that must be controlled. Once this has been accomplished, the source of food, entry, and attraction must be identified and eliminated. This includes the following:

- Check exterior doors. If you can see light under the door, this is a potential problem.
  - Install door thresholds.
  - Seal around windows.
- Check for windows that do not fit properly or have holes in the screens.
  - Seal around windows.
  - Use mesh or screens to fix holes in the screens.
- Check for openings around any objects that penetrate the building's foundation such as plumbing, electrical service, telephone wires, HVAC, and so on.
  - Seal around these objects.
- Do not store materials against the foundation of a building. This could be a nesting place for bugs.
- Do not leave outside lights on all of the time. Light attracts bugs.
  - Use motion sensor lights.
- Maintain a plant-free zone of about 12 inches around the building to deter insects from entering.
- Fix problem areas in the structure of the building that provide a nesting place for birds or rodents.
- Place outdoor garbage containers away from the building and on concrete or asphalt slabs and keep the area clean.
- Keep facilities and area around facilities clean.

If these measures do not control and/or eliminate the pest in a structure, consider selecting a **pesticide** that will control the pest and have the least possible harmful effects on the occupants and the environment.

Nonpesticidal Pest Control

Although pesticides can be successfully used to eliminate pests, they can be harmful to the occupants and pets. Therefore, it is recommended researching and understanding nonpesticidal pest control methods that can be applied to a particular type of pest. Nonpesticidal pest control methods include the following:

- Trapping
- Hoeing
- Hand weeding
- Installing pest barriers
- Sanitizing the area
- Eliminating food, water, and/or cover for the pest

- Using natural predators (spiders in some cases)
- Natural insecticides (neem-tree products)
- Ultrasonic pest control (use ultra-high-frequency sound waves to repel insects and rodents)

Pest Control with Pesticides

Pesticides should only be applied in areas in which there is currently a threat. Also preventative application of a pesticide should be avoided. When applying a pesticide, use the least toxic product(s) available that will effectively and safely control the pest.

Effects of Pesticides on Pests

Not all pesticides eliminate pests the same way. The most common methods of eliminating pests used by pesticides are:

- **Ingested poison**—Kills when swallowed.
- **Contact poison**—Sprayed directly on pest.
- **Fumigants**—Gas that is inhaled or absorbed.
- **Systemics**—Will kill the pest when it eats the host, but does not harm the host.
- **Protectants**—Prevent the entry of pests.

CAUTION

Always follow the manufacturer's instructions, and never mix pesticides with other chemicals. Because pesticides can be harmful to the environment, never dump them into a drain or dumpster. Always follow the manufacturer's disposal instructions as well as any local, state, and federal regulations.

Pesticide Safety

Because pesticides are a low-tech approach to controlling and eliminating pests, they do not discriminate against whom they will kill and/or affect. Therefore, when working with pesticides, it is important to use the following safety procedures:

- Select a pesticide with the lowest toxicity that can be employed legally on the target area, crop, or plant and that will safely and effectively control the pest.
- Buy only the amount of pesticide necessary to complete a particular extermination task.
- Keep pesticides in a secure and separate location from other items.
- Before attempting to use pesticides, ensure that you have the proper safety and application equipment available and know how to use it.
- Read, understand, and follow the pesticide's instructions before using it. This includes all safety information associated with the product.
- Inspect the area to be treated and its surroundings. Because plants and animals can be injured by the chemical used, always read and follow the manufacturer's instructions. If you cannot guarantee their safety, then do not spray the pesticide.
- In most communities, it is against the law to dispose of hazardous waste such as pesticides directly in the garbage or to pour remaining chemicals down the drain. Check if your community has a waste collection program.

- Store pesticides out of reach of children—in locked cabinets or in cabinets with childproof latches.
- Store pesticides only in their original containers with labels visible and intact.
- Never mix pesticides together or with other chemicals.

Applying Pesticides

The success or failure of an extermination job depends greatly on the type of pesticide application equipment used. In addition to selecting the proper application equipment, using it correctly as well as properly maintaining it will ensure the success of the extermination job as well. Remember, pests of any kind are a nuisance to tenants. When working on a complaint about pests, be as nonintrusive as possible, but do whatever is necessary to eliminate the pests. Once the elimination or preventive measures have been performed, check back with the tenant to make sure the pests have been eliminated.

Sprayers

The most common of the pesticide application equipment is the sprayer. In fact, it has become standard issue for most professional extermination crews. Sprayers range in complexity and size from simple, handheld models to intricate machines weighing several tons.

Hand Sprayers

When small amounts of pesticides are needed, a hand sprayer is typically employed. They can be used indoors as well as outdoors as well as in hard-to-reach areas. Most use a hand pump to supply compressed air to the unit. Hand sprayers include:

- **Pressurized can (aerosol sprayer)**—Pesticide contained in a sealed container under pressure (Figure 14-1).
- **Trigger pump sprayer**—When the trigger is squeezed, pressure propels the pesticide and diluents through the. These units range in size from 1 pint to 1 gallon (Figure 14-2).

FIGURE 14-1: Pesticide aerosol can.

FIGURE 14-2: Trigger pump pesticide sprayer.

- **Hose-end sprayer**—Typically consists of a hose attached to a unit that mixes and dispenses a fixed amount of pesticide (Figure 14-3).

- **Push–pull hand pump sprayer**—Air is forced out of a cylinder using a hand operated plunger, thus creating a vacuum at the top of a siphon tube. The suction draws pesticide from a small tank and forces it out with the air flow. Capacity is usually 1 quart or less (Figure 14-4).

- **Compressed air sprayers**—Usually operated with pressure that has been created from a self-contained manual pump (Figure 14-5).

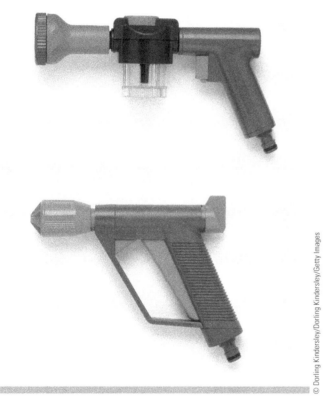

FIGURE 14-3: Spraying with hose-end pesticide sprayer.

FIGURE 14-4: Push–pull hand spray pump.

FIGURE 14-5: Compressed-air pesticide sprayer.

Small Motorized Sprayers

Small motorized sprayers are usually not self-propelled. They may be mounted on wheels, so they can be pulled manually, mounted on a small trailer for pulling behind a small tractor, or skid-mounted for carrying on a small truck. They may be low-pressure or high-pressure sprayers, according to the pump and other components with which they are equipped. Small motorized sprayers include:

- **Estate sprayers**—Attached on a two-wheel cart with handles for pushing (Figure 14-6).
- **Power backpack sprayer**—This usually has a small gas-powered engine. This model can generate high pressure and is best suited for low-volume applications of diluted or concentrated pesticides (Figure 14-7).

FIGURE 14-6: Estate pesticide sprayer.

FIGURE 14-7: Power backpack sprayer.

SUMMARY

- The first step to controlling pests is to determine the type of pest that is infesting a facility.
- Controlling cockroaches is seldom easy because getting the insecticide to the insect is difficult.
- To control rodents, seal off entry points into the facilities.

- Like ants, termites are social insects that mostly feed on dead plant material (wood) and animal waste.
- Pesticides are considered a low-tech approach to controlling and eliminating pests, they do not discriminate against whom they will kill and/or affect.
- Sprayers are the most common pesticide application equipment.
- Hand sprayers are often used to apply small quantities of pesticides.
- For facilities in which the pest control is contracted out, the facility maintenance technician is usually the primary contact person for the facility.

REVIEW QUESTIONS

1   Describe the difference between pest control with pesticides and nonpesticidal pest control.

2   Describe how to control the following pests:

- Cockroaches
- Ants
- Spiders
- Termites

3   List the way in which pesticides kill pests.

4   What is the difference between a pesticide and an insecticide?

5   How do ultrasonic pest controls manage pest?

Name: _____

Date: _____

Pest Prevention

IDENTIFY AND REMOVE PESTS

Upon completion of this job sheet, you should be able to identify pests that may be causing problems at your facility and develop a plan to remove them.

1. Inspect your facility or training area, and make a list of the pests that you observe or pests that the tenants report to you. If possible, use a digital camera to photograph the pests for easier identification.

2. Document where the pests were observed and what they were doing at the time of observation.

3. Estimate the level of infestation to determine if there needs to be a plan in place to remove the pests.

4. Research the appropriate method for removing the pests, if deemed necessary, and develop a plan.

5. Implement your plan if applicable.

INSTRUCTOR'S RESPONSE:

_____

_____

_____

© iStockphoto.com/kkgas

CHAPTER 15

Landscaping and Groundskeeping

OBJECTIVES

By the end of this chapter, you will be able to:

KNOWLEDGE-BASED

- Identify the various parts of a plant.
- Discuss the proper procedure for installing a retaining wall.
- Discuss the proper procedure for removing snow.
- Discuss the procedure for removing a swimming pool.
- Discuss the inspections that must take place when removing a swimming pool.

SKILL-BASED

- Maintain and police grounds, including mowing, edging, planting, mulching, leaf removal, and other assigned tasks.
- Perform basic small engine repair and preventive maintenance according to the manufacturer's specifications.
- Perform basic swimming pool maintenance that does not require certification.
- Remove refuse and snow as required.
- Maintain public areas, including hallways, kitchens, and lobbies.
- Repair asphalt using cold-patch material.

Groundskeeping the activity of maintaining an area for the purpose of aesthetics and functionality; this includes mowing grass, trimming hedges, pulling weeds, planting flowers, and so on

Groundskeeper a technician who maintains the functionality and appearance of landscaped grounds and gardens

Landscaper an individual who modifies the features of an area (land) such as stone walls, wooden fences, brick pathways, statuary, fountains, benches, trees, flowers, shrubs, and grasses

Belly deck a mower that is built like a car and has a blade deck underneath the driver and engine

Organic mulch natural substances placed over soil such as bark, wood chips, leaves, pine needles, and grass clippings

Inorganic mulch gravel, pebbles, black plastic, and landscape fabrics placed over soil

Introduction

Groundskeeping is the activity of maintaining an area for the purpose of aesthetics and functionality. It includes mowing grass, trimming hedges, pulling weeds, planting flowers, and so on. A **groundskeeper** is a technician who maintains the functionality and appearance of landscaped grounds and gardens, whereas a **landscaper** is an individual who modifies the features of an area (land) such as stone walls, wooden fences, brick pathways, statuary, fountains, benches, trees, flowers, shrubs, and grasses.

Typically, these technicians use manual and power equipment to maintain the facility's grounds. Technicians who normally work in this are perform the following tasks: laying sod, mowing, trimming, planting, watering, fertilizing, digging, raking, sprinkler installation, and installation of mortarless segmental concrete masonry wall units (retaining walls).

Mowing

Mowing is the process of maintaining the height of the lawn by using a lawnmower to cut the grass. Proper lawn maintenance is more involved than simply starting a power mower and making a few laps around the grounds. Although it is not rocket science, there are a few basic rules that should be followed:

1. Never cut grass too short (see Table 15-1).

2. Never remove more than one-third of the grass's height in one session.

3. Never mow the grass when it is wet.

4. Change the direction in which the grass is cut with each mowing.

5. On average a lawn should be mowed once a week.

6. Growing season is usually from May to October.

Type of Grass	Desired Height (Inches)
Centipede	2.5
Common Bermuda	1.5
Hybrid Bermuda	1
Kentucky Blue	2
St. Augustine	2.5
Zoysia japonica	1.5
Zoysia matrella	1

TABLE 15-1: Desired Grass Height for Common Grasses.

Lawn Mowing

Depending on the size of the lawn to be maintained, most facilities maintenance technicians use either a push mower (which may be self-propelled) or a riding mower (Figure 15-1).

Riding mowers are available in different sizes and types. The most common riding mower for residential use is one with a "belly" deck. A **belly deck** is a mower that is built like a car (four wheels, a driver's seat, a steering wheel, engine located in the front) and has a blade deck underneath the driver and engine (Figure 15-2).

FIGURE 15-1: Push mower.

FIGURE 15-2: Riding mower.

Edging and Trimming

Traditionally, edging and/or trimming were done as the final step in mowing a lawn. However, depending on the size of the groundskeeping crew, it should be done before or while the lawn is being mowed. This allows the clipping generated to be either mulched and left on the lawn or mulched and bagged by the mower. Trimming and edging is a simple, but all-important, practice that gives a lawn a finished, manicured look. When edging:

- Use a handheld edger between pavement and grass. Place the wheel on the pavement with the blade over the edge and push and pull. For large lawns, use a power-driven model (Figure 15-3).

FIGURE 15-3: Handheld edger.

- Edge along walkway to prevent grass from growing out past the borders.
- Edge along the edges of turf and landscape beds to prevent grass from growing into the beds.

When trimming:

- Use grass shears around delicate plants and trees that could be damaged using string trimmers (Figure 15-4).
- Use a string trimmer in areas that are difficult or impossible to reach with a lawn mower (Figure 15-5).

FIGURE 15-4: Grass shears.

Introduction to Botany

Botany is a branch of science that is concerned with the study of plant life and its development. Plants are vital to all life forms on this planet because they provide food, fuel, and medicine, and they provide oxygen. Therefore, the study of plants is critical for:

- The production of food.
- The production of medicine.
- Controlling environmental changes due to changes in plant life.

FIGURE 15-5: String trimmer.

Parts of a Plant

To understand how to take care of plants, one must understand the different parts of plants and their function. Although there are thousands of different plant varieties, all plants produce seeds, and most produce their own food by means of photosynthesis. The basic parts of a plant are roots, stems, leaves, flowers, and fruits (see Figure 15-6).

Roots

Plants' roots absorb water and minerals from the soil as well as help stabilize and anchor the plants to the ground.

Stems

Stems are the plant's transportation system; they help transport the water and minerals taken in by the roots. In addition, stems are used to transport the food produced by the leaves to the other parts of the plant.

Leaves

Leaves are the tiny plant food factories, in which sunlight is captured and combined with minerals and water to produce sugar that the plant uses to produce food.

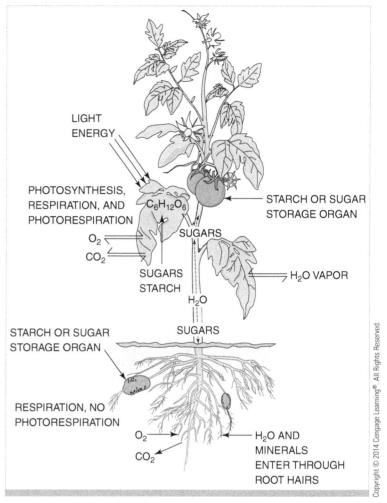

FIGURE 15-6: Parts of a plant.

Flowers

Flowers are the reproductive elements of the plant. Flowers contain both pollen and tiny eggs, called ovules, which when fertilized (pollinated) turn into fruit.

Fruits

Fruit is a protective covering for the plant's seed. In other words, fruit is a ripened ovary.

Seeds

Seeds are embryonic plants that contain stored food enclosed in a covering.

Mulching

Mulch is defined as any loose material placed over soil to control weeds and conserve soil moisture by providing 10% to 25% reduction in soil moisture loss from evaporation. In addition, mulch helps keep the soil well-aerated by reducing soil compaction that results when raindrops hit the soil. It also reduces water runoff and soil erosion. Mulching trees and shrubs is a good method of reducing landscape maintenance and keeping plants healthy.

Large quantities of mulch are generated naturally from such sources as fallen leaves, pine needles, twigs, branches, dead plants, pieces of bark, spent flower blossoms, fallen fruit, and other organic material. There are basically two types of mulches: organic and inorganic. Both types are used in landscaping.

Organic mulch is composed of natural materials such as bark, wood chips, leaves, pine needles, or grass clippings. Organic mulches have a tendency to provide a food source for insects, slugs, cutworms, and the birds. As organic mulch decomposes over time, thus creating a need to replace the mulch over time. **Inorganic mulches** are composed of such materials as gravel, pebbles, black plastic, and landscaping fabrics. Unlike organic, inorganic mulch will not attract insects nor will it decompose. A 2–4-inch layer (after settling) is adequate to prevent most weed seeds from germinating. Mulch should be applied to a weed-free soil surface (Figure 15-7).

FIGURE 15-7: Mulch around plants.

Aeration

As organic material dies, it tends to leave a layer on the lawn known as "thatch; this thatch can deprive the lawn of oxygen. In addition, soil that has been compacted is difficult for plants to grow roots in as well as hard to provide proper irrigation in. To help remedy the situation, small soil plugs or cores can be removed from the lawn in a process known as "aeration." Currently, there are two types of aerators used: spike and core aerators.

FIGURE 15-8: Aeration machine.

Spike aerators can be pulled by hand or behind a tractor. These aerators use a solid spike to penetrate the soil, leaving a small hole in the ground. This is accomplished by pushing the soil away from the spike. For soils that are heavily compacted, this type of aeration is not recommended, because it can contribute to the compaction problem. Because the spikes simple push the soil away from them, the benefits of aeration are short term. Core aeration, on the other hand, uses hollow or spoon shaped tines to extract a small soil plug (also known as a core) from the lawn. On average these types of aerators create a hole between $\frac{1}{2}$ to $\frac{3}{4}$ inch in diameter having a depth of approximately 1 to 6 inches in depth. The holes are typically spaced between 2 and 6 inches apart (Figure 15-8).

To be effective, aeration should be performed twice a year, early spring and late fall.

Mark sprinkler heads with flag or paint before aeration.

1. Rake all yard debris from the lawn.

2. Make sure the lawn has been mowed (to approximately 2–2$\frac{1}{2}$ inches in height) and is free of obstacles.

3. When aerating, the soil should be moist. If the soil is dry, then water the lawn the night before. Soil that is moist will be easier to work with.

4. Work in a north to south or east to west direction.

5. Fertilize the lawn after aerating it.

FROM EXPERIENCE

Lawn aerator shoes can be an effective way to aerate small lawn or areas.

FROM EXPERIENCE

To maintain a healthy lawn, core aeration should be performed every three years. In between core aeration, spike aeration should be performed.

Amending Soil Prior to Installation

A source of confusion for most people is the distinction between dirt and soil. Basically, dirt is the soil that has been depleted of the nutrients needed to support and maintain plant life (its fertility). The primary nutrients required to support life are nitrogen, phosphorus, and potassium. Amending the soil is the process of changing its ability to sustain plant life, that is, adding necessary nutrients and/or changing its texture (the ratio of sand, silt, and clay). The type of plant life that is being planted will determine the type and amount of nutrients as well as the soil texture needed.

Nitrogen

Plants use nitrogen to generate protein (in the form of enzymes) and nucleic acids. Nitrogen is present in chlorophyll, a green pigment responsible for photosynthesis. It is distributed throughout the plant, starting with the older tissue and working its way to the new tissue. A plant that displays yellowish older leaves is typically deficient in nitrogen.

Phosphorus

Phosphorus is another essential element in the process of photosynthesis. It is also used in the development of the plant's root system as well as the production of the plant's flowers.

Potassium

Potassium is essential for the overall health of the plant by aiding in the production of proteins, photosynthesis, high-quality fruit, and the reduction of diseases.

Using Bushes and Shrubs in Landscapes

Both flowering shrubs and evergreen bushes are used extensively in landscaping. Because many flowering shrubs and plants produce berries, they have a tendency to attract wildlife as well as provide beautiful fall foliage colors Evergreen bushes can be used as privacy screens or even pruned into hedges. A flowering shrub or evergreen bush can be used as a stand-alone piece to provide a focal point (see Figures 15-9 and 15-10).

FIGURE 15-9: Flowering shrubs.

FIGURE 15-10: Evergreen bushes.

Planting Bushes and Shrubs

When planting bushes and shrubs, always follow the instruction provided with the plant or by the retailer from which the plant was purchased. To plant a shrub or a bush, follow these guidelines:

1. Select a location for the plant, keeping the following in mind:
   - Growing condition (temperature, water conditions, sunlight, etc.)
   - Potential hazards (to the plant)
2. When digging the receiving hole for the plant, be sure that it is twice as wide and deep as the pot or rootball of the plant.
3. Add a layer of mulch that is approximately one-fourth the depth of the receiving hole.
4. Add a cup of slow-release fertilizer to the receiving hole.
5. Add two shovelfuls of dirt to cover the fertilizer.
6. Fill the hole with water and allow it to be absorbed into the compost and surrounding soil.
7. Remove the plant roots from the pot or burlap wrap.
8. Gently spread out the roots.
9. Place the rootball centrally in the hole and backfill to the soil line on the plant.
10. Gently tamp the soil to hold the bush upright.

 FROM EXPERIENCE

Mulching the beds is done midspring, usually with bark or woodchips 2–3 inches in depth to help keep the weeds to a minimum.

Retaining Walls

A retaining wall is a structure that is designed to hold back soil and/or rock from a facility or an area. Retaining walls are used to control erosion, extend usable grounds, and control changes in grade (the slope of the land). When built correctly, they can add beauty and value to any property.

Installing a Cinder-Block Retaining Wall

Concrete block retaining walls, also known as cinder-block retaining walls, are more extensive to construct and are usually longer lasting than other forms of retaining walls (landscaping timbers, see Figure 15-11). However, with a little planning, these types of retaining walls can be easily constructed. Before attempting to construct a cinder-block retaining wall, always consult local and state building codes. To construct a cinder-block retaining wall:

1. Determine where the retaining wall must be located.

2. Excavate for the footing. The excavation for the footer should be done so that the footer is below the frost line; in some locations this can be as deep as 36 inches. The width of the footer should be twice the width of the wall. So for a wall that is 3 feet high built with 12-inch concrete block, the footer will be 24 inches. Also, the frost line will aid in determining of the depth of the footer.

3. Place three rows of $\frac{1}{2}$-inch steel rebar 6 inches from the outside edges, and one in the center should be placed in the footer about $\frac{1}{3}$ the depth of the footer from the bottom.

4. Using 3,000-psi concrete, pour the footer.

5. Place $\frac{1}{2}$-inch rebar perpendicular to the footer at the location in which the row of blocks are to be installed. Ensure that the position of the rebar will match the cavities of the blocks!

FIGURE 15-11: Cinder-block retaining wall.

© Steven Frame/www.Shutterstock.com

6. Using a chalk line, snap the position of the first row of the cinder blocks. Lay a $\frac{3}{8}$-inch-thick mortar bed, and apply mortar to the blocks; tap the blocks into position, and level.

7. Alternate the block joints for the second course to maximum the strength of the wall.

8. For retaining walls over 3-feet tall, deadmen should be installed every 4 feet.

9. Attach galvanized deadmen cables parallel to the retaining wall and buried in undisturbed soil.

10. On the uphill side of the footing, place preformed weeping pipe in a layer of gravel or crushed stone.

11. Using gravel, backfill and pack (using a tamper) into place the first two courses of the wall.

12. Fill the first two courses with concrete.

13. Install all additional courses, filling all of them with concrete and using deadmen supports needed.

14. To give the wall a finished look, install a wall facing, using flagstone, brick, or stucco.

15. Cap the final course as desired and complete the backfilling.

Installing a Castlerock Retaining Wall

Castlerock, or mortarless block, retaining walls are much simpler to install and maintain than cinder-block retaining walls (see Figure 15-12). In addition, if you make a mistake and most likely you will, castlerock retaining walls can be easily dismantled and adjusted or corrected. Finally the skill level required to construct a cinder-block retaining wall is much greater than that of a castlerock retaining wall.

To construct a castlerock retaining wall:

1. Determine where the retaining wall must be located.

2. Excavate the footing. Normally for a castlerock wall 3-feet high, the footing should be 18 inches wide and 6 inches deep. Be sure to place the foundation course of the wall below the surface or ground level. Add a layer of leveling sand or paver base (1–2 inches) and tamp.

3. Set the first castlerock into place, checking it to ensure that it is level. If the block is not level, then using a hammer, tap the block to adjust it.

4. Continue this until the entire foundation course has been set.

5. After the foundation course has been set, starting with the end of the retaining wall, offset the second block (as you would for cinder blocks) and set it into position.

6. Attaching a line between the two blocks and a line level, check to see if the blocks are level with one another. If the blocks are not level, adjust them until they are level.

7. Once the blocks are level continue laying the course.

8. For each course, repeat Steps 6 and 7 until the desired height has been reached.

9. Crushed rock or gravel should be used to prevent the soil from washing through spaces in the retaining wall. The gravel should be protruding out from the retaining wall a minimum of 8–12 inches.

FIGURE 15-12: Castlerock retaining wall.

Installing a Landscaping Timber

Like castlerock retaining walls, landscaping timber retaining walls are much simpler to install and maintain than cinder block ones (see Figure 15-13). To construct a landscaping timber retaining wall:

1. Determine where the retaining wall must be located.

2. Excavate the footing. Normally for a landscaping timber retaining wall 3 feet high, the footing should be 18 inches wide and 6 inches deep. The footer should be filled with crushed stone or gravel and tamped flat.

3. Using a chainsaw, cut the landscaping timber to size as needed.

4. Using a drill, drill holes through the timbers for the rebar. The holes should be spaced about 4-feet apart.

5. Place the first course of landscaping timber into position, and drive the 24–36-inch rebar through the timber into the ground.

6. Before adding subsequent courses, check each course to ensure that it is level. If the course is not level, adjust the course using a sledge hammer.

7. Lay the next course into position and connect it to the previous course using 12-inch galvanized spikes about every 24 inches.

8. To every third course add a deadmen anchored to the course using 12-inch galvanized spikes.

9. Once the desired height has been achieved, backfill the retaining wall with crushed rock or gravel from the retaining wall to 8–12 inches out from the retaining wall.

FIGURE 15-13: Landscaping timbers retaining wall.

Paving Stone Walks

Paving stones can offer an attractive, quick, and easy alternative to concrete in constructing a patio or walkway for a fraction of the cost.

Installing Paving-Stone Walks

The process for installing a paving-stone walk is similar to that for a castlerock retaining wall (see Figure 15-14). To install a paving-stone walk:

1. Determine where the walk will be located.

2. Excavate the walk area. If the walk is designed for pedestrian traffic, then the area should be excavated 7–9 inches with a 4–6-inch gravel base. For vehicular traffic, the area should be excavated 9–11 inches with a 6–8-inch gravel base. The base should be tamped flat and level.

3. Install the edge restraints in their desired location and shape. Edge restraints are used to prevent the pavers from moving.

FIGURE 15-14: Paving-stone walkway.

4. Install the bedding sand. The bedding sand is installed by laying 1-inch conduit 6–8-feet apart in the area where the pavers are to be installed, followed by screening the sand into place. Once the bedding sand is screened into place, remove the 1-inch conduit.

5. Starting in one of the corners, lay the pavers in the desired pattern.

6. Clear the surface of any debris, then spread masonry sand over the paver's surface. Using a broom, sweep the sand into the joints, leaving surplus sand on the pavers. Using a plate compactor, tamp the pavers down.

7. Sweep the remaining excess dry sand over the surface filling the joints.

Irrigation Systems

An irrigation system is a network of piping, valves, and controls used to artificially maintain the moisture level in the soil. The most widely used types of irrigation are ditch irrigation, terraced irrigation, drip irrigation, sprinkler system, rotary systems, and center pivot irrigation.

Winterizing the Irrigation System

Every year, before the first freeze, winterization should be performed on all irrigation systems located in climates in which freezing can occur. This is especially necessary in climates in which the frost level is below the piping system. To minimize the possibility of damage to the system due to freezing, winterization is a critical step and, therefore, should be performed every winter (depending upon local climate conditions).

Manual Drain Method

This method is used when the manual valves are located at the end and low points of the irrigation piping (Figure 15-15).

To drain these systems:

1. Shut off the irrigation water supply. The shutoff will be located in the basement and will be a gate/globe valve, ball valve, or stop–waste valve (see Figures 15-16 through 15-18).

2. Open all the manual drain valves.

FIGURE 15-15: Valves on the irrigation piping.

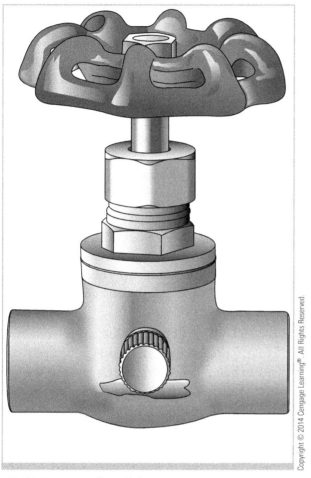

FIGURE 15-16: Gate Valve.

FIGURE 15-17: Ball Valve.

FIGURE 15-18: Stop/waste valve.

3. Once the main line has been drained of water, open the boiler drain valve or the drain cap on the stop–waste valve and drain all the remaining water that is present between the irrigation water shutoff valve and the backflow device (Figure 15-19).

4. Open the test cocks on the backflow device. If check vales are present on the sprinkler system, you will need to pull up on the sprinklers to allow the water to drain out the bottom of the sprinkler body (Figure 15-20).

FIGURE 15-19: Boiler drain valve/drain cap.

FIGURE 15-20: Test cocks.

Automatic Drain Method

This method is used when the automatic drain valves are located at the end and low points of the irrigation piping (Figure 15-21). These valves will automatically open and drain water if the pressure in the piping is less than 10 psi.

FIGURE 15-21: Valves at the end of irrigation piping.

1. To activate the automatic drain valves, shut off the irrigation water supply and activate a station to relieve the system pressure. Normally, the shutoff valve will be a gate/globe valve, ball valve, or stop–waste valve located (Figures 15-18, 15-22, and 15-23).

2. Once the water has drained out of the main line, open the boiler drain valve or the drain cap on the stop–waste valve and drain the remaining water that is present between the irrigation water shutoff valve and the backflow device (Figure 15-19).

3. Open the test cocks on the backflow device. If your sprinklers have check valves, you will need to pull up on the sprinklers to allow the water to drain out the bottom of the sprinkler body (Figure 15-20).

"Blow-Out" Method

It is recommended that a qualified licensed contractor perform this type of winterization method. The blow-out method utilizes an air compressor with a cubic foot per minute (CFM) rating of 125–185 for any mainline of 2 inches or less and a PSI of 50–80.

1. Open the test cocks on the vacuum breaker (Figure 15-20).

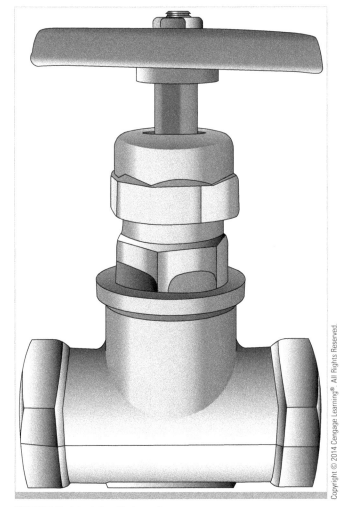

FIGURE 15-22: Gate valve.

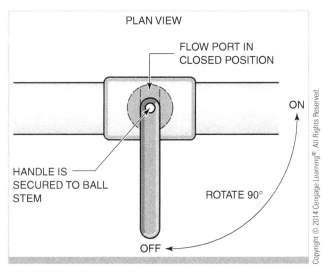

PLAN VIEW

FLOW PORT IN CLOSED POSITION

HANDLE IS SECURED TO BALL STEM

ON

ROTATE 90°

OFF

FIGURE 15-23: Ball valve.

2. Shut off the irrigation water supply and open the drain on the supply line.

3. Once the line is drained, close the drain.

4. Attach the compressor to the main line via a quick coupler, hose bib, or other type of connection, which is located before the backflow device (Figure 15-24).

5. Activate the station on the controller that is the zone or sprinklers highest in elevation and the farthest from the compressor (Figure 15-25).

6. Do not close the backflow isolation or test cock valves. Slowly open the valve on the compressor; this should gradually introduce air into the irrigation system. The air pressure should be constant at 50 psi. If the sprinkler heads do not pop up and seal, increase the air until the heads do pop up and seal. The air pressure should NEVER exceed 80 psi.

7. Activate each station/zone, starting from the station/zone farthest from the compressor, slowly working your way to the closest station/zone. Each station/zone should be activated until no water can be seen exiting the heads. This should take approximately 2–4 minutes per station/zone.

FIGURE 15-24: Compressor attached to the mainline.

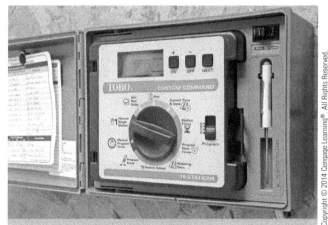

FIGURE 15-25: Station controller.

Spring Irrigation Startup

If the system was correctly winterized in the fall, the chances of cracks and breaks due to freezing were greatly reduced. However, even properly winterized systems are subject to damage from extreme conditions. Therefore, the spring startup procedure should be done in four stages:

1. Safely reintroducing water to each zone.

2. Checking winter damage, making repairs as needed, and cleaning or replacing nozzles.

3. Examining the entire system to see that it is still operating the way it is supposed to and providing even coverage.

4. Resetting the controls.

FIGURE 15-26: Main drain valve.

FIGURE 15-27: Rain sensor.

Starting Up the Irrigation System in the Spring

Follow this procedure when first starting an irrigation system in the spring:

1. Make sure that all the manual drains valves are closed before turning on any water to the system (Figure 15-26).

2. The pipes should be allowed to fill slowly by opening the main water valve slowly. Opening the main sprinkler line to quickly can subject the system to an extremely high pressure surge that can lead to water hammer or uncontrollable flow. A water hammer can result in damage to the piping system.

3. Each zone's proper operation should be verified by manually activating it from the controller.

4. Each station can be activated on the controller, thus allowing you to check for proper operation of the zone. Check the operating pressure of the system. Low pressure will result typically indicates a break in the line or even a missing sprinkler. Check the sprinkler head for proper rotation and adjustment as well as coverage. All filters should be clean and if necessary replaced. Clogged filters generally result in poor sprinkler performance.

5. Adjust the controller programming for automatic watering, if necessary.

6. Check the controller's backup battery and, if necessary, replace it.

7. If the system has a rain sensor, uncover and/or clean the sensor (Figure 15-27).

8. For a drip irrigation zone clean and/or replace any inline filters.

Pool Maintenance

Routine pool maintenance is an unavoidable fact of life if you want to keep your pool looking clean, sparkling, and inviting day after day.

Cleaning the Pool Deck

In addition to maintaining the pool, most facilities maintenance technicians are responsible for the upkeep of the pool deck. When maintaining the pool deck area:

1. Eliminate as much debris from the pool as possible.

2. Sweep or use a hose to remove the debris near the pool. Cover pool if necessary (Figure 15-28).

Cleaning the Surface of the Pool

Floating debris can be easily removed by using a skimmer or a pool leaf rake and a telescopic pole. To keep the debris from getting back into the pool, empty the skim or pool leaf rake into a trash can of plastic garbage bag. Emptying the skimming debris into the garden or on the lawn usually leads to the debris getting back into the pool once it has dried out.

There is no set method for skimming debris, but as you do, scrape the tile line, which acts as a magnet for small bits of leaves and dirt. The rubber–plastic edge gasket on the professional pool leaf rake will prevent scratching the tile (Figure 15-29).

If there is scum or general dirt on the water surface, squirt a quick shot of tile soap over the length of the pool; doing so will cause the scum to spread toward the edges of the pool, making it more concentrated and easier to skim off.

FIGURE 15-28: Pool deck.

Maintaining the Pool

In addition to maintaining the appearance of the pool (removing trash, dirt, etc.), you must also maintain the quality of the water. You can maintaining the integrity of the water in the pool by doing the following:

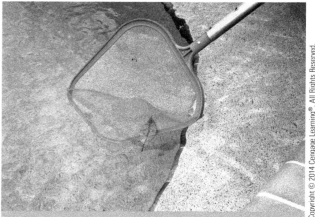

FIGURE 15-29: Skim debris from the pool.

1. You should use a stabilized chlorine product to sanitize the pool, providing protection against bacteria. These products are generally available in stick, tablet or granular form and are fed through a distribution container near the pump and filter system.

2. Algae preventative products should be used to help prevent the more than 15,000 kinds of algae from ever getting started. You apply the liquid version of the product simply by pouring it into the water near the skimmer intake so that the pump system can distribute it.

3. Consult a pool professional or supply dealership for water testing and computer analysis of the water sample.

4. Check for chlorine and pH levels daily. To check pH in pool water, use a kit from your local pool store. It contains tubes that hold about 4 ounces of pool water. You then put in drops of liquid that comes with the test kit, shake for about 30 seconds, and match the colors with the accompanying chart to check the chlorine and pH levels (yellow for chlorine and red for pH).

Snow Plowing

Unfortunately, as with rainfall, you cannot always predict when or how much snow will fall in any one spot. When it does fall, it is the responsibility of the facilities maintenance technician to remove the snow. This does not mean that the technician has to remove the snow himself or herself; the snow removal can be contracted out.

FIGURE 15-30: Clear walkways.

FIGURE 15-31: Snow on side of parking lot.

Recommended Snow Removal Procedures

The following procedure should be followed when removing snow:

- For the initial opening of steps, large walk-through areas, and large entry ways a partially path along the railings should be shoveled. Full access should be given to handicap areas (Figure 15-30).

- After a storm, the cleanup operations should include completely opening all walks and entry ways and deicing.

- Snow should be removed (pushed back) from roadway sides, walks and parking lots (Figure 15-31).

- Any remaining snow should be removed from all stairs and entryways.

- Any remaining ice can be removed by using ice choppers and ice-melt applications.

- It is the grounds supervisor who will determine the appropriate application of ice-control products, based on current weather conditions or when freezing occurs.

- The removal of melted ice and slush from the surface completes the cleaning of all surfaces.

- As necessary, return trips for sanding and salting equipment should be made as determined by the grounds supervisor.

When parking lots are plowed, snow should be piled so that it does not interfere or block any thoroughfares and sidewalk areas. Avoid pushing snow back up over a curb. However, if it is absolutely necessary to push snow over a curbed area, it must be positioned so it will not fall back into the lot or obstruct any adjacent sidewalks.

Small Engine Repair

The facilities maintenance technician should be able to perform basic small engine repair and preventive maintenance according to manufacturer's specifications. However, to avoid voiding the warranty of a lawnmower, trimmer, chain saw, or the like, never work on any equipment that is still under warranty.

Common Problems with Small Engines

Some common problems associated with small engines that a facilities maintenance technician should be able to address are engines that will not start, engines that smoke, and engines that sputter.

Engine Will Not Start

One of the most common problems that a facilities maintenance technician will have to troubleshoot is engines that will not start.

- **Ensure that the equipment's fuel level is sufficient**—If necessary, fill the gas tank to just below the fill neck, leaving enough space for the fuel to expand and/or slosh around. Filling the tank all the way to the top will cause fuel to leak and the equipment will only run until the fuel level drops.

- **Check the fuel valve**—Check to see if the fuel valve is in the ON position.

- **Check the spark plug**—If it is "carbon shorted" (carbon is present between the electrode gap) clean or replace it. If the plug is pitted, is burned, or has cracked porcelain, replace it with an identical spark plug (Figure 15-32).

- **Check for sparking**—With a commercially available spark tester, test for sparking by putting the spark plug wire (high-tension lead) on one side of the tester and clipping the other side of the tester to the shroud, fins, or head bolts (anything metallic). Or ground the plug, with high tension lead on it, to the head of the engine and crank on the engine. Sparking is present when you see blue sparks jump the electrode gap.

FIGURE 15-32: Spark plug.

- **Prime or choke the engine**—If the engine is cold (has not been started in a recently) and has a primer or choke, use them! Read the manufacturer's instructions for priming and/or choking the engine. If the engine has a primer bulb, typically pushing it three times will be sufficient. If the engine contains a choke, choke the engine until it starts to fire and then open the choke.

If the engine still won't start after going through these steps, take it in to a repair center to be repaired.

Engine Is Smoking

The second common problem that a facilities maintenance technician will have to troubleshoot is engines that smoke.

- **Dirty or plugged air filter**—Replace or clean air filter (Figure 15-33).

- **Wrong grade of oil**—Oil that is less than 30w HD will vaporize if used for a long time. Use only a good grade of 30w HD oil for all four-cycle lawn equipment.

- **Worn valve guides**—A valve guide that is worn out due to excessive wear on the engine must be taken to the shop for a replacement guide bushing. Common causes are mowing on a hillside for excessive amounts of time and debris build-up on the cooling fins (Figure 15-34).

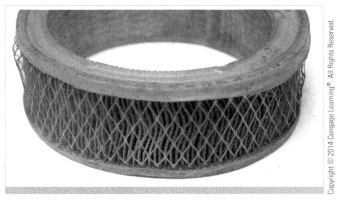

FIGURE 15-33: Dirty air filter.

FIGURE 15-34: Valve guide.

- **Choke is still on**—Open the choke as soon as the engine fires and continues to run without faltering.
- **Too much oil**—Correct the amount of oil in the crankcase. Fill only to the FULL mark.

If the engine still will not start after going through these steps, take it in to a repair center to be repaired.

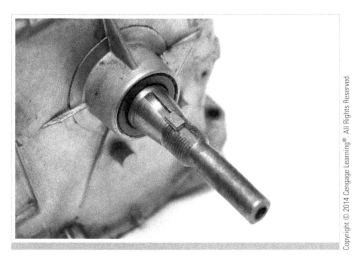

FIGURE 15-35: Flywheel key.

FIGURE 15-36: Flywheel.

Engine That Sputters

Finally, engines that sputter are something that a facilities maintenance person will have to troubleshoot. Causes include:

- **Water in the gas**—Drain the old gas from the fuel tank and replace it with fresh gas.
- **The electrical system is grounding on the equipment's frame (not the engine)**—This is usually caused by cracked or worn wires. Locate them and replace them.
- **Flywheel key is sheared**—Replace it as necessary (Figure 15-35).
- **Stop wire is rubbing on the flywheel**—Remove the flywheel, tape up wire if not severed, and reroute it so that it will not touch flywheel (Figure 15-36).
- **Bad spark plug (Figure 15-37)**—Replace the spark plug with an identical new one.

If the engine still will not start after going through these steps, take it in to a repair center to be repaired.

FIGURE 15-37: Bad spark plug.

Service Recommendations

Up to this point, the focus has been on repairing small engines. However, preventive maintenance is the most critical task that a facilities maintenance technician will perform on a small engine. Proper maintenance will extend the life of the equipment and reduce the repair cost. Preventive maintenance includes the following:

FIGURE 15-38: Air filter.

- **Change the oil**—This is the most important thing you can do to extend the life of any small engine. Always follow the manufacturer's recommendation and instructions for changing the engine oil. Generally, a SAE 30w oil is sufficient for most small engines. Never over fill the engine oil; doing so will cause the engine to smoke or, in some cases, blow the oil out of the engine's breather. On an engine with a dipstick, fill to the top line of the crosshatch mark. On engines without dipsticks, fill to the top of the hole. On horizontal engines without dipsticks, fill until oil comes out of the fill hole.

- **Clean or replace the air filter**—This is also a very important part of engine maintenance. Clean the foam filters with soap and hot water. Re-oil and squeeze out excess. Paper elements must be replaced if extremely dirty (Figure 15-38).

- **Change spark plugs**—Replace them once a season instead of trying to clean and reuse them (Figure 15-39).

- **Remove fuel before storing the engine**—Doing this will prevent you from having to do maintenance in the spring time.

If the engine still will not start after you have completed all of these steps, take it in to a repair center to be repaired.

Maintaining Public Areas

FIGURE 15-39: Spark plug.

Public areas are defined as areas that are not assigned to specific individuals. This includes kitchens, hallways, lobby areas, and stairwells—areas open to the public.

Recommended Maintenance Procedures for Public Areas

All facility must be maintained for both cosmetic and safety reasons. In some cases, dirty facilities can be a health violation that could result in a fine and even the closing of the facility. The following are recommended maintenance procedures for common public areas.

Bathroom

The maintenance of restrooms is mostly a safety concern. Improperly maintained restrooms can spread disease.

- Must be disinfected a minimum of once a day, possibly more depending upon their usage

- Paper products must be restocked a minimum of once a day according to the restrooms' usage.

- Must be thoroughly cleaned once a week
- Hand soap should be replaced as needed.

Stairways

The maintenance of stairways is a safety issue. In North America, the accident rate of stairs is rather high, that is, thousands of people are injured each year from falls on stairways.

To maintain stairs:

- Thorough cleaned them at least once a week.
- Sweep them on a daily basis.
- Make sure there is adequate lighting.

Hallways

Although hallways are not as dangerous as stairways, they should also be maintained on a regular basis.

- Once a week they should be buffed to a shine, twice time permitting.
- General cleaning should occur on a daily basis.
- On a weekly basis, the walls should be cleaned and spots should be removed as needed.
- Alcoves and shelves should be thoroughly cleaned.
- Water fountains should be disinfected daily.
- Make sure there is adequate lighting.

Trash Cans

If trash cans are not properly maintained, they can be used as a food source by insects and pests. Therefore, it is imperative that they be properly maintained. Trash cans should be:

- Emptied daily or as needed.
- Thoroughly clean and disinfected once a week.

Lobbies

Lobbies are often the first impression that a person will have of a facility; therefore, it is critical that they be properly maintained. Lobbies should:

- Be thoroughly cleaned weekly.
- Be given a general cleaning daily.

Repairing Asphalt Using Cold-Patch Material

The most common types of materials used on driveways and parking lots today are asphalt and concrete. Even though asphalt is a durable building material, it does require periodic maintenance. This is especially true in colder climates, where freezing and thawing cycles frequently occur. Before minor damage due to climatic conditions becomes a major problem, problems should be addressed. Crack prevent can be accomplished by applying asphalt sealer periodically. However, sealing will not prevent damage caused by the ground under the asphalt settling due to improper installation.

Cold patch is a fast, permanent, easy-to-use repair material for asphalt and concrete surfaces.

Repairing Small Cracks

Repairing small cracks will usually postpone and/or prevent major cracks and problems from developing later. Therefore, it is important to repair cracks as they are found regardless of their size.

1. Fill any cracks in a blacktop drive as soon as possible to keep water from getting under the slab and causing more serious problems. Asphalt cold patch is typically used to fill cracks that are $\frac{1}{2}$ inch and wider. Crack filler is usually used for narrow cracks. Crack filler is available in cans, plastic pour bottles, and handy caulking cartridges.

2. Chunks of loose or broken material should be removed by using a masonry chisel, wire brush (Figure 15-40).

3. Using a stiff-bristled brush sweep out the crack (Figure 15-41).

4. Clean off all dust using a garden hose with a pressure nozzle.

5. Apply the crack filler. For cold patch applications deeper than 2 inches, crack filler should be applied in 1–2-inch layers. Again, apply enough cold patch material as to create a crown once the area has been tamped. Usually over time a patch will compact below the roadway's surface, thus requiring additional material to be added. If this occurs, then the area to be refilled and properly cleaned and additional materials added (Figure 15-42).

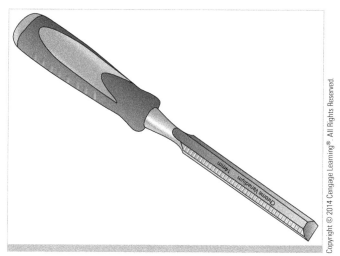

FIGURE 15-40: Use chisel or sharp object to clean out the crack.

FIGURE 15-41: Sweep out the cracks with stiff-bristled broom.

FIGURE 15-42: Apply crack filler to the crack.

Repairing Large Holes with Cold Patch

Cold patch commonly used on large holes and potholes has a larger-diameter aggregate than paste patch. It is available in quantities ranging from 60–70-pound bags. Although cold patch resembles regular asphalt, it is specifically treated with chemicals that will permit it to remain workable for temperature above 50°F. For applications in which the temperature is below 50°F, the cold patch will have to be heated before applying. Also, cold patch will have to be rolled. To repair a large hole:

1. Remove any loose debris, as well as any jagged edges from around the hole with a hammer and cold chisel (Figures 15-43 and 15-44).

2. Apply enough cold patch to leave a slight mound after tamping (Figure 15-45).

3. Fill the hole using a two-step process in which you fill half the hole and tamp it down, then fill and tamp the rest of the hole. This will help you avoid leaving a water-collecting depression.

4. For best results, the asphalt should be compacted using a vibrating plate tamper or compactor; however, if these are not available, then the weight of your service vehicle can be used. To use the weight of the service vehicle, start by placing a piece of $\frac{3}{4}$-inch plywood or a layer of sand spread over the cold patch, then drive your vehicle over it (Figure 15-46).

FIGURE 15-43: Clear loose debris.

FIGURE 15-44: Use a hammer and cold chisel to remove jagged edges around a hole.

FIGURE 15-45: Cold patch piled in a hole.

FIGURE 15-46: Drive a car over the cold patch to spread and compact it.

REMODELING

Remodeling and Landscaping

As mentioned in the introduction, when a structure is remodeled that contains a swimming pool, most locations require that a permit be obtained before the remodeling begins. The general procedure for remodeling a structure that contains swimming pool is as follows.

Demolishing an In-Ground Pool

Most locations require two inspections to remove an in-ground pool. The first inspection ensures that either the bottom of the pool has been removed or that proper drainage has been applied (holes punched in the bottom of the pool). The second is typically conducted after the pool has been backfilled.

> *Always check with your local and state agencies before starting a remodeling project. Before beginning the demolition or deconstruction process, the proper permits must be obtained.*

- Once the proper permits and paper work have been completed, if the pool is constructed from steel or fiberglass, then the materials must be removed and properly disposed of before the pool area can be backfilled.

- Some areas allow for pools constructed from concrete/gunite to be crushed and used as part of the fill. (Check with the local building department.)

- A section of the bottom of the pool must be broken and removed to allow the water table in the area of the pool to rise and lower naturally.

- Some areas allow for the concrete pavers and decking surrounding the pool to be crushed as used as backfill also. (Check with the local building department.)

- Crushed stone can also be used to backfill the pool area within 16 inches from the top.

- The remaining pool area can be backfilled using top soil, compacting as more and more soil is added. Compacting the soil will help prevent settling.

SUMMARY

- Groundskeeping is the activity of maintaining an area for the purpose of aesthetics and functionality. It includes mowing grass, trimming hedges, pulling weeds, planting flowers, and so on.

- Mowing is the process of maintaining the height of the lawn using a lawnmower to cut the grass; there is more to mowing than simply starting a power mower and making a few laps around the grounds.

- Trimming and edging is a simple but necessary exercise that gives a lawn a manicured look.

- Botany is a branch of science that is concerned with the study of plant life and its development; the study of plants is critical for the production of food and medicine, and controlling environmental changes due to changes in plant life.

- The roots of a plant are used to absorb water and minerals from the soil as well as to help stabilize and anchor the plants to the ground.
- Stems are the plant's transportation system and help transport the water and minerals taken in by the roots.
- Leaves are the tiny plant food factories, in which sunlight is captured and combined with minerals and water to produce sugar that the plant uses to produce food.
- Flowers are the reproductive portion of the plant. Flowers contain both pollen and tiny eggs, called "ovules," which when fertilized (pollinated) turn into fruit.
- Fruit is a protective covering for the plant's seed. In other words, fruit is a ripened ovary.
- Seeds are small embryonic plants that contain stored food enclosed in a covering.
- Any loose material that is placed over soil to control weeds and moisture is known as "mulch." Mulch can reduce the evaporation of moisture by as much as 10% to 25%.
- As organic material dies it tends to leave a layer on the lawn known as "thatch"; thatch can deprive the lawn of oxygen.
- Amending the soil is the process of changing its ability to sustain plant life, that is, adding necessary nutrients and/or changing the texture (the ratio of sand, silt, and clay).
- Plants use nitrogen to generate protein (in the form of enzymes) and nucleic acids.
- Phosphorus is another essential element in the process of photosynthesis.
- Potassium is essential for the overall health of the plant, aiding in the production of proteins, photosynthesis, fruit quality, and reduction of diseases.
- Retaining walls are used to control erosion, extend usable grounds, and control changes in grade (the slope of the land).
- Before attempting to construct a cinder-block retaining wall, always consult local and state building codes.
- Paving stones can provide an attractive, quick, and easy alternative to concrete when constructing a patio or walkway, for a fraction of the cost.
- The most common types of irrigation systems currently in use are ditch irrigation, terraced irrigation, drip irrigation, sprinkler system, rotary systems, and center pivot irrigation.
- Routine pool maintenance is an unavoidable fact of life if you want to keep your pool looking clean, sparkling, and inviting day after day.
- Some common problems associated with small engines in which a facilities maintenance technician should be able to address are engines that will not start, engines that smoke, and engines that sputter.
- Most locations require two inspections to remove an in-ground pool.

REVIEW QUESTIONS

1   List the steps for planting brushes and shrubs.

2   List three methods for winterizing an irrigation system.

3   List the steps for starting up the irrigation system in the spring.

4   List the steps for maintaining a swimming pool.

5   If the lawnmower's engine does not start, what should the technician check?

6   What should the technician check if the engine on the lawnmower starts to sputter?

7   Typically, what inspections are required when removing a swimming pool?

Name: _____

Date: _____

Groundskeeping

LAWNMOWER MAINTENANCE

Upon completion of this job sheet, you should be able to identify a need for mower maintenance.

Task Completed		
____	1	Use the gas and oil recommended by the manufacturer.
____	2	Blades are sharp.
____	3	Blades and crankshaft are tight.
____	4	Underside of mower is cleaned after each use.
____	5	Check grass-catcher bag for wear and tear or deterioration.
____	6	Check mower wheels, bearings, and axles for wear and lubrication.
____	7	Mower is thoroughly inspected every year.

INSTRUCTOR'S RESPONSE:

_____

_____

_____

Name: _____

Date: _____

Botany

PLANT IDENTIFICATION

Upon completion of this job sheet, you should be able to research the web to determine the growing conditions for a plant.

TASK COMPLETED

Using the Internet, determine the growing conditions for the following plants:

Little Jamie white cedar

Dwarf fothergilla

Blaauw's juniper

Rosegold pussy willow

Ginny Bruner holly

Mary Nell holly

Yoshino cryptomeria

Swamp white oak

INSTRUCTOR'S RESPONSE:

_____
_____
_____

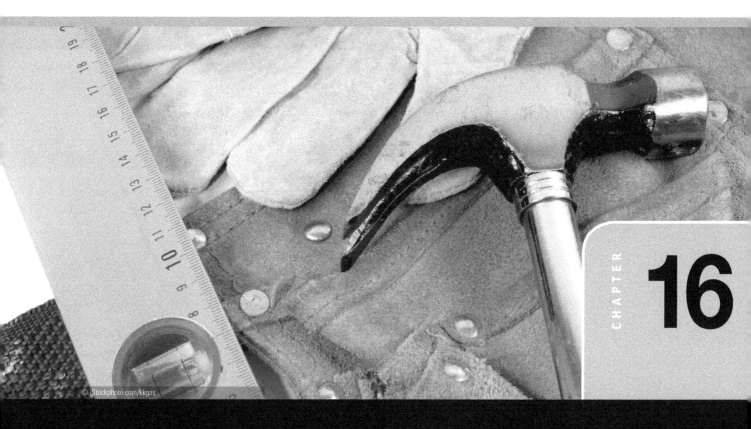

Basic Math for Facilities Maintenance Technicians

OBJECTIVES

By the end of this chapter, you will be able to:

KNOWLEDGE-BASED

- State the difference between a real number and a whole number.
- State the difference between an integer and a whole number.

SKILL-BASED

- Add whole and real numbers and fractions.

- Subtract whole and real numbers and fractions.
- Multiply whole and real numbers and fractions.
- Divide whole and real numbers and fractions.
- Solve problems involving multiple operations with whole numbers.

Integer a positive or negative whole number that is factorable or nonfactorable; includes zero

Real number a positive or negative number that is rational or irrational; includes zero.

Whole numbers counting numbers, including zero

Fraction a number that can represent part of a whole

Introduction

Regardless of the assigned task, it is imperative that the facilities maintenance technician have a good grasp of basic math principles and techniques for dealing with whole numbers, decimals, fractions, percentages, and area and volume. In the following sections, we will be discussing some of the principles and techniques associated with these.

Integers

An **integer** is defined as positive or negative whole number that is factorable or nonfactorable, including zero. For example: -3, -2, -1, 0, 1, 2, 3, 4, 125, and so on. In other words, an integer is any number between negative infinity and positive infinity.

Real or Real Numbers

A **real number** is a positive or negative number that is rational or irrational, including zero. A rational number is any number that is the quotient (the outcome of division) and/or ratio of two integers, whereas an irrational number is any number that cannot be—for example, -3, -2, 0, 5.35, 10, and so on are real numbers. In other words, a real number ranges from negative infinity and positive infinity.

Whole Numbers

Whole numbers are counting numbers, including zero. Whole numbers are an important part of our everyday life. They are used in every aspect of modern living. This includes everything from ordering lunch to filling out an employment application. To be an effective facilities maintenance technician, you should have a good understanding of the basic operation of whole numbers. In other words, you must be able to add, subtract, multiple, and divide to properly order supplies, maintain inventory, and so on.

Basic Principles of Working with Whole Numbers

When working with numbers, there are four basic operations of which the facility maintenance technician should have a good working knowledge: addition, subtraction, multiplication, and division. With the exception of division, these operations can usually be done by lining up the numbers in columns. When working with columns, it is important to correctly line up columns of numbers properly. The numerals on the right should line up over one another for addition, subtraction, and multiplication. Also, keep long involved problems simple by breaking them down and solving them one step at a time. Finally, when working with whole numbers that have units always remember to carry the units to the answer.

Addition and Subtraction of Whole Numbers

As stated earlier, when performing addition and/or subtraction of whole numbers, it is easy to arrange the numbers in a table format, and then starting from the right-hand side of the table add or subtract each column while moving to the left.

Examples:

		270 pounds	2,768	112 ft
		814 pounds	814	96 ft
13		58 pounds	644	40 ft
+8		+ 9 pounds	+ 555	+57 ft
21		1,151 pounds	4,781	305 ft

29	57 ft	28 cu yd
−5	−18 ft	−13 cu yd
24	39 ft	15 cu yd

2,768	114 lb	120 in
−814	−102 lb	−106 in
1,954	12 lb	14 in

Multiplication of Whole Numbers

Multiplication can be thought of as repeated addition, that is, 2 × 4 is the same as saying 2 + 2 + 2 + 2 or 4 + 4.

Examples:

9	22	427 ft	377 gal
× 5	× 3	× 23	× 14
45	66	9,821 ft	5,278 gal

54	24	$12.00
× 4	× 7	× 37
216	168	$444.00

Division of Whole Numbers

When a whole number division problem is arranged in a tabular format, the bottom number is the divisor and the top number is the dividend.

Examples:

12	144	96 ft	5,012 lb
÷ 4	÷ 6	÷ 8	÷ 14
3	24	12 ft	358 lb

14,400 V	352 oz	$720.00
÷ 120	÷ 16	÷ 14
120 V	22 oz	$51.43

Decimals

As mentioned earlier whole numbers are all the counting numbers, including zero. However, in reality, not all numbers are considered to be whole numbers. For example, how do you express quantities that are less than one? In these situations, we can express the quantity as either a fraction (which we will be discussing later in this chapter) or a decimal number. In the decimal system, numbers positioned on the left side of the decimal point represent whole numbers, whereas those on the right side represent the portion of the whole numbers that is less than one (Figure 16-1). When a number does not have a whole number portion associated with it (i.e., the whole number portion is equal to zero), it is said to be a decimal fraction.

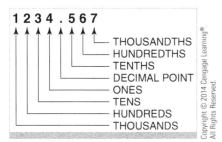

1 2 3 4 . 5 6 7

— THOUSANDTHS
— HUNDREDTHS
— TENTHS
— DECIMAL POINT
— ONES
— TENS
— HUNDREDS
— THOUSANDS

FIGURE 16-1: Decimal number.

Addition and Subtraction of Decimals

To add or subtract decimal numbers, line up the decimal points and perform the operation exactly as you would do for whole numbers. Be careful to always vertically align the decimals. Always make sure that the decimal point is positioned in the same place in the answer.

Examples:

41.12	56.92		
5.25	7.80	9.4	24.3
+2.60	+61.23	−5.2	−5.6
48.97	125.95	4.2	18.7

13.52	47.58
+8.41	−5.31
21.93	42.27

Multiplication of Decimals

When multiplying decimals, don't align the decimals as you would in an addition or subtraction problem. Instead make sure that the number are aligned starting on the left. In other words treat the problem as you would a whole number problem. Next, multiply each number in the top row of the equation by each digit in the bottom row of the equation (see Figure 16-2). Finally, find the location of the decimal in the answer by totaling the numbers to the right of the decimal of the numbers being multiplied. The total number of decimal places is then transformed into the answer by moving the decimal the equivalent number of places starting on the right side of the answer and moving toward the left side of the answer

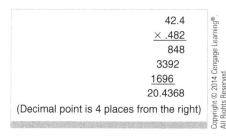

FIGURE 16-2: Multiplication of decimals.

Division of Decimals

Similarly to the division of whole numbers, the division of decimal numbers involves making sure that the divisor is a whole number. If the divisor is not a whole number, then it must first be made a whole number by moving the decimal to the right until the divisor is a whole number. However many places you move the decimal to the right in the divisor, you must move the decimal the same number of places to the right in the dividend. Next, place the decimal above the dividend and perform the division as if it were a typical division problem (see Figure 16-3).

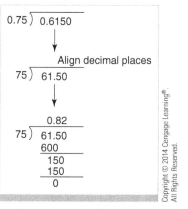

FIGURE 16-3: Division of decimals.

Fractions

By definition a **fraction** is a number that can represent part of a whole. Fraction problems are best worked longhand. Fractions can be converted to decimal numbers and thus worked on a calculator and/or by hand, using the methods described for decimals. Common fractions, frequently just called "fractions," are used for many other things, such as parts of volume measure ($\frac{1}{2}$ cup) or parts of a dollar (a quarter). A whole inch may be divided into equal parts in many ways. For instance, an inch may be divided into eight equal parts. Each part is an eighth of an inch ($\frac{1}{8}$ inch). If five of these parts were needed in measuring a length, the quantity would be five-eighths of an inch and would be written as $\frac{5}{8}$ inch. Each fraction is made up of two numbers:

$$\frac{\text{Numerator}}{\text{Denominator}}$$

Sometimes it is necessary to change fractions to equivalent fractions. An equivalent fraction is one that contains a denominator and numerator equivalent to those of another fraction. For example, $\frac{2}{3}$ is an equivalent fraction to $\frac{4}{6}$ and $\frac{8}{12}$. In other words, the overall value generated by the fraction is the same. For example, $\frac{2}{3} = 0.6666$, $\frac{4}{6} = 0.6666$, and $\frac{8}{12} = 0.6666$. This is so because the relative value of the numerator and the denominator doesn't change when they are multiplied by the same number. For example, $\frac{2}{3}$ is the same as $\frac{4}{6}$ because the numerator and the denominator were both simply multiplied by 2. More often than not, it is necessary to find a common denominator in a group of fractions. This is referred to as finding the least common denominator of that group. For example, the least common denominator of $\frac{2}{3}$ and $\frac{5}{14}$ is 42. This would produce the equivalent fraction

$\frac{28}{42}$ (multiply the numerator and the denominator by 14), and $\frac{15}{42}$ (multiply the numerator and the denominator by 3).

Addition and Subtraction of Fractions

The addition and subtraction of fractions is done by first determining the least common denominator of the fractions to either add or subtract. For example, $\frac{2}{3} + \frac{3}{4}$ would have the least common denominator of 12, producing: $\frac{8}{12} + \frac{9}{12}$. Next for addition, the numerators are added together. For example, in this example $8 + 9 = 17$. This is followed by writing down the least common denominator. For example, in the above example $\frac{8}{12} + \frac{9}{12} = \frac{17}{12}$. Finally the fraction is reduced to its lowest form. For example, in the previous example $\frac{17}{12}$ will reduce to $1\frac{5}{12}$.

For subtraction, after determining the least common denominator, subtract the numerators and carry the result over. Next, carry over the least common denominator and reduce the fraction to its lowest form.

Examples:

$$\frac{3}{1} - \frac{17}{5} = -\frac{2}{5} \qquad \frac{1}{17} - \frac{3}{7} = -\frac{44}{119} \qquad \frac{6}{13} - \frac{8}{9} = -\frac{50}{117}$$

$$\frac{5}{20} - \frac{19}{17} = -\frac{59}{68} \qquad \frac{18}{17} + \frac{6}{14} = 1\frac{59}{119}$$

Multiplication of Fractions

The multiplication of two or more fractions is accomplished simply by multiplying the numerators. followed by multiplying the denominators. For example:

$$\frac{5}{6} \times \frac{3}{4}$$

$5 \times 3 = 15$ (numerators)

$6 \times 4 = 24$ (denominators)

Therefore $\frac{5}{6} \times \frac{3}{4} = \frac{15}{24}$

Finally, the answer is expressed in the lowest form possible. In the previous example, the lowest form of the faction is $\frac{5}{8}$.

Examples:

$$\frac{2}{20} \times \frac{16}{4} = \frac{2}{5} \qquad \frac{19}{19} \times \frac{14}{19} = 1\frac{5}{9} \qquad \frac{4}{19} \times \frac{18}{16} = \frac{9}{38}$$

Division of Fractions

The opposite of multiplication is division. When you divide a number by one-half, it is the same as multiplying the number by the faction $\frac{1}{2}$. When working with fractions, division is done simply by inverting the second fraction and then performing multiplication. For example: $\frac{1}{4}$ divided by $\frac{5}{12}$ is the same as $\frac{1}{4} * \frac{12}{5} = \frac{12}{20}$. The final step to division of fractions is to reduce the answer to its lowest form. In the previous example, the answer can be reduced to $\frac{3}{4}$.

Examples:

$$\frac{4}{3} \div \frac{18}{6} = \frac{4}{9} \qquad \frac{2}{2} \div \frac{17}{11} = \frac{11}{17} \qquad \frac{14}{20} \div \frac{15}{17} = \frac{119}{150}$$

Percent and Percentages

Percentages are a way of expressing a number as a fraction of 100. Percent means "per hundred," just as miles per hour means miles traveled each hour. Percentages can be expressed as either a percentage or a decimal. When converting a percentage to a decimal equivalent, start by dividing the percentage by 100 and then removing the percentage symbol. Dividing the number by 100 will determine the decimal equivalent. For example, 25% can be converted to a decimal as follows:

1. Drop the percentage symbol.
2. Divide the number by 100; $(\frac{25}{100})$.
   a. 0.25 is the decimal equivalent.

Working with Percentages

When working with percentages, remember to specify the relationship. Figure 16-4 illustrates the relationship between percentages and whole numbers. For example, when determining percentage (%), cover the percentage section of the diagram; the fractional parts P (part) and W (whole) remain with P above W. Thus, the formula for percentage is % = P/W.

Example:

60 is 80% of _____.

60/80% = whole number

60/0.80 = whole number

60/0.80 = 75

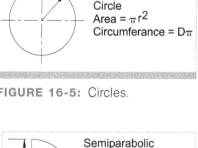

FIGURE 16-4: Relationship between percentages and whole numbers.

Areas and Volumes

Area is always measured in square units—for instance, square inches, square feet, and square yards. Volume is the space enclosed by a three-dimensional figure. Volume can be calculated by first determining the area of an object and then multiplying the area by the object's depth (Figures 16-5 through 16-12).

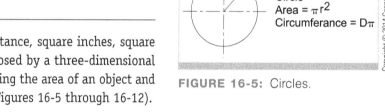

FIGURE 16-5: Circles.

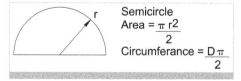

FIGURE 16-6: Semicircles.

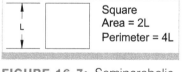

FIGURE 16-7: Semiparabolic figure.

FIGURE 16-8: Square.

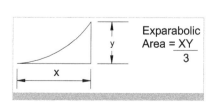

FIGURE 16-9: Exparabolic figure.

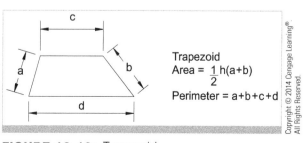

FIGURE 16-10: Trapezoid.

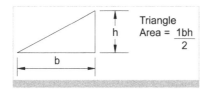

FIGURE 16-11: Triangle.
Copyright © 2014 Cengage Learning®. All Rights Reserved.

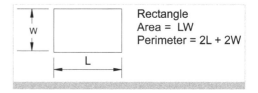

FIGURE 16-12: Rectangle.
Copyright © 2014 Cengage Learning®. All Rights Reserved.

Combined Operations

There are two keys to solving a combined-operations problem: First, analyze the problem to determine what information is given and what you must calculate. Second, write down the steps you will go through. One of the steps very often will be to get everything into the same units. When solving problems that combine operations, the following general rule of thumb should be followed:

Rule 1: Always do the calculations within the parentheses first.
Rule 2: Starting on the left side of the problem, perform all multiplication and/or division operations as they are encountered.
Rule 3: Starting on the left side of the problem, perform all addition and subtraction operations as they are encountered.

> **Example:**
>
> Evaluate $4 + 5 \times (6 + 2) \div 3 - 5$ using the order of operations.
>
> **Solution:**

Step 1: $4 + 5 \times (6 + 2) \div 4 - 5 = 4 + 5 \times 8/4 - 5$ Parentheses
Step 2: $4 + 5 \times 8 \div 4 - 5 = 4 + 40/4 - 5$ Multiplication
Step 3: $4 + 40 \div 4 - 5 = 4 + 10 - 5$ Division
Step 4: $4 + 10 - 5 = 14 - 5$ Addition
Step 5: $14 - 5 = 9$ Subtraction

Applying Basic Math in the Building Trades

In most situations, the facility maintenance technicians can use the basic math skills outlined in the previous sections to solve everyday technical problems associated with their trade. The following are some practical examples demonstrating how the basic math skills can be used to solve construction problems.

One area in which facility maintenance technicians will have to use multiplication is when ordering concrete (sidewalks, driveways, footer, concrete pads). For example, to determine the amount of concrete needed to create a side walk 4 feet wide × 30 feet long × 4 inch thick the facility maintenance technician would do the following:

> 4 feet × 30 feet × 0.33 feet (0.33 feet is determined by dividing 4 inches by 12 inches):
> 120 feet × 0.33 feet
> 39.6 cubic feet

To determine the amount of concrete needed, the number of cubic feet of concrete is divided by 27 to determine the cubic yards needed. Thus, the amount of concrete needed is:

$39.6 \text{ ft}^3/27 = 1.48$ cubic yards

Determining the Number of Acoustical Ceiling Tiles Needed

To determine the Number of acoustical ceiling tiles needed for a project, first calculate the room's square footage by multiplying the dimensions of the room (length × width). Next, divide the room's square footage obtained in the previous step by the square footage of the ceiling tile. For example:

Step 1: Calculate the total area of the ceiling to be covered.

Ceiling Area = Length of Room × Width of Room
Ceiling Area = 10 ft × 15 ft
Ceiling Area = 150 sq. ft

Step 2: Calculate the total area of the ceiling tile.

Ceiling Tile Area = Length of Ceiling Tile × Width of Ceiling Tile
Ceiling Tile Area = 2 ft × 4 ft
Ceiling Tile Area = 8 sq. ft

Step 3: Determine the number of ceiling tile to use.

Number of Ceiling Tiles = Ceiling Area/Ceiling Tile Area
Number of Ceiling Tiles = 150 sq. ft/8 sq. ft
Number of Ceiling Tiles = 18.75, or 19

Determining Aluminum and Vinyl Siding Needed

Aluminum and vinyl siding panels are typically sold by the square. To determine the amount of siding needed for a project, first determine the amount of wall area to be covered. Second, add 10 percent of the wall area calculated, for waste. Third, divide the amount calculated in the previous step by 100. For example, to determine the amount of siding needed for the building shown in Figure 16-13 (assuming the structure has an exterior wall height of 9 feet 6 inches):

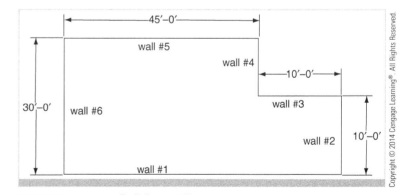

FIGURE 16-13: Building outline.

Step 1: Determine the wall area to be covered.

Section #1

Length = length of wall #5 + length wall #3
Length = 45 feet 0 inches + 10 feet 0 inches
Length = 55 feet 0 inches

Because the height of the wall is 9 feet 6 inches, or 9.5 feet, the area is calculated as:

Area1 = 55 feet × 9.5 feet
Area1 = 522.5 square feet

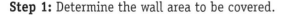

Section #2

Area 2 = 10 feet 0 inches × 9.5 feet

Area 2 = 95 square feet

Section #3

Area 3 = 10 feet 0 inches × 9.5 feet

Area 3 = 95 square feet

Section #4

Area 4 = 20 feet 0 inches × 9.5 feet

Area 4 = 190 square feet

Section #5

Area 5 = 45 feet 0 inches × 9.5 feet

Area 5 = 427.5 square feet

Section #6

Area 6 = 30 feet 0 inches × 9.5 feet

Area 6 = 285 square feet

Total Square Footage

Area Total = Area 1 + Area 2 + Area 3 + Area 4 + Area 5 + Area 6

Area Total = 522.5 square feet + 95 square feet + 95 square feet
+ 190 square feet + 427.5 square feet + 285 square feet

Area Total = 1,615 square feet

Step 2: Determine the amount of waste.

Allowing for waste

Waste Allowance = Area Total × 0.10

Waste Allowance = 161.5 sq. ft

Step 3: Determine the Number of Squares.

Total Wall Area with Waste = 1615 sq. ft + 161.5 sq. ft

Total Wall Area with Waste = 1,776.5 sq. ft

Number of Square = Total Wall Area with Waste/100

Number of Square = 1778.5/100

Number of Square = 17.765, or 18

SUMMARY

- An integer is defined as positive or negative whole number that is factorable or nonfactorable, including zero.

- A real number is a positive or negative number that is rational or irrational, including zero. Whole numbers are counting numbers, including zero.

- When working with numbers, there are four basic operations of which the facility maintenance technician should have a good working knowledge.

- Addition and/or subtraction of whole numbers is done by arranging the numbers in a table format, and then starting from the right side of the table add or subtract each column while moving to the left.

- Multiplication can be thought of as repeated addition, that is, 2×4 is the same as saying $2 + 2 + 2 + 2$ or $4 + 4$.

- When a whole number division problem is arranged in a tabular format, the bottom number is the divisor and the top number is the dividend.

- In the decimal system, numbers are placed on either side of a decimal point. Numbers positioned on the left side of the decimal point are whole numbers, whereas numbers located to the right of the decimal point represent the portion of the whole numbers that is less than one.

- To add or subtract decimal numbers, start by vertically aligning the decimals, and perform the operation exactly as you would for whole numbers.

- The multiplication of decimal numbers is done the same way as it is for whole numbers.

- Similarly to the division of whole numbers, the division of decimal numbers involves making sure that the divisor is a whole number. A fraction is a number that represents part of a whole.

- Each fraction is made up of two numbers:

$$\frac{\text{Numerator}}{\text{Denominator}}$$

- The addition and subtraction of fractions is done by first determining the least common denominator of the fractions to either add or subtract.

- The multiplication of two or more factions is accomplished simply by multiplying the numerators, followed by multiplying the denominators.

- When working with fractions, division is accomplished simply by inverting the second fraction and then performing multiplication. Percentages are a way of expressing a number as a fraction of 100.

- The area of a shape is the size of the inside of the shape. Area is always measured in square units—for instance, square inches, square feet, and square yards.

- Volume is the space enclosed by a three-dimensional figure.

- There are two keys to solving a combined-operations problem: First, analyze the problem to determining the amount and type of information that is provided as well as the information that must be calculated. Second, write down the steps you will go through. One of the steps very often will be to get everything into the same units.

REVIEW QUESTIONS

1   Complete the following:

$$58 + \frac{13 + 32 - 9}{3} - 17 = \underline{\hspace{3cm}}$$

2   The lowest common denominator for $\frac{7}{12}$, $\frac{2}{3}$, and $\frac{1}{2}$ is _____.

3   The lowest common denominator for $\frac{3}{4}$, $\frac{7}{10}$, $\frac{1}{3}$, and $\frac{8}{15}$ is _____.

4   Using the prime factors, the lowest common denominator for $\frac{2}{3}$, $\frac{7}{10}$, $\frac{5}{9}$, and $\frac{11}{12}$ is _____.

5   Add the following decimal fractions:

     1478.35
     4362.31
     +116.96
     ―――――――

     a. 5,960.62
     b. 6,191.54
     c. 5,957.62
     d. 5,879.65

6   Add the following decimal fractions:

     3,425.26
     1,334.63
     +1,606.53
     ―――――――

     a. 6,368.42
     b. 6,366.41
     c. 6,364.42
     d. 5,563.16

7   Add the following decimal fractions:

     1,270.54
     1,798.54
     +3,400.99
     ―――――――

     a. 6,473.33
     b. 13,272.31
     c. 6,470.07
     d. 4,203.00

8   Add the following decimal fractions:

     1,014.19
     2,567.85
     +1,093.83
     ―――――――

     a. 4,677.86
     b. 4,675.87
     c. 4,673.86
     d. 4,128.95

REVIEW QUESTIONS

9   Add the following decimal fractions:

   2,656.21
   5,235.27
   +5,091.55

   a. 12,986.03
   b. 23,166.13
   c. 12,983.03
   d. 9,588.67

11 Subtract the following decimal fractions:

   3,721.43 − 3,633.047 = _____

   a. 88.38
   b. 795.24
   c. 79.36
   d. 09.82

12 Multiply the following decimal fractions:

   4,228.43
   ×3,883.72

   a. 16,422,165.01
   b. 39,181.90
   c. 16,422,025.47
   d. 1,824,670.50

10 Subtract the following decimal fractions:

   4,722.57 − 3,889.43 = _____

   a. 837.14
   b. 3332.56
   c. 829.14
   d. 833.14

13 Multiply the following decimal fractions:

3,035.03
×5,870.45

a. 17,816,975.90
b. 55,869.05
c. 17,816,991.86
d. 1,979,663.99

14 Multiply the following decimal fractions:

102.53
×1,448.38

a. 5,896.03
b. 148,502.40
c. 148,495.03
d. 148,499.03

15 Multiply the following decimal fractions:

3,523.46
×5,998.62

a. 27,517.95
b. 27,519.42
c. 21,135,897.62
d. 21,135,919.74

16 Divide the following decimal fractions:

216.53 ÷ 06.07

17 Divide the following decimal fractions:

397.27 ÷ 33.14

REVIEW QUESTIONS

18 Add the following fractions:

$$\frac{3}{4} + \frac{7}{8} + \frac{5}{16}$$

20 Multiple the following:

$$\frac{3}{4} \times 12 \times 1\frac{5}{8}$$

19 Subtract the following fractions:

$$9\frac{3}{8} - \frac{3}{4}$$

21 Divide the following:

$$5\frac{3}{5} \div \frac{15}{16}$$

22 **Add the following:**

362 + 1,491 + 73 + 29,248

24 **Multiply the following:**

189 × 9

23 **Subtract the following:**

4,793 − 404

25 **Add the following:**

13,328 + 238

Name: _____

Date: _____

Whole Number Addition

Upon completion of this job sheet, you should be able to add whole numbers.

PROCEDURE

Add the following whole numbers:

3,887	1,308	5,041
5,223	3,783	179
+5,602	+3,951	+5,578

2,974	3,270	3,615
4,784	5,489	1,268
+948	+4,833	+2,309

2,373	1,027	4,552
2,181	218	50
+2,758	+1,052	+5,962

1,821
2,245
+2,565

INSTRUCTOR'S RESPONSE:

_____

_____

_____

Name: _____

Date: _____

Whole Number Subtraction

Upon completion of this job sheet, you should be able to subtract whole numbers.

PROCEDURE

Subtract the following whole numbers:

2,291 − 585   = _____
3,341 − 3,173 = _____
4,430 − 1,162 = _____
3,957 − 561   = _____
4,174 − 1,931 = _____
4,879 − 2,660 = _____
5,816 − 4,598 = _____
2,251 − 1,699 = _____
1,763 − 1,571 = _____
2,147 − 417   = _____

INSTRUCTOR'S RESPONSE:

_____
_____
_____

Name: _____

Date: _____

Whole Number Multiplication

Upon completion of this job sheet, you should be able to multiply whole numbers.

PROCEDURE

Choose the correct answer for the following multiplication:

793
× 449

  a. 2,139   c. 355,954

  b. 356,057   d. 118,651

5,600
× 2,435

  a. 12,906   c. 13,638,102

  b. 13,636,000   d. 4,546,033

2,667
× 4,013

  a. 10,694   c. 10,705,399

  b. 10,705,401   d. 10,702,671

88
× 2,564

  a. 225,632   c. 226,626

  b. 226,628   d. 113,313

3,821
× 4,959

  a. 23,657   c. 18,948,944

  b. 18,948,339   d. 18,948,948

1,775
× 3,565

  a. 6,254,950   c. 6,327,875

  b. 6,256,575   d. 6,327,875

2,051
× 4,270

  a. 19,130   c. 8,756,491

  b. 8,757,770   d. 8,756,495

2,396
$\times$ 49

a. 117,404   c. 116,392

b. 2,541   d. 38,797

1,154
$\times$ 961

a. 4,038   c. 1,109,057

b. 1,108,994   d. 369,685

5,808
$\times$ 4,999

a. 29,034,192   c. 29,032,560

b. 20,805   d. 9,677,519

INSTRUCTOR'S RESPONSE:

_____
_____
_____

Name: _____

Date: _____

Whole Number Division

Upon completion of this job sheet, you should be able to divide whole numbers.

PROCEDURE

Divide the following whole numbers:

$931 \div 19$ = _____

$1{,}150 \div 46$ = _____

$306 \div 34$ = _____

$2{,}597 \div 49$ = _____

$858 \div 39$ = _____

$1{,}232 \div 56$ = _____

$2{,}065 \div 59$ = _____

$28 \div 2$ = _____

$2{,}450 \div 50$ = _____

INSTRUCTOR'S RESPONSE:

_____

_____

_____

Name: _____

Date: _____

Adding Fractions

Upon completion of this job sheet, you should be able to add fractions.

PROCEDURE

Add the following common fractions:

$\dfrac{6}{13} + \dfrac{8}{9}$ = _____

$\dfrac{19}{9} + \dfrac{6}{14}$ = _____

$\dfrac{18}{17} + \dfrac{6}{14}$ = _____

$\dfrac{4}{1} + \dfrac{19}{4}$ = _____

$\dfrac{10}{16} + \dfrac{4}{19}$ = _____

$\dfrac{3}{12} + \dfrac{10}{1}$ = _____

$\dfrac{8}{4} + \dfrac{2}{5}$ = _____

$\dfrac{8}{4} + \dfrac{16}{19}$ = _____

$\dfrac{19}{19} + \dfrac{1}{10}$ = _____

$\dfrac{12}{15} + \dfrac{7}{12}$ = _____

INSTRUCTOR'S RESPONSE:

_____

_____

_____

Name: _____

Date: _____

Subtracting Fractions

Upon completion of this job sheet, you should be able to subtract fractions.

PROCEDURE

Subtract the following common fractions:

$$\frac{3}{1} - \frac{17}{5} = \underline{\hspace{3cm}}$$

$$\frac{1}{17} - \frac{3}{7} = \underline{\hspace{3cm}}$$

$$\frac{20}{20} - \frac{15}{10} = \underline{\hspace{3cm}}$$

$$\frac{20}{9} - \frac{14}{19} = \underline{\hspace{3cm}}$$

$$\frac{12}{19} - \frac{13}{15} = \underline{\hspace{3cm}}$$

$$\frac{5}{20} - \frac{19}{17} = \underline{\hspace{3cm}}$$

$$\frac{4}{3} - \frac{12}{20} = \underline{\hspace{3cm}}$$

$$\frac{9}{12} - \frac{10}{15} = \underline{\hspace{3cm}}$$

$$\frac{16}{2} - \frac{10}{10} = \underline{\hspace{3cm}}$$

$$\frac{14}{18} - \frac{16}{7} = \underline{\hspace{3cm}}$$

INSTRUCTOR'S RESPONSE:

_____

_____

_____

Name: _____

Date: _____

Multiplying Fractions

Upon completion of this job sheet, you should be able to multiply fractions.

PROCEDURE

Multiply the following common fractions:

$\dfrac{2}{20} \times \dfrac{16}{4} =$ _____

$\dfrac{20}{16} \times \dfrac{2}{6} =$ _____

$\dfrac{19}{19} \times \dfrac{14}{9} =$ _____

$\dfrac{19}{8} \times \dfrac{13}{18} =$ _____

$\dfrac{11}{18} \times \dfrac{12}{14} =$ _____

$\dfrac{4}{19} \times \dfrac{18}{16} =$ _____

$\dfrac{3}{2} \times \dfrac{11}{19} =$ _____

$\dfrac{9}{11} \times \dfrac{9}{14} =$ _____

$\dfrac{15}{1} \times \dfrac{9}{10} =$ _____

$\dfrac{9}{12} \times \dfrac{10}{2} =$ _____

INSTRUCTOR'S RESPONSE:

_____

_____

_____

Name: _____

Date: _____

DIVIDING FRACTIONS

Upon completion of this job sheet, you should be able to divide fractions.

PROCEDURE

Divide the following common fractions:

$\dfrac{4}{3} \div \dfrac{18}{6} =$ _____

$\dfrac{3}{19} \div \dfrac{4}{8} =$ _____

$\dfrac{2}{2} \div \dfrac{17}{11} =$ _____

$\dfrac{2}{10} \div \dfrac{15}{1} =$ _____

$\dfrac{14}{20} \div \dfrac{15}{17} =$ _____

$\dfrac{7}{2} \div \dfrac{1}{19} =$ _____

$\dfrac{6}{5} \div \dfrac{13}{2} =$ _____

$\dfrac{11}{13} \div \dfrac{12}{16} =$ _____

$\dfrac{18}{13} \div \dfrac{12}{12} =$ _____

$\dfrac{16}{20} \div \dfrac{18}{9} =$ _____

INSTRUCTOR'S RESPONSE:

_____
_____
_____

© iStockphoto.com/kkgas

CHAPTER 17

Blueprint Reading, Building Codes, and Permits

OBJECTIVES

By the end of this chapter, you will be able to:

KNOWLEDGE-BASED

- Identify the various views of an orthographic drawing.
- Identify the various symbols used on plumbing plans to represent piping types, fittings, and symbols on a plumbing plan.
- Identify the various symbols used on HVAC plans to represent HVAC line types, ducts, and equipment.
- Identify the various symbols used on electrical plans to represent wiring, switches, fixtures, and so on.
- Understand standard abbreviations and symbols used on blueprints.

- Explain when it is necessary to obtain and building permit.
- Discuss some of the various types of building permits that are required.

SKILL-BASED

- Determine the length of objects presented on a blueprint using an architect's scale and/or tape measure.
- Determine the angle of a line on a blueprint using a protractor.
- Understand standard abbreviations and symbols used on a blueprint.

Linear measurement the measurement of two points along a straight line

English system of measure the system of measure currently used in the United

Metric system of measure a system of measurement that uses a decimal units for physical quantities

Auxiliary view A view that uses orthographic techniques to

produce a projection on a plane other than one of the three principal planes

Introduction

Because technology is evolving at an ever-increasing rate and systems are becoming more and more complex, it is becoming ever more important that the technician have a basic understanding of blueprint reading. Blueprints are drawn plans of homes and buildings that are used in the construction of the structure. The contractor and technicians doing the work must have a good understanding of the portion of the prints that apply to their individual trade as well as the trades of others and even equipment locations. For example, the location of the electrical and electrical wiring, plumbing and pipe locations, HVAC duct work, and equipment locations. In addition, understanding the basics of blueprint reading can save a technician numerous hours troubleshooting a problem with a particular piece of equipment or a system.

In addition to having a good understanding of blueprint reading, it is critical that the facility maintenance technician have a good understand of the process of obtaining a building permit, as well as when and where it obtain a building permit. Not getting a permit before starting a remodeling project can lead to costly fines and project overruns.

Linear Measurement

Linear measurement is defined as the measurement of two points along a straight line (Figure 17-1). All objects, whether they are man-made or the result of natural conditions and/or forces, consist of points and lines. A point can be the center location of a gas line (Figure 17-2) or one of the edges of a rectangular duct (Figure 17-3).

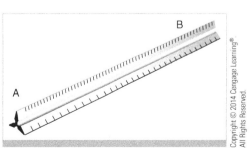

FIGURE 17-1: Linear measurement.

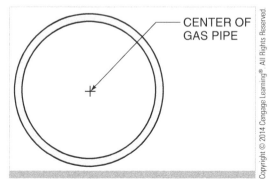

CENTER OF GAS PIPE

FIGURE 17-2: The center line of a gas pipe.

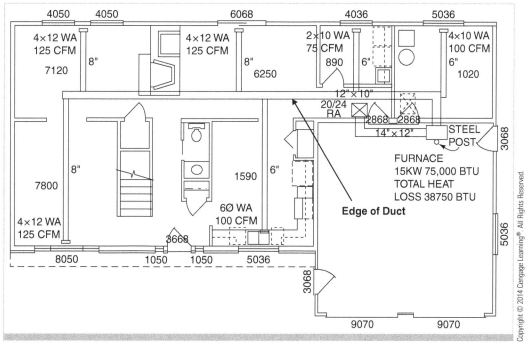

FIGURE 17-3: Edge of rectangular duct supplying air to a room in a facility.

Whatever the application of points and/or lines, the fact remains the same; any person entering a technical field must understand how to locate them.

Currently two basic systems, **English system of measure** and **metric system of measure,** are used to make measurements in the world today. The English system of measure is the unit of measure currently used in the United States by most technicians (see Table 17-1). The base units for the English system for determining the length of an object are:

- Inches: in
- Foot: ft
- Yards: yd
- Miles: mi

Unit English	Conversion Factors
1 inch (in)	
1 foot (ft)	12 in
1 yard (yd)	3 ft
1 mile (mi)	5,280 ft

TABLE 17-1: English Conversion Factors.

For the metric system the base unit of measurement for determining the length of an object is the meter. This base unit is divided into larger and smaller units (in multiple of 10) by adding prefixes. Common metric prefixes are deci (10), centi (100), and milli (1,000) (see Table 17-2). For example a meter contains 10 decimeters.

For example, to convert from 5 centimeters to millimeters, the technician would start by locating the cm row on the left side of the table and the mm column at the top of the table. Where the two intersect will be the conversion factor. Thus, to convert from 5 cm to mm: 5 × 10 = 50 mm.

Although the English system of measure is commonly used in the United States, a facility maintenance technician should still be able to recognize and convert from one unit to another in the metric system of measure. Table 17-3 provides the conversion factors for converting from one system of units to another.

Metric Symbols	
Symbol	Meaning
mm	millimeter
cm	centimeter
dm	decimeter
m	meter
dam	dekameter
hm	hectometer
km	kilometer

Metric Equilivants

Converting To

Converting From		mm	cm	dm	m	dam	Hm	km
	mm	1	0.1	0.01	0.001	0.0001	0.00001	0.000001
	cm	10	1	0.1	0.01	0.001	0.0001	0.00001
	dm	100	10	1	0.1	0.01	0.001	0.0001
	m	1,000	100	10	1	10	0.01	0.001
	dam	10,000	1,000	100	10	1	0.1	0.01
	hm	100,000	10,000	1,000	100	10	1	0.1
	km	1,000,000	100,000	10,000	1,000	100	10	1

TABLE 17-2 Metric Conversion Factors.

Starting with	Multiply	To Find
Inches	2.5	centimeters
Feet	30	centimeters
Yards	0.9	meters
Miles	1.6	kilometers
Centimeters	0.3937	inches
Meters	1.1111	yards
Kilometers	0.625	miles

For example:

Converting From		mm	cm	Dm	m	dam	hm	km
					Converting To			
	Inches	25.4	2.54	0.254	0.0254	0.00254	0.000254	0.0000254
	Feet	304.8	30.48	3.048	0.3048	0.03048	0.003048	0.0003048
	Yards	914.4	91.44	9.144	0.9144	0.09144	0.009144	0.0009144
	Miles	1609344	160934.4	16093.44	1609.344	160.9344	16.09344	1.609344

TABLE 17-3: Converting from Metric to English and English to Metric.

For example, to convert from 5 inches to millimeters, the technician would start by locating the inches row on the left hand side of the table and the mm column at the top of the table. Where the two intersect will be the conversion factor. Therefore to convert from 5 inches to mm: $5 \times 25.4 = 127$ mm

To convert 12'6" to centimeters the technician would perform the following steps:

Convert from Inches to Centimeters

Step 1: Converting 12 feet into inches:

12 ft $\times$ 12 in/ft $= 144$ in

Step 2: Next add the inch portion of the original measurement:

144 in $+$ 6 in $= 150$ in

Step 3: Find the conversion factor using Table 17-3:

120 in $\times$ 2.5 cm/in $= 300$ cm

Convert from Centimeters to Inches

Step 1: Find the conversion factor using Table 17-3:

300 cm $\times$ 0.4 in/cm $= 120$ in

Step 2: Convert from inches to feet and inches by dividing by 12:

120 in / 12 ft/in $= 10$ ft

Therefore the measure would be written as 10 ft.

Reading Linear Measurement on a Blueprint

When a mechanical drawing is created on the computer, typically it is drawn at a scale of 1:1 (also known as full scale). However, in most cases when the file is printed, it is printed at a reduced scale so that it can fit onto a single sheet of paper (typically $8\frac{1}{2} \times 11$, 24×36, or 30×42). When a drawing is printed at full scale, determining the length of a line is easy to accomplish. This is done be simply placing a ruler along the length of the line and reading the dimensions (Figure 17-4).

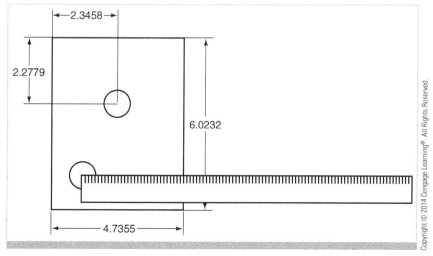

FIGURE 17-4: Reading the length of a line drawn at full scale.

However, if the drawing is printed at a reduced scale (a scale smaller then 1:1), then a scale factor must be applied to determine the length of a line not dimensioned (Figures 17-5, 17-6, and 17-7).

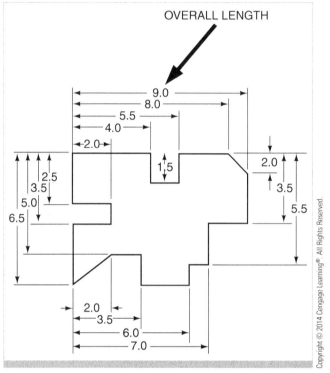

FIGURE 17-5: Determining the overall length of the object.

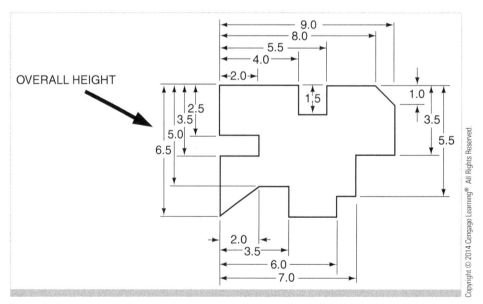

FIGURE 17-6: Determining the overall height of the object.

When a drawing is printed to a scale $\frac{1}{4}$ times smaller than its actual size, then that drawing is said to be drawn at $\frac{1}{4}$ scale and is therefore read as every $\frac{1}{4}"$ is equal to 1'-0". For example, if a portion of a building footing plan measures 2" using a ruler, then the actual length of the footing can be determined by multiplying the measured length by the inverse (meaning swap the top and bottom numbers) of the drawings scale factor. For a scale factor of $\frac{1}{4}$ the inverse or multiplier would be 4 and therefore the true length of the portion of the footing would be 8ft (Figure 17-8).

Table 17-4 gives the conversion factors for various scales commonly used on engineering drawings.

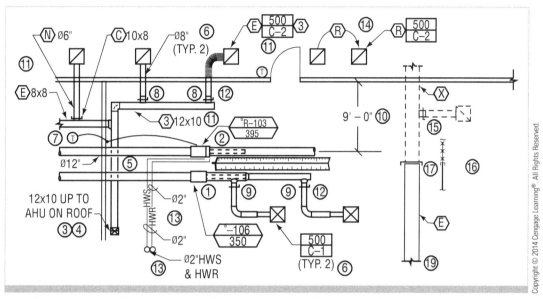

FIGURE 17-7: Reading the length of a line that is not dimensioned.

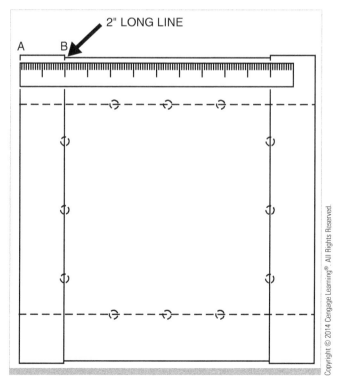

FIGURE 17-8: Determining the length of a portion of a footing using a ruler.

Drawing Scale	Multiplier
$\frac{1}{16}$" = 1' 0"	16
$\frac{3}{32}$" = 1' 0"	10.66
$\frac{1}{8}$" = 1' 0"	8
$\frac{3}{16}$" = 1' 0"	5.33
$\frac{1}{4}$" = 1' 0"	4
$\frac{3}{8}$" = 1' 0"	2.66
$\frac{1}{2}$" = 1' 0"	2
$\frac{3}{4}$" = 1' 0"	1.33
1" = 1' 0"	1
$1\frac{1}{2}$" = 1' 0"	1.5
3" = 1' 0"	0.33
Half-size	2
Full size	1

TABLE 17-4: Drawing Scale Conversion Factors.

Determining Linear Measure Using a Scale

When a drawing is created, one of the responsibilities of the draftsman is to note the scale at which the drawing was created. Usually, this information is listed in the drawing title block region. The title block region is usually located along the bottom edge or lower-right corner of the drawing (Figure 17-9). The length and position of items not dimensioned on a drawing can be determined using a ruler and the scale factor as discussed in the previous section. However, an easier method of determining the length and position of objects that are not dimensioned is to measure them on the drawing using a scale (Figures 17-10 and 17-11). The scale most commonly used in residential and light commercial is the architect's scale.

Henderson Engineering
Little Rock, Arkansas

Tolerances	Date:	Drawn By:	Checked By:
	Scale:		
	Engineer:	Designed By:	Approved By:
	Department:		
	Division:		
	Part Name		Part Number

FIGURE 17-9: Typical engineering title block.

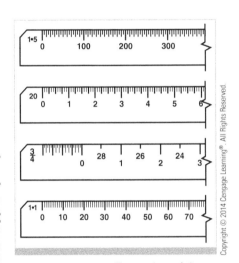

FIGURE 17-10: Examples of the engineer's, architect's, and metric scales.

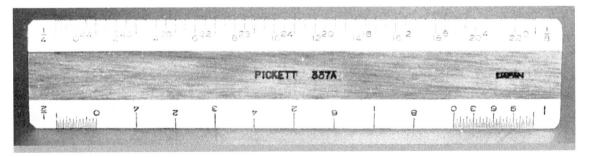

FIGURE 17-11: Six-inch architectural scale.

The Architect's Scale

The architect's scale is a type of ruler in which ranges of precalibrated ratios are illustrated. These scales can be made of a variety of materials and contain as few as two scales (ratio) and as many as eleven (on ten of them, each 1 inch' represents a foot and is subdivided into multiples of 12). The most commonly used architect's scale used is the triangular scale. This scale receives it name because its cross section is in the shape of a triangle (Figures 17-11). Architect's scales are standardized into feet and inches ranging from $\frac{1}{16}$" = 11 0" through 3" = 11 0". The flat scale architectural scale commonly used by technicians, because it requires less space when stored in a toolbox (Figure 17-12).

FIGURE 17-12: Twelve-inch triangular architect's scale.
Copyright © 2014 Cengage Learning®. All Rights Reserved.

Reading the Architect's Scale

If you understand how to read one of the ratios (scales) on an architect's scale, then reading the remaining ones will be simple. Because these scales are used to represent feet and inches, the scale is divided into two sections. The first section represents feet whereas the second section represents inches and fractions of an inch graduated in $\frac{1}{16}$, $\frac{1}{8}$", $\frac{1}{4}$", and $\frac{1}{2}$" increments. This will be different from an engineer's scale that is graduated in $\frac{1}{1,000}$", $\frac{1}{100}$", and $\frac{1}{10}$ of an inch measured with caliper type devices and is used for manufacturing machine products. For example, using the scale shown in (Figure 17-13), the marks on the right side of the zero line indicate 1-foot increments, whereas the marks on the left side represent inches and fraction of an inch.

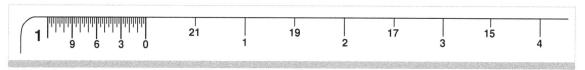

FIGURE 17-13: Architect's scale 1"=1' 0".
Copyright © 2014 Cengage Learning®. All Rights Reserved.

To find the length of a line using the 1-inch scale, the following procedure should be used:

Step 1: Position the scale zero mark indicator on the beginning of the line or object to be measured (see Figure 17-14).

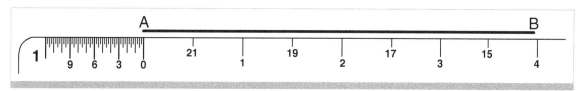

FIGURE 17-14: Placing the architectural scale on the beginning of the line to be measured.
Copyright © 2014 Cengage Learning®. All Rights Reserved.

Step 2: Moving from left to right, count the number of feet marks until you reach the end of the line or the foot mark nearest to the end (Figure 17-15).

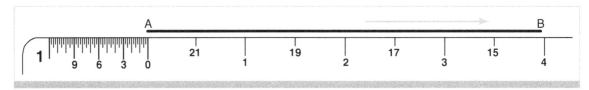

FIGURE 17-15: Reading the architectural scale, moving from left to right.
Copyright © 2014 Cengage Learning®. All Rights Reserved.

Step 3: Slide the scale to the right until the end of the line is on the foot mark indicated in Step #2 (Figure 17-16).

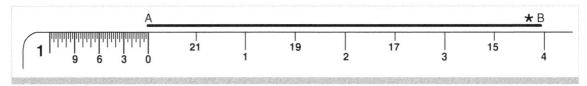

FIGURE 17-16: Finding the foot mark nearest to the end of the line to be measured.
Copyright © 2014 Cengage Learning®. All Rights Reserved.

Step 4: Starting at the zero, count the number of inch marks to the end of the line. See Figure 17-17.

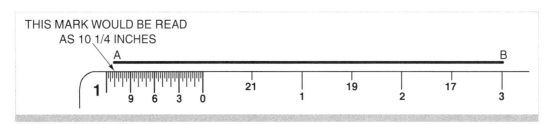

FIGURE 17-17: Determining the inches and fraction of an inch.
Copyright © 2014 Cengage Learning®. All Rights Reserved.

Angular Measurement

Understanding and being able to read linear measurements are only a part of the skills needed to interrupt a blueprint; the facility maintenance technician must also be able to read and understand angular measurements (measuring angles) as well.

Angles

Three types of units are used to express an angle: angular degrees, radians, and gradients, but typically only angular degrees are used on blueprints. Since the circumference of any circle contains 360 degrees, an angular degree is equal to $\frac{1}{360}$th of the circumference of a circle. This means that by drawing a circle (of any size) and dividing its circumference into 360 equal segments (called arc lengths), the angle formed by constructing a line from the center of the circle to the endpoints of one arc length would produce a wedge equal to 1 degree (Figure 17-18).

Reading and Measuring Angles

Angles are usually given on a blueprint with either a dimension or a leader; however, from time to time it may be necessary to determine the angle on a blueprint using a protractor. A protractor is an instrument consisting of a half circle with a midpoint (center) marked on the horizontal position (base) of the protractor. This midpoint, also called the "reference point," is marked on the protractor. The half-circle is divided into degrees (180). The degrees are labeled from right to left and vice versa, allowing for angles to be measured from either direction (Figure 17-19).

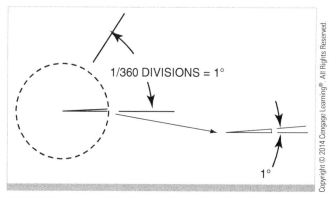

FIGURE 17-18: Wedge produced by dividing a circle 360 times.

To measure an angle using a protractor, first place the end of the protractor's base angle on the vertex of the angle to be measured (Figure 17-20). Next, align the baseline with one of the sides of the angle to be measured (Figure 17-21). Finally, count the number of degrees of the side adjacent to the baseline.

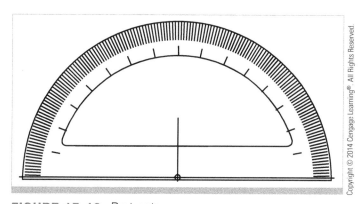

FIGURE 17-19: Protractor.

FIGURE 17-20: Vertex of angle to be measured.
Copyright © 2014 Cengage Learning®. All Rights Reserved.

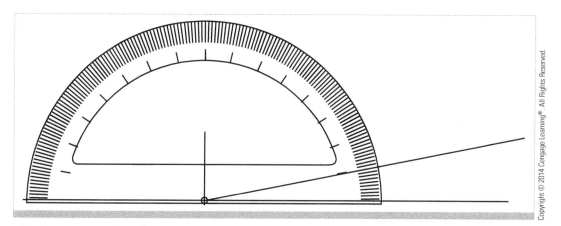

FIGURE 17-21: Reading the protractor.

Standard Abbreviations and Symbols

Standard abbreviations and symbols have been developed for the engineering and architectural community that not only facilitate the development of blueprints but also ensure consistency in their interpretation. Therefore, a good understanding of what these symbols are is critical in developing blueprint reading skills. An example of some of the abbreviations used in the HVAC industry is shown in Table 17-5.

Like abbreviations, symbols are used to speed up the drawing process by providing a shorthand method of representing commonly used equipment. Typically, an engineering or architectural firm will employ a standard set of symbols commonly used and accepted in the building trade. However, if a nonstandard symbol is used, it is typically identified in the drawing legend. A sample of commonly used symbols is illustrated in Figure 17-22.

Abbreviations	Definition
A/D	Analog to digital
AC	Alternating current
ACH	Air changes per hour
BD	Boiler blow-off
CFC	Chlorofluorocarbon
CFM	Cubic feet per minute
CHWR	Chilled water return
D	Drain
FOS	Fuel oil suction
H	Humidification line
HPR	High pressure return
LPR	Low pressure return
MPR	Medium pressure return
MU	Makeup water
PPD	Feedwater Pump Discharge
RD	Refrigerant discharge
RL	Refrigerant liquid
RS	Refrigerant suction
V	Air relief line
VPD	Condensate or vacuum

TABLE 17–5: HVAC Abbreviations

Architectural Symbols

As stated earlier, drawing symbols are used as a shorthand method of communicating information about arrangements, appliances, and applications in a standard way from drawing to drawing as well as from design firm to design firm. When architectural drawings are created using a computer-aided drawing and design (CADD) applications, walls are drawn to their exact thickness. Exterior wood-framed walls are typically drawn as 6 inches thick, exterior masonry walls are typically drawn as 5 inches thick. Interior walls are typically drawn to a thickness of 4 inches, and soil stack walls are typically drawn as 8 inches thick (see Figure 17-23). On architectural drawings, walls are usually shaded so that they stand out from the rest of the drawing.

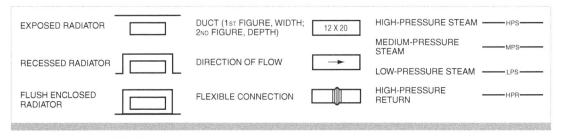

FIGURE 17-22: HVAC symbols.

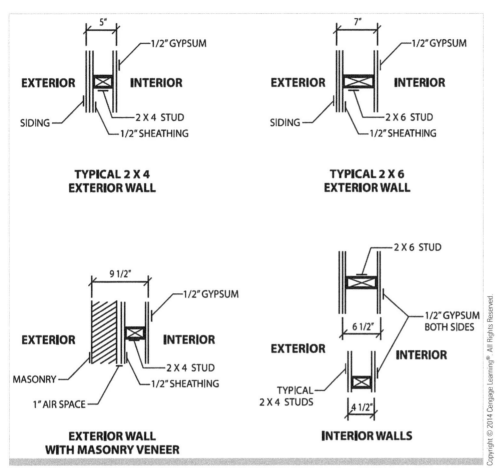

FIGURE 17-23: Typical wall sections.

On architectural drawings, partial walls are at least 36 inches above the floor and defined with a note; while guardrails are noted and at least 36 inches tall with no more than 4 inches between rails (see Figure 17-24).

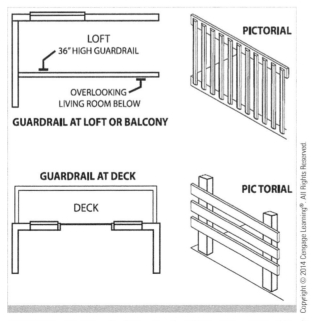

FIGURE 17-24: Architectural railing shown on drawings.

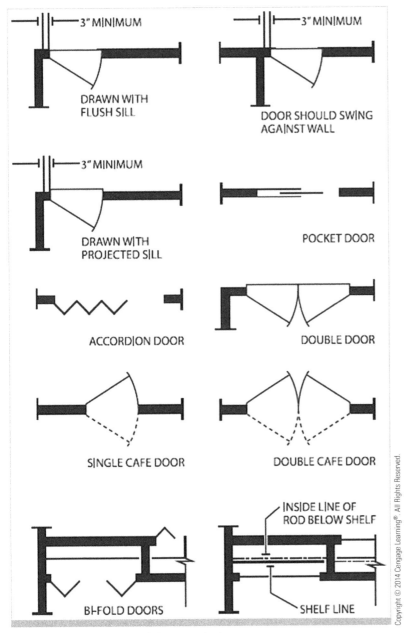

FIGURE 17-25: Typical door symbols.

Door Symbols

Like walls, exterior doors are drawn on a CADD system to actual size and reduced to a scale when they are printed. Most exterior doors are drawn to 3' 0" wide × 6' 8" height. Exterior doors are typically drawn to 2' 8" wide; utility rooms are 2' 8" wide × 6' 8" height; bathroom doors are typically 2' 4" to 2' 6" wide. Other rooms are 2' 6" to 2' 8" tall; closets are 2' 0" to 2' 4" tall (see Figure 17-25).

Window Symbols

For windows, the sill is drawn on both the inside and outside of the window and can range in size from 2' 0" to 12' 0" at 6" intervals. The size of the window typically depends on the purpose of the window. Casement windows have the ability to open to 100 percent. Pictorial windows have two windows that slide vertically, while awning windows are used in basements or below a fixed window. Fixed windows are larger and do not open or close; bay windows extend beyond the wall and can extend from the floor to the ceiling or contain a bench. Garden windows are usually in a kitchen or utility room, while skylights add natural light and are drawn with dashed lines on the floor plan.

Written Specifications

It is said that a picture is worth a thousand words, and the overall purpose of a set of construction drawings is to provide the user with a basic set of instructions that can be easily interpreted by almost anyone with or without a technical background. All that is required is a little training in the art of blueprint reading and practice. However, for the most part architectural drawings can be created that will cover almost all aspects of construction. When information necessary to complete a project cannot be illustrated in a schedule or a drawing, then a construction specification must be created. Construction specifications are legal documents that are created by one or all of the following:

- The architect
- The engineer
- The contractor
- The client

Typically, specifications are created that will cover or describe in detail: product requirements or details, material, and workmanship requirements. It is an exact statement that describes a certain aspect or characteristic of a project. They are used to describe when the project is to start, how the owner is to make payments, where equipment is to be stored (until needed), and the types of insurance that the contractors are to carry, as well as other information necessary for the completion of the project.

As stated earlier, written specifications are legal documents, therefore, they must be: concise, clear, well written, and to the point. The sentences must be well formed and free of grammatical errors. Upon careful completion of the specifications, the contractor must derive one and only one conclusion. There cannot be multiple conclusions reached from reading the specifications. In short, construction specifications:

- Must be clear, concise, technically and grammatically correct
- Must not contain ambiguous, misleading or any wording that will lead to a misinterpretation
- Should use short sentences with simple words that can be easily understood
- Should avoid using slang technically incorrect or field words and terms
- Should avoid repeating or stating conflicting requirements
- Should state construction requirement sequentially whenever possible

Specifications are typically divided into various division and sections. The number of division and sections is determined by the complexity of the project. Each section is denoted with a content-specific heading. For architectural projects, specification are typically divided into the following divisions:

- **General Requirements**—This section contains information regarding the general administrative and technical provisions of the project. This include the contractual and legal requirements, a summary and explanation of work to be done, and a description of project reporting requirements (meetings, documentation), quality control, and submittal procedures.

- **Site Work**—This section contains information regarding work to be performed on the site. This includes soil testing, core drilling, standard penetration testing, and seismic testing and exploration.

- **Concrete**—This section contains information regarding any and all concrete work, form work, expansion and concrete joints, cast-in-place concrete, specially placed concrete, and recast concrete.

- **Masonry**—This section contains information regarding any brick, stone, glass brick, clay backing tile, and ceramic veneer, as well as information regarding joint reinforcement, mortar, and anchors.

- **Metal**—This section contains information regarding the metal in a structure; this includes: structural members, metal roofing, metal floor decking, and the like.

- **Wood and Plastics**—This section contains information regarding the methods of carpentry used.

- **Thermal and Moisture Protection**—This section contains information regarding roofing materials, waterproofing materials, flashing and sheet metal trim, insulation, roof accessories, and sealants.

- **Doors and Windows**—This section contains information regarding the windows, doors and storefronts used, including the glazing as well the hardware used.

- **Finishes**—This section contains information regarding wall, ceiling, and floor finishing.

- **Specialties**—This section contains information regarding specialty items such as chalkboards and tackboards, louvers and vents, grilles and screens, pest control, fireplaces, flagpoles, lockers, storage shelving, directional signage, and sun control devices.

- **Equipment**—This section contains information regarding maintenance equipment, bank and vault equipment, food service equipment, vending equipment, athletic equipment, laundry equipment, library equipment, medical equipment, waste-handling equipment, and loading dock equipment.

- **Furnishings**—This section contains information regarding the window treatments, furniture and furnishing accessories, and the like.

- **Special Construction**—This section contains the necessary information for such areas as clean rooms, operating rooms, incinerators, instrumentation rooms, nuclear reactors, radiation treatment rooms, sound and vibration rooms, vaults, and swimming pool spaces

- **Conveying Systems**—This section contains the necessary information for elevators, escalators, moving walks, pneumatic tube systems, and the like.

- **Mechanical**—This section contains information regarding the HVAC and plumbing as well as other mechanical systems. It contains information on basic materials and methods, equipment, piping systems and services, and insulation. Plumbing also includes fire protection (sprinklers) equipment and special equipment and materials such as medical gases, fuel gases, compressed air, and process piping systems.

- **Electrical**—This section contains information on general provisions, basic materials and methods, power generation, power transmission, power service and distribution, lighting, special systems, communications, controls, and instrumentation.

- **Responsibility and Liability**—As its name implies, this section provides information regarding the responsibility for the design, including the calculations of the architect, engineer, contractor and subcontractor. By law, registered architects and engineers assume the responsibility for all work done by employees working under their guidance.

Another form of written specifications is the Statement of Objective. Used by government agencies, this document contains general statements or guidelines for the use of the contractor when developing the specifications and statement of work.

Two-Dimensional Views

The way in which an object is presented or viewed is extremely important. If a part is to be manufactured and the angle or view shown does not provide the necessary information, then the part could never be manufactured to design expectations.

The following section examines the most common method of representing objects on engineering and architectural drawings.

Multiview Drawings

When something is to be constructed or created using a blueprint, it is the job of the draftsperson to supply all the necessary information regarding the object's sizes and the location of its features. All this can easily be accomplished when the object is relatively simple and all the essential data can be provided in one view. For example, suppose that a 2-inch square piece of steel plating that is $\frac{1}{2}$ inch' thick is to be fabricated with a $\frac{1}{2}$-inch diameter hole drilled at its center. All the required information about this part can be contained in one drawing consisting of a single view (Figure 17-26). When a more complex part is produced, a single view is not sufficient to clearly show all its features. In other words, a series of grooves on the opposite side of this same part would be shown as hidden features, resulting in a drawing that is difficult to interpret (Figure 17-27). To solve this problem, additional views must be created that will reveal all the hidden attributes of the part. In this instance, an additional view of the side is required (Figure 17-28A). Very complex parts might contain views showing the front, sides, back, bottom, and top of the object. A drawing containing two or more of these views is called a multiview drawing. The different views in an orthographic projection are created by projecting the lines from one view to another. In other words, by projecting the edge of the object down from the top view, the width of the object can be established in the front view (Figure 17-28B). Another way to look at this is to suppose that the object is placed in a glass box (Figure 17-29), with the object's surfaces and features being projected onto the surface of the glass (Figure 17-30). Once the object's features have been transferred to the glass box, the cube is then unfolded to reveal six different views of the object (Figure 17-31). The primary views used on an engineering drawing are: left side view, front view, right side view, rear view, top view and bottom view.

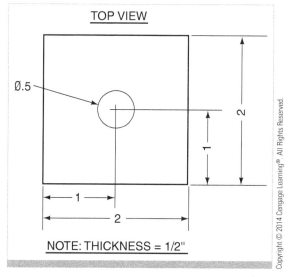

FIGURE 17-26: Single view of a 2" × 2" × $\frac{1}{2}$" steel plate with a $\frac{1}{2}$"-hole drilled in the center.

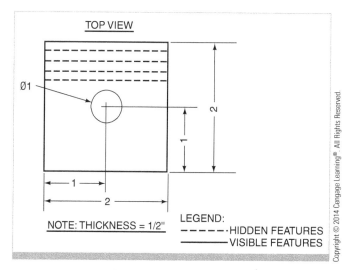

FIGURE 17-27: Single view of a 2" × 2" × $\frac{1}{2}$" steel plate with a $\frac{1}{2}$"-hole drilled in the center and containing a series of grooves as hidden features.

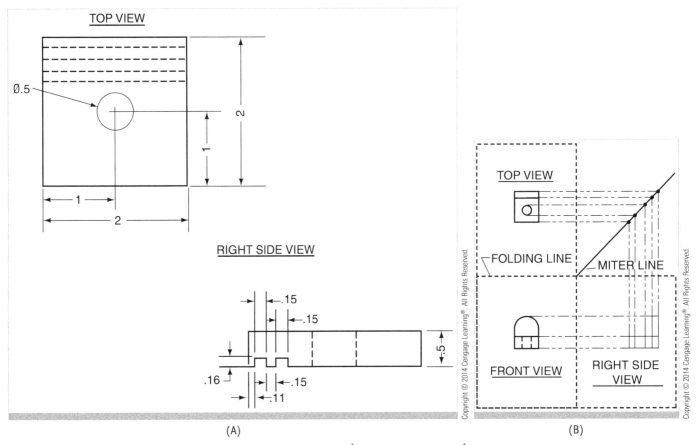

(A)

(B)

FIGURE 17-28: (A) Orthographic projection of a 2" × 2" × $\frac{1}{2}$" steel plate with a $\frac{1}{2}$"-hole drilled in the center and containing a series of grooves. (B) Orthographic projection showing how the right side view was created.

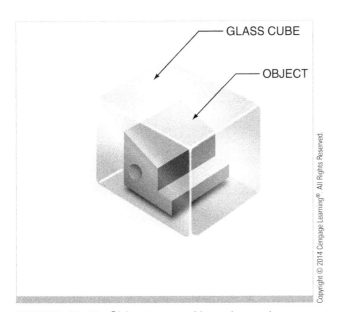

FIGURE 17-29: Object encased in a glass cube.

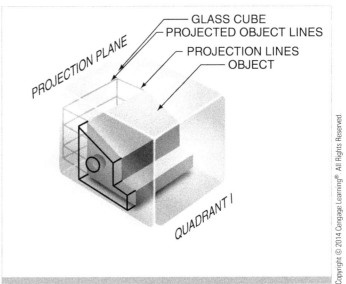

FIGURE 17-30: Object encased in glass cube with object lines projected onto surface of glass cube.

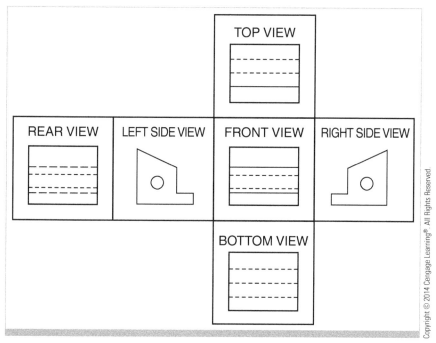

FIGURE 17-31: Glass cube unfolded to reveal the six different views.

Axonometric Projections

Axonometric projections are pictorial drawings in which the object is rotated so that all three dimensions are seen at once (length, width, and depth). This type of projection is especially helpful when a nontechnical person is trying to visualize what a part is going to look like.

Oblique Projections

Oblique projections are another form of pictorial drawing. In this form of drawing, the projection plane is set parallel with the front surface of the object. The line of sight is at an angle to the projection plane, producing a view that reveals all three axes (length, width, and depth). This is similar to an isometric because all three axes are shown, but the front of the object is not rotated. As a result, the front is drawn at full scale.

Perspective Projections

Perspective projections give the object a more natural appearance by allowing the projection lines to converge to a single point. This type of drawing is more commonly used in the art world, because the illusion produced by the lines makes the image appear more realistic. However, perspective projections have found their way into the engineering and architectural world as presentation drawings. These drawings are especially helpful if a presentation is to be given to a group of nontechnical persons by providing a better visualization of how the object will appear in reality.

Auxiliary Views

We have already discussed the fundamentals of viewing two- and three-dimensional objects using orthographic, axonometric, oblique, and perspective projection techniques. The purpose of these projections, except the orthographic, is to illustrate to a nontechnical person exactly how the object will appear. The orthographic projection shows the exact location of lines, surfaces, and contours, and projects them from one view to another. Recall that in an orthographic projection, the object's features are projected onto three principal planes—horizontal, frontal, and profile. It is from these three principal planes that the six principal views are produced. However, when an object contains features that are inclined or oblique, then a view other than one of the six principal views must be derived to show those features in their true form. In other words, in an orthographic projection these features would appear foreshortened (not as the actual size or shape), as shown in Figure 17–32.

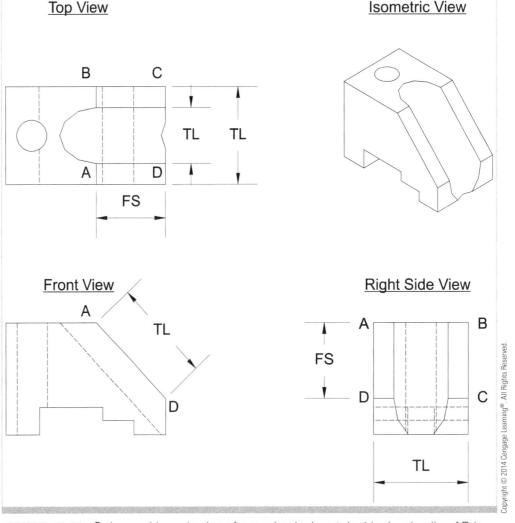

FIGURE 17-32: Orthographic projection of a mechanical part. In this drawing line AD is shown in true length in the front view, while lines AB and DC are shown in true length in the top and right side views. The true shape of the inclined surface is never revealed.

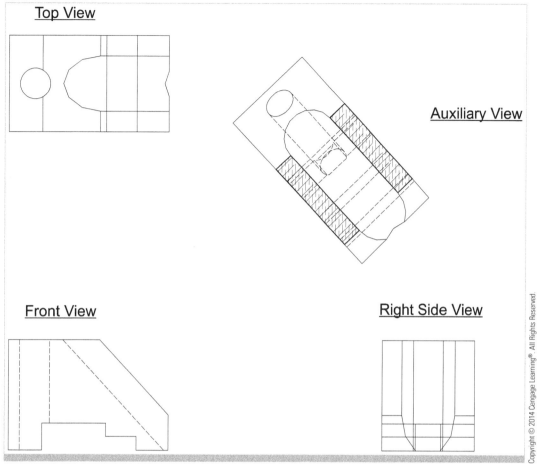

FIGURE 17-33: Auxiliary view showing plane ABDC in true shape and size.

The new view shown in Figure 17–33 is necessary to fully describe the sloping face of this part; this type of projection is called an auxiliary view. An **auxiliary view** uses orthographic techniques to produce a projection on a plane other than one of the three principal planes. Its purpose is to show the true shape, size and/or length of features that would otherwise be foreshortened in the principal views, see Figure 17-34.

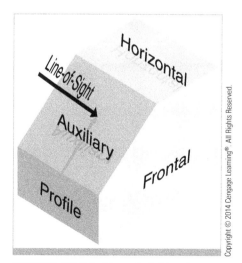

FIGURE 17-34: Line of sight set perpendicular to the auxiliary plane.

Orthographic Projection and Architectural Drawings

Architectural drawings are typically not labeled the same way as engineering drawings; however, they are created based on the same principles. In an architectural drawing, the top view is not referred to as a top view, but instead it is called a "plan." In addition, architectural drawings do not refer to left and right side views as left and right side views, instead they are referred to as "elevations."

Floor Plans

The floor plan provides a representation of the intended location of the major items in a home or other building. The plan shows the location of walls, doors, windows, cabinets, appliances, and plumbing fixtures. The drawing allows the owner to evaluate the project to ensure that the building will meet current and future needs once it is constructed. The floor plan also serves as a key tool in the communication process between the design team and the building team.

Types of Plans

The floor plan is the skeleton of framework for the development of other drawings required to complete the construction of the structure. These drawings include electrical plans, fire protection plans, framing plans, plumbing plans, and HVAC or mechanical plans.

Plumbing Plans

The plumbing plan contains the size and location of each piping system contained in the proposed structure. Typically, piping on a plumbing plan is represented using a single line and the different functions of that piping system are represented using different line types (Figure 17-35). For example, drain lines are typically shown on plumbing plans using a heavier line, whereas the domestic cold water lines are presented as light lines that consist of long dashes and short dashes. Domestic hot water lines, on the other hand, are shown as a series of long dashes followed by two short dashes. All vent piping is shown as a series of dashes.

In addition to recognizing the types of plumbing lines used in a facility, the technician must also be able to recognize various plumbing fittings and valves used in plumbing drawings. The most commonly used symbols are shown in Figures 17-36 and 17-37.

Soil and Waste, Above Grade...
Soil and Waste,
   Below Grade.........................
Vent...
Cold Water.................................
Hot Water..................................
Hot Water Return.....................
Fire Line...................................
Gas Line...................................
Acid Waste...............................
Drinking Water Supply..............
Drinking Water Return..............
Vacuum Cleaning.....................
Compressed Air.......................

FIGURE 17-35: Plumbing Limetypes.

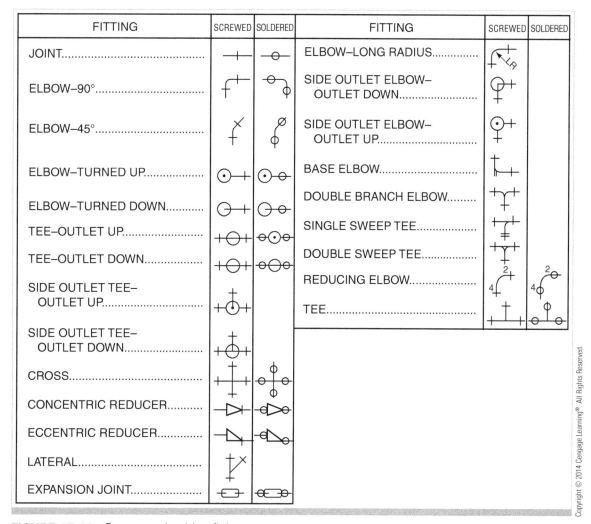

FIGURE 17-36: Common plumbing fittings.

HVAC Plans

The HVAC plan contains the size and location of each HVAC system contained in the proposed structure. Typically, piping on a HVAC plan is represented using a single line in which the different functions of that piping system are represented using different line types (Figures 17-38 and 17-39). For example, drain lines are typically shown on HVAC plans using a single line with the letter "D" inserted evenly throughout the line.

In addition to recognizing the types of HVAC lines used in a facility, the technician must also be able to recognize equipment and duct symbols used in HVAC drawings. The most commonly used symbols are shown in Figures 17-40 and 17-41.

Elevations

An elevation is an orthographic drawing that shows one side of a building. In true orthographic projection, the elevations would be displayed as shown in Figure 17-42. The true projection is typically modified as shown in Figure 17-43 to ease viewing.

VALVE	SCREWED	SOLDERED
GATE VALVE.....................................		
GLOBE VALVE...................................		
ANGLE GLOBE VALVE......................		
ANGLE GATE VALVE........................		
CHECK VALVE..................................		
ANGLE CHECK VALVE......................		
STOP COCK.....................................		
SAFETY VALVE.................................		
QUICK-OPENING VALVE..................		
FLOAT VALVE..................................		
MOTOR-OPERATED GATE VALVE...		

FIGURE 17-37: Common plumbing valves.

HIGH-PRESSURE STEAM	——HPS——
MEDIUM-PRESSURE STEAM	——MPS——
LOW-PRESSURE STEAM	——LPS——
HIGH-PRESSURE RETURN	——HPR——
MEDIUM-PRESSURE RETURN	——MPR——
LOW-PRESSURE RETURN	——LPR——
BOILER BLOW OFF	——BO——
CONDENSATE OR VACCUUM PUMP DISCHARGE	——VPD——
FEEDWATER PUMP DISCHARGE	——FPD——
MAKEUP WATER	——MU——
AIR RELIEF LINE	——V——
FUEL OIL SUCTION	——FOS——
FUEL OIL RETURN	——FOR——
FUEL OIL VENT	——FOV——
COMPRESSED AIR	——A——
HOT WATER HEATING SUPPLY	——HW——
HOT WATER HEATING RETURN	——HWR——

FIGURE 17-38: Heating line types.

REFRIGERANT LIQUID	——RL——
REFRIGERANT DISCHARGE	——RD——
REFRIGERANT SUCTION	——RS——
CONDENSER WATER SUPPLY	——CWS——
CONDENSER WATER RETURN	——CWR——
CHILLED WATER SUPPLY	——CHWS——
CHILLED WATER RETURN	——CHWR——
MAKEUP WATER	——MU——
HUMIDIFICATION LINE	——H——
DRAIN	——D——

FIGURE 17-39: Air-conditioning line types.

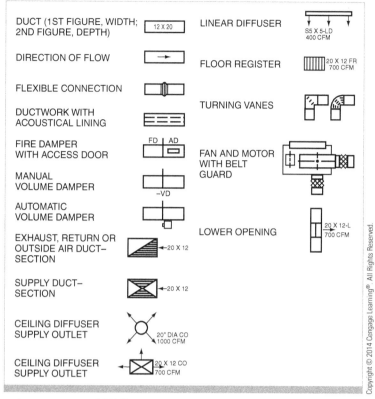

FIGURE 17-40: HVAC duct.

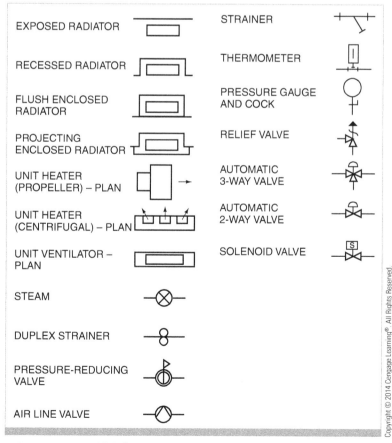

FIGURE 17-41: HVAC equipment symbols.

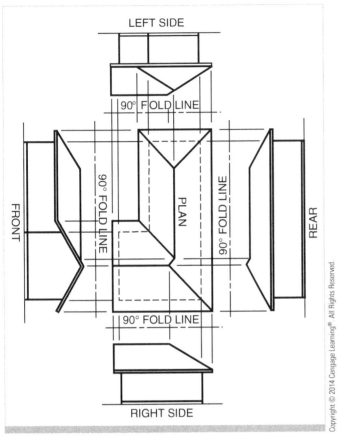

FIGURE 17-42: Elevations are orthographic projections showing each side of a structure.

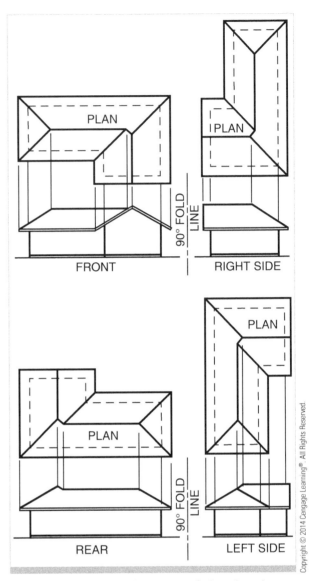

FIGURE 17-43: The placement of elevations is usually altered to ease viewing. Group elevations so that a 90° rotation exists between views.

No matter how they are displayed, it is important to realize that between each elevation, projection, and the plan view is an imaginary 90° fold line. An imaginary 90° fold line also exists between elevations in Figure 17-43. Elevations are drawn to show exterior shapes and finishes, as well as the vertical relationships of the building levels. By using the elevations, sections, and floor plans, the exterior shape of a building can be determined.

Dimensioning

It is essential for the technician to have a good understanding of the basics of manual drafting and dimensioning. Currently, there are five different styles of dimensioning used in the engineering community today. They are: unidirectional, aligned, tabular, arrowless, and chart dimensioning.

Unidirectional Dimensioning

This style of dimensioning is used widely in the mechanical and manufacturing engineering fields. In this style of dimensioning, all numerical text, dimensioning figures, and notes are placed on a drawing in a horizontal position and read starting from the bottom of the drawing (see Figure 17-44).

Aligned Dimensioning

This style of dimensioning dictates that all numerical text, dimensioning figures, and notes are placed on a drawing so that they are aligned with the dimension lines. For horizontal dimensions, the numerical text is read from the bottom. For vertical dimensions, the numerical text is read from the right side (see Figure 17-45).

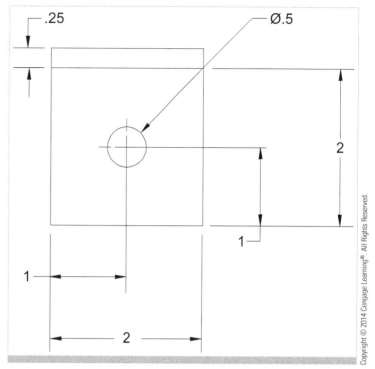

FIGURE 17-44: Unidirectional dimensioning.

Tabular Dimensioning

In this style of dimensioning, the dimensions are placed in a table and reference to them is made on the drawing (see Figure 17-46).

Arrowless Dimensioning

In this style of dimensioning, dimensions are identified on the drawing using a system of letters that are linked to a table. However, location dimensions are established with extension lines as coordinates from determined datums (see Figure 17-47).

Chart Dimensioning

The illustrations produced with this style of dimensioning is also known as "chart drawings." They are used extensively in manufacturing and in catalogs in which a part or assembly of parts has one or more dimensions that change, depending on the specific application (see Figure 17-48). Typically, the geometry of these parts remains the same and only a few of the dimensions change.

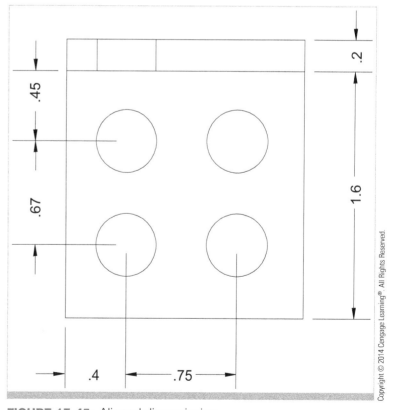

FIGURE 17-45: Aligned dimensioning.

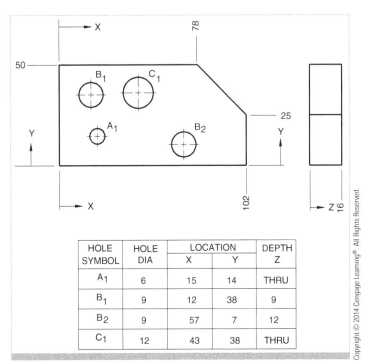

| HOLE SYMBOL | HOLE DIA | LOCATION | | DEPTH |
		X	Y	Z
A_1	6	15	14	THRU
B_1	9	12	38	9
B_2	9	57	7	12
C_1	12	43	38	THRU

FIGURE 17-46: Tabular dimensioning.

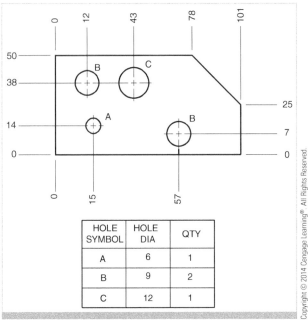

HOLE SYMBOL	HOLE DIA	QTY
A	6	1
B	9	2
C	12	1

FIGURE 17-47: Arrowless dimensioning.

Baseline Dimensioning (Datum)

In manufacturing and engineering, a common method used to place dimensions on a drawing is the baseline or datum dimensioning method. In this method, a base point is established as a reference (which can be a common surface, axis, point, center plane) with each dimension originating from the established reference point (datum). Because each dimension is independent of the others, tolerances from a previous dimension do not interfere with the exact placement of a parts feature as they do with other styles of dimensioning (see Figure 17-49).

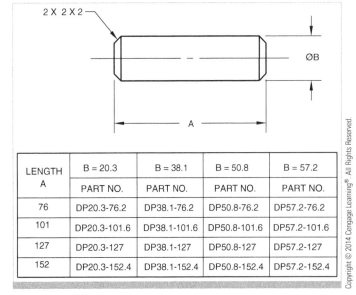

| LENGTH A | B = 20.3 | B = 38.1 | B = 50.8 | B = 57.2 |
	PART NO.	PART NO.	PART NO.	PART NO.
76	DP20.3-76.2	DP38.1-76.2	DP50.8-76.2	DP57.2-76.2
101	DP20.3-101.6	DP38.1-101.6	DP50.8-101.6	DP57.2-101.6
127	DP20.3-127	DP38.1-127	DP50.8-127	DP57.2-127
152	DP20.3-152.4	DP38.1-152.4	DP50.8-152.4	DP57.2-152.4

FIGURE 17-48: Chart dimensioning.

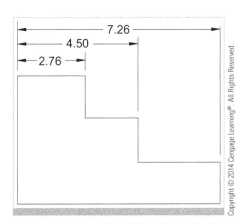

FIGURE 17-49: Baseline dimensioning.

Chain Dimensioning

Chain dimensioning is a method of dimensioning in which dimensions are placed end to end so that where one dimension ends another dimension begins (see Figure 17-50). In this method of dimensioning, each dimension is dependent on the previous dimension. Even though this is a common method of dimensioning, in most cases, it produces an undesirable result when tolerances are applied to dimensions that are chained together. This result is known as "tolerance stacking" or "tolerance build-up"; for example, a part containing chained dimensions in which individual tolerances of $\pm.25$ are applied to each dimension. If the part is manufactured at the upper or lower limit of the $\pm.25$ tolerance, then it is possible for the part to be distorted overall by an amount of .75, or three times the specified tolerance ($3 \times .25$).

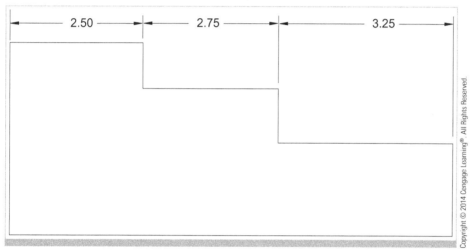

FIGURE 17-50: Chain dimensioning.

Material Take-Offs

For engineering and construction professionals, a material take off is nothing more than a list of materials, including the quantities and types needed to complete a project. The material list is generated from the construction drawings and documents.

When estimating the quantity of materials needed to complete a construction project:

- Convert concrete volumes into cubic yards by using the formula (Length $\times$ Width $\times$ Height) / 27.

- Use linear feet measurements for dimensional lumber. Typically, it is easiest to start with the sill plates, working upward and allowing for framing members to be placed on 16-inch centers (unless otherwise noted on the drawing or in the specifications). Allow for extra framing materials for windows and doors

- For electrical count all the switches, outlets, light fixtures, wiring runs. The key to successfully determining the amount of wire need is to start away from the panel, working toward it, going point to point. Use a scale to measure distances. Never estimate the ceiling height, and always add 2 to 4 feet for wire termination.

- Count all the windows and doors, grouping them by type, style, and their measurements.
- Determine the roofing materials needed:
  - For shingles
    - ➤ Determine the roof's square footage. This is accomplished by multiplying the length times the width of each section of the roof.
    - ➤ The number of squares needed for the roof is determined by dividing the total square footage by 100.
    - ➤ To determine the number of bundles needed, multiple the number of squares by 3 (for a three-tab shingle).
  - Underlayment
    - ➤ Determine the roof's square footage. Again, this is accomplished by multiplying the length times the width of each section of the roof.
    - ➤ The number of squares needed for the roof is determined by dividing the total square footage by 100.
    - ➤ The most common type of underlayment used today is 15-pound underlayment; for 15-pound underlayment divide the number of square by 4.

Building Codes

Regardless of the type of structure that is created, all buildings are constructed in accordance with the building codes. These building codes define the standards that regulate the construction of a building. These include, the facility's overall ability to react to certain load conditions, its overall height and even overall size, and the methods and the number of egresses (exits), as well as the permissible number of occupants in relation to the size, location, and exits, just to name a few aspects. In short, building codes are designed to protect the occupants.

REMODELING

Remodeling Permits and Zoning

Most remodeling projects require that the project manager/facility maintenance technician to acquire the proper permits. In general, the following remodeling projects may require a permit:

- Structural additions and/or modifications
- Electrical (high-voltage Systems, lighting systems, etc.)
- Mechanical (heating, ventilation, and air-conditioning systems; exhaust systems, etc.)
- Plumbing (potable water systems, sewage systems, drainage systems, etc.)
- Alarm systems

The exact type of permit(s) that might be required for a remodeling project will be determined by the local and state building jurisdictions where the structure is located. Always check with local and state agencies before starting a remodeling project. For example, some locations require a permit for any home improvement project (construction, additions, remodeling, repairs, replacements, and/or upgrades) that exceed a certain dollar amount.

In addition to checking to ensure that the proper permits have been obtained before starting a remodeling project, always check the local zoning ordinances. In most cases, the permit process involves:

1. Completing a permit application

2. Providing a set of construction documents

Permit requirements vary from state to state and city to city. Always check with the local building departments before attempting to obtain a permit.

Remodeling Building Inspections

Once the building/remodeling permits have been obtained and the work has started, various stages of the project may require field inspection from building inspectors before other phases can be completed. Check with the local building department regarding what inspections are required, when inspections should be performed, and how to schedule an inspection. Examples of various types of building inspections are:

- Footing(s)
- Slab
- Foundations
- Waterproofing
- Framing (rough)
- Insulation
- Electrical rough-in
- Mechanical rough-in
- Gas piping
- Fire alarms
- Plumbing rough-in

Demolition Permits

Before starting any demolition phase of a project always check to determine if a demolition permit is required. The requirements for a demolition permit will vary from location to location and typically include:

- Demolition permit application
- Asbestos affidavit
- Certificate of appropriateness if the structure is in an historic district
- Address of the facility
- Structure's owner's information (name, address, city, state, zip code, and phone number)
- Utility termination letters
- Certificate of liability insurance

Before attempting to obtain a demolition permit, always verify the requirements. If the demolition involves any of the following, additional requirements may be necessary:

- Septic system
- In-ground pool
- Spa
- Basement

Sheetmetal

Products today are manufactured from a combination materials of different types and thicknesses. These materials can be divided into five categories: plastics (thermoplastics, thermoset plastics, and elastomers), ceramics (oxides, nitrides, carbides, glass, graphites), composites (reinforced plastics, metal-matrix, ceramic-matrix, laminates), pulp (wood and paper), and metals (ferrous and nonferrous).

Once designed, metal products must have a development created before they can be manufactured (Figures 17-51 and 17-52). A development is the pattern produced when a three-dimensional geometric shape is unfolded onto a flat surface. Developments are also called "flat patterns" or "flat pattern layouts." Their purpose is to show the true shape and size of each area of an object before it is folded into its final shape. Products manufactured from sheet metal are not the only ones that

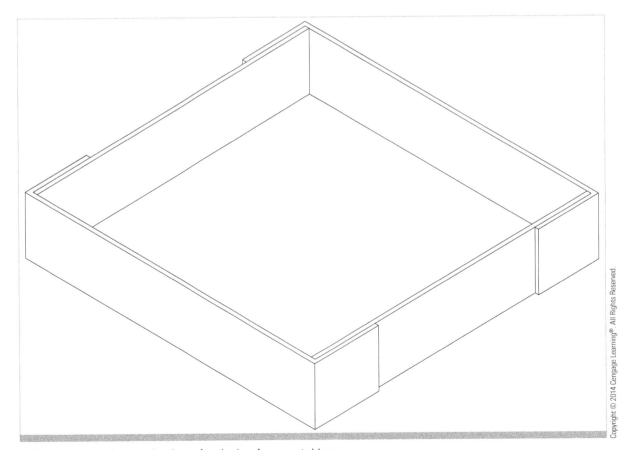

FIGURE 17-51: Isometric view of a design for a metal box.

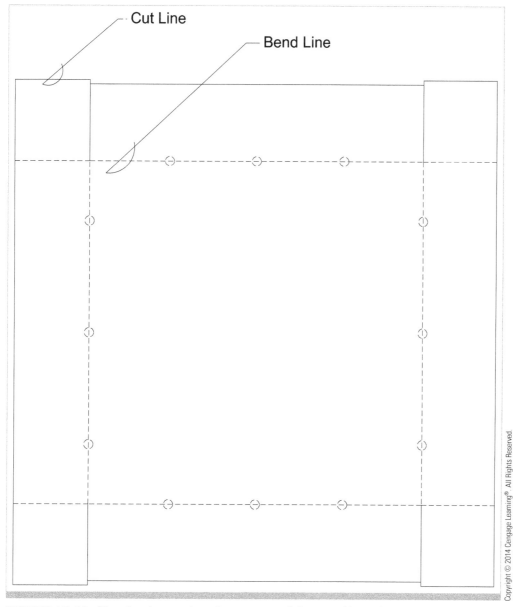

FIGURE 17-52: The development, or flat pattern, of the metal box shown in Figure 17–51.

require the use of developments; any product that receives its shape from a bending operation needs a development. Heating and air conditioning, aerospace, automobile manufacturing, and packaging are just a few of the industries that require the extensive use of developments. For this reason, it is easy to see why conceptualizing and producing the development of a part is an integral part of the engineering process.

When sheet metal products are manufactured, special consideration must be given to the thickness of the material and the angle at which it is bent. This consideration is important because the material, when subjected to a bending operation, will experience tension on one side, causing that side to stretch, while the other side is exposed to compressive forces, causing it to shrink (see Figure 17–53). The result is an angle (or a bend) that does not contain sharp edges or corners. Additional length, called "bend allowance," must be added to the stock material to compensate for these rounded corners.

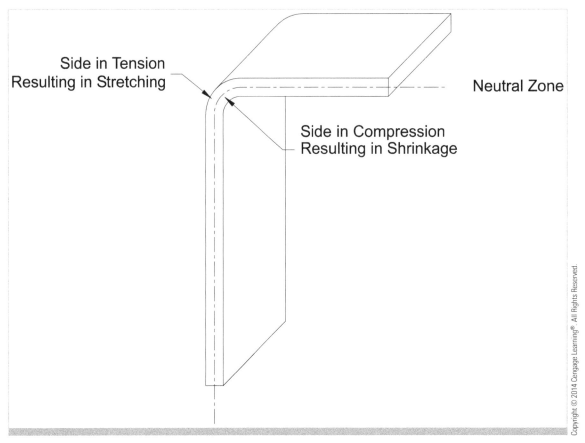

FIGURE 17-53: Forces experienced by a piece of steel subjected to a 90° bend.

For materials thinner than 0.65 mm (0.025"), the allowance can be omitted, but for any material thicker than 0.65 mm (0.025"), the allowance must be calculated to ensure that the part is produced within design tolerances. Bend allowances can be found using the allowance tables in the *Machinery's Handbook* (Industrial Press) or *ASME Handbook*. If neither of these reference books are available, then the allowance can be calculated by using the following formulas. For soft copper or soft brass, the formula is: Allowance = (0.55 × Thickness) + (0.5 × 3.14 × Radius of Bend). For aluminum, medium-hard copper, medium-hard brass, and soft steel, the formula is: Allowance = (0.64 × Thickness) + (0.5 × 3.14 × Radius of Bend). For bronze, hard copper, cold-rolled steel, and spring steel, the formula is: Allowance = (0.71 × Thickness) + (0.5 × 3.14 × Radius of Bend). In all three of these formulas, the calculations are based on a bend angle of 90°; when an angle other than 90° is to be used, the modifier (angle°/90°) must be applied to these formulas. For example, using soft steel and a bend angle of 56°, the formula now becomes: Allowance = [(0.64 × Thickness) + (0.5 × 3.14 × Radius of Bend)] × $(\frac{56°}{90°})$. If the same sheet of soft metal has a thickness of 0.125 and a bend radius of 0.25 at an angle of 56°, the complete equation would be: Allowance = [(0.64 × 0.125) + (0.5 × 3.14 × 0.25)] × $(\frac{56°}{90°})$. After plugging the information into a calculator, a bend allowance of 0.2871 is computed.

Once the bend allowance is found, it is then used to calculate the overall **length of the stock** needed to manufacture the part. This is accomplished by adding the inside distance from either allowance to allowance (Distance1 in Figure 17–54), or

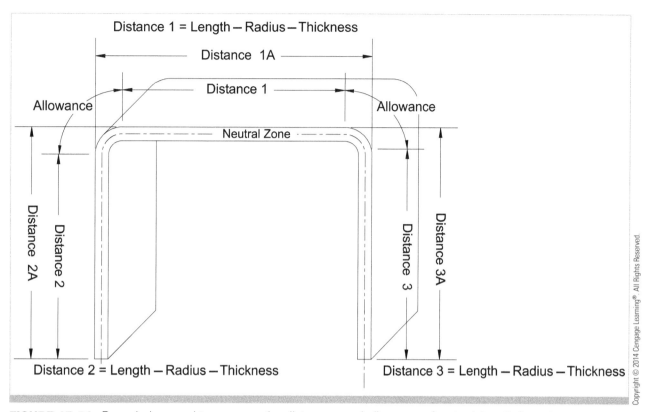

FIGURE 17-54: Boundaries used to measure the distances and allowances for stock length formulas. $Distance_1 + Distance_2 + Diatance_3 + allowance_1 + Allowance_2$ or Stock Length $= [(Distance_{2a} - Radius - Thickness) + (Distance_{1a} - Radius - Thickness) + allowance_1 + allowance_2].$

from edge of part to allowance (Distance2 in Figure 17–54). This particular part would yield the formula: Stock length $= Distance_1 + Distance_2 + Distance_3 + Allowance_1 + Allowance_2$. Naturally, the complexity of the part determines the number of distances and allowances in the formula. For this equation to work, the distances supplied must be minus the bends. If the distances supplied do contain the bends, the equation would become: $(Distance_{1a} - Radius - Thickness) + (Distance_{2a} - Radius - Thickness) + (Distance_{3a} - Radius - Thickness) + Allowance_1 + Allowance_2$.

In the preceding example, the formula could have been written as Distance + Distance + Distance + allowance − (2R + 2T). This practice is not recommended for technicians with little experience, because it can lead to confusion and possible mistakes when calculating large problems.

Another variable that must be taken into consideration when measuring for stock length is the method used to fasten sheets together. The most common methods used are: welding, brazing, soldering, adhesive bonds, and mechanical joining. For each of these methods, a different amount of material must be allowed to make an adequate connection. The method used is determined by the application of the final product, the strength of material, and the forces to which the part will be subjected. Among the most common mechanical joining methods used today are seaming and hemming (examples of this method are illustrated in Figure 17–55). When calculating the length of stock required to construct a seam, the formulas discussed earlier must be used for each bend or fold in a particular seam. This is shown in Figure 17–56.

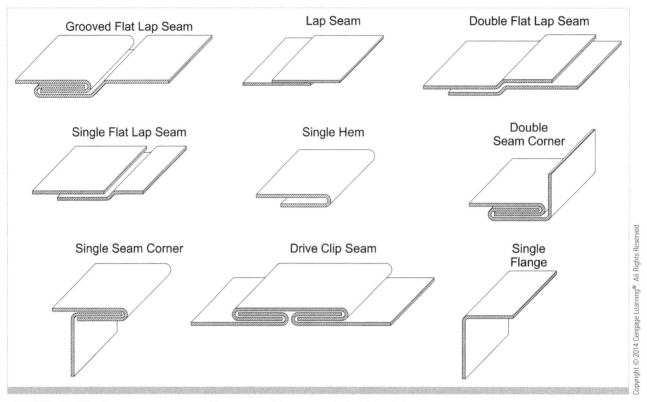

FIGURE 17-55: Common seams and hems used to connect sheet metal.

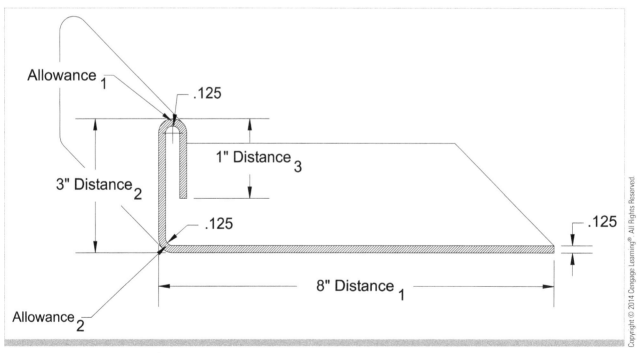

FIGURE 17-56: Material = Soft steel.

Calculation for Allowance$_1$

$$\text{Allowance}_1 = [(0.64 \times \text{Thickness}) + (0.5 \times 3.14 \times \text{Radius of Bend})] \times \text{Angle}°/90°$$

$$\text{Allowance}_1 = [(0.64 \times 0.125) + (0.5 \times 3.14 \times 0.125)] \times 180°/90°$$

Allowance$_1$ = [0.08 + 0.19625] × 2

Allowance$_1$ = 0.5525

Calculation for Allowance$_2$

Allowance$_2$ = [(0.64 × Thickness) + (0.5 × 3.14 × Radius of Bend)] × Angle°/90°

Allowance$_2$ = [(0.64 × 0.125) + (0.5 × 3.14 × 0.125)] × 90°/90°

Allowance$_2$ = [0.08 + 0.19625] × 1

Allowance$_2$ = 0.27625

Calculation for Stock Length

Stock Length = (Distance − (0.125 + 0.125)) + (Distance − (0.125 + 0.125)) + (Distance − (0.125 + 0.125)) + Allowance + Allowance

Stock Length = 7.75 + 4.25 + 0.75 + 0.5525 + 0.27625

Stock Length = 12.82875

SUMMARY

- Linear measurement is defined as the measurement of two points along a straight line.

- Currently, two basic systems are used to make measurements throughout the world—the English system of measure and metric system of measure.

- The English system of measure is the unit of measure currently used in the United States by most technicians.

- The base unit for the metric system is the meter.

- When a mechanical drawing is created on the computer, typically it is drawn at a scale of 1:1.

- However, in most cases when the file is printed, it is printed at a reduced scale so that it can fit onto a single sheet of paper

- When a drawing is printed to a scale $\frac{1}{4}$ times smaller than its actual size, then that drawing is said to be drawn at $\frac{1}{4}$ scale

- The scale most commonly used in residential and light commercial applications is the architect's scale.

- Three types of units are used to express an angle: angular degrees, radians, and gradients, but typically only angular degrees are used on blueprints.

- Since the circumference of any circle contains 360 degrees, an angular degree is equal to $\frac{1}{360}$th of the circumference of a circle.

- Specifications are typically divided into various divisions and sections. The number of divisions and sections is determined by the complexity of the project.

- The floor plan is a representation showing where to locate the major items of a home or other building.

- The plumbing plan contains the size and location of each piping system contained in the proposed structure.

- The HVAC plan contains the size and location of each HVAC system contained in the proposed structure.
  - An elevation is an orthographic drawing that shows one side of a building.
  - In general, the following remodeling projects may require a permit:
    - ➤ Structural additions and/or modifications
    - ➤ Electrical (high-voltage systems, lighting systems, etc.)
    - ➤ Mechanical (heating, ventilation, and air-conditioning systems; exhaust systems, etc.)
    - ➤ Plumbing (potable water systems, sewage systems, drainage systems, etc.)
    - ➤ Alarm systems
- The exact type of permit(s) that might be required for a remodeling project will be determined by the local and state building jurisdictions where the structure is located.

REVIEW QUESTIONS

1 _____ is defined as the measurement of two points along a straight line.

2 All objects, whether they are man-made or the result of natural conditions and/or forces, consist of _____ and _____.

3 Currently, there are two basic systems used to make measurements in the world today: _____ and _____.

4 What are the base units of the English system of measurement?

5 The _____ is based on the meter.

6 When a mechanical drawing is created on the computer, it is typically drawn at a scale of 1:1 (also known as _____ scale).

REVIEW QUESTIONS

7 The title block region is typically located along the _____ edge or _____ of the drawing.

8 The length and position of items not dimensioned on a drawing can be determined using a _____.

9 Three types of units are used to express an angle: _____, _____, and _____.

10 Angles are typically given on a blueprint with either a _____ or a leader.

11 The primary views used on an engineering drawing are _____.

12. What is a bend allowance and why is it important?

13. Given the following information, calculate the bend allowances:

Material = Soft Brass, Radius = 0.125, Angle = 45°, Thickness = 0.0625

Material = Cold Roll Steel, Radius = 0.25, Angle = 34°16', Thickness = 0.125

Material = Soft Steel, Radius = 0.375, Angle = 90°, Thickness = 0.0313

Name: _____

Date: _____

Unit Conversion

Upon completion of this job sheet, you should be able to convert from one unit to another for both metric and English.

PROCEDURE

Convert the following:

144 feet	= _____	inches
234 inches	= _____	feet
234,539 inches	= _____	miles
5 miles	= _____	inches
23 miles	= _____	yards
100 feet	= _____	meters
245.9 centimeters	= _____	feet
123 millimeters	= _____	feet
256 yards	= _____	meters
452.1 inches	= _____	centimeters

INSTRUCTOR'S RESPONSE:

_____
_____
_____

Name: _____

Date: _____

Reading a Scale

Upon completion of this job sheet, you should be able to read a scale.

PROCEDURE

Using the following scale, determine the length of lines "A," "B," "C," "G," and "H."

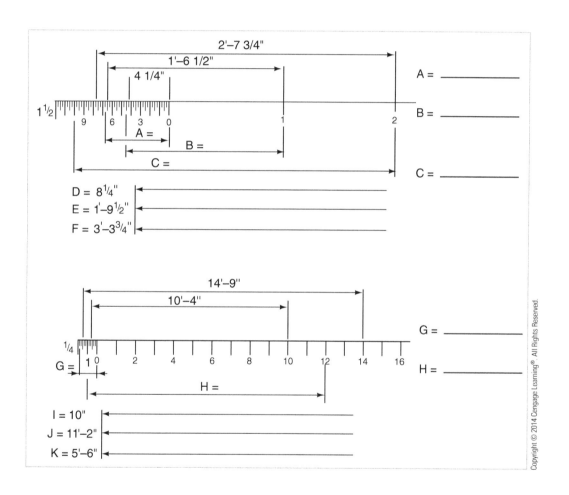

INSTRUCTOR'S RESPONSE:

_____

_____

_____

Name: _____

Date: _____

Determine Angles

Upon completion of this job sheet, you should be able to identify angles.

PROCEDURE

Determine the angles of the following:

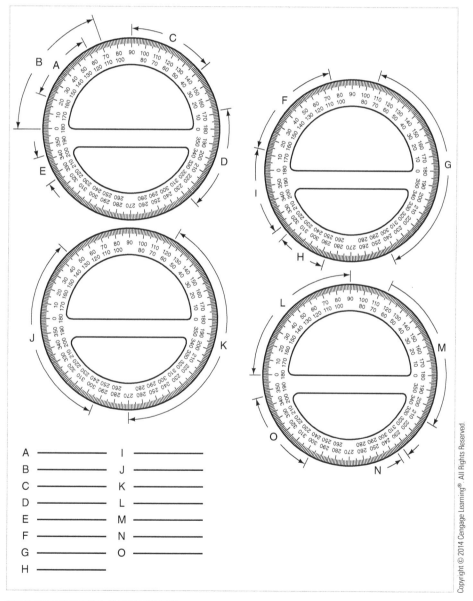

A —————— I ——————
B —————— J ——————
C —————— K ——————
D —————— L ——————
E —————— M ——————
F —————— N ——————
G —————— O ——————
H ——————

INSTRUCTOR'S RESPONSE:

_____

_____

_____

Name: _____

Date: _____

Valve Identification

Upon completion of this job sheet, you should be able to identify various valves on a blueprint.

PROCEDURE

Identify the following valve:

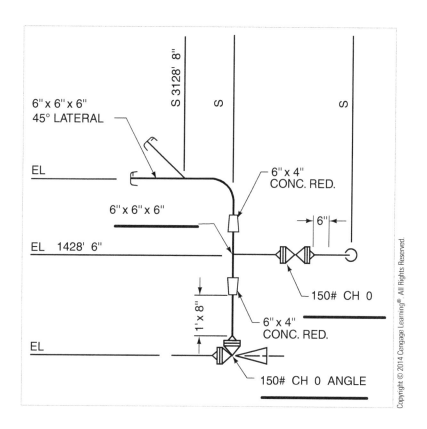

INSTRUCTOR'S RESPONSE:

_____

_____

_____

Name: _____

Date: _____

Line Type Identification

Upon completion of this job sheet, you should be able to identify various line types used on blueprints.

PROCEDURE

Identity the following line types:

————————————	——— RL ———
————————————	——— RD ———
————————————	——— RS ———
————————————	——— CWS ———
————————————	——— CWR ———
————————————	——— CHWR ———
————————————	——— MU ———
————————————	——— H ———
————————————	——— D ———

INSTRUCTOR'S RESPONSE:

_____

_____

_____

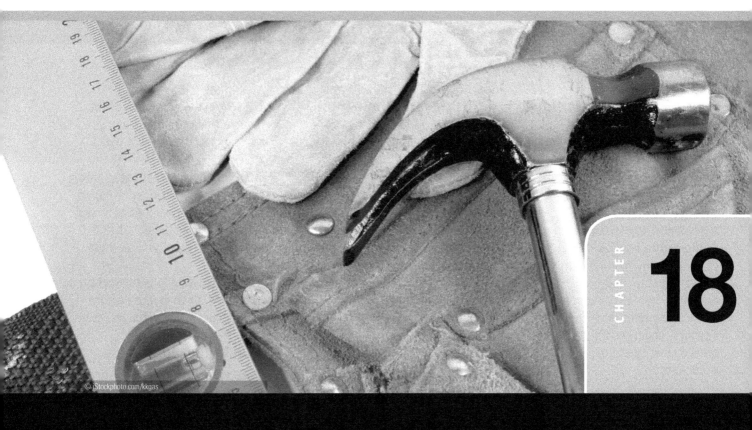

© iStockphoto.com/kkgas

CHAPTER 18

Weatherization Concepts

OBJECTIVES

By the end of this chapter, you will be able to:

KNOWLEDGE-BASED

- Identify sources of infiltration in a structure.
- Discuss the factors that influence a base load.
- Discuss the methods of heat transfer as well as the first and second laws of thermodynamics.
- Discuss the major area in a structure in which energy is used.
- Discuss the potential hazards that OSHA has identified for weatherization workers.
- Define the term *energy audit*.
- Discuss the importance of performing visual inspections.
- Explain the concept of "house as a system."
- Explain the basics of an energy audit and *Manual J*.

- Identify characteristics of energy-efficient home.
- Explain how to visually inspect exterior and interior of home.
- Identify historic preservation requirements.

SKILL-BASED

- Calculate the infiltration loads of a structure.
- Calculate the base load energy consumption of a structure and it occupants.
- Use the basic formula Btu = $\Delta T \times W$ to determine the amount of heat required to raise or lower the temperature of a given amount of fluid.
- Use the basic formula q = $A \times U \times TD$ to determine the amount of heat transfer through a given building material.
- Properly apply a sealant to an outdoor application.

Return on investment (ROI) the rate at which the total investment is recovered from by the savings of energy conservation

Energy audit a survey of a residence or a structure conducted to analyze the current energy requirements of the structure and its occupants as well as how that energy flow can be reduced without having a negative effect on the structure and/or occupants

Backdraft occurs when an exhaust fan(s) and combustion style heating devices compete for air; most often the exhaust fan pulls the combustion gases back into the structure as a result

Thermodynamics the branch of physics that deals with the conversion of energy

British thermal unit (Btu) the amount of heat needed to raise the temperature of one pound of water one degree Fahrenheit

Conduction the transfer of heat from one molecule to another within a substance or from one substance to another

Convection the transfer of heat from one place to another using a fluid

Radiation heat that travels through air, heating solid objects that in turn heat the surrounding area

Insulator a material that has a stable atomic structure and does not permit the free flow of electrons from atom to atom

Heat of transmission the amount of heat transferred through building materials

U factor the reciprocal of the R value (measure of a material's resistance to the flow of heat)

Infiltration air that enters into a structure through cracks, windows, doors, or other openings as a result of the indoor air pressure being lower than the outdoor air pressure

Ventilation planned induction of outside air to create a slight positive pressure on the structure

Base-load energy consumption the amount of energy consumed by a structure and its occupants under the best possible conditions

Shoulder months the months in which the heating and cooling systems are not used

Energy management is the process of monitoring, controlling, and ultimately conserving energy in a structure

Introduction

For home owners the cost of energy, regardless of the source is a major expense. However, with proper energy conservation techniques, these costs can easily be controlled and, in most cases, substantially reduced. One major misconception regarding energy conservation is that it will alter or reduce the comfort level of the occupants of the facility. In most cases, the energy conservation techniques will have an opposite effect; in fact, they usually will enhance the quality of life. Another misconception is that energy conservation techniques are cost prohibitive. In some cases, the cost of home improvement can be quite steep. However, some of the most effective energy conservation techniques cost very little to implement and generate the most savings. Other energy conservation techniques, such as cleaning air filters turning back thermostats, and changing general energy usage patterns, do not cost anything. Finally, energy conservation is a personal responsibility and cannot be legislated nor can it be forced upon someone. In other words, it is up to the occupants of a facility to choose and implement the most effective energy conservation techniques.

The Cost of Energy Conservation

Energy conservation techniques that require a substantial investment can be evaluated based on their **return on investment**, or **ROI**. ROI is the rate at which profit on an investment occurs. In the case of energy conservation, ROI is the rate at which the total investment is recovered by the savings the customer realizes in their utility bill from the energy conservation.

The cost of energy for any residential dwelling is based upon the following factors:

- The geographic location of the dwelling

- The square footage of the dwelling

- The lifestyle of the occupants of the dwelling

- The construction of the dwelling

- The design of the dwelling

- The condition and age of the appliances that are currently in service

While the lifestyle of the occupants should be addressed when considering energy conservation techniques, it is the construction and design of the dwelling that are the most important factors. For example, the quantity and type of insulation used in the ceiling, exterior walls, and floor will have a greater impact on the residential home owners than whether or not they unplug their phone charger when not in use.

Before facility maintenance technicians can start an energy conservation endeavors, they should understand a few basic concepts, the first of which is the major areas in which energy is most often used. The average residential consumer energy usage can be summarized as follows (see Figure 18-1):

- Heating accounts for 29%.

- Cooling accounts for 17%.

- Appliances account for 13%.

- Hot water generation accounts for 14%.

- Lighting, electronics, and other accounts for 28%.

In addition, facility maintenance technicians should have a good basic understanding of thermodynamics before attempting an energy conservation project.

FIGURE 18-1: The average residential consumer energy usage.

Weatherization Safety and OSHA

Weatherization is an important aspect of becoming energy efficient and helping conserve our natural resources. However, like every other building activity, this area has its dangers, which can cause severe injury and/or even death, not to mention property damage and loss. Therefore, safety should be the utmost concern of any facility maintenance technician working in this area.

When working with power tools, always ensure that all hand tools and power tools are inspected before and after use. Look for any defects or damage that may cause injury while the hand tool or power tool is in use.

Working with power tools in confined areas magnifies the noise levels generated. Hearing protection may be needed.

When installing fiberglass insulation, it is important to wear protective clothing (long sleeves and gloves).

According to the American Lung Association (www. lungusa.org), skin, eye, and nose irritation can be a direct result of exposure to fiberglass materials as well as fiberglass airborne particles.

Hazards associated by OSHA with SPF are:

- **Chemical Hazards—isocyanates**
- **Fire**

Work-related asthma is a incapacitating and sometimes fatal. It is also estimated that one in seven workers suffer from work-related asthma. Currently, there are more than 300 substances used in the workplace that are known to either cause or aggravate asthma.

Facility maintenance technician working in the weatherization area can be exposed to the following:

- Confined spaces
- Fall hazards
- Electrical hazards
- Airborne pollutants
- Mechanical hazards
- Exposure to hazardous materials
- Exposure to hazardous chemicals

As in other building activities previously discussed, facility maintenance technicians working in the area of weatherization must use personal protection equipment (PPE). This equipment includes, but is not limited to:

- **Chemical-resistant gloves**—Gloves made of chloroprene rubber (Neoprene), nitrile rubber (Buna N), chlorinated polyethylene, polyvinylchloride (Pylox), or butyl rubber, fluoroelastomer (Viton)
- **Tight-fitting safety glasses**—If there is the possibility of a splashing or splatter hazard, then wear a face shield or a full face mask.
- **Full body suit with hood**—Saran-coated material
- Face protection

OSHA is a part of the U.S. Department of Labor that ensures the safety of workers as well as controlling working conditions by setting and enforcing regulations, and providing training and education, outreach, and assistance. Some potential hazards that OSHA has identified for weatherization workers include:

- **Exposure to fiberglass**—This can lead to skin and eye irritation and inflammation as well as respiratory problems. When fiberglass is used in a blow-in application, such as blown-in insulation, employers are required to furnish employees with the proper respiratory protection. OSHA, the International Agency for Research on Cancer (IARC), and the National Toxicology Program (NTP) have linked the inhalation of fiberglass material to cancer.
- **Exposure to cellulose**—Employees who work with this type of insulation should be provided with the proper dust mask in order to prevent exposure. Cellulose is used in insulation and is considered to be an irritant to the respiratory system; irritation can occur when an employee is exposed without proper safety equipment (NTP, 2006; 29 CFR 1910. 1000 Subpart Z; 29 CFR 1926. 55 Appendix A).
- **Exposure to spray polyurethane foam (SPF)**—Employees who work with this type of insulation should follow the manufacturer's application instructions as well as the manufacturer's safety instructions. This type of insulation contains Isocyanates. According to researchers and the National Institute for Occupational Safety and Health, isocyanates are the leading chemical that has been linked to work-related asthma
- **Exposure to polystyrene**—Polystyrene is used in insulation. When inhaled, it has been linked to several health issues, ranging from respiratory irritation to neurological effects. Employees that work with this type of insulation should be provided with the proper personal safety equipment (the same as that used for

SPF applications) to limit exposure (29 CFR 1910. 1000 Subpart Z; 29 CFR 1926. 55 Appendix A – Air contaminants; EPA IRIS). Because styrene is flammable the same precautions used for SPF should also be used for polystyrene (29 CFR 1910. 106 – Flammable and combustible liquids).

- **Exposure to latex sealant**—Often used with fiberglass batting, latex is a known allergy sensitizer. Exposure to latex can cause minor to severe health effects in some individuals. Therefore, employees should be provided with proper personal protection equipment to limit skin, eye, and respiratory exposure.

⚠ CAUTION ⚠

OSHA has identified that polystyrene is a potential fire hazard; use extreme caution when using it around open flames or heat.

The House as a System

More information regarding exposure to insulation, as well as about OSHA in general, can be found at www.osha.gov.

An important concept to keep in mind before starting a weatherization project is the concept of the "house as a system." This principle states that, although the home might be made up of individual components, these components come together as a whole (see Figure 18-2). For example, the human body is made up of individual parts (fingers, toes, ears, eyes, etc.); it is the combination of these part that make up the whole human body.

As with the human body, when a change is made to one component of a structure, it will often affect the performance of the remaining components. Therefore, this should be taken into consideration when starting a weatherization project. If the systems are treated merely as individual systems and not as part of the overall system, then performance, balance, comfort, indoor air quality, and even safety can be compromised.

When a home is treated as a system the following features are considered:

- The physical location and its terrain
- Climatic conditions
- Solar orientation
- Solar gain
- The structure's design
- The structure's components
- The structure's environmental impact on the occupants
- The occupants' impact on the building's environment

FIGURE 18-2: The house as a system.

When the structure is treated as a whole system, the overall objectives should be to:

- Improve the overall indoor environment
- Increase the comfort level in the structure
- Reduce the facility's overall utility cost
- Reduce maintenance costs associated with the facility
- Improve the facility's overall sustainable aspects

Manual J and Energy Audits

Before a heating, ventilation, and air-conditioning (HVAC) system can be selected for a facility, a heating/cooling load calculation must be generated. To be comfortable, the air temperature must be within the acceptable comfort range and the air-distribution system must not create drafts, either hot or cold, in the conditioned space. A typical comfort level for residential and light commercial buildings is 70°F with a relative humidity level between 40% and 60%.

The quality of the air in an environment is an important aspect of HVAC. If the air is too dry, the environment can be a health hazard. The same holds true for an environment in which the humidity is too high. When the humidity is too high, dangerous mold may be produced.

As stated earlier, the design sequence starts with a load calculation of the structure. A heating load calculation is used to tell the technician how much heat is escaping from the structure while a cooling load tells the technician how much heat is entering into the structure. The load calculation is important from another standpoint. Many lending institutions do not lend money for new construction unless they are assured that the heating system will be adequate. These institutions do not want a poor heating system in a building that they will have to sell if the building owner defaults on the loan.

The most common method for performing a heating/cooling load calculation is the Air Conditioning Contractors of America's (ACCA) *Manual J*. In many locations, the building code requires the use of the *Manual J* techniques. *Manual J* is an accepted industry standard that has been accepted by American National Standards Institute ANSI for accurately sizing and selecting HVAC equipment in residential structures. The latest version of *Manual J*, version eight (see Figure 18-3), provides a comprehensive approach to ensure that environmental systems are efficient, safe, and promote healthy environments.

Some of the key factors that must be taken into consideration when sizing HVAC equipment are:

- The weather conditions
- Size, shape, and orientation of the structure
- Amount of insulation
- Square footage of the windows and exterior doors as well as their location, and type
- Air infiltration rates
- The number and ages of occupants
- Occupant comfort preferences
- The types and efficiencies of lights and major home appliances (which give off heat)

Energy Audits

An **energy audit** is a survey of a residence conducted to analyze the current energy requirement of the structure and its occupants as well as how that energy flow can be

ACCA. Air Conditioning Contractors of America

Residential Plans Examiner Review Form for HVAC System Design (Loads, Equipment, Ducts)

Form RPER 1.01 8 Mar 10

County, Town, Municipality, Jurisdiction
Header Information

Contractor: ABC Heating and Air Conditioning Company
Mechanical License #: MCL# 123456789
Building Plan #: Model P54321-987, dated 13 June 2010
Home Address (Street or Lot#, Block, Subdivision): 123 Elm Street, Beatrice, Nebraska

REQUIRED ATTACHMENTS[1]	ATTACHED
Manual J1 Form (and supporting worksheets):	Yes [X] No [ ]
or MJ1AE Form[2] (and supporting worksheets):	Yes [ ] No [X]
OEM performance data (heating, cooling, blower):	Yes [X] No [ ]
Manual D Friction Rate Worksheet:	Yes [X] No [ ]
Duct distribution system sketch:	Yes [X] No [ ]

HVAC LOAD CALCULATION (IRC M1401.3)

Design Conditions

Winter Design Conditions
- Outdoor temperature (1) -2 °F
- Indoor temperature (2) 70 °F
- Total heat loss (14) 59,000 Btu

Summer Design Conditions
- Outdoor temperature (3) 95 °F
- Indoor temperature (4) 75 °F
- Grains difference (5) 35 Δ (6) @ 50 % Rh
- Sensible heat gain (15) 22,000 Btu
- Latent heat gain (16) 5,000 Btu
- Total heat gain (17) 27,000 Btu

Building Construction Information

Building
- Orientation (Front door faces) (7) South
  North, East, West, South, Northeast, Northwest, Southeast, Southwest
- Number of bedrooms (8) 3
- Conditioned floor area (9) 1,773 Sq Ft
- Number of occupants (10) 4

Windows
- Eave overhang depth (11) 2 Ft
- Internal shade (12) Blinds, light, 45 Angle
  Blinds, drapes, etc
- Number of skylights (13) 1

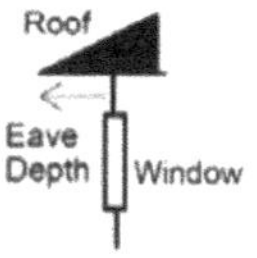

HVAC EQUIPMENT SELECTION (IRC M1401.3)

Heating Equipment Data
- Equipment type (18) Gas Furnace
  Furnace, Heat pump, Boiler, etc
- Model (19) XYZ 080-14
- Heating output capacity (20) 64,000 Btu
  Heat pumps - capacity at winter design outdoor conditions
- Auxiliary heat output capacity (21) N/A Btu

Cooling Equipment Data
- Equipment type (22) Air Conditioner
  Air Conditioner, Heat pump, etc
- Model (23) XYZ 030 Condenser 030 Coil
- Sensible cooling capacity (24) 21,200 Btu
- Latent cooling capacity (25) 6,500 Btu
- Total cooling capacity (26) 28,700 Btu

Blower Data
- Heating CFM (27) 1,117 CFM
- Cooling CFM (28) 1,000 CFM

HVAC DUCT DISTRIBUTION SYSTEM DESIGN (IRC M1601.1)

- Design airflow (29) 1,117 CFM
- External Static Pressure (ESP) (30) 0.75 IWC
- Component Pressure Losses (CPL) (31) 0.40 IWC
- **Available Static Pressure (ASP)** (32) 0.35 IWC
  ASP = ESP - CPL
- Longest supply duct: (33) 278 Ft
- Longest return duct: (34) 110 Ft
- **Total Effective Length (TEL)** (35) 388 Ft
- **Friction Rate:** (36) 0.09 IWC
  Friction Rate = (ASP × 100) ÷ TEL

Duct Materials Used (circle)
Trunk Duct: Duct board, Flex, Sheet metal, Lined sheet metal, Other (specify)
Sheet metal (insulated R-8) (37)
Branch Duct: Duct board, Flex, Sheet metal, Lined sheet metal, Other (specify)
Flex duct (insulated R-8) (38)

I declare the load calculation, equipment selection, and duct system design were rigorously performed based on the building plan listed above. I understand the claims made on these forms will be subject to review and verification.

Contractor's Printed Name: Bartholomew J. Simpson Date: 1 April 2010

Contractor's Signature:

Reserved for use by County, Town, Municipality, or Authority having jurisdiction.

[1] The AHJ shall have the discretion to accept Required Attachments printed from approved ACCA software vendors, see list on page 2 of instructions.
[2] If abridged version of Manual J is used for load calculation, then verify residence meets requirements, see Abridged Edition Checklist on page 13 of instructions.

Courtesy of Air Condition Contractors of America (ACCA)®

FIGURE 18-3: ACCA's *Manual J*.

reduced without having a negative effect on the structure and/or occupants. The key to an energy audit is that it will not decrease the quality of the structure or quality of life of the occupants. Although energy audits should be conducted by a certified professional, a facility maintenance technician can still do a diligent walk-through and spot many of the problems. When conducting an energy audit (or in the case of a facility maintenance technician a walk through), keep a list of the areas that you have already examined as well as any issues connected to those areas. Problem areas that a facility maintenance technician should be looking for include, but are not limited to:

- **Air leaks**—The reduction of drafts in a dwelling can reduce the energy usage by as much as 30% on an annual basis as well as increase the comfort level. Common sources for air leaks include:
  - Receptacles
  - Switch plates
  - Window frames
  - Baseboards
  - Weather stripping around doors
  - Fireplace dampers
  - Attic hatches
  - Wall- or window-mounted air conditioners

When sealing a dwelling, keep in mind the dangerous situation known as backdraft. A **backdraft** occurs when an exhaust fan(s) and combustion-style heating devices compete for air. The result is most often the exhaust fan pulling in the combustion gases back into the structure. As a general rule of thumb, heating devices that use combustion (natural gas, fuel oil, propane, or wood) require one square inch of ventilation for every 1,000 Btu of heat output.

- **Insulation**—Depending upon the age of the dwelling, heat transfer through the ceiling and walls could be a very substantial.

- **HVAC equipment**

- **Lighting**

The following is an example of an energy audit checklist. A more extensive energy checklist can be obtained from the visiting a cooperative extension:

Name _____

Daytime Phone _____

E-mail _____

Fax _____

Building Information

Building Name _____

Address _____

City/State/Zip _____

Floors _____

Utility Rates: Electricity Billing Information (or include copy of bill)

Energy Cost per KWh $ _____

Demand Cost per KW $ _____

Fixed Costs per Month $ _____

(Continued)

Window Area Construction Thickness Appearance

North sq. ft _____ **Single** _____ **Double Pane** _____ $\frac{1}{4}$" _____ $\frac{1}{8}$" _____ **Clear** _____ **Colored** _____

South sq. ft _____ **Single** _____ **Double Pane** _____ $\frac{1}{4}$" _____ $\frac{1}{8}$" _____ **Clear** _____ **Colored** _____

East sq. ft _____ **Single** _____ **Double Pane** _____ $\frac{1}{4}$" _____ $\frac{1}{8}$" _____ **Clear** _____ **Colored** _____

West sq. ft _____ **Single** _____ **Double Pane** _____ $\frac{1}{4}$" _____ $\frac{1}{8}$" _____ **Clear** _____ **Colored** _____

Attics and Ceilings

Insulated Yes _____ **No** _____ **Type of Insulation** _____ **Thickness** _____ **Rating** _____

Walls (Outside exposure walls)

Thickness _____ **Construction** _____ **Insulated Yes** _____ **No** _____ **Type** _____ **Rating** _____

Attic Walls (Outside exposure walls)

Thickness _____ **Construction** _____ **Insulated Yes** _____ **No** _____ **Type** _____ **Rating** _____

Floors (Crawlspaces)

Thickness _____ **Construction** _____ **Insulated Yes** _____ **No** _____ **Type** _____ **Rating** _____

Duct in Un-heated Areas)

Insulated Yes _____ **No** _____ **Type** _____ **Rating** _____ **Sealed Yes** _____ **No** _____

Exterior Doors

Insulated Yes _____ **No** _____, **Contains Glass Yes** _____ **No** _____ **sq ft** _____ **Single** _____ **Double Pane** _____
$\frac{1}{4}$" _____ $\frac{1}{8}$" _____ **Clear** _____ **Colored** _____

Overall Structure Construction

Well Sealed Yes _____ **No** _____

Water Heaters

Insulated Yes _____ **No** _____, **Energy Star Yes** _____ **No** _____

Water Heaters

Age _____ **SEERS Rating** _____,

Overall Structure Construction

Yes _____ **No** _____

Comments

_____

_____

_____

_____

_____

_____

_____

Performing Visual Inspections

Before an energy audit can be effectively preformed, the person performing the audit must become oriented to and acquainted with the facility. A visual inspection of the facility is not only required when performing the audit, but can be extremely helpful in assisting the auditor in becoming acquainted with the structure, its orientation, the building design, state of maintenance, site issues, floor plan, and main utilities.

Inspecting the Exterior

The visual exterior inspection of a facility consists of evaluating the efficiency of the structure as well as identifying any safety issues that might exist. If an unsafe condition is detected, the home owner should be notified at once. When inspecting the exterior of a structure:

When performing an exterior inspection, always be aware of your surroundings. If the home owner has a pet, determine the location of the pet before entering an area.

- Inspect the characteristics and condition of the foundation, roof, siding, windows, doors, and overhangs.
- Inspect the foundation, and note the amount of exposure.
- Evaluate the site drainage, and look for evidence of moisture accumulation and damage.
- Note any additions to the dwelling.
- View the building through an infrared scanner, if available, to identify thermal flaws.
- Assess the roof and window shading from trees, awnings, and other buildings.
- Determine the orientation each side of the structure faces to assess the effect of solar gain and the opportunity to use solar energy or to block it as appropriate.
- Inspect the chimney(s) and exhaust vents.

Inspecting the Interior

Although, inspecting the interior of a structure consists of examining a different set of conditions than when inspecting the exterior, the overall outcome is still the same. You evaluate the efficiency of the structure, its equipment, and its occupants as well as identifying any safety issues that might exist. When inspecting the interior of a structure:

FROM EXPERIENCE

When conducting an interior inspection of a structure, always start with the areas of the home in which there is sufficient opportunity for energy reduction. This will allow you to spend more time and conduct a more intense evaluation in those areas.

- Pinpoint and classify components of the thermal boundary.
- Appraise the type, thickness, and condition of insulation in the attic, walls, floors, and foundation.
- Pinpoint large air leaks.
- Measure the structure's floor space and interior volume.
- Check for signs of moisture problems such as mold, water stains, or musty smells.
- Inspect the wiring in the attic or other areas affected by weatherization measures.
- Identify other health and safety issues.

Characteristics of an Energy-Efficient Home

Even though there are considerable discussions involved in obtaining an energy-efficient structure, the basic element or characteristics are the same. They are:

- A tightly sealed and well-constructed facility consisting of:
  - A weather barrier (the outer skin)
  - An air barrier (limiting air leakage and infiltration)
  - A thermal barrier (prevents the movement of heat)
  - A vapor barrier (prevents movement of moisture)
  - Interior finish
- Controlled ventilation
- Properly sized high-efficiency heating and cooling systems (systems that are not properly sized can actually decrease the efficiency of the unit as well as drive up the operating cost).
- Energy-efficient windows and doors as well as appliances

Heat Transfer

To understand how to control the flow of heat during the various seasons, the facility maintenance technician must have a good basic understand of thermodynamic as well as the principle methods of heat transfer. **Thermodynamics** is the branch of physics concerned with the conversion of different forms of energy. One of these laws states that energy can be neither created nor destroyed, but is simply changed from one form to another. However, it is the second law of thermodynamics on which we will center out discussion. This law states that heat will travel from hot to cold and not in the reverse. This is because the molecules of the warmer substance will transfer energy to the molecules of the colder substance. This meaning that in the winter time heat is trying to leave the structure, while in the summer time heat is always trying to enter the structure. It is the temperature difference that is the driving force behind heat transfer. Therefore, the greater the temperature difference, the greater the heat transfer rate.

Because we are dealing in energy, the unit of measure most often used to describe the amount of heat transfer in residential system in the United States is the **British Thermal Unit (Btu)**. The Btu is the amount of heat needed to raise the temperature of one pound of water one degree Fahrenheit. In other words the amount of heat required to raise the temperature of 1 pound of water from 40°F to 41°F would be 1 Btu. The formula for the Btu can be expressed as follows: Btu = $\Delta T \times W$. This is where ΔT is equal to the temperature difference and W is equal to the weight of the water. For example, to raise the temperature of 55 pounds of water from 56°F to 75°F would require 950 Btu. This is determined as follows:

Btu = (75°F − 56°F) × 50pounds
Btu = 19°F × 50pounds
Btu = 950 Btu

FIGURE 18-4: Placing a branding iron in a fire will cause the heat from the fire to be transferred to the metal branding iron.

There are three different methods by which heat is transferred from one location or substance to another. These three methods are: conduction, convection, and radiation. In a structure, heat transfer is typically accomplished through a combination of the three rather than one single method.

Conduction is a form of heater transfer in which the heat is transferred from within a substance from one molecule to another. It is transferred from the faster-moving molecules to the slower-moving molecules. For example, applying heat (in the form of a flame) to the metal rod will cause heat to be transferred from the flame to the metal rod. It will also cause heat to be transferred along the length of the rod. Depending upon the length of the rod, the end opposite the flame could become too hot to handle without gloves (see Figure 18-4).

Convection is heat transfer from one place to another using a fluid (for example, the heat transformation of water, see Figure 18-5). For HVAC the convection is the most common method of heat transfer. Figure 18-5 illustrates the changing of water from a solid to a liquid and then to a gas. It is the amount of heat that is absorbed by the water while it is in a particular state (also known as a phase) that is referred to as "sensible heat." That is not the case when the water is in the process of changing states while absorbing heat. In this particular instance, the heat that is absorbed is

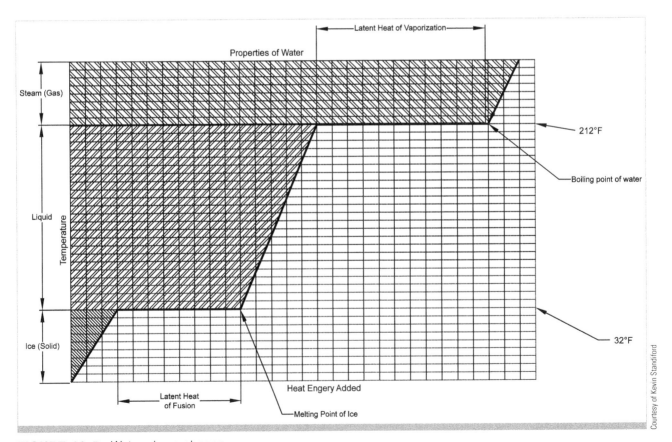

FIGURE 18-5: Water phase change.

referred to as "latent heat." It is latent heat that is responsible for changing the water from a liquid to a gas as well as changing it from a solid to a liquid. It is latent heat that is responsible for why it will snow sometimes when the temperature is 32°F and at other times it will rain at the same temperature (see Figure 18-5).

Radiation is heat that passes through air, heating solid objects that in turn heat the surrounding area. This is the principle behind a radiant heater.

Heat Transfer and Building Materials

One of the determining biggest factors as to the amount of heat that a building will transmit to the outside environment is the type of materials that are used in the construction of the facility. Other factors that help determine the amount of heat transfer from a structure to its surrounding and vice versa is the orientation of the structure, its geographical location, and even the building's design. However, as stated earlier, it is the materials that are being used in the construction of the building that will have the biggest impact on the transmission of heat. Some materials are better at transmitting heat, whereas other materials are not. Any material that is a good conductor of electricity will be a good conductor of heat. Recall from Chapter 5 that a material that is a good conductor of electricity typically has one or two electrons in the outermost shield. These electrons are considered to be free electrons because they move easily from one atom to another. However, atoms with several electrons in their outer orbits are poor conductors. It is difficult to free these electrons, and materials containing such atoms are considered to be **insulators**. Generally speaking, anything that is a good conductor of electricity is a good conductor of heat. For example, a hollow-core steel door without any insulation would conduct more heat than a wooden door. If the same hollow-core steel door was filled with Styrofoam insulation, it would conduct less heat than a solid steel door (Figure 18-6A).

The amount of heat transferred through building materials is known as **heat of transmission**. Heat of transmission uses conduction as it mechanism for transmitting heat from one building material to another. It can be calculated by using the formula $q = A \times U \times TD$.

Where

q = Heat transfer in Btu/h
A = Area of the building material in square feet
U = Overall heat-transfer coefficient
TD = Temperature difference from one side to the other in °F

The overall heat transfer coefficient, also referred to as the **U factor**, is the reciprocal of the R value of a material. The R value is the material's resistance to the flow of heat. For example, the U-factor of R-30 insulation can be determined as follows:

$$U = \frac{1}{R}$$

$$U = \frac{1}{30}$$

$$U = 0.03$$

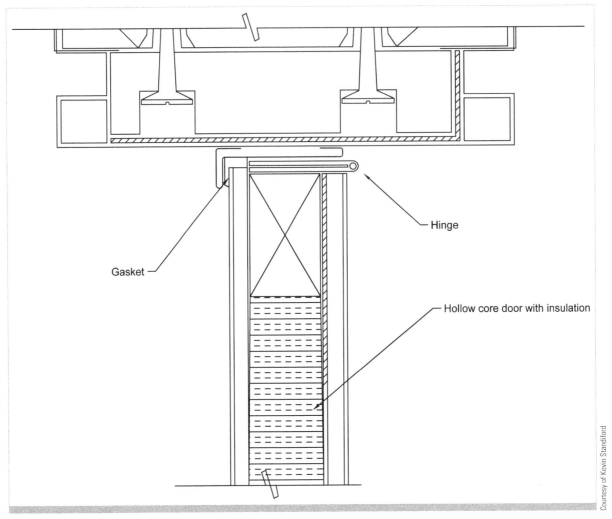

FIGURE 18-6A: Steel door with insulation inside.

It should be noted that the higher the R value, the lower the U-factor and the less heat that is allowed to transfer. For example:

	R-15 Insulation	R-23 Insulation	R-30 Insulation
R Value	15	23	30
U-Factor	0.066666667	0.043478261	0.033333333

To determine the amount of heat transfer from one side of a piece of building materials to the other (Figure 18-6B) is calculated using the formula q = A × U × TD. If the building material is 1 foot by 1 foot and has a R-factor of 30, then the heat of transmission is determined as follows:

Step 1: Determine the U factor of the material:

$$U = \frac{1}{R}$$
$$U = \frac{1}{30}$$
$$U = 0.0333$$

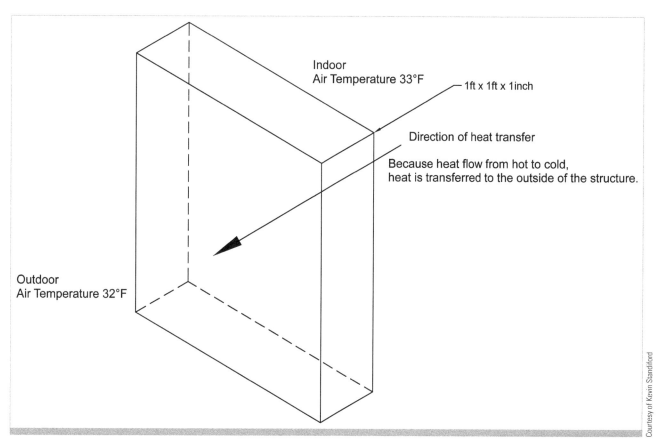

Indoor
Air Temperature 33°F

1ft x 1ft x 1inch

Direction of heat transfer

Because heat flow from hot to cold,
heat is transferred to the outside of the structure.

Outdoor
Air Temperature 32°F

Courtesy of Kevin Standiford

FIGURE 18-6B: Overall heat-transfer coefficient, or U factor.

Step 2: Determine the square footage:

> $A = L \times W$
> $A = 1 \text{ ft} \times 1 \text{ ft}$
> $A = 1 \text{ sq. ft}$

Step 3: Determine the heat of transmission:

> $q = A \times U \times TD$
> $q = 1 \text{ sq. ft} \times (0.0333 \text{ Btu/h} \times °F \text{ TD} \times \text{sq. ft}) \times 1°F \text{ TD}$
> $q = 0.0333 \text{ Btu/h}$

Determining the area of the building material can typically be accomplished by using one of the following formulas:

Square $= \text{Length}^2$
Rectangle $= \text{Length} \times \text{Width}$
Parallelogram $= \text{Base} \times \text{Height}$
Trapezoid $= \text{height}/2 \, (\text{base}_1 \times \text{base}_2)$
Circle $= \pi \, \text{radius}^2$
Ellipse $= \pi \, \text{radius}_1 \times \text{radius}_2$
Triangle $= \frac{1}{2} (\text{base} \times \text{height})$

The temperature difference is determined by subtracting the inside design temperature (the desired temperature) against the outside design temperature (typically, the coldest of the hottest day of the year). The outside design temperature can sometimes be determined by contacting your local HVAC distributor or a HVAC company. It can also usually be determined by contacting your local building department. Once

The summer TD subtracts the outdoor design temperature from the indoor design temperature. This is because of the second law of thermodynamics that states that heat flows from hot to cold.

the outside design temperature has been determined the temperature difference can calculated. For example, for a structure located in Omaha NE having an outside design temperature of $-5°F$ (winter) and $91°F$ (summer) and an inside design temperature of $72°F$ would be determined as follows:

Summer

$$TD = 91°F - 72°F$$
$$TD = 19°F$$

Winter

$$TD = 72°F - (-5°F)$$
$$TD = 77°F$$

When calculating the heat of transmission, it is the overall all U-factor that is typically the most difficult to determine. In most cases, the U-factor is determined by finding the R values of the various components of the composite wall, summing them up, and then taking the reciprocal. For example, to determine the U-Factor of the composite wall shown in Figure 18-6C the following table is created:

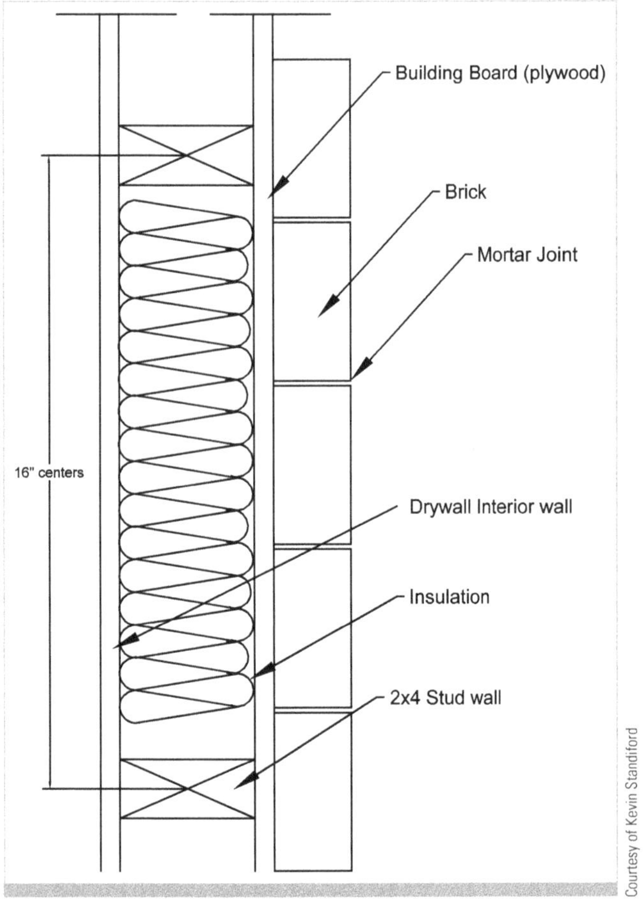

FIGURE 18-6C: Composite stud wall.

Building Board (plywood)

Brick

Mortar Joint

16" centers

Drywall Interior wall

Insulation

2x4 Stud wall

Courtesy of Kevin Standiford

Building Material	R Value
Brick	0.44
Insulation R-30	30
Plywood ($\frac{5}{8}$)	0.77
Drywall ($\frac{1}{2}$)	0.45
2 × 4	4.38
Total R value	36.04
Total U Factor	0.027746948

Heat Loss through Walls

In a structure, the walls make up the vast majority of the structure in which exposure to the outside in which heat is permitted to exit or enter a building. As stated earlier, the heat of transmission can be determined by finding the R values of the various wall components, adding them together, and then taking the reciprocal. Once the overall U factor has been determined, the formula $q = A \times U \times TD$ is then used to calculate the total Btu/hr heat loss (winter) or gain (summer) for that particular wall.

Also, when calculating the rate of heat transfer through a wall, all windows and doors have to be taken into consideration as well as their square footage must be subtracted from the square footage of the wall (see Figure 18-7). Regardless of the orientation of the walls (above or below ground), the rate of heat transfer must be determined in order to accurately determine the size of heat/air-conditioning unit needed to maintain the structure's indoor temperature (see Figure 18-8).

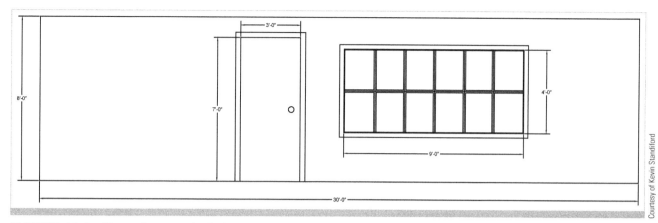

FIGURE 18-7: Net wall area calculation.

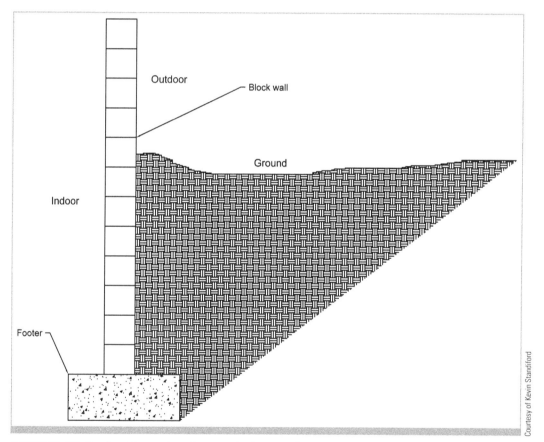

FIGURE 18-8: A wall may be above or below grade.

Walls located below grade are basement walls, and their heat loss is just as important as the heat loss for above-grade walls.

The total R value of a wall is easy to determine. Determining the resistance of the various components, on the other hand, can be a little more difficult. The R values of some of the more common building materials is provided in Figure 18-9. However, a more complete listing can be obtained from the American Society of Heating Refrigeration and Air Conditioning Engineers.

There is another factor that must be taken into consideration before the total resistance of a wall can be determined. This is the fact that an air film on each side of a wall offers resistance to the heat flow. This film is very still on the inside of the house, where the air in the room is not moving very fast. Heat is not readily transferred through still air. The resistance to heat flow from the film on the outside depends on the wind speed. This resistance is called the film factor. These two film factors are expressed in the form of resistances in this text and are as follows:

Inside film resistance = 0.68 + 1.47 = 2.15/2 = 1.1
Outside film resistance for 15-mph wind speed = 0.25

Material	R/Inch	R/Thickness
Fiberglass Batt	3.14	
Fiberglass Blown (attic)	2.2	
Fiberglass Blown (wall)	3.2	
Rock Wool Batt	3.14	
Rock Wool Blown (attic)	3.1	
Rock Wool Blown (wall)	3.03	
Cellulose Blown (attic)	3.13	
Cellulose Blown (wall)	3.7	
Vermiculite	2.13	
Air-entrained Concrete	3.9	
Urea Terpolymer Foam	4.48	
Rigid Fiberglass (Greater Than 4lb/ft3)	4	
Expanded Polystyrene (Beadboard)	4	
Extruded Polystyrene	5	
Polyurethane (Foamed-in-Place)	6.25	
Polyisocyanurate (Foil-Faced)	7.2	
Concrete Block 4"		0.8
Concrete Block 8"		1.11
Concrete Block 12"		1.28
Brick 4" Common		0.8
Brick 4" Face		0.44
Poured Concrete	0.08	
Soft Wood Lumber	1.25	
2" Nominal ($1\frac{1}{2}$")		1.88
2 × 4 ($3\frac{1}{2}$")		4.38
2 × 6 ($5\frac{1}{2}$")		6.88

FIGURE 18-9: R values for many common building materials.

Material	R/Inch	R/Thickness
Cedar Logs and Lumber	1.33	
Plywood	1.25	
$\frac{1}{4}$"		0.31
$\frac{3}{8}$"		0.47
$\frac{1}{2}$"		
$\frac{25}{32}$"		2.06
Fiberglass ($\frac{3}{4}$")		3
(1")		4
($1\frac{1}{2}$")		6
Extruded Polystyrene ($\frac{3}{4}$")		3.75
(1")		5
($1\frac{1}{2}$")		7.5
Foil-faced Polyisocyanurate ($\frac{3}{4}$") ($\frac{3}{4}$")		5.4
(1")		7.2
($1\frac{1}{2}$")		10.8
Hardboard ($\frac{1}{2}$")		0.34
Plywood ($\frac{5}{8}$")		0.77
($\frac{3}{4}$")		0.93
Wood Bevel Lapped		0.8
Aluminum, Steel, Vinyl (Hollow Backed)		0.61
(w/$\frac{1}{2}$" Insulating Board)		1.8
Brick 4"		0.44
Gypsum Board (Drywall $\frac{1}{2}$")		0.45
($\frac{5}{8}$")		0.56
Paneling ($\frac{3}{8}$")		0.47
Plywood	1.25	
($\frac{3}{4}$")		0.93
Particle Board (underlayment)	1.31	
($\frac{5}{8}$")		0.82

FIGURE 18-9: (*Continued*)

Material	R/Inch	R/Thickness
Hardwood Flooring	0.91	
$(\frac{3}{4}")$		0.68
Tile, Linoleum		0.05
Carpet (Fibrous Pad)		2.08
(Rubber Pad)		1.23
Asphalt Shingles		0.44
Wood Shingles		0.97
Single Glass		0.91
w/storm		2
Double Insulating Glass ($\frac{3}{16}"$ Air Space)		1.61
($\frac{1}{4}"$ Air Space)		1.69
($\frac{1}{2}"$ Air Space)		2.04
($\frac{3}{4}"$ Air Space)		2.38
($\frac{1}{2}"$ w/Low-E 0.20)		3.13
(w/Suspended Film)		2.77
(w/2 Suspended Films)		3.85
(w/Suspended Film and Low-E)		4.05
Triple Insulating Glass ($\frac{1}{4}"$ Air Spaces)		2.56
($\frac{1}{2}"$ Air Spaces)		3.23
Addition for Tight-Fitting Drapes or Shades, or Closed Blinds		0.29
Wood Hollow-Core Flush ($1\frac{3}{4}"$)		2.17
Solid-Core Flush ($1\frac{3}{4}"$)		3.03
Solid-Core Flush ($2\frac{1}{4}"$)		3.7
Panel Door w/$\frac{7}{16}"$ Panels		1.85
($1\frac{3}{4}"$)		
Storm Door (wood 50% glass)		1.25
(Metal)		1
Metal Insulating (2" w/Urethane)		15

FIGURE 18-9: (*Continued*)

Also, the framing member used in the construction of a wall will reduce the wall overall R value or its effective total R value. This is because the thermal resistance of wood is less than the thermal resistance of insulating material in which the framing members have displaced. It is possible to get a more accurate total resistance of a wall by calculating the R value of the wall between the framing members as well as the R value of the wall at the framing members. The resulting R values are then weighted according to the percentage of the solid framing in the wall assembly. The formula for calculating the effective R value is as follows:

$$R_{effective} = (R_i)\,(R_f)/p(R_i - R_f) + R_f$$

Where

$R_{effective}$ = Effective R value of panel (°F × ft² × hr/Btu)
P = Percentage of panel occupied by framing (decimal %)
R_f = R value of panel at framing (°F × ft² × hr/Btu)
R_i = R value of panel at insulation cavities (°F × ft² × hr/Btu)

For example, using the construction information provided below, determine the heat loss of room #1 shown in the structure illustrated in Figures 18-10 and 18-11.

Note for this example, assume the following:

R value for all windows will be R-3.0.
R value for all exterior doors will be R-5.0.
R value for the foundation edge will be R-11.
Outdoor design temperature will be 5°F.
Indoor design temperature will be 68°F.

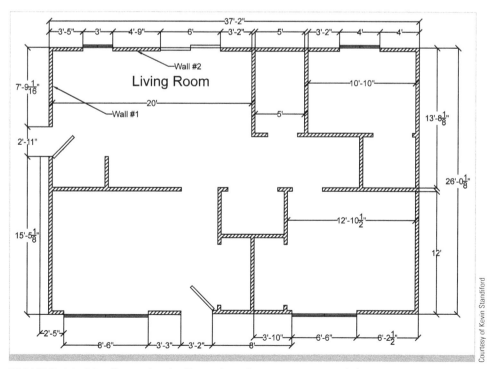

FIGURE 18-10: Example of a floor plan of a structure containing a room to calculate the total heat of transmission.

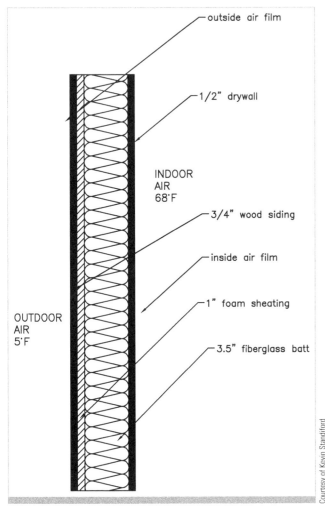

outside air film

1/2" drywall

INDOOR
AIR
68°F

3/4" wood siding

inside air film

1" foam sheating

OUTDOOR
AIR
5°F

3.5" fiberglass batt

FIGURE 18-11: Typical wall section for example.

Window height: 4'
Door Height: 6' 8"
Wall height will be 8'.

Step 1: Using the table shown in Figure 18-09 the R values of the walls can be determined as follows:

Material	R value between framing	R value at framing
Inside air film	0.68	0.68
$\frac{1}{2}$-inch drywall	0.45	0.45
Stud cavity	21	5.5
3.5" Fiberglass Batt R12	12	0
1" Foam sheeting	5	5
$\frac{3}{4}$" Wood siding	2.06	2.06
Outside air film	0.17	0.17
Total	41.36	13.86

Step 2: Calculate the effective overall resistance of the wall:

Assuming 15 percent of wall is solid framing, the effective total R value of the wall is:

$$R_{effective} = (R_i)\ (R_f)/p(R_i - R_f) + R_f$$
$$R_{effective} = (41.36 \times 13.86)/0.15(41.36 - 13.86) + 13.86$$
$$R_{effective} = 31.873$$

Step 3: Determine the area of the walls and windows in room #1:

Room #1 Calculation:

Wall One Area

   7' − 9 1/16" × 8' = 62.0417 ft²

Wall Two Area

   20' − 0" × 8 = 160 ft²

Window One Area

   3 × 4 = 12 ft²

Door Area One

   6' 0" × 6' 8" = 40 ft²

Subotal Wall

   62.0417 ft² + 160 ft² = 222.0417 ft²

Total Wall (Subotal Wall − Windows − Doors)

   222.0417 ft² − 12 ft² − 40 ft² = 170.0417 ft²

Step 4: Calculate the heat of transmission:

Heat of transmission for walls with outside exposure:

$q = A \times U \times TD$

$q = 170.0417\,\text{ft}^2 \times (1/31.873)(68°F - 5°F)$

$q = 336.1035\ \text{Btu/hr}$

Heat of transmission for window:

$q = A \times U \times TD$

$q = 12\ \text{ft}^2 \times (1/3) \times (68°F - 5°F)$

$q = 252\ \text{Btu/hr}$

Heat of Transmission for door.

$q = A \times U \times TD$

$q = 40\ \text{ft}^2 \times (1/3) \times (68°F - 5°F)$

$q = 504\ \text{Btu/hr}$

The windows and the doors in this particular area of the house are allowing the majority of the room's heat to escape. Increasing the overall R value of these features will reduce the amount to heat that is allowed to escape and, thus, lower the home owner's utility bill. If the windows were replaced with a more efficient window with an R value of 4.05, then the amount of heat that will be permitted to escape from the windows would be reduced.

$q = A \times U \times TD$

$q = 12\ \text{ft}^2 \times (1/4.05) \times (68°F - 5°F)$

$q = 186.6666\ \text{Btu/hr}$

Roof and Ceiling Heat Loss

Heat is also lost through the top of a building by conduction. In a structure, the ceiling may either have a vented attic above or be part of the roof. Either way, the ceiling will transmit heat to the cooler surroundings in the winter months and heat into the structure in the summer months. The building materials used will not have much resistance to heat transfer. Therefore, the key to stopping heat from being transferred into and out of the structure will be the insulation used in this area.

Heat Loss through Floors

In many areas, houses are located over crawlspaces or basements. If the crawlspace is not vented, it is treated like an unheated basement. If the crawlspace is vented and the vents are closed in the winter months, the space is also treated as an unheated basement. In other words, the outside design temperature will be a lot less than an area exposed to the outside. The temperature difference can be calculated the same way as for a garage; use an average between the inside and outside temperatures. When the structure is located on a slab that is positioned on the ground, heat is permitted to exit the structure from the slab's perimeter as well as through the bottom of the. However, most of the heat that leaves a structure through a slab does so through the perimeter. Therefore, in colder climates insulating the perimeter will help reduce the heat transfer. In warmer climates, the perimeter is not insulated.

Heat Loss through Windows and Doors

Typically, heat loss through windows and doors makes up the majority of the heat that escapes from a structure. This is because these opening are connected to the outside using tracks and hinges. These openings are extremely difficult to effectively seal because they are designed to be opened.

The basic window has a single glaze, commonly called the pane. These windows offer very little resistance to the transmission of heat. Double-glazed windows use two pieces of glass separated by a dead vapor space. The vapor in the space is filled with an inert gas, such as argon. Triple-glazed windows are also available. When the heat loss for windows is calculated, the U factor is based on the rough window opening, including the frame.

Air infiltration

When air enters a structure as the result of prevailing winds and breezes, it is referred to as **infiltration**. When the air from a breeze impacts one side of a structure, it creates a slightly higher pressure on that side and a slightly lower pressure on the other side. The result is air passing through opening in the structure. In other words, air enters from one side and exits through the other. Some common places in a structure in which air can enter and/or escape a structure are:

- Around the windows and the doors
- Around plumbing fixtures where they enter the structure, if the fixtures are above grade
- Where electrical wires penetrate the structure

Calculating the infiltration load on a structure is probably the most difficult part of a load calculation because there is no way of knowing how much air will move through the structure without an actual test. Typically, infiltration is determined using one of two methods: a blower door test or estimating the infiltration through cracks around windows and doors. Modern windows and doors are tested by blowing wind of a specific velocity across them. The typical test calls for a wind velocity of 25 miles per hour (mph) across the window or door and rates the window or door according to how much air leaks through per foot of crack. The best window typically leaks 0.25 cubic foot of air per minute per foot of crack.

Weatherization Sealants

As you may have seen in the "Plumbing" and "Surface Treatments" chapters, caulks and sealants are used in a wide variety situations for a wide range of uses. When selecting a weatherization sealant, there are many factors to take into consideration. These include application location, physical requirements, appearance, and service life. Also, when selecting a sealant for weatherization purposes, consult with the American Society for Testing and Materials (ASTM) for standards and for the expected performance various construction applications.

For step-by-step instructions about applying sealant outdoors, see Procedure 18-1.

Ventilation

In many climates, air must be conditioned for people to be comfortable. One way to condition air is to use a fan to move the air over the conditioning equipment. This conditioning equipment can be a cooling coil, a heating device, a device to add humidity, or a device to clean the air. Because a fan is used to move the air, the system is a forced-air system. The forced-air system reuses the same room air many times. Air from the room enters the system and is conditioned and returned to the room. Fresh air enters the structure either by infiltration around the windows and doors or by ventilation from a fresh-air inlet connected to the outside. The forced-air system is different from a natural-draft system, in which the air passes naturally over the conditioning equipment. **Ventilation** is planned induction of outside air to create a slight positive pressure on the structure. When more air is induced into the house than will leak in, the air will leak out the cracks where infiltration would occur. When the ventilation is known, there is no infiltration calculation.

As mentioned earlier, in many climates the air must be conditioned before it can be released into the structure. Failure to condition the air could raise or lower the overall temperature of the structure, raise or lower the humidity level of the structure, or increase the amount of pollutants (pollen and particulates) in the structure (thus making the structure unhealthy).

However if the climate permits natural ventilation can help reduce the structure's electrical requirements by allowing air to enter the structure by a natural means.

Natural ventilation can:

- Reduce operating costs
- Reduce construction costs
- Reduce the building's energy usage
- Potentially increase the productivity of the occupants

Basically, there are three types of natural ventilation. They are:

- Ventilation caused by air entering into the structure
- Ventilation caused by temperature differences (warmer air inside and cooler air outside can cause air to exit through ceiling vents)
- Humidity differences; this form of natural ventilation is only effective when the outdoor humidity is very low

Finishes and Preservation Requirements

When performing weatherization on a structure that is managed by the federal government and considered to be historic, the facility maintenance technician must take into consideration the characteristics of the windows before making any alterations that could affect the overall appearance. In fact, according to the secretary of the interior, *windows that make a significant contribution cannot be destroyed*. In other words, the appearance of the facility cannot be altered from its original state.

In most historic structures, metal windows were used. These windows are generally not energy efficient; however, they can be made more energy efficient. Applying caulking around the masonry openings as well as adding weather stripping will reduce the air infiltration around the windows. Other weatherization treatments that can be applied to historic windows are: adding layers of glazing over the windows, installing operable storm windows, and replacing existing glass with thermal glass.

Building Shell Measurements

The purpose of the building's shell is to protect the occupants from the extreme conditions of the outdoors (temperature swings and moisture content). However, as stated earlier, most buildings lose energy through cracks and leaks. The air tightness of a structure can be tested using a blower door. A blower door is a diagnostic tool designed to test the air tightness of a structure and assist in locating its leaks. Blower door test kits vary greatly in design and price.

A typical blower door testing unit is composed of three basic components: a calibration fan, a door-panel system, and an instrument used to measure the building pressure and the fan flow. Using the door-panel system, the blower door system is temporarily located in an exterior doorway. Once secure, a small air pressure difference is created between the interior and the exterior of the structure by either pumping air into or out of the structure. This air pressure difference forces air through all cracks and penetrations in the building's shell. The tighter a structure, the less air less air that is required to create and maintain the difference in air pressure.

Before you conduct a blower door test, you should check the following items:

- Ensure that all exterior windows, doors, storm windows, and storm doors are closed.
- Ensure that all interior doors are open.
- Disable all nonelectric heating equipment.
- Shut all wood-stove dampers and fireplace dampers.
- Turn off the clothes dryer and all bathroom and kitchen exhaust fans.
- Fill any plumbing traps that you suspect to be dry.

Base-Load Measurements

Determining the base load energy consumption of a structure can be a tricky at best. The **base-load energy consumption** of a structure is the amount of energy consumed by a structure and its occupants under the best possible conditions. In other words, it is the amount of energy required to operate the structure when neither the heating nor the air-conditioning systems are in use. The months in which the heating and cooling systems are not used are referred to as the **shoulder months**. Typically, these months are in the spring and fall.

To calculate the base load of the structure, take several years of energy bills (electricity and gas—if the structure uses gas). Next, take the three lowest months' bills and average them, for each year. For example, if the three lowest bills are 113.35, 115.00, and 116.52, then the base load is determined as follows:

	Electric (All Electric)						
	Year One	Year Two	Year Three	Year Four	Year Five	Year Six	Year Seven
Month 1	$125.00	$124.32	$113.35	$136.63	$125.85	$119.85	$120.00
Month 2	$115.00	$116.52	$132.52	$125.63	$125.74	$128.65	$135.25
Month 3	$123.35	$115.00	$125.10	$134.23	$119.52	$124.56	$135.52
Average	$121.12	$118.61	$123.66	$132.16	$123.70	$124.35	$130.26
						Base load	$124.84

To calculate the amount of money spent on the heating and cooling for the structure, simply multiple the base load by 12 and subtract it from the total electric bill for one year.

$2,719.90   Total electric one year

$1,508.27   Total base load one year

Amount spent on heating and cooling for one

$1,211.63   year

For example, when determining the base load as check for the following conditions that could affect the base load of the structure:

- Measure the refrigerator's electrical consumption as well as the temperature setting for the fridge and the freezer. The freezer should be set between 0°F and 5°F, while the refrigerator should be set between 36°F and 40°F.

- Determine if any lighting is continuously lit.

- Check to ensure at the settings on the domestic water heater is set at 120°F.

- Check for sinks, water lines, laborites, water lines, and tubs for leaks that could cause an excessive base load.

- Check for large energy users (hot tubs, fish tanks, freezers, etc.)

Electricity	
January	$211.71
February	$210.83
March	$116.69
April	$118.72
May	$149.38
June	$287.45
July	$287.45
August	$305.00
September	$305.00
October	$287.45
November	$195.58
December	$244.64
Total	$2,719.90

Energy Management and Monitoring

Energy management is the process of monitoring and controlling, and ultimately conserving, energy in a structure. As mentioned earlier, because of the increasing cost of energy, as well as the effects on the environment, energy conservation is an important aspect of facility maintenance and typically involves the following steps:

1. Monitor the energy consumption by collecting the data.

2. Search for opportunities to increase energy efficiency, save energy, and reduce the overall footprint left on the environment. Analyze the data collected from

the monitoring phase; problem areas can be detected and recommendations developed to address the problems detected.

3. Resolve problem areas and track progress.

One tool that is available to monitor energy consumption is the Google PowerMeter. The Google PowerMeter is a web application that allows the facility maintenance technician to see energy consumption broken down by appliance.

SUMMARY

- The sources of infiltration in a structure are the areas around windows and doors, and where plumbing fixtures and electrical wires penetrate the structure.

- The base load energy consumption of a structure is the amount of energy consumed by a structure and its occupants under the best possible conditions.

- Natural ventilation occurs as a result of air entering into the structure, temperature differences (warmer air inside and cooler air outside can cause air to exit through ceiling vents), and humidity differences (only effective when the outdoor humidity is very low).

- The second law of thermodynamics states that heat naturally moves from hot to cold and never in the opposite direction.

- The cost of energy for any residential dwelling is based upon the dwelling's geographic location, the square footage, they occupants' lifestyle, the building's construction and design, and the age and condition of the appliances used in the dwelling.

- Average residential consumer energy usage can be summarized as follows:
  - Heating accounts for 29%.
  - Cooling accounts for 17%.
  - Appliances account for 13%.
  - Hot water generation accounts for 14%.
  - Lighting, Electronics, and Other accounts for 28%.

- OSHA has identified potential hazards for weatherization workers pertaining to exposure, chemicals, and fire.

- When sizing HVAC equipment for a house, consider the local weather conditions; the size, shape, and orientation of the structure; insulation levels; square footage of the windows and exterior doors; air infiltration rates;, the number and ages of occupants and their comfort preferences; and the types and efficiencies of lights and major home appliances.

- An energy audit is a survey conducted of a residence or a structure to analyze the current energy requirements of the structure and its occupants as well as how that energy flow can be reduced without having a negative effect on the structure and/or occupants.

- A visual inspection of the facility is not the only process required in performing the audit, but is important.

- The principle known as "house as a system" states that, although the home might be made up of individual components, these components come together as a whole.

- The most common method of performing a heating/cooling load calculation uses the ACCA's *Manual J. Manual J* is a comprehensive approach that ensures that environmental systems are efficient and safe and promotes healthy environments.

- One should consult with the ASTM when selecting a weatherization sealant.

- When inspecting the exterior of a structure, inspect the characteristics and condition of the foundation, roof, siding, windows, doors, and overhangs. Inspect the foundation, the site drainage, and any additions to the dwelling; identify thermal flaws, shading from trees, awnings, and other buildings, the home's orientation, and the effect of solar heat gain; and inspect the chimney(s) and exhaust vents.

- When inspecting the interior of a structure, locate, identify, and evaluate the components of the thermal boundary.

PROCEDURE 18-1

Applying Sealant Outdoors

When applying a sealant in an outdoor situation, the following guideline should be followed:

1. The surface should be clear and clean from dirt and debris, oils, and grease as well as old caulk and sealants.

2. When filling a gap greater than $\frac{1}{2}$ inches wide and/or deeper than $\frac{1}{4}$ inch, use a backer rod or bond breaker tape

3. Apply sealants only when the surface is above freezing and less than 120°F.

4. Do not apply a sealant if snow and/or rain is expected.

5. A cured sealant bead should be between $\frac{1}{8}$ inch and $\frac{1}{2}$ inch wide.

6. One tube of sealant should produce a bead 56 feet long and $\frac{3}{16}$ inch wide.

7. Always consult the manufacturer for curing times.

REVIEW QUESTIONS

1   List the sources of infiltration in a structure.

2   List the factors that influence a base load.

3   What are the types of natural ventilation?

4   List the areas of an HVAC system that should be checked to air leaks.

5   What are the first and second laws of thermodynamic?

6   What the factors on which the cost of energy for a residential structure is based?

REVIEW QUESTIONS

7   Discuss the major areas in a structure in which energy is used.

8   What are the potential hazards that OSHA has identified for weatherization workers?

9   Explain the concept of "the house as a system" and how it is important to weatherization.

10   What are the characteristics of an energy-efficient home?

Name: _____

Date: _____

Load Calculations

Upon completion of this job sheet, you should be able perform a
load calculation.

PROCEDURE

A wall in Oklahoma City, Oklahoma, has a U value of 0.046 and an area of 400 square
feet. What is the estimated heat loss for this wall if the 97.5% value is used and the
indoor design temperature is 70°F?

INSTRUCTOR'S RESPONSE

_____

_____

_____

Name: _____

Date: _____

2

Load Calculations

Upon completion of this job sheet, you should be able perform a
load calculation.

PROCEDURE

A wall has a U value of 0.51, and the owner wants a U value of 0.09. How many
inches of expanded polystyrene, which is extruded and has a smooth surface, must
be added?

INSTRUCTOR'S RESPONSE

_____

_____

_____

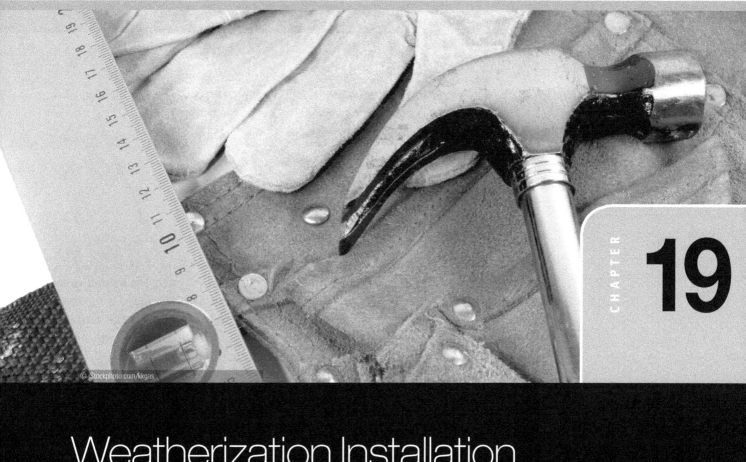

© iStockphoto.com/kkgas

CHAPTER 19

Weatherization Installation, Maintenance, and Repair

OBJECTIVES

By the end of this chapter, you will be able to:

KNOWLEDGE-BASED

- Discuss the factors involved in selecting weather stripping.
- List the various types of insulation used in residential construction.

SKILL-BASED

- Properly install a water heater blanket.

- Properly insulate gas water heaters.
- Properly insulate switches and outlets.
- Properly apply weather stripping.
- Properly insulate an unventilated crawlspace.
- Properly insulate a ventilated crawlspace.

Crawlspaces a area under a house that provides a means of raising the structure off the ground

Unconditioned space design space that is not air conditioned

Mold a plant that is a member of the fungi family

Introduction

Once you have a basic understanding of electricity and mechanical systems (plumbing and HVAC), determining and correcting situations in which energy is wasted can become easier. In addition, these skills will help you in determining if a recommended energy conservation method will deliver the expected results and the desired return on investment. Having a good understanding of thermodynamics will help the technician decide what R value of insulation will better ensure the comfort level of the occupants. However, selecting the amount of insulation to use in a structure is only the first step. The type of insulation must also be determined for the application and area in which it is intended to be used. In addition to insulation, leak detection and weather stripping will go a long way in helping lower utility bills.

Air Leaks and Barriers

As mentioned earlier, insulation is one of the single most important weatherization techniques that a facility maintenance technician can perform to help lower the facility's utility bills and conserve energy. However, other techniques such as leak detection and barrier installation are extremely effective and will have a major impact on the structure's energy consumption.

A drafty home is not a healthy dwelling. When unconditioned air enters into the structure and bypasses the HVAC equipment, this allows pollution and humidity to enter the structure. Controlling airflow in to the structure and controlling airflow from room to room are both very important. One way to control airflow in a structure is to incorporate an air barrier. An air barrier is a material that prevents air from an unconditioned space from entering into the conditioned space. For an air barrier to be effective, it must:

- Resist air movement
- Completely and continuously surround the building envelope
- Be strong, durable, and easy to install

A common material used for air barriers is gypsum board, also known as drywall. When drywall seams are properly taped, the material becomes air tight (Figure 19-1). To achieve airtight barriers using drywall, a continuous bead of sealant must be applied to the all

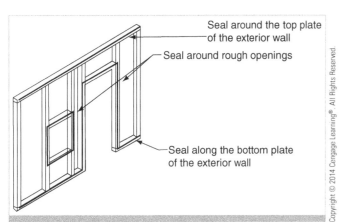

Seal around the top plate of the exterior wall

Seal around rough openings

Seal along the bottom plate of the exterior wall

FIGURE 19-1: Sealing drywall.

exterior wall top and bottom plates, all top plates at insulated ceilings, penetrations and rough openings, and both sides of the first interior stud of a partition wall.

Weather Stripping

One way to seal air leaks is to use weather stripping around windows and doors. Determining the amount of weather stripping needed to seal around a window(s) and/or door(s) is accomplished by adding the perimeters of all the windows and doors together, plus 10% for waste. For example, determine the amount of weather stripping needed to seal four windows that have the dimensions listed below:

Window/Door	Location	Left Jamb Length Inches	Right Jamb Length Inches	Header Length Inches	Sill Length Inches
1	Living Room	45	45	24	24
2	Living Room	45	45	24	24
3	Living Room	65	65	36	36
4	Living Room	85	85	48	48

Step 1: Determine the perimeter of each window:

The perimeter is determined by adding the "header length" + "sill length" + ""right jamb length" + "left jamb length."

Window #1

   45" + 45" + 24" + 24" = 138"

Window #2

   45" + 45" + 24" + 24" = 138"

Window #3

   65" + 65" + 36" + 36" = 202"

Window #4

   85" + 85" + 48" + 48" = 266"

Step 2: Determine the extra 10%:

Window #1

   138" × 0.10 = 13.8"

Window #2

   138" × 0.10 = 13.8"

Window #3

   202" × 0.10 = 20.2"

Window #4

   266" × 0.10 = 26.6"

Step 3: Determine the total inches needed for each window:

Window #1

   138" + 13.8" = 151.8"

Window #2

138" + 13.8" = 151.8"

Window #3

202" + 20.2" = 222.2"

Window #4

266" + 26.6" = 292.6"

Step 4: Determine the grand total of weather stripping needed:

Window #1 + Window #2 + Window #3 + Window #4

151.8" + 151.8" + 222.2" + 292.6" = 818.4"

Step 5: Determine the total amount of weather stripping needed in feet:

818"/12 = 68.2 ft

The following form can be used to determine the perimeter of windows and doors in a structure:

Window/ Door	Location	Left Jamb Length Inches	Right Jamb Length Inches	Header Length Inches	Sill Length Inches	Total Perimeter Inches (L Jamb + R Jamb + Header + Sill)	Extra 10% Inches (Total Perimeter *0.10)	Total Inches (Total Perimeter + Extra)
							Total Inches	
							Total Feet	

For example, using the previous example:

Window/ Door	Location	Left Jamb Length Inches	Right Jamb Length Inches	Header Length Inches	Sill Length Inches	Total Perimeter Inches (L Jamb + R Jamb + Header + Sill)	Extra 10% Inches (Total Perimeter *0.10)	Total Inches (Total Perimeter + Extra)
Window #1	Living Room	24	24	45	45	138	13.8	151.8
Window #2	Living Room	24	24	45	45	138	13.8	151.8
Window #3	Living Room	36	36	65	65	202	20.2	222.2
Window #4	Living Room	48	48	85	85	266	26.6	292.6
							Total Inches	818.4
							Total Feet	24.38333333

When selecting a type of weather stripping, also take into consideration the following factors and select a type of weather stripping accordingly:

Temperature differences and changes

- Friction
- Weather
- Durability
- Overall wear and tear
- Cost

Crawlspaces

In residential construction, homes built using a crawlspace have several advantages over homes that were built having a basement or those built on a slab. **Crawlspaces** provide a means of raising the structure off the ground in areas where termites and surface water pose a problem. Crawlspaces also provide a location for plumbing, ducts, and other mechanical systems, so they can be easily serviced. If not properly insulated and ventilated, crawlspaces can be a source of mold and undesirable heat transfer.

Because moisture can damage insulation as well as structural components, crawlspaces must be properly ventilated and a drainage system installed (Figure 19-2). The drainage system usually consists of a trench, perforated pipe, and gravel going around the perimeter of the structure.

FIGURE 19-2: Crawlspace.

Attics

Of all the different locations in a home, the attic is typically the easiest area to check the amount of insulation. If necessary, add additional insulation to improve the comfort level, as well as the overall efficiency, of the dwelling (Figure 19-3). As a general rule of thumb, if you can see the joist, then you should add more insulation. The attic is well insulated if the insulation covers the tops of the joist and is evenly spaced. For most attics, it is recommended that you insulate to a depth of 10–14 inches. This should be equivalent of using R-38 insulation. However, this is dependent on the type of insulation that is to be used. Insulation can be mixed; therefore, when applying additional insulation to an attic, it is not necessary to use the same exact type as the existing insulation. In other words, it is a common practice to add loose fill insulation over batt insulation. The recommended level for most attics is to insulate to R-38, or about 10 to 14 inches, depending on the type of insulation. When adding additional insulation, it is not necessary to use the same type of insulation that is currently used in the attic; loose-fill insulation can be added on top of fiberglass batts or blankets.

FIGURE 19-3: Insulating the attic.

Insulating the Attic

When insulating an attic, start by plugging all the holes in the attic. Because the attic is typically such a large area, a greater return on investment will be realized when the larger holes and openings are contained.

Step 1: Cut 16-inch long pieces fiberglass insulation and place them into any opening of stud cavities.

Step 2: Sealing cover interior dropped soffits:

a. Remove the insulation from the dropped soffit.

b. Using reflective foil backed insulation, cut a length of reflective foil slightly greater in length then the opening being covered.

c. Go around the opening with a bead of caulk. Finally, apply the insulation to the opening attaching it in place using staples.

Step 3: Sealing behind kneewalls:

a. Cut 24-inch long pieces of fiberglass insulation and place into any joist space.

Step 4: Sealing furnaces flues:

a. Before attempting to insulate around a flue pipe, check your area local building code as well as the equipment manufacturer's instructions. As a general rule of thumb, always allow for at least 1 inch of clearance between the flue and any material that is combustible. In most locations, this is a building code requirement. Always check the local building codes before applying insulation.

b. Cut and fit an aluminum flashing to encapsulate the furnace's flue (two pieces that overlap by 3 inches in the center (Figure 19-4).

c. Use high-temperature caulk around the seams of the flashing as well as the edges.

d. Press the flashing into the caulk, stapling, or nailing it into place.

e. The gap between the flashing and the flue should be sealed using high-temperature caulk.

f. Using the same aluminum used for the flashing, cut out insulation dam with 1-inch slots cut into the top of the flashing and 2-inch slots cut into the bottom of the flashing.

g. Secure the insulation dam to the flashing; afterward insulate up to the insulation dam.

Vented soffits should be kept clear of insulation to allow for proper ventilation of the attic and roof.

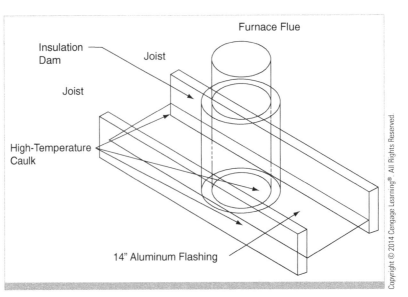

FIGURE 19-4: Aluminum flashing around a furnace flue.

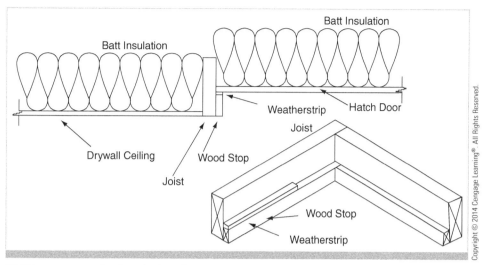

FIGURE 19-5: Coving the hatch door with batt or foam insulation.

Step 5: Sealing the attic hatch door:
   a. Apply a self-adhesive foam weather stripping to the attic entrance.
   b. Cover the hatch door with batt or foam insulation (Figure 19-5).

Duct and Piping Systems

Regardless of the type of climate in which a structure is located, the air inside the structure must undergo some type of conditioning before the occupants of the structure will be comfortable and safe. Conditioning air maybe nothing more than moving the air over an air-conditioning coil for the purpose of cooling the air or modifying the humidity (either adding or removing) or simply cleaning the air using filtration; in most of these systems, a fan is used to move the air. A system that uses a fan to move air is referred to as a "forced-air system." These systems reuse the contained within the structure several times by pulling the air from the rooms applying some form of conditioning and returning it back to the room. Fresh air is permitted to enter the structure either by infiltration around the windows and doors or by ventilation from a fresh-air inlet connected to the outside.

As stated earlier, the forced-air system uses a fan to force air through the air-conditioning system. This is different from a natural draft system, which relies on air to pass over the conditioning equipment naturally. For example, in a heating system the air passes over the heating pipes. This causes the air to expand and rise, thus causing a vacuum effect and pulling in cooler air to replace it.

The Forced-Air System

The forced-air system is composed of the following components:

• An air handler (this includes the blower—fan)

• A heat source (gas, oil, electric)

• A means of delivering the air to the heating unit (return)

• A means of delivering the air back to the rooms (ductwork and registers)

- A thermostat
- Filtration system

When these components are correctly selected and properly maintained, the occupants will not feel any air movement, nor will they feel any temperature swings or hear any system noise. In short, the occupant will not be aware that the system is either on or off.

The Supply Air Duct System

The conditioned air is transported to the space in which it is needed via supply air ducts, registers, and diffusers. The supply air duct must be designed to permit air to move freely throughout the duct system without the duct being oversized (Figure 19-6). The types of duct systems commonly used in residential structures are:

- Plenum
- Extended plenum
- Reducing extended plenum
- Perimeter loop

The Plenum System

The plenum systems are perfect for applications in which the rooms are close to the unit that services them. This type of system typically easy to install and has a low installation cost. Normally, these systems have supply lines that are located on interior walls.

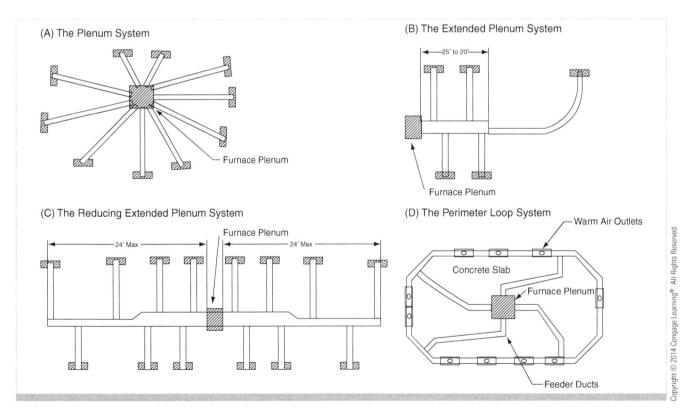

FIGURE 19-6: Supply air duct systems: (A) plenum duct system, (B) extended plenum system, (C) reducing extended plenum system, (D) perimeter loop system with feeder and loop ducts in concrete slab.

The Extended Plenum System

The extended plenum system is the most common duct system used in residential construction today. In this type of system, a large main supply duct is connected to the supply plenum. This acts as an extension to the furnaces supply plenum, thus allowing small ducts, called "branches" to connect to the trunk duct and thus supply air to the individual rooms. The trunk duct can be round, square, or rectangular. These systems typically restrict the main supply trunk to 24 feet in length. A length longer than 24 feet tends to allow pressure to build in the end of the duct, thus allowing too much airflow into the room. This also results in the rooms closer to the equipment having insufficient airflow.

The Reducing Extended Plenum System

Another approach to keeping the systems balanced is to use a reducing extended plenum system. This system reduces the trunk duct size as branch ducts are added. It has an advantage of saving materials and maintaining the same pressure from one end of the duct system to the other. Each branch duct has approximately the same pressure pushing air into its takeoff from the trunk duct.

The Perimeter Loop System

The perimeter loop system is particularly well suited for structures that incorporate a concrete slab construction in colder climates. Because this system is run under the slab, the slab is kept at a more even temperature. The loop has a constant pressure around the system and provides the same pressure to all outlets.

Duct Insulation

HVAC design space that is not air conditioned is referred to as **unconditioned space**. When a heating and air-conditioning duct passes through unconditioned space, the result is that in the summer time heat is transferred from the surrounding space into the conditioned air, thus reducing the effectiveness and the output air entering the room (Figure 19-7). In the wintertime this is reversed, that is, heat is transferred from the duct into the unconditioned space. Again, this lowers the effectiveness of the conditioned air entering into the room. In either case, when the effectiveness of the supply air is decreased, the result is the same: an increase in the running time of the HVAC equipment. If there is 15°F temperature difference between the surrounding air of the unconditioned space and the supply air, then it is necessary to insulate the duct.

FIGURE 19-7: HVAC duct passing through unconditioned space.

If fiberglass ducts are used in the structure, then the insulation is built into the duct. However, if the structure employs metal ducts, then insulation must be applied to the ducts. For metal ducts, this is accomplished using one of two different techniques—insulation is either applied to the outside or the inside of the duct.

FIGURE 19-8: Insulating air-conditioning ducts.

The easier of the two methods is applying insulation to the outside of the duct. This can be done at any time during the process of installing the system or afterward. When insulation is applied to the outside of an air conditioning duct, a foil- or vinyl-backed insulation is used (see Figure 19-8). This insulation is available in several different thicknesses depending upon the climate and the situation. The backing acts as a moisture barrier between the air duct and the unconditioned space. The insulation is joined by lapping and stapling, and then taped to prevent moisture from entering the seams. When the insulation is to be applied to the inside of the duct, this is done during the fabrication of the duct and the insulation is either glued or fastened to tabs mounted on the duct by spot welding or glue.

HVAC Duct and Air Leaks

Having a well-insulated duct system is only half of the story; if the duct system is not airtight, then it can waste energy as well as become a potential health hazard. Poorly sealed HVAC duct can increase the homeowners' energy cost between 10 and 30 percent. Poorly sealed ductwork is a potential health hazard because it can draw unconditioned and unfiltered air into the duct system from the attic, basement, and crawlspace areas. This air could be filled with toxic chemicals, pollen, and mold.

If there is any question of the air leaking out, special tape can be applied to the connections to ensure that no air will leak. However, the best material to use to seal ductwork is a thick white paste known as "mastic." Mastic is preferable because it can be applied relatively easily with a paintbrush and dries to a hard but pliable finish.

In the duct system, every seam must be airtight (Figure 19-9). In an existing structure, this can be difficult to verify and correct. The following areas of a HVAC duct system should be inspected:

FIGURE 19-9: Sealing an air-conditioning duct.

1. All seams in the air handler

2. Both the supply plenums and the return plenums

3. The main trunk lines

4. Ducts that are formed using materials such as wood

5. Joints between sections of the branch

6. All connections to the registers and grills

Mold and Indoor Air Quality

The quality of the air in a structure is an important aspect of HVAC. If the air is too dry, then the environment can be a health hazard. The same holds true for an environment in which the humidity is too high. When the humidity is too high, there is a possibility that dangerous mold will be produced.

Mold is plant that is a member of the fungi family (Figure 19-10). It is one of the catalysts responsible for the breakdown of organic matter. However, unlike most plants, mold products spores instead of seeds. These spores become airborne particles that, when inhaled, can produce adverse health effects.

There are more than 100,000 species of mold that can be currently found growing in a variety of different materials (food, soil, etc,). However, according to the United States Centers for Disease Control, in the United States the most common species of mold are *cladosporium*, *penicillium*, *aspergillus*, and *alteraria*. Of these species, *aspergillus* and *penicillium* are considered toxin-producing species. One of the causes of the sick building syndrome is considered to be the toxin-producing mold *Stachybotrys chartarum*.

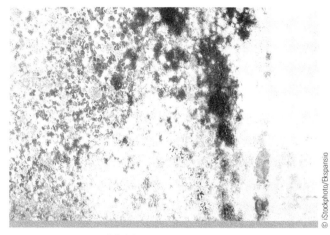

FIGURE 19-10: Mold.

Replacing the Filters on an HVAC System

There are a number of thing that a technician can do to help keep an air-conditioning system operating efficiently. However, making sure that the air filters are replaced on a regular schedule is one of the most important. Failure to replace the air filters on regular basis can result in unfiltered air entering the air supply and being transported into other area of the structure. In addition, restricted airflow will affect the efficiency of the equipment. When replacing the filters on a system, you should these follow guidelines should be followed:

- Inspect the channel that holds the filters to be sure that the channels are in good shape and that the filter is supported on at least two sides (see Figure 19-11).

- Make certain that the replacement filter is the same size as the filter channel, not necessarily the size of the filter that came out of the unit.

- Install the filter in the channel with the arrow on the filter pointing in the direction of airflow, which is toward the blower.

- Mark the edge of the filter with the date and your initials.

- Once the filter has been installed, inspect the filter and the channel to be certain that no air is able to bypass the filter.

- Seal any and all air leaks to prevent air from bypassing the filter.

FIGURE 19-11: Inspecting a filter.

Air Duct and Air-Handling Equipment Cleaning and Maintenance

When a structure becomes so badly polluted that an air filtration system does not properly work, often a thorough cleaning of the duct is necessary to correct the problem (Figure 19-12). Duct cleaning involves cleaning the return and supply ducts as well as the coils, drain pans, and motors and housings, in addition to any part with

FIGURE 19-12: Air duct inspection.

which supply or return air might come into contact. When cleaning a duct system, it is necessary to inspect and, if at all possible, clean each component. Failure to do so will negate the benefits of having the system cleaned by allowing the system to become recontaminated. Also, if mold is present in the air system, then this is a good indication that moisture is present in the air system. If this is the case, then the source of that moisture should be identified and corrected. Failure to correct moisture problems in a forced-air system will result in the mold returning, thus requiring the system to be cleaned again at a future time.

When a system is cleaned, special brushes and a vacuum are used to dislodge dirt, dust and other debris. Also chemicals such as biocides are used to neutralize microorganism. In addition, an ultraviolet light can be installed on the supply line to neutralize microorganism that are traveling through the ductwork. However, before using any chemical cleaning agents, always read and follow all instructions. Failure to carefully read and follow the instructions on a chemical cleaning agent can result in damage to the air system, ineffective cleaning, and/or personal injury. Also, always check all local building codes before attempting to clean an air system.

Air Duct Cleaning Associations and Regulations

Clean air is an important topic in HVAC and is often debated and argued; it is also becoming a very large industry that is establishing its own requirements, standards, and regulations. Some local and state governments have adapted these regulations and are now requiring compliance with them when cleaning an air system, while others have not. Regardless, a good understanding of the building codes governing HVAC is critical. In addition, if a technician is going to start cleaning duct systems, then it is strongly recommended that the technician become acquainted with the National Air Duct Cleaning Association (NADCA). Formed in 1989, this nonprofit organization was originally founded to promote the removal of HVAC of contaminated equipment as the only acceptable method of cleaning. However, it mission has changed, and now NADCA's goal is to be the number one source for HVAC cleaning and restoration. In other words, removal of the contaminated equipment now the last line of defense.

Measuring Duct Leakage

Because air duct leaks cannot be seen , a performance test must be administered that will measure the airflow through the fan and its effects on the pressure in the duct system. It stands to reason that the tighter the air ducts, the less work the fan will have to perform to deliver the desired amount of air. A duct system is pressurized by static pressure, velocity pressure, and total pressure. Static pressure is the same as the pressure on an enclosed vessel, such as a tank of air that is pushing outward.

Currently there are two types of tests that can be performed to determine if a HVAC duct system is leaking. Testing only the duct that is connected to the outside or testing the ducts that are connected to the outside and inside of the home. The two most commonly used types of duct test are: duct blast and blower door.

A duct blast unit works on the same principle as a plumbing pressure test. The duct system is directly pressurized to determine the amount of leakage currently: associated with the system. The fan of the duct blast unit it directly connected to either a grill or register and in some cases to the air handler cabinet. The remaining grill and registers are then sealed, and the duct blast unit's fan is brought up to speed until a standard pressure has been achieved. Measurements can then be made using airflow and pressure gauges that are connected directly to the duct blast unit. In addition, a nontoxic fog can be introduced into the system to help you visualize the location and extent of any leaks in the system.

Installing the Blower Door Testing System

When installing a blower door system, it is important to select an exterior doorway. Typically, an external doorway to large open rooms works better. If at all possible, avoid installing the system in a doorway containing stairs or large obstructions that might impede the airflow. If the exterior door opens into a garage or onto a porch, the connecting space should be open to the outside (Figure 19-13).

When installing a blower door system, the primary door should be removed and the blower door system frame installed in its place. After the primary door has been removed, install the aluminum frame and nylon panel in the primary door opening. After securing the blower door frame system in place, attach the gauge-mounting

© Sacramento Bee/McClatch-Tribune/Getty Images

FIGURE 19-13: Installing a blower door testing unit.

board either to a door or to the aluminum blower door frame gage hanger bar. Once the gauge mounting and fan speed control have been installed, connect the blower door fan with the flow rings and no-flow plate to the blower door opening in the nylon panel. Ensure that the exhaust side of the blower fan is positioned on the exterior side of the building. Always follow the blower door system manufacturer's installation instructions.

Operating Blower Door for Air Sealing

Before conducting a blower system test, always ensure that the following activities have been completed:

1. All exterior doors and windows have been closed.
2. All interior door are open.
3. All combustion appliances are set not to turn on during the test.
4. All fireplaces and wood stove fires have been extinguished and their doors are closed.
5. All exhaust fans, air conditioners, and dryers are shut off.

Once the structure has been prepped, the door test is started by bringing the fan up to speed slowly. This gives you time to inspect the structure being tested to ensure that nothing unexpected is happening in the structure (Figure 19-14). Once the you have ensured that all is well, increase the fan speed until the pressure difference between the house and the outdoors is 50 pascals. Always follow the manufacturer's instructions before starting a blower door test.

© Melanie Stetson Freeman/Christian Science Monitor/Getty Images

FIGURE 19-14: Inspecting the structure before performing the blower door test.

Inspecting the Structure before Performing the Blower Door Test

Depending upon the age of the structure and its geographic location, the removal and disposal of building insulation may be restricted to licensed technicians. Therefore, before attempting to perform any repair, and/or removal of insulation always check the local and state regulations regarding the removal and disposal of insulation. If asbestos is present, then check the local building codes, as well as with the air pollution control board and any other local agency responsible for worker safety before attempting to remove the asbestos. The Environmental Protection Agency also has regulations that oversee the removal and disposal of asbestos. These regulations are outlined in two different federal laws: the Clean Air Act (CAA) and the Toxic Substances Control Act (TSCA).

Regardless of the type of insulation being used, always read and follow the manufacturer's instructions regarding the application and installation of the product.

Types of Insulation

Recall from Chapter 18 that heating and cooling loads are based on the occupants lifestyle as well as the building materials used in the structure. Some building materials allow less heat transfer and some allow more. For example, the Styrofoam used in coolers to keep cold drinks cold is a good insulator. You can keep drinks cold much longer in a Styrofoam cooler than in a metal washtub because Styrofoam is an insulator, and metal is a conductor of heat. Choosing the best insulation for a home is worthwhile for the homeowner.

Currently, there are several types of insulation available for residential construction. The type of insulation that should be used in a residential construction project depends upon its particular application. The most common types of insulation used in residential construction and their uses are described in the following sections.

When handling insulation wear proper eye protection, a particle mask, and protective clothing. Exposure to fiberglass insulation will not cause long-term negative heath effects. However, it can cause irritation to the skin, eyes, and upper respiratory tract.

Blanket: Batts and Rolls

One of the most versatile forms of insulation is the batt or roll insulation (Figure 19-15). Batt insulation is typically made from a mineral fiber or rock wool and can be purchased in several widths, thicknesses, and R values. Blanket insulation is often used in unfinished walls (foundation walls, floors, and ceilings).

Foam Board Insulation

Foam board, also referred to as "rigid insulation," is a form of insulation made from such materials as polystyrene, polyisocyanurate, and polyurethane (Figure 19-16). In residential construction, this type of insulation can be installed in practically any location in a structure. In addition to offering excellent heat transfer properties, foam board insulation will provide small degree of strength and stability not offered by batt insulation.

FIGURE 19-15: Batt insulation.

FIGURE 19-16: Foam board insulation.

Currently, there are three types used in residential construction—molded expanded polystyrene foam board, extruded expanded polystyrene foam board, and polyisocyanurate and polyurethane foam board.

- **Molded expanded polystyrene (MEPS) foam board**—This is used in everything from coffee cups to insulation. It typically ranges from 3.8-R per inch of thickness to approximately 4.4-R per inch of.

- **Extruded expanded polystyrene foam board**—This is also a closed-cell material. This is a more expensive insulation than MEPS; however, it has a greater R-values and is an excellent choice for roofs and walls panels.

> **CAUTION**
>
> Before using any insulation, always consult the manufacturer and check local building codes to determine any restrictions on types and locations.

- **Polyisocyanurate and polyurethane foam board**—This offers the greatest R values, as much as R8 per inch of insulation thickness. It has the ability to remain stable when subjected compressive loads as well as temperatures ranging from (−100°F to +250°F). It is an excellent choice for roofing insulation.

Blown-In Loose-Fill Insulation

Blown-in loose-fill insulation is an excellent choice for new or existing structures because it has the ability to conform to any shape or space without interference from existing structural members and without damaging any finishes (Figure 19-17). This form of insulation is made up of tiny particles of recycled materials such as cellulose, fiberglass, or rock wool. The typical R value for loose-fill insulation is:

- **Cellulose**—3.2 to 3.8 per inch; 10–12 inches are needed to achieve the same effect as R-38

FIGURE 19-17: Blown-in loose-fill Insulation.

- **Fiberglass**—2.2 to 2.7 per inch; 14–17 inches are needed to achieve the same effect as R-38.
- **Rock wool**—3.0 to 3.3 per inch; 11.5–13 inches are need to achieve the same effect as R-38.

All three insulations are perfectly fine to use with $\frac{1}{2}$-inch drywall on 16 inch centers or $\frac{5}{8}$-inch drywall on 24-inch centers. However, if you are using $\frac{1}{2}$-inch drywall on 24-inch centers, then you can only use fiberglass loose-fill insulation.

Sprayed-Foam and Foamed-in-Place Insulation

FIGURE 19-18: Spray-foam insulation.

This form of insulation is the most versatile form of insulation on the market today (Figure 19-18). It can be injected or poured as well as sprayed. This also means that this type of insulation, when applied properly, will penetrate the smallest area. This form of insulation also typically has a R value that is approximately twice per inch as that of batt insulation.

Water Heater Blankets

The installation of a water heater blanket has the potential to reduce annual utility bills because of a reduction in energy used to heat hot water of approximately 4%–9%. The average life span of a water heater is 13 years and the production of hot water accounts for approximately 25% of each entity dollar spent. For the average American family that spends approximately $1,900 a year on utility bills, this constitutes a savings of $3.56 per month. This alone will produce an annual savings of approximately $42.72. The average life expectancy of a water heater is approximately 13 years. If a water heater blanket is installed with the water heater the savings over the life span of the water heater would equal $555.36. Depending upon the make and model of the water heater, this could more than pay for the water heater itself.

Water heater blankets can be installed on either electric water heaters or combustion water heaters. For both types of water heaters, the installation process is basically the same.

Drain Water Heat Recovery

Up to 90% of the energy used in the production of hot water in a residential structure goes down the drain in the form of drain water (graywater). A recovery system (Figure 19-19) uses the heat from the drain water to heat the water entering the water heater, thus reducing the amount of energy necessary to heat and maintain the water in the water heater storage tank. These systems work extremely well with on-demand water heaters and solar water heaters.

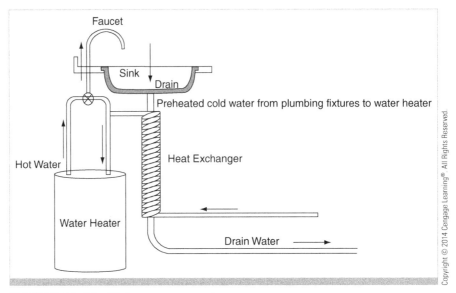

FIGURE 19-19: Water recovery system.

Determine Air Leaks without a Blower Door Test

The most efficient means of determining air leaks in a structure is by the use of a blower door test; however, if a blower door testing unit is not available, then the facility maintenance technician can still find air leaks if the technician knows or general areas to check. This includes:

- Locations where different materials come in contact with one another
- Between the foundations and the walls
- Between chimney and siding
- Doors and window frames
- Mail chutes
- Utility service entrances and penetrations
- Cable and wire penetrations
- Plumbing penetrations
- Vent penetrations
- The exterior siding and foundation
- HVAC systems
- Fans

Air leaks may be determined by depressurizing the structure and performing a smoke test. The steps to perform a smoke test are as follows:

Step 1: Turn off the HVAC equipment.
Step 2: Close all exterior and interior windows and exterior and interior doors.
Step 3: Turn on all exhaust fans.
Step 4: Light an incense stick or use a smoke equipment, pass around the common leak areas listed above.

Installing a Water Heater Blanket

Step 1: Turn off electrical power to the water heater before starting.

Step 2: (optional step): bridges Adjust the water heaters thermostat to 120°F. Once the temperature has been adjusted replace water heater thermostat cover. Lowering the water temperature to 120°F, can result in the average homeowner saving approximately $30–$40 per year.

Step 3: Starting from the water heater, wrap insulation around the side of the water heater taping it into place.

Step 4: Locate the thermostat doors and cut an opening into the insulation to provide access to the unit's thermostats. Replace the insulation in the opening(s).

Step 5: When insulating the top, cut the insulation to fit, providing holes and slots for the water pipes extending from the top of the water heater.

Step 6: Install the insulation on the top of the water heater, taping the seams.

Before attempting to install the water heater blanket, read the manufacturer's instructions carefully.

It is extremely important that the airflow to the burner not be obstructed. Never place insulation over the top of the water heat or over the thermostat.

Basic Installation for Gas Water Heaters

Step 1 (optional step): Readjust the thermostat to 120°F. Once the temperature has been adjusted, replace the water heater cover (Figure 19-20).

Step 2: Starting in the front of the water heater, apply the insulation to the side of the water heater, tapping it into place.

Step 3: Leave an opening for the pilot light and the pressure–temperature relief valve (Figure 19-21).

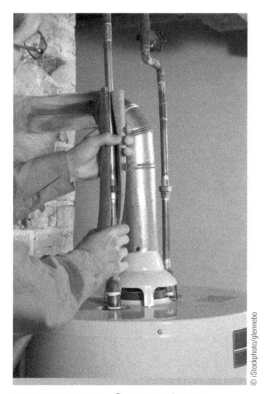

FIGURE 19-20: Gas water heater.

FIGURE 19-21: Insulating a gas water heater.

Insulating Hot Water Pipes

Insulate the hot water supply line to reduce the possible amount of heat loss from hot water supply lines, which can save as much as 2°F–4°F. Over time, this will add up to a noticeable difference in the structure's utility bills. In addition, it reduces the lag time to get hot water commonly experienced in the wintertime. When using polyethylene or neoprene foam pipe sleeves, always place them so that the sleeve is facing down on the pipe. Once installed use tape every foot to secure the insulation.

Sealing Switches and Outlets

As stated earlier switches and outlets along exterior walls are one source of heat transfer. This source of heat transfer can be easily eliminated by applying outlet and switch sealers. The following steps outline how to install switch and outlet sealers.

Step 1: Disconnect the power to the outlet(s) and/or switches.
Step 2: Using a screwdriver remove the device cover (see Figure 19-22).
Step 3: Place the precut sealer around the device.
Step 4: Replace the device cover (see Figure 19-23).
Step 5: Reenergize the circuit.

FIGURE 19-22: Removing an outlet cover.

FIGURE 19-23: Replacing the outlet cover.

Applying Weather Stripping

For windows:

Step 1: Determine the amount of weather stripping needed for the application, using the weather stripping form previously provided.
Step 2: From a building supply distributor, select the correct shape and size of the weather stripping to be used (this is often determined by the application of the weather stripping).
Step 3: Remove old worn-out weather stripping before applying the new weather stripping. In most cases, this can be done with a knife and/or sandpaper.
Step 4: Clean the surface and surrounding areas in which the weather stripping is to be applied be attempting to install the new weather stripping.
Step 5: Cut the weather stripping to fit the jambs, header, and sill of the window.
Step 6: Following the manufacturer's instructions, install the weather stripping.

For Doors:

Step 1: Tighten any loose hinges.
Step 2: Determine the amount of weather stripping needed for the application, using the weather stripping form previously provided.
Step 3: From a building supply distributor, select the correct shape and size of the weather stripping to be used (this is often determined by the application of the weather stripping).
Step 4: Remove old warn out weather stripping before applying the new weather stripping. In most cases, this can be done with a knife and/or sandpaper.
Step 5: Clean the surface and surrounding areas in which the weather stripping is to be applied be attempting to install the new weather stripping.
Step 6: Cut the weather stripping to fit the sides and top of the door.
Step 7: Following the manufacturer's instructions, install the weather stripping.
Step 8: For the threshold, install a door sweep.

Insulating an Unventilated Crawlspace

When insulating an unventilated crawlspace, it is recommended that the foundation walls be sealed and insulated. That way, insulating an unventilated crawlspace will require less insulation, and it typically will allow for uninsulated pipe and ductwork. because they are considered to be in a conditioned space. To insulate an uninsulated crawlspace:

Step 1: Seal all air leaks and penetrations through the exterior wall, including but not limited to the band joist.
Step 2: Locate and insulate the access door to the crawlspace. This is typically done by using a weather stripping that will retain its shape overtime and after repeated usage.

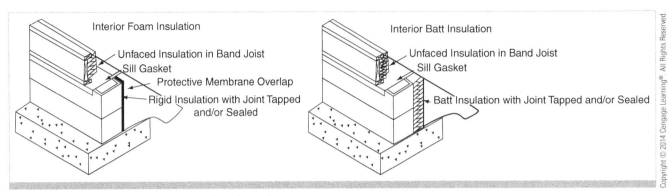

FIGURE 19-24: Insulating an unventilated crawlspace.

Step 3: Insulate the masonry foundation wall—this can be done using either rigid foam board insulation or batt insulation (see Figure 19-24). In addition to insulating the foundation walls, insulate the access door using foam board or batt insulation. Finally insulate the band joist using batt insulation,

Step 4: If possible, install a termite barrier, also known as a termite shield, between the foundation wall and the band joist. This is typically done during the construction phase of a facility. If a termite shield cannot be installed, then a gap (2–4 inches) is commonly left to allow for termite inspection.

Insulating a Ventilated Crawlspace

Step 1: Inspect the interior of the facility for crack and holes in the floor. Once you have located any crack or holes, repair them to prevent air infiltration into the structure.

Step 2: Check the space between floor joists. If it is not insulated, apply fiberglass insulation to this area. This is typically done by stapling fiberglass insulation to the floor joist so that it is tight against the subfloor. Seal any seams in the insulation to achieve a seamless coverage.

Step 3: Using a house wrap (this can be purchased at any home improvement store) cover the insulation and the floor joists to provide a vapor barrier. The seams should be sealed and taped to provide a seamless vapor retardant over the insulation and subfloor system.

SUMMARY

- When selecting a weather stripping consider temperature differences and changes, friction, weather, durability, overall wear and tear, and cost.

- Residential construction homes built using a crawlspace have several advantages over homes that were built having a basement or those built on a slab.

- Of all the different locations in a home, the attic is the typically the easiest to estimate the amount of insulation for; if necessary, add additional insulation to improve the comfort level as well as the overall efficiency of the dwelling.

- Regardless of the type of climate in which a structure is located, the air inside the structure must undergo some type of conditioning before the occupants of the structure will be comfortable and safe.
- A forced-air system uses a fan to force air through the air-conditioning system.
- The forced-air system is composed of the following components:
  - An air handler (this includes the blower-fan)
  - A heat source (gas, oil, electric)
  - A means of delivering the air to the heating unit (return)
  - A means of delivering the air back to the rooms (ductwork and registers)
  - A thermostat
  - A filtration system
- The conditioned air is transported to the space via supply air ducts, registers, and diffusers.
- The plenum systems are perfect for applications in which the rooms are close to the unit that services them.
- The extended plenum system is the most commonly used duct system in residential construction today.
- Another approach to keeping the systems balanced is to use a reducing extended plenum system.
- The perimeter loop duct system is particularly well suited to structures that incorporate a concrete slab construction in colder climates.
- Having a well-insulated duct system is only half of the story; if the duct system is not airtight, then it can also waste energy as well as become a potential health hazard.
- A poorly sealed HVAC duct can increase the homeowners' energy cost between 10 and 30 percent.
- Poorly sealed ductwork is a potential health hazard because it can draw unconditioned and unfiltered air from the attic, basement, and crawlspace areas into the duct system.
- The quality of the air in an environment is an important aspect of HVAC.
- Mold is plant that is a member of the fungi family.
- The installation of a water heater blanket can potentially reduce annual utility bills because of a reduction in the energy required to heat hot water by approximately 4%–9%.
- If a water heater blanket is installed with the water heater, the savings over the life span of the water heater would equal approximately $555.36.
- When installing a water heater blanket, always follow the manufacturer's instructions.
- When installing a gas water heater blanket, always leave an opening for the pilot light and the pressure–temperature relief valve.
- One source of heat transfer can be easily eliminated by applying outlet and switch sealers.

REVIEW QUESTIONS

1   List the types of duct systems commonly used in residential construction.

3   List the guidelines for replacing an air filter.

2   List the components of a forced-air system:

4   Describe the methods used to determine an air leak in a duct system.

5   List the steps to determining an air leak without a blower door test system.

7   What are the basic steps for insulating a water heater?

6   Determine the amount of weather stripping for the following windows:

Window #1: 36 × 36 × 48 × 48
Window #2: 60 × 60 × 72 × 72
Window #3: 24 × 24 × 36 × 36

Note: All dimensions are given in inches.

8   What are the basic steps for insulating a crawlspace? An attic?

Name: _____

Date: _____

PERFORMING AIR
LEAK DETECTION
WITHOUT USING
A BLOWER DOOR
SYSTEM
JOB SHEET

Leak Detection

1

PERFORM A LEAK TEST WITHOUT A BLOWER DOOR.

Upon completion of this job sheet, you should be able to identify air leaks without using blower door system.

1. Using your training facility, document the locations in which different materials come in contact with one another. For example, stucco meets rock.

2. Document the locations of mail chutes in the training facility.

3. Document the locations of windows and doors in your training facility.

4. Document the location of air-conditioning equipment in the training facility.

5. Document the locations of outdoor water faucets.

Using the smoke test outlined in "Determine Air Leaks without a Blower Door Test" test your training facility, and record your findings in the space provided below.

INSTRUCTOR'S RESPONSE:

_____

_____

_____

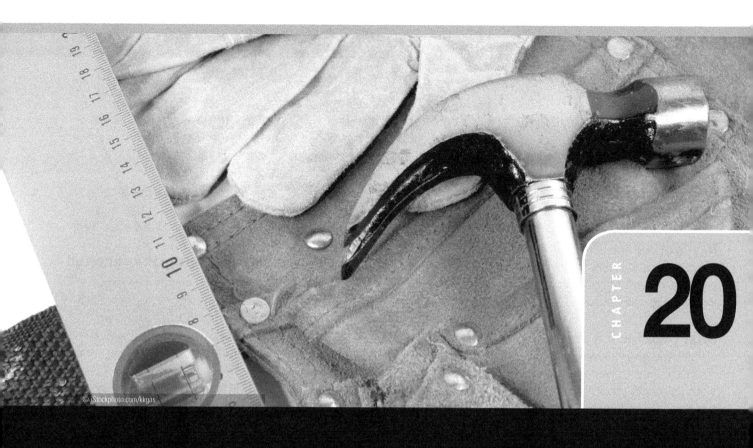
© iStockphoto.com/kkgas

20

Solar Energy Systems: Maintenance and Repair

OBJECTIVES

By the end of this chapter, you will be able to:

KNOWLEDGE-BASED

- Identify various types of renewable energy sources.

- Define the terms: electromagnetic radiation, and photovoltaic.

- Explain the difference between a stand-alone grid connected system,

an off-grid system, and a grid-connected system.

- Discuss the biggest challenges associated with the use of solar energy.

- Explain the difference between passive solar design and active solar heating.

Renewable energy any form of energy that is derived from sources that can be regenerated

Solar constant the amount of power available from the sun

Daylighting the placement of windows in a location that will promote natural lighting as well as a view to the outside

Solar collector a device that utilizes the greenhouse effect and direct sunlight to heat up a copper plate containing heat transfer tubing (heat exchanger)

Indirect gain process in which heat is collected in a storage device and later permitted to be distributed via conduction, convection, and radiation

Passive solar design the use of daylighting, landscaping thermal mass, and proper building orientation to provide an efficient building design

Photovoltaics the production of electricity from direct sunlight

Light the range of electromagnetic radiation that is detectable by the human eye

Wave–particle duality the fact a photon exhibits the properties of a wave as well as a particle.

Photons the tiny packet that makes up light

Solar array multiple solar panels that are connected together

Energy density the amount of energy stored in a region of space per unit volume

Grid-connected system a system that is uses both energy from the power grid and solar energy

Stand-alone grid-connected system basically the same as a grid-connected system, with the exception of having an added storage device, typically in the form of a battery system

Off-grid system a stand-alone non-grid-tied system in which the peak demand must be taken into consideration because the system is not connected to the electrical grid

Stand-Alone Off-Grid Hybrid System a standalone system that incorporates a diesel generator

Introduction

As stated in Chapter 18, because of the increasing price of energy and the side effects of burning fossil fuels, the search for a substitute has become ever more important. When a substitute has been determined, it must be reliable, renewable, clean, and easily obtained. It must be clean, meaning it cannot have a negative impact on the environment or be a contributor to the greenhouse effect. It must be renewable, meaning it is derived from sources that can be easily regenerated (replenished). For the energy can be considered reliable, the process used to generate the energy must be repeatable, and it must produce the same results every time. Finally, to be considered easily obtainable the energy source cannot be a huge financial burden.

Renewable Energy

Renewable energy is any form of energy that is derived from regenerative sources such as wind (wind mills), hydro (derived from moving water, also known as hydro-electric or dams), fuel cells (hydrogen), the burning of wood (also known as biomass), geothermal (water that has been heated by the earth's interior), and tidal power (energy generated by ocean tides). Of all the renewable energy sources, solar energy has the most potential because it could supply thousands of times more energy than the world currently consumes. It is estimated that the sun delivers enough energy to the Earth an hour to power the planet for an entire year.

In reality, the sun is a thermonuclear furnace that is approximately 93,000,000 miles from Earth. It has a surface temperature of approximately 10,000 °F and generates 380 billion billion megawatts of power. In comparison, the Hoover dam only generates 2,080 megawatts.

The **solar constant** is the amount of power available from the sun. If the solar constant is measured in terms of Btu (British thermal units), then the output of the sun is 442 Btu per ft^2 per hour.

Daylighting Concept

Windows when placed in the proper location will not only provide natural lighting and a view to the outside, but they can also reduce the amount of energy consumed by a structure, thus lowering the facility's utility bills by reducing the structure daytime lighting requirements. This concept is known as **daylighting**. Daylighting takes into consideration type of window, window placement and interior design as well as many other factors to utilize the benefits of natural sunlight. In the United States, windows placed along the southern exposure of a structure will provide maximum solar radiation in the winter months while limiting the effects of the sun in the summer months. North-facing windows let in more natural light while limiting the exposure to radiation in the summer months. Window with western and eastern exposures will provide very little benefit to the structure. The light that is allowed to enter the structure from windows placed along the east and west slide of a structure typically produce higher glare and heating effects from the sun in the morning and evening hours. In addition to windows, skylights and light tubes are an excellent way of allowing sunlight to enter a structure.

Like a skylight, light tubes are designed to allow sunlight to enter into a room via an opening in the roof. However, unlike a skylight, light tubes do not have to be located directly over the room they are intended to service. A light tube consists of an opening covered with framed glass and a flexible tube lined with highly reflective foil.

Studies have indicated that adding natural lighting to a structure will not only reduce the amount of artificial light needed but will improve the overall physical and mental health of the facility's occupants.

Different Types of Solar Energy Equipment

Currently, there are two basic types of equipment used with solar energy. They are equipment that captures the photons in the solar energy as they strike the surface of the solar cell and transform them directly into electricity and equipment that captures the heat generated by the solar radiation and stores it in thermal storage equipment for facility heating and hot water production. The energy stored in the thermal storage can also be used to power a heat engine. A heat engine is a thermal system that converts heat into mechanical work by transforming a liquid from a higher temperature to a lower temperature.

Active and Passive Solar Space Heating Systems

Chapter 18 discusses how to calculate the heating load of a structure in order to maintain a desired comfort level in a structure during the winter months. However, this amount can be reduced without altering the target comfort level by using either active or passive solar heating designs. In an active solar design, heat is transferred from the sun using one of two different types of active systems, based on either liquid or air using a solar collector. For a liquid-based system, the **solar collector** is device that uses the greenhouse effect and direct sunlight to heat up a copper plate containing heat transfer tubing (heat exchanger; see Figure 20-1). A fluid (water or antifreeze) is circulated through the collector and either used to heat the structure or stored in a storage device (water tank, etc.). Heat stored in a storage device can be used at a later time for space heating.

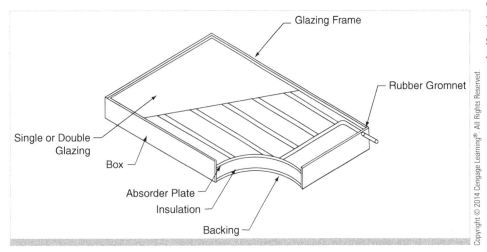

FIGURE 20-1: Solar collector.

In an active solar heating system, the heat is absorbed using solar radiation and the transferred directly to a storage device or to the interior of the structure (Figure 20-2). Liquid-based active solar heating systems are particularly well suited for radiant heating systems. However, they can be used as a supplement to a forced-air system.

In a passive solar design, the structure is used as the collection and storage device. A passive solar design does not use mechanical or electrical devices in order to collect and distribute the heat. In the winter months the walls, windows, and even the floors are used to store and distribute the heat; while in the summer months, the roof overhang plays an important factor in shading windows and walls and floors from

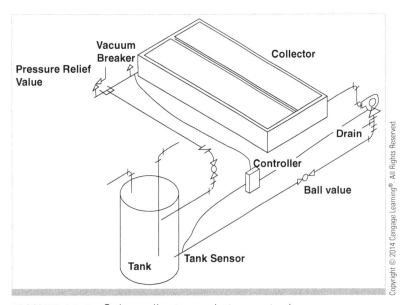

FIGURE 20-2: Solar collector and storage tank.

direct solar gains. In a passive solar design, the local climate is the key to an effective design. In a passive solar design, the heat that is permitted to enter and is stored in the floors and walls is known as "direct gain." When heat is collected in a storage device (Trombe walls) and later permitted to be distributed via conduction, convection, and radiation, the process is known as **indirect gain**. In a system in which the heat is collected in one area of the structure and redirect to another, the process is known as "isolated gain."

Passive Solar Design

When a facility correctly uses daylighting, along with landscaping, thermal mass, and proper building orientation, this is known as **passive solar design**. Proper building orientation consists of aligning the facility's length along a plane that is 15° from due south. Doing this will maximize the structure's natural ability to utilize solar heat gain in the winter months.

The initial construction cost of a passive solar building will be higher than the initial construction cost of a standard structure. Over time, the reduction in the structure's utility bill will justify the initial construction cost. A common myth associated with passive solar building design is that passive solar buildings will look much different from standard buildings. In fact, most passive solar buildings will not look any different from conventional buildings.

In a passive solar design, the area of the windows (length × width) should not exceed 10 percent of the total floor area of the structure in which they are installed. For example, in a 2,300 square foot structure there should be 230 square feet of glass with a southern exposure. This typically means that the south side of the structure will have more windows than the remaining sides.

If an existing structure is to be retrofitted to become a passive solar structure, then the cost of the changes and their expected return on investment (ROI) must be taken into account before the project is started. In addition, each change should be evaluated to determine the effect on the structures visual appearance. Finally, always consult the local building code before starting any construction project.

Photovoltaic Solar System Concepts

One of the biggest challenges associated with the use of solar energy is that the production of electricity from direct sunlight, as known as **photovoltaics**. While many people might think of this as a relatively new concept, in fact it was in 1839 that the French physicist Edmund Bequerel noted that some materials would generate small amounts of electricity when they were exposed to sunlight. It wasn't until 1905 that the production of electricity took another leap forward when Albert Einstein completely described the photoelectric effect and the nature of light. The photoelectric effect states that some materials release small amounts of electrons when exposed to light.

It was at this point that photovoltaic technology was born. However, it took another 49 years before the first photovoltaic module (solar cell) was developed by Bell Laboratories. It wasn't until the space race of the 1960 that this technology became more than just a curiosity and was taken as a serious method of producing electricity. The space industry needed a lightweight method of producing electricity that was renewable.

Light is nothing more than a term used to describe a range of electromagnetic radiation that is detectable from the human eye. Electromagnetic radiation is a form of energy that is has both wave and particles like properties—this is known as the **wave–particle duality.** Light ranges in wavelength from 7×10^{-5}cm (red) to 4×10^{-5}cm (violet) and travels at a speed of 186,282 miles per second. It has a frequency ranging from about 405 THz to 790 THz. It is made up of tiny packets known as **photons.**

When a photon collides with the surface of a solar cell, some of the electrons of the semiconductor material that the solar cell is made of become loose and produce a small electric charge. The semiconductor material used in a solar cell is the same as that used in the production of microelectronic chips for computers, gaming systems, and the like. Photovoltaic modules contain a number of solar cells connected to one another and mounted in a structural frame (see Figure 20-3). Photovoltaic modules are designed to supply a specific voltage—for example, 12 volts. The amount of voltage produced by a photovoltaic module can be determined by the amount of light that is permitted to strike the surface to the solar cells. For example, solar cells located in a shady area will

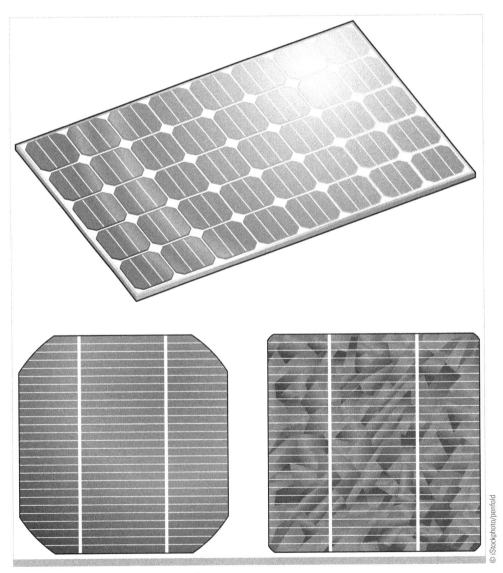

FIGURE 20-3: Photovoltaic cells.

not produce as much electricity as those placed in direct sunlight (see figure 20-4).

Photovoltaic modules can be wired in together to form an array. The more arrays that are connected ,the more electricity that will be produced. All electricity that is produced by a **solar array** is in the form of direct current (DC). Photovoltaic modules can be connected in series to generate more voltage and/or parallel to produce more current. The final arrangement will be determined by the voltage/current requirements of the system the array is servicing.

Solar panels are rated by their DC power output under normal conditions. The surface area of the panels is determined by their efficiency. A panel with an efficiency rating of 5% will have five times the surface area of a panel with an efficiency rating of 25%. The efficiency of solar panels is affected by high temperatures; therefore, having good ventilation behind the solar panels is critical.

FIGURE 20-4: Photovoltaic celliinstallation.

The **energy density** (the amount of energy stored in a region of space per unit volume) for a solar panel is the energy density is measured in watts per ft^2 and is the product of the peak power that is produced for a given surface area. Photovoltaic technology has several advantages over the other forms of renewable energy sources. They are:

- It produces no emissions that are harmful to the environment, plants, or animals.
- It is quiet.
- Modules can be installed in unused spaces (rooftops).
- Modules are designed to operate for long periods of time with little to no maintenance.

As stated earlier, energy produced by solar cell is in the form of direct current (DC). If you are dealing with consumer electronics or spacecraft, then direct current is preferable. However, for residential, commercial, and industrial applications, the power must be transformed from direct current to alternating current. For a solar energy system, this is achieved by using a solar inverter.

Grid-Connected Systems

A **grid-connected system** is a system that is uses both energy from the power grid and a solar energy system. These systems generally use the solar energy during daylight hours and at night they tap into the power from the conventional power grid. These systems have several advantages over other types of solar power systems. Some of these advantages are:

- The surplus energy can be exported back to the grid.
- The solar energy systems do not have to be sized the meet the facility's peak demand. During peak demand times, overages can be drawn from the conventional power grid.
- They are typically less expensive because they don't have to be designed to meet the peak demand load requirements.

However, grid-connected systems must meet all local building codes as well as the requirements of the local power companies.

Stand-Alone Grid-Tied Systems

A **stand-alone grid-connected system** is basically the same as a grid-connected system, with the exception that they have an added storage device; typically in the form of a battery system. This allows the system to provide power even in times of electrical grid failure. Some of their advantages are:

* The solar energy systems do not have to be sized the meet the facility's peak demand. During peak demand, overages can be drawn from the conventional power grid.
* The system can be designed to provide uninterrupted power to the facility.

Off-Grid Systems

An **off-grid system** is more expensive than a stand-alone grid-tied system or a grid-tied system in that the peak demand must be taken into consideration because the system is not connected to the electrical grid. Therefore, the facility is entirely dependent upon the sun to provide all of the facility's power requirements.

Stand-Alone Off-Grid Systems

When it is necessary for a system to provide a guaranteed output (day or night), an additional backup power source (other than batteries) must be combined with the solar power system. These types of systems are referred to as "stand-alone off-the-grid systems." When a stand-alone off-the-grid-system is combined with a diesel generator, the system is known as a **stand-alone off-grid hybrid system**.

Principles of Solar Hot Water Heating

The production of domestic hot water can be a rather complex process, depending upon the type of equipment selected to generate the hot water. However, solar hot water production is a rather simple process that typically involves the use of three major components: a solar collector, a transfer medium (water), and a storage device (water tank; Figure 20-5). The production of hot water using solar energy is accomplished by means of two basic principles of physics. First, heat is more easily absorbed by dark darker colors (black). Lighter colors, such as white, have a tendency to reflect heat. Second, when water is heated above 39.1°F, its volume expands. To generate hot water using solar energy a solar collector is generally mounted on the roof of the facility facing the sun. Currently, there are three main system configurations used in the production of solar hot water. They are: open hot water/radiant system, the drain back system, and the open solar hot water/radiant system.

FIGURE 20-5: Solar water heater.

When water is pumped through a system, the system is referred to as an "active system," (Figure 20-6). When water is drawn through the system using natural convection, this is known as a "passive system."

The open hot water/radiant system utilizes a heat exchanger and antifreeze solution design. The heat exchanger is used to transfer energy from the antifreeze solution to the storage tank containing the potable water. Heat is transferred by conduction. This type of system not only provides energy for the heating of domestic water but also can be used to provide radiant floor heating. These types of systems are typically powered using a low-wattage (wattage = voltage * amps) pump.

Drain-back systems also provide heating and domestic hot water, but these types of systems require a larger pump because of the location of the collector to the storage tank. In this type of system, the all the water in the collector is drained back into the storage tank. When the water is stored inside the collector itself, it is referred to a "batch collector."

When the same water (output from the collector) is used for dual purposes, typically heating and hot water production, this is referred to as an "open system." This type of system is typically more efficient because it employs a single heat source multiple purposes (Figure 20-7).

Antifreeze is typically used in a solar collector be it will ensure that the fluid will not freeze in the winter months thus protecting the equipment from damage. It also provides an additional benefit in which the boiling point of the water is increased. This allow for higher temperatures to be generated by the solar collector. The amount of antifreeze used in a collector is determined by the region of the country in which the collector is to service.

FIGURE 20-6: Active solar water heater.

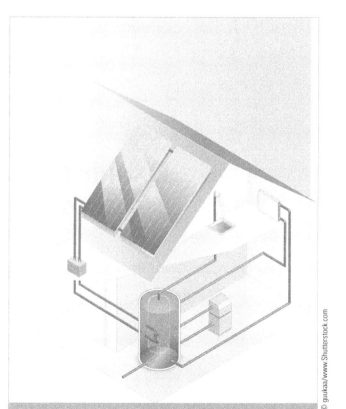

FIGURE 20-7: Solar water heater system.

The NEC® and Building Codes

Most towns, cities, and counties have building codes. A building code is a set of regulations (usually in the form of a book) that ensure that all buildings in that jurisdiction (area covered by a certain government agency) are of safe construction. Building codes specify such things as minimum size and spacing of lumber for wall framing, steepness of stairs, and fire rating of critical components. The local building department enforces the local building codes. States usually have their own building codes, and state codes often require local building codes to be at least as strict as the state code. Most small cities and counties adopt the state code as their own, meaning that the state building code is the one enforced by the local building department.

Because antifreeze can be hazardous if consumed, always follow the manufacturer's instruction regarding the disposal of the antifreeze. In addition, always check with the department of health before releasing the antifreeze into a public waste water system.

Until recently three major model codes were published by independent organizations. (A model code is a suggested building code that is intended to be adopted as is or with revisions to become a government's official code.) Each model code was widely used in a different region of the United States. By themselves model codes have no authority. They are simply a model that a government agency can choose to adopt as their own or modify as they see fit. In 2009, the International Code Council published a new model code called the *International Building Code* (2012 edition now available). They also published the *International Residential Code* to cover home construction (2012 edition now available). Since publication of the first *International Building Code,* states have increasingly adopted it as their building code.

Other than the building code, many codes govern the safe construction of buildings: plumbing codes, fire protection codes, and electrical codes. Most workers on the job site do not need to refer to the codes much during construction. It is the architects and engineers who design the buildings who usually see that the code requirements are covered by their designs. Plumbers and electricians do, however, need to refer to their respective codes frequently. Especially in residential construction, it is common for the plans to indicate where fixtures and outlets are to be located, but the plumbers and electricians must calculate loads and plan their work so it meets the requirements of their codes. The electrical and plumbing codes are updated frequently, so the workers in those trades spend a certain amount of their time learning what is new in their codes.

For electrical installations, local, state, and often municipal governments often use the *National Electric Code* (NEC) as the basis of their codes. The NEC is an effort to standardize the installation and maintenance of electrical equipment and devices.

Solar Pool-Heating Systems

The cost of heating a swimming pool can be greatly reduced by the use of a solar pool heater. Solar pool-heating system are typically inexpensive (as low as $100.00) to install because they employ the pool's filter pump to push the water from the pool through the solar collectors. These systems usually also have a low operating cost because they employee the pool's filter pump to circulate the water (see Figure 20-8). A solar pool heater typically consists of:

- A solar collector
- A filter
- A pump (usually the pool's filter pump)
- A flow control valve

Water is pumped from the pool, using the pool's filter system, and then through the solar collector. Collectors used for solar pool heating are manufactured from different materials, depending upon the climate in which the pool heater is to be used. If the unit is installed in a location in which the temperature is above freezing, the collector is usually constructed as unglazed (without a glass covering). Unglazed collectors are constructed using a heavy-duty plastic that has been treated with a ultraviolet light inhibitor.

FIGURE 20-8: Solar pool heater.

© luiggi33/www.Shutterstock.com

Glazed collectors are more expensive and typically consists of copper tubing that is placed between and an aluminum plate (in the back) and a tempered glass cover. These types of systems are generally used in colder climates and employ a heat exchanger and a transfer fluid.

Installation and Maintenance of a Solar Pool Heater

These systems should be installed by a professional solar pool heater technician. Before these systems can be installed, the pool-heating professional must take into consideration the following

- Solar radiation (See Chapter 18)
- Weather patterns
- User safety
- Building codes
- Manufacturer's instructions and guidelines

Once the system has been installed, check the manufacturer's instructions for recommended maintenance procedures and schedules. Also, maintenance should include ensuring that the pool's chemistry and filtering systems are properly maintained. If a glazed collector is used, the glass on the collector might have to be cleaned.

Maintaining Solar Water Storage Tank

An integral part of the part of a solar heating system is the storage tank. In reality, these tanks function the same as a traditional water heater tanks. Their efficiency can be increased by installing a water heater blanket. As water is circulated through the system, depending upon the quality of the water, sediments and various materials in the water can settle in the bottom of the water storage tank. This can be resolved by shutting off the water supply to the tank and draining the water from the storage tank. The drain is typically located on the bottom of the storage tank. If the tank is equipped with a pressure relief valve, then it might be necessary to open the pressure relief valve. Once the water has been drained from the system, it should be flushed using the makeup water. Once the tank has been flushed, close the drain and the pressure relief valve and refill the system.

Testing and Recharging Heat Transfer Fluid

The heat that has been collected by the solar collectors is transferred to the storage tank via the heat exchanger and the heat transfer fluid. It is the responsibility of the heat transfer fluid to move the heat from the collector to the heat exchanger (located in the storage tank). There are many factors involved in the selection of a fluid; therefore, when selecting a heat transfer fluid always consult the equipment manufacturer as well as the supplier.

When testing and recharging a heat transfer fluid, always check the manufacturer's instructions and guideline for testing and recharging it. Depending upon the type of fluid used by the system, an EPA certification might be required to handle the heat transfer fluid.

Types of Heat Transfer Fluids

Some commonly used heat transfer media used in solar collectors are:

- **Air**—Air has very low heating capability; however, it will not freeze and is noncorrosive.
- **Water**—At standard atmospheric pressure, water has a boiling point of 212°F and a freezing point of 32°F. It is nontoxic and inexpensive. It also has the capability of containing a high material content (also known as heavy water); this can cause mineral deposits in the collector as well as in the storage tank.
- **Glycol/water mixture**—Ethylene and propylene glycol are "antifreezes."
- **Hydrocarbon oils**—These require more energy to pump.
  - **Silicones**—These are longlasting, are noncorrosive, and have a very low heating capability.
  - **Refrigerants (R22, R13a, etc.)**—These are used in air-conditioning equipment as well some types of heat pumps.

Cleaning Collectors

Regardless of its purpose, a properly functioning solar collector is critical for maintaining the integrity of a solar panel system. To ensure that the collector is working properly, the solar collector must be regularly cleaned and kept free of any dirt or grime as well as bird feces (Figure 20-9). This is can be done by using the same chemical used in window washing. The collector should never be cleaned using anything that might etch the glass. Also, the cleaning solution should not be harmful to the environment. In the north section of the country, snow cover is an issue that must be addressed (Figure 20-10). Never clean the collectors when they are hot. Doing so can possibly break the glass.

FIGURE 20-9: Inspecting a PV cell.

CAUTION

Because solar panels are usually installed on rooftops, extra care should be taken to guard against the possibility of falls when installing or maintaining them.

FIGURE 20-10: Cleaning a PV cell.

SUMMARY

- Renewable energy is any form of energy that is derived from sources that can be regenerated.

- Renewable energy is typically derived from sources such as: water (hydro), geothermal, biomass, solar, wind, hydrogen, and tidal power.

- Daylighting requires taking into consideration the type of window, window placement, and interior design as well as many other factors to utilize natural sunlight.

- In the United States, windows placed along the southern exposure of a structure will provide maximum solar radiation in the winter months while limiting the effects of the sun in the summer months.

- A light tube consists of an opening covered with framed glass and a flexible tube lined with highly reflective foil.

- Currently, there are two basic types of equipment used with solar energy. Equipment that captures the photons in the solar energy as they strike the surface of the solar cell and transform them directly into electricity and equipment that captures the heat generated by the solar radiation and stores it in thermal storage equipment for facility heating and hot water production. In an active solar heating system, the heat is absorbed using solar radiation and the transferred directly to a storage device or to the interior of the structure.

- In a passive solar design, the structure is used as the collection and storage device.

- When a structure incorporates an energy-efficient design, proper usage of daylighting, along with landscaping, thermal mass, and proper building orientation, this is known as passive solar design.

- One of the biggest challenges associated with the use of solar energy is the production of electricity from direct sunlight, known as photovoltaics).

- Electromagnetic radiation is a form of energy that is has both wave- and particle-like properties. This is known as "wave–particle duality."

- The semiconductor material used in a solar cell is the same as that used in the production of microelectronics (computers, gaming systems, etc.).

- The energy density the amount of energy stored in a region of space per unit volume and is expressed in watts per ft^2.

- Photovoltaic technology has several advantages over the other forms of renewable energy sources. They are:
  - It produces no emissions that are harmful to the environment, plants, or animals.
  - It is quiet.
  - Modules can be installed in unused spaces (rooftops).
  - Modules are designed to operate for long periods of time with little or no maintenance.

- A grid-connected system is a system that is uses both energy from the power grid and solar energy.

- A stand-alone grid-connected system is basically the same as a grid-connected system, with the exception of an added storage device; typically in the form of a battery system.

- An off-grid system is more expensive than a stand-alone grid-tied system or a grid-tied system in that the peak demand must be taken into consideration because the system is not connected to the electrical grid.

REVIEW QUESTIONS

1   List the types of heat transfer fluid commonly used in solar collectors as well as advantages and disadvantages of each.

2   What is the difference between a passive system and an active system?

3   What should a facility maintenance technician do when selecting a heat transfer fluid?

4   Calculate the window area need to create a passive solar building design for the following square footages:

   a. 1,200 sq. ft
   b. 2,000 sq. ft
   c. 3,600 sq. ft
   d. 7,250 sq. ft

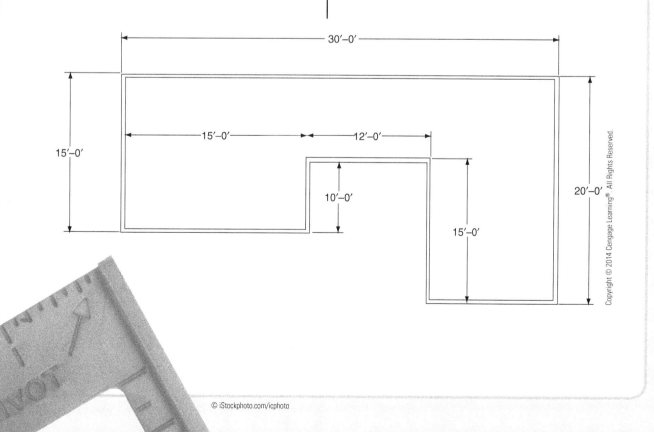

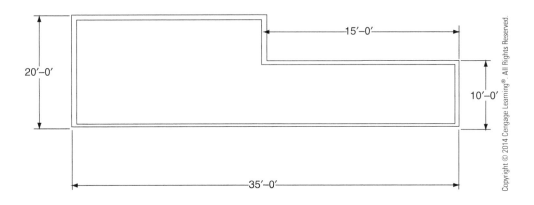

5   How does a photovoltaic cell
work?

Name: _____

Date: _____

Grid-Connected Solar Energy System

1

INSTALLING A GRID-CONNECTED SOLAR ENERGY SYSTEM

Upon completion of this job sheet, you should be able to state the building code and utility requirements for a grid-connected solar energy system.

TASK

Consult your local building codes and utility requirements. In the space provided below, list the requirements for installing a grid-connected solar system.

INSTRUCTOR'S RESPONSE

_____

_____

_____

Name: _____

Date: _____

Grid-Connected Solar Energy System

INSTALLING A STAND-ALONE GRID-TIED SOLAR ENERGY SYSTEM

Upon completion of this job sheet, you should be able to state the building code and utility requirements for a stand-alone grid-tied solar energy system.

TASK

In the space provided below, list the local building requirement and utility requirements for the installation of a stand-alone grid-tied solar energy system.

INSTRUCTOR'S RESPONSE

_____

_____

_____

Name: _____

Date: _____

Solar Water-Heating System

INSTALLING A SOLAR WATER-HEATING SYSTEM

Upon completion of this job sheet, you should be able to state the building code and utility requirements for a solar water-heating System.

TASK

In the space provided below, list the local building requirement and utility requirements for the installation of a solar water-heating system.

INSTRUCTOR'S RESPONSE

_____

_____

_____

Appendix A

Remodeling and Demolition Tips

Remodeling Tips

Remodeling is no different from any other construction project. That is, permits must be obtained, financing must be secured and contractors must be selected in many cases. The following tips should make remodeling a little easier:

1. Make a complete comprehensive remodeling plan before starting the remodeling process. In most cases, financing for the project cannot be obtained without one.

2. Obtain as many comprehensive proposals from contractors as possible.

3. If the contractor is supplying the materials, make sure they meet the requirements of the remodeling project.

4. Obtain references from the contractors submitting the bids.

5. Check the contractor's references and if possible inspect and examine previous projects completed by the contractors.

6. Examine the contractor's safety records as well as obtain any information that can be furnished by the Better Business Bureau.

7. Make sure that the contractor is licensed and bonded and/or insured. When a contractor is bonded, the facility manager or owner is protected against such issues as the contractor not completing the project or loss/damage to property. When a contractor is insured, the facility manager or owner is protected from liabilities such as worker injuries.

Demolition Tips

Regardless of the type of remodeling that is underway (with the exception of constructing a stand-alone structure), there will be a need to remove or demolish part of the existing structure in-order to complete the remodeling process. This is typically done by either demolishing or deconstructing part of the area to be remodeled. When *demolition* is involved, little or no attention is given to preserving the elements that are being removed from the structure. This is in contrast to *deconstruction* in which the overall goal is to preserve the elements that are being removed for possible reuse.

Before demolition can be performed, there are certain steps that must be performed. These steps include:

- Acquiring the necessary permits

- Submitting notifications to the residents and occupants as well as the surrounding neighbors

- Identifying and removing hazardous materials within the affected area(s). This include but is not limited to asbestos (was used in the production of insulation before it was banned in 1978), lead (was used in the production of paint before it was banned in 1977) and mercury (is still used in some forms of lighting). If it is determined that hazardous materials are present in the area that is to be affected, then proper removal and disposal techniques must be performed by licensed technicians. If the demolition includes the removal of HVAC equipment, then a licensed HVAC contractor must recover the refrigerant in the unit before it can be disassembled.

- Having the utilities in the affected area(s) disconnected and or removed.

- Developing and posting a site-specific safety procedure and work plan(s).

- If the demolition involves any load bearing walls or structural members then a structural engineer must be consulted and in most cases a demolition plan must be developed by the engineer.

© iStockphoto.com/Gencay M. Emin

Fasteners

Screws and Bolts

Wood screws

Phillips flat head
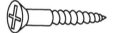

Slotted flat head

Slotted oval head
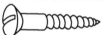

Slotted round head

Hex bolts

Carriage bolt

Full thread tap bolt

Standard bolt

Socket bolts

Socket head

Socket button head

Socket flat head

Socket set screw
with cup point

Sheet metal screws

Phillips flat head

Phillips oval head

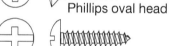
Phillips truss head

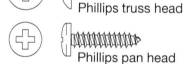
Phillips pan head

Phillips pan head self drilling

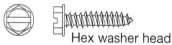
Hex washer head

Hex washer head self drilling

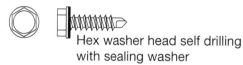
Hex washer head self drilling
with sealing washer

Machine screws

Phillips flat head

Phillips pan head

Slotted flat head

Phillips oval head

Slotted oval head

Combination round head

Combination truss head

Slotted round head

Lag bolt

Washers and Nuts

Washers

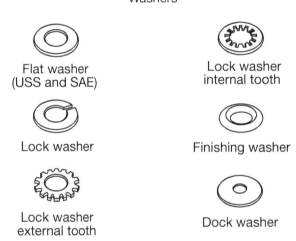

Flat washer
(USS and SAE)

Lock washer
internal tooth

Lock washer

Finishing washer

Lock washer
external tooth

Dock washer

Nuts

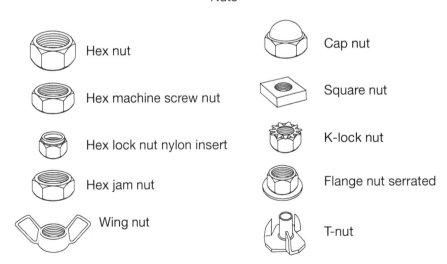

Hex nut

Hex machine screw nut

Hex lock nut nylon insert

Hex jam nut

Wing nut

Cap nut

Square nut

K-lock nut

Flange nut serrated

T-nut

Nails and Other Fasteners

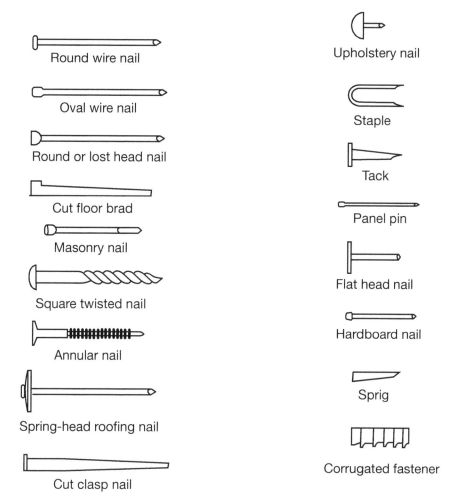

Round wire nail

Oval wire nail

Round or lost head nail

Cut floor brad

Masonry nail

Square twisted nail

Annular nail

Spring-head roofing nail

Cut clasp nail

Upholstery nail

Staple

Tack

Panel pin

Flat head nail

Hardboard nail

Sprig

Corrugated fastener

© iStockphoto.com/Gencay M. Emin

Wire Connectors

Crimp connectors used to splice and terminate 20-AWG to 500-kcmils aluminum-to-aluminum, aluminum-to-copper, or copper-to-copper conductors

A

PROPERLY CRIMP
THEN TAPE

Connectors used to connect wires on combinations of 18-AWG through 6-AWG conductors. They are twist-on, solderless, and tapeless.

•Wire-Nut® and Wing-Nut® are registered trademarks of Ideal Industries, Inc. Scotchlok® is a registered trademark of 3M.

B

WIRE CONNECTORS VARIOUSLY KNOWN AS
WIRE-NUT,® WING-NUT,® AND SCOTCHLOK.®

Connectors used to connect wires in combinations of 16-, 14-, and 12-AWG conductors. They are crimped on with a special tool, then covered with a snap-on insulating cap.

C

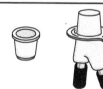

CRIMP-TYPE WIRE CONNECTOR
AND INSULATING CAP

Solderless connectors are available for 14 AWG through 500 kcmil conductors. They are used for one solid or one stranded conductor only, unless otherwise noted on the connector or on its shipping carton. The screw may be of the standard screwdriver slot type, or it may be for use with an Allen wrench or socket wrench.

D

SOLDERLESS CONNECTORS

Compression connectors are used for 8-AWG through 1,000-kcmil conductors. The wire is inserted into the end of the connector, then crimped on with a special compression tool.

E

COMPRESSION CONNECTOR

Split-bolt connectors are used for connecting two conductors or for tapping one conductor to another. They are available in sizes 10 AWG through 1,000 kcmil. They are used for two solid and/or two stranded conductors only, unless otherwise noted on the connector or on its shipping carton.

F

SPLIT-BOLT CONNECTOR

Appendix D

Conversion Tables

Fraction	Decimal	Fraction	Decimal	Fraction	Decimal
1/64	0.0156	11/32	0.3438	23/32	0.7188
1/32	0.0313	23/64	0.3594	47/64	0.7344
3/64	0.0469	**3/8**	**0.3750**	**3/4**	**0.7500**
1/16	0.0625	25/64	0.3906	49/64	0.7656
5/64	0.0781	13/32	0.4063	25/32	0.7813
3/32	0.0938	27/64	0.4219	51/64	0.7969
7/64	0.1094	7/16	0.4375	13/16	0.8125
1/8	**0.1250**	29/64	0.4531	53/64	0.8281
9/64	0.1406	15/32	0.4688	27/32	0.8438
5/32	0.1563	31/64	0.4844	55/64	0.8594
11/64	0.1719	**1/2**	**0.5000**	**7/8**	**0.8750**
3/16	0.1875	33/64	0.5156	57/64	0.8906
13/64	0.2031	17/32	0.5313	29/32	0.9063
7/32	0.2188	35/64	0.5469	59/64	0.9219
15/64	0.2344	9/16	0.5625	15/16	0.9375
1/4	**0.2500**	37/64	0.5781	61/64	0.9531
17/64	0.2656	19/32	0.5938	31/32	0.9688
9/32	0.2813	39/64	0.6094	63/64	0.9844
19/64	0.2969	**5/8**	**0.6250**		

Conversion Factors

Two basic systems, the **English system** and the **metric system**, are used to make measurements in the world today. The English system of measure is the one currently used in the United States today by most technicians. Its base units are inch (in), foot (ft), yard (yd), and mile (mi).

Unit	Divisions
English	
1 inch (in)	6 picas
1 foot (ft)	12 inches
1 yard (yd)	3 feet
1 mile (mi)	5,280 feet

English Conversion Factors

The base unit for the metric system is meter. This base unit is further divided into larger and smaller units (in multiples of 10), indicated by prefixes. Common metric prefixes are deci- (10), centi- (100), and milli- (1,000). For example, a meter contains 10 decimeters.

Unit	Divisions
Units of Length	
10 millimeters (mm)	1 centimeter (cm)
10 centimeters	1 decimeter (dm), 100 millimeters
10 decimeters	1 meter (m), 1,000 millimeters
10 meters	1 dekameter (dam)
10 dekameters	1 hectometer (hm), 100 meters
10 hectometers	5.1 kilometer (km), 1,000 meters

Metric Conversion Factors

The metric system of measure is occasionally used in the United States and, therefore, it is important that the technician be able to recognize and convert from one unit to another. The following table gives conversion factors for converting from one system to another.

Starting With	Multiply	To Find
inches	2.5	centimeters
feet	30	centimeters
yards	0.9	meters
miles	1.6	kilometers
centimeters	0.4	inches
centimeters	0.0333	inches
meters	1.1111	yards
kilometers	0.625	miles

Converting from metric to English and English to metric

For example, to convert 12 feet 6 inches to centimeters, perform the following steps.

Convert from Inches to Centimeters

Step 1 Converting 12 feet into inches.

$$12 \text{ ft} \times 12 \text{ in/ft} = \textbf{144 in}$$

Step 2 Next add the inch portion of the original measurement.

$$144 \text{ in} + 6 \text{ in} = \textbf{150 in}$$

Step 3 Find the conversion factor using the previous table.

$$150 \text{ in} \times 2.5 \text{ cm/in} = \textbf{375 cm}$$

Convert from Centimeters to Inches

Step 1 Find the conversion factor using the previous table.

375 cm × 0.4 in/cm = **140 in**

Step 2 Convert from inches to feet and inches by dividing by 12.

150 in/12 ft/in = 12 ft with a remainder of 6

Therefore, the measure would be written as **12 ft 6 in** or **12' 6"** .

Electrical Wire Gauge and Current Chart

AWG Gauge	Diameter (inches)	Ohms/1,000 ft	Maximum Amps
0000	0.46	0.049	302
000	0.4096	0.0618	239
00	0.3648	0.0779	190
0	0.3249	0.0983	150
1	0.2893	0.1239	119
2	0.2576	0.1563	94
3	0.2294	0.197	75
4	0.2043	0.2485	60
5	0.1819	0.3133	47
6	0.162	0.3951	37
7	0.1443	0.4982	30
8	0.1285	0.6282	24
9	0.1144	0.7921	19
10	0.1019	0.9989	15
11	0.0907	1.26	12
12	0.0808	1.588	9.3
13	0.072	2.003	7.4
14	0.0641	2.525	5.9
15	0.0571	3.184	4.7
16	0.0508	4.016	3.7
17	0.0453	5.064	2.9
18	0.0403	6.385	2.3

Small Engine Recommended Preventive Maintenance Charts

Recommended Chainsaw Preventive Maintenance				
	Daily	**Weekly**	**Monthly**	**As Needed**
Sprocket	Inspect			Replace
Fuel filter		Clean		
Muffler			Clean	
Muffler screen	Clean			Replace
Fuel tank			Clean	
Spark plug		Clean and adjust		Replace
Fuel, oil, and hoses	Check			
Air filter	Clean	Replace		
Screws, nuts, and bolts	Inspect and tighten			
Chain	Inspect and sharpen			

Recommended Lawnmower Preventive Maintenance				
	Daily	**Weekly**	**Monthly**	**As Needed**
Fuel filter		Clean		
Muffler			Clean	
Fuel tank			Clean	
Spark plug		Clean and adjust		Replace
Fuel, oil, and hoses	Check			
Air filter	Clean	Replace		
Screws, nuts, and bolts	Inspect and tighten			

Recommended String Trimmer Preventive Maintenance				
	Daily	**Weekly**	**Monthly**	**As Needed**
Fuel filter		Clean		
Muffler			Clean	
Fuel tank			Clean	
Spark plug		Clean and adjust		Replace
Fuel, oil, and hoses	Check			
Air filter	Clean	Replace		
Screws, nuts, and bolts	Inspect and tighten			

Appendix E

Using Sine, Cosine, and Tangent to Determine the Angles and Lengths of the Lines of a Triangle

As indicated by its name, a triangle contains three angles. When these three angles are added, their sum is equal to 180°, that is, angle A + angle B + angle C = 180°. Using this fact, a missing angle can be calculated if the other two angles are given. For example, the missing angle of a triangle containing a 70° and a 55° angle is found by adding the two known angles, and then subtracting their product from 180° (180° − [70° + 55°]). After performing the calculation, the missing angle is found to be 55°.

Each side of a triangle has a name: the hypotenuse, opposite, and adjacent sides. In a right triangle the hypotenuse is always the longest side. The other two sides, opposite and adjacent, are labeled relative to the acute angle (any angle less than 90°) being focused on in a given calculation (Figures E-1 and E-2).

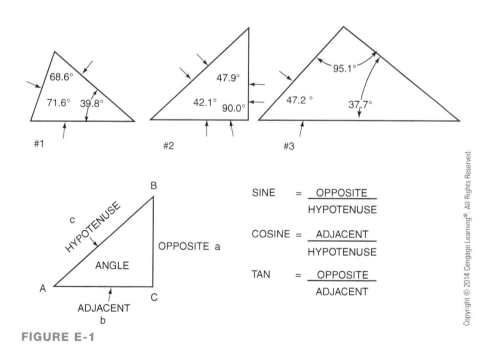

FIGURE E-1

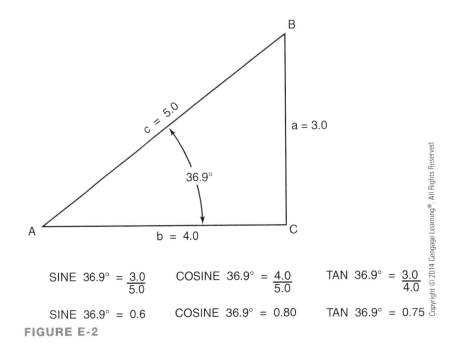

$$\text{SINE } 36.9° = \frac{3.0}{5.0} \qquad \text{COSINE } 36.9° = \frac{4.0}{5.0} \qquad \text{TAN } 36.9° = \frac{3.0}{4.0}$$

$$\text{SINE } 36.9° = 0.6 \qquad \text{COSINE } 36.9° = 0.80 \qquad \text{TAN } 36.9° = 0.75$$

FIGURE E-2

The adjacent side is the side adjacent to the angle in question. The opposite is the side opposite to that angle.

In a right triangle, there is a direct relationship between the angles and the lengths of the sides. This relationship can be summed up in three fundamental trigonometric functions: sine, cosine, and tangent. The sine function is defined as the ratio of the side opposite to an acute angle divided by the hypotenuse or (Sine A = a/c), as shown in Figures E-1 and E-2. The cosine function is defined as the ratio of the side adjacent to an acute angle divided by the hypotenuse or (Cosine A = b/c), and the tangent function is defined as the ratio of the side opposite to an acute angle divided by the side adjacent (Tan A = a/b). Another important concept in trigonometry is the Pythagorean theorem. It states that the square of the hypotenuse of a right triangle is equal to the sum of the square of the other two sides or $R^2 = X^2 + Y^2$. In this equation R is equal to the hypotenuse, and X and Y are equal to the adjacent and opposite sides of the triangle. With this equation, the missing side of a right triangle can be determined if the other two sides are given by applying the Pythagorean theorem. To calculate the hypotenuse, you would have to use algebra to isolate the variable R. In other words, the hypotenuse is found by solving for R, not R^2. To isolate R, the square root of both sides of the equation must be taken, and doing so would yield the equation $R = \sqrt{(X^2 + Y^2)}$. For example, given a right triangle that contains an adjacent side (b) equal to 3 inches and an opposite side (a) equal to 4 inches, the hypotenuse can be found by using the formula: ($R = \sqrt{3^2 + 4^2}$, $R = \sqrt{25}$, $R = 5$), as shown in Figure E-3.

#1 Acute triangle—all angles are less than 90°.

#2 Right triangle—one angle is 90°.

#3 Obtuse triangle—one angle is greater than 90°.

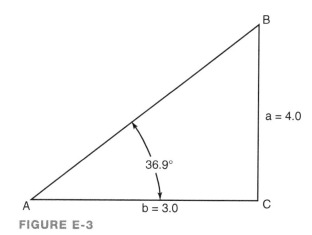

Using the equation:

$$R^2 = A^2 + B^2$$

Solving for R not R^2

the equation becomes

$$R = \sqrt{B^2 + A^2}$$

$$R = \sqrt{9 + 16}$$

$$R = \sqrt{25}$$

$$R = 5$$

a = 4.0

b = 3.0

36.9°

FIGURE E-3

Glossary

A

Acoustics the ability of a material to absorb and reflect sound is referred to as acoustical.

Acoustical analysis an analysis to determine the level of reverberation or reflection of sound in a space as it is influenced by the building materials used in it construction.

Acoustical consultant a consultant who is experienced in providing advice on the acoustical requirements and noise control of a facility or a room.

Acoustical insulation insulation that reduces the passage of sound through the section of a building.

Alternating current (AC) electron flow that flows in one direction and then reverses at regular intervals.

Americans with Disabilities Act (ADA) a law that prohibits discrimination based on disability.

Ampere (amp) unit of current flow.

Appreciation the expression of gratitude toward your customers.

Asphyxiation loss of consciousness that is caused by a lack of oxygen or excessive carbon dioxide in the blood.

Assigning tasks giving a task to someone to complete.

Atom the smallest particle of an element.

B

Auxiliary view A view that uses orthographic techniques to produce a projection on a plane other than one of the three principal planes.

Backdraft occurs when an exhaust fan(s) and combustion style heating devices compete for air; most often the exhaust fan pulls the combustion gases back into the structure as a result.

Back miter an angle cut starting from the end and coming back on the face of the stock.

Base-load energy consumption the amount of energy consumed by a structure and its occupants under the best possible conditions.

Belly deck a mower that is built like a car and has a blade deck underneath the driver and engine.

Booking the process used to activate the paste.

Boring jig a tool frequently used to guide bits when boring holes for locksets.

Bounty program a program offered by most utilities and municipalities whereby they purchase (in the form of a rebate) an old appliance when a person purchases any appliance that qualifies as an energy star appliance.

British thermal unit (Btu) the amount of heat needed to raise the temperature of one pound of water one degree Fahrenheit.

C

Carbon dioxide by-product of natural gas combustion that is not harmful.

Carbon monoxide a poisonous, colorless, odorless, tasteless gas generated by incomplete combustion.

Cardiopulmonary resuscitation (CPR) an emergency first-aid procedure used to maintain circulation of blood to the brain.

Caulk used to fill cracks as well as other defects in interior wall as well as exterior walls and foundation cracks.

Class A fire extinguishers fire extinguishers used on fires that result from burning wood, paper, or other ordinary combustibles.

Class B fire extinguishers fire extinguishers used on fires that involve flammable liquids such as grease, gasoline, or oil.

Class C fire extinguishers fire extinguishers used on electrically energized fires.

Class D fire extinguishers fire extinguishers typically used on flammable metals.

Competence having the skills, knowledge, ability, or qualifications to complete a task.

Compressor the system component that raises the pressure and temperature of the refrigerant in the system.

Condenser the system component that is used to eject heat from the refrigerant to the surrounding outdoor environment.

Conduction the transfer of heat from one molecule to another within a substance or from one substance to another.

Confidence having a belief in yourself and your abilities.

Continuous load a load in which the maximum current is expected to continue for 3 hours or more.

Convection the transfer of heat from one place to another using a fluid.

Conventional current flow theory states that current flows from positive to negative in a circuit.

Coulomb the amount of electric charge transported in one second by one ampere of current.

Courtesy acting respectfully toward your customers.

Crawlspace an area under a house that provides a means of raising the structure off the ground.

Current (or amperage) the flow of electrons through a given circuit, which is measured in *amps*.

D

Daylight the placement of windows in a location that will promote natural lighting as well as a view to the outside.

Diagnostics the process of determining a malfunction.

Direct current (DC) electron flow that flows in only one direction; used in the industry only for special applications such as solid-state modules and electronic air filters.

Dumpster a large waste receptacle.

E

Electron theory states that current flows from negative to positive in a circuit.

Element any of the known substances (of which ninety-two occur naturally) that cannot be separated into simpler compounds.

Elevator platform the floor of an elevator.

Empathy the capacity to understand your customers' state of mind or emotion.

Energy audit a survey of a residence or a structure conducted to analyze the current energy requirements of the structure and its occupants as well as how that energy flow can be reduced without having a negative effect on the structure and/or occupants.

Energy density the amount of energy stored in a region of space per unit volume.

Energy management is the process of monitoring, controlling, and ultimately conserving energy in a structure.

Engineered panels human-made products in the form of large reconstituted wood sheets.

English system of measure a system of measurement primarily used in the United States based on the dimensions of the human body.

Evaporator a device used to transfer heat from the inside of a structure to the outside by passing air over its coils.

F

Faceplate markers a tool used to lay out the mortise for the latch faceplate.

Fittings devices used to connect pipes together in a plumbing system.

Fraction a number that can represent part of a whole.

Frostbite injury to the skin resulting from prolonged exposure to freezing temperatures.

Frostnip the first stage of exposure, which causes whitening of the skin, itching, tingling, and loss of feeling.

G

Garbage waste or food that has spoiled as well as other household refuse.

Green lumber lumber that has just been cut from a log.

Grid-connected system a system that uses both energy from the power grid and solar energy.

Ground fault circuit interrupter (GFCI) electrical device designed to sense small current leaks to ground and deenergize the circuit before injury can result.

Groundskeeper a technician who maintains the functionality and appearance of landscaped grounds and gardens.

Groundskeeping the activity of tending an area for aesthetic or functional purposes. It includes mowing grass, trimming hedges, pulling weeds, planting flowers, and so on.

H

Hardwood deciduous trees.

Heating element a device used to transform electricity into heat through electrical resistance.

Heat of transmission the amount of heat transferred through building materials.

Honesty acting truthfully with your customers.

I

Indirect gain process in which heat is collected in a storage device and later permitted to be distributed via conduction, convection, and radiation.

Infiltration air that enters into a structure through cracks, windows, doors, or other openings as a result of the indoor air pressure being lower than the outdoor air pressure.

Illumination lighting.

Impact noise sound that is caused by the dropping of object onto a floor.

Inorganic mulch constitutes gravel, pebbles, black plastic, and landscape fabrics.

Insecticide chemical used to kill insects.

Insulator a material that has a stable atomic structure and does not permit the free flow of electrons from atom to atom.

Integer any whole number, including zero that is positive or negative, factorable or nonfactorable.

K

Kill spot an inconspicuous area used to start and end wallpaper in a room.

L

Landscaper an individual who modifies the feature of an area (land) such as construct stone walls, wooden fences, brick pathways, statuary, fountains, benches, trees, flowers, shrubs, and grasses.

Law of centrifugal force states that spinning object has a tendency to pull away from its center point and that the faster it spins, the greater the centrifugal force will be.

Law of charges states that like charges repel and unlike charges attract.

Light the range of electromagnetic radiation that is detectable by the human eye.

Linear measurement the measurement of two points along a straight line.

Litter waste that is unlawfully disposed of outdoors.

M

Mask a faux finish used to apply a new color over a dry base coat to create an image or shape.

Matter a substance that takes up space and has weight.

Megger another name for a megohmmeter.

Megohmmeter an instrument used to measure high electrical resistance.

Metering device the system component that removes from the system the remaining heat and pressure previously introduced by the compressor.

Metric system of measure a system of measurement that uses a single unit for any physical quantity.

Mold a plant that is a member of the fungi family.

Multispur bits a power-driven bit, guided by a boring jig, that is used to make a hole in a door for the lockset.

Mural a faux finish used to give the illusion of scenery or architectural elements.

O

Occupants (elevator) people riding on an elevator.

Occupational Safety and Health Administration (OSHA) branch of the U.S. Department of Labor that strives to reduce injuries and deaths in the workplace.

Off-grid system a stand-alone non-grid-tied system in which the peak demand must be taken into consideration because the system is not connected to the electrical grid.

Ohm's law relationship between voltage and current and a material's ability to conduct electricity.

Organic mulch constitutes natural substances such as bark, wood chips, leaves, pine needles, or grass clippings.

P

Passive solar design the use of daylighting, landscaping thermal mass, and proper building orientation to provide an efficient building design.

Personal protective equipment (PPE) any equipment that will provide personal protection from a possible injury.

Pesticide chemical used to kill pests (as rodents or insects).

Plies layers of wood used to build up a product such as plywood.

Photons the tiny packet that makes up light.

Photovoltaics the production of electricity from direct sunlight.

Potable water that is safe for human consumption.

Power the electrical work that is being done in a given circuit, which is measured in wattage (watts) for a purely resistive circuit or volt-amps (VA) for an inductive/capacitive circuit.

Power tool a tool that contains a motor.

Priority giving a task precedence over others.

Q

Qualified technician a person that has the necessary knowledge and skills directly related to the operation of electrical equipment.

R

Radiation heat that travels through air, heating solid objects that in turn heat the surrounding area.

Real number any number including zero that is positive or negative, rational or irrational.

Reliability the quality of being dependable.

Renewable energy any form of energy that is derived from sources that can be regenerated.

Resistance the opposition to current flow in a given electrical circuit, which is measured in *ohms*.

Return on investment (ROI) the rate at which the total investment is recovered from by the savings of energy conservation.

S

Safety interlock a device used to help prevent a machine from causing damage to itself or injury to the operator by disengaging the device when the interlock is tripped.

Screws used when stronger joining power is needed.

Self-talk what you say silently to yourself as you go through the day or when you are faced with difficult situations.

Shoulder months the months in which the heating and cooling systems are not used.

Softboard a low-density fiberboard.

Softwood coniferous, or cone-bearing, trees.

Solar array multiple solar panels that are connected together.

Solar collector a device that utilizes the greenhouse effect and direct sunlight to heat up a copper plate containing heat transfer tubing (heat exchanger).

Solar constant the amount of power available from the sun.

Sound transmission class (STC) a rating used to describe the resistance of a building section to the passage of sound.

Spark ignition a device used to ignite a gas stove.

Stand-alone grid-connected system basically the same as a grid-connected system, with the exception of having an added storage device, typically in the form of a battery system.

Stand-alone off-grid hybrid system a standalone system that incorporates a diesel generator.

Striker plate a plate installed on the door jamb against which the latch on the door engages when the door is closed.

T

Task an activity that needs to be performed to complete a project.

Thermodynamics the branch of physics that deals with the conversion of energy.

Thermostat a device used to control the temperature of water by controlling the heat source.

Troubleshooting the process of performing a systematic search for a resolution to a technical problem.

U

Unconditioned space design space that is not air conditioned.

U factor the reciprocal of the R value (measure of a material's resistance to the flow of heat).

V

Valence shell the outermost shell of an atom.

Valves devices used to control the flow of water, waste, gas, and so on.

Ventilation planned induction of outside air to create a slight positive pressure on the structure.

Voltage the potential electrical difference for electron flow from one line to another in an electrical circuit.

W

Watt a unit of power applied to electron flow. One watt equals 3.414 Btu.

Wave–particle duality the fact a photon exhibits the properties of a wave as well as a particle.

Whole numbers counting numbers, including zero.

Glosario

A

Acoustics (Acústica)
la capacidad de un material de absorber y reflejar el sonido se denomina acústica.

Acoustical analysis (Análisis acústico) un análisis para determinar el nivel de reverberación o reflexión del sonido en un espacio al ser influenciado por los materiales de construcción utilizados en su edificación.

Acoustical consultant (Consultor acústico)
un consultor experto en prestar asesoramiento sobre los requisitos acústicos y el control del ruido de una instalación o una sala.

Acoustical insulation (Aislamiento acústico)
aislamiento que reduce el paso del sonido a través de la sección de un edificio.

Alternating current (AC) (Corriente alterna) flujo de electrones que fluye en una dirección y luego se invierte a intervalos regulares.

Americans with Disabilities Act (ADA) (Ley de Protección para Ciudadanos Estadounidenses con Discapacidades) ley que prohíbe la discriminación por razones de discapacidad.

Ampere (amp) (Amperio (A))
unidad de flujo de corriente.

Appreciation (Apreciación)
la expresión de gratitud hacia sus clientes.

Asphyxiation (Asfixia)
pérdida de conciencia que es causada por falta de oxígeno o exceso de dióxido de carbono en la sangre.

Assigning tasks (Asignación de tareas) dar una tarea a una persona para que la complete.

Atom (Átomo) la partícula más pequeña de un elemento.

Auxiliary view (Vista auxiliar) Una vista que utiliza técnicas ortográficas para producir una proyección en un plano distinto de uno de los tres planos principales.

B

Backdraft (Contracorriente)
ocurre cuando uno o más ventiladores de escape y dispositivos de calefacción estilo combustión compiten por el aire; con mayor frecuencia, como resultado, el ventilador de escape atrae los gases de la combustión hacia el interior de la estructura.

Back miter (Inglete) ángulo cortado desde el extremo hacia la superficie del material.

Base-load energy consumption (Consumo de energía de carga base) la cantidad de energía consumida por una estructura y sus ocupantes en las mejores condiciones posibles.

Belly deck (Tractor cortacésped) cortadora de césped que está construida como un automóvil y tiene una cubierta con cuchilla debajo del conductor y el motor.

Booking (Reserva) el proceso utilizado para activar la pasta.

Boring jig (Matriz de taladrado) herramienta que se usa con frecuencia para guiar las brocas cuando se perforan orificios para juegos de cerraduras.

Bounty program (Programa de recompensas) un programa ofrecido por la mayoría de las empresas de servicios públicos y municipalidades mediante el cual estos compran (en forma de reembolso) un viejo electrodoméstico cuando una persona compra cualquier electrodoméstico que califique como un electrodoméstico Energy Star.

British thermal unit (Btu) (Unidad térmica británica (Btu)) la cantidad de calor necesaria para elevar la temperatura de una libra de agua un grado Fahrenheit.

C

Carbon dioxide (Dióxido de carbono) producto derivado de la combustión del gas natural que no es nocivo.

Carbon monoxide (Monóxido de carbono) gas tóxico, incoloro, inodoro e insípido generado por la combustión incompleta.

Cardiopulmonary resuscitation (CPR) (Resucitación cardiopulmonar (RCP)) procedimiento de primeros auxilios de emergencia utilizado para mantener la circulación de la sangre al cerebro.

Caulk (Sellador) se utiliza para llenar grietas y otros defectos tanto en paredes interiores como en paredes exteriores y en grietas de los cimientos.

Class A fire extinguishers (Extinguidores de incendios clase A) extinguidores utilizados en incendios que se producen de la quema de madera, papel u otros combustibles comunes.

Class B fire extinguishers (Extinguidores de incendios clase B) extinguidores utilizados en incendios que involucran líquidos inflamables como grasa, gasolina o petróleo.

Class C fire extinguishers (Extinguidores de incendios clase C) extinguidores utilizados en incendios energizados por electricidad.

Class D fire extinguishers (Extinguidores de incendios clase D) extinguidores que se usan por lo general en metales inflamables.

Competence (Competencia) contar con habilidades, conocimientos, capacidad o cualificaciones para completar una tarea.

Compressor (Compresor) el componente del sistema que eleva la presión y la temperatura del refrigerante en el sistema.

Condenser (Condensador) el componente del sistema que se utiliza para expulsar el calor del refrigerante al entorno exterior circundante.

Conduction (Conducción) transferencia de calor de una molécula a otra por dentro de una sustancia, o de una sustancia a otra.

Confidence (Confianza) creer en uno mismo y en sus aptitudes.

Continuous load (Carga continua) carga en la cual se prevé que la corriente máxima continúe por 3 horas o más.

Convection (Convección) transferencia de calor de un lugar a otro usando un fluido.

Conventional current flow theory (Teoría de flujo de la corriente convencional) afirma que la corriente fluye de positivo a negativo en un circuito.

Coulomb (Culombio) la cantidad de carga eléctrica transportada en un segundo por un amperio de corriente.

Courtesy (Cortesía) actuar de manera respetuosa hacia los clientes.

Crawlspace (Sótano de poca altura) área debajo de una casa que proporciona un medio de elevar la estructura del suelo.

Current (Corriente) (o amperaje) el flujo de electrones a través de un circuito determinado, que se mide en amperios.

D

Daylight (Iluminación natural) colocación de ventanas en un lugar que promueve la iluminación natural así como una vista del exterior.

Diagnostics (Diagnóstico) proceso de determinación de un desperfecto.

Direct current (DC) (Corriente continua) flujo de electrones que fluye en una sola dirección; se utiliza en la industria solamente para aplicaciones especiales, como módulos de estado sólido y filtros de aire electrónicos.

Dumpster (Contenedor) receptáculo grande para desperdicios.

E

Electron theory (Teoría de electrones) manifiesta que la corriente fluye de negativo a positivo en un circuito.

Element (Elemento) cualquiera de las sustancias conocidas (de las cuales noventa y dos se producen naturalmente) que no pueden separarse en compuestos más simples.

Elevator platform (Plataforma de ascensor) el piso de un ascensor.

Empathy (Empatía) la capacidad de entender el estado de ánimo o la emoción de sus clientes.

Energy audit (Auditoría de energía) inspección de una residencia o una estructura realizada para analizar los requisitos de energía de corriente de la estructura y sus ocupantes, así como de la manera en que ese flujo de energía puede reducirse sin tener un efecto negativo en la estructura y/o los ocupantes.

Energy density (Densidad de la energía) la cantidad de energía almacenada en una región de espacio por volumen de unidad.

Energy management (Administración de la energía) es el proceso de monitoreo, control y, en última instancia, de conservación de la energía en una estructura.

Engineered panels (Paneles industrializados) productos hechos por el hombre en la forma de grandes hojas de madera reconstituidas.

English system of measure (Sistema inglés de medición) sistema de medición que se usa principalmente en Estados Unidos, con base en las dimensiones del cuerpo humano.

Evaporator (Evaporador) dispositivo utilizado para transferir calor desde el interior de una estructura al exterior, pasando aire por sus bobinas.

F

Faceplate (Placa frontal) herramienta utilizada para disponer la muesca para la placa frontal del pestillo.

Fittings (Accesorios) dispositivos utilizados para conectar tubos entre sí en un sistema de tuberías.

Fraction (Fracción) número que puede representar parte de un entero.

Frostbite (Congelación) lesión en la piel causada por exposición prolongada a temperaturas bajo cero.

Frostnip (Congelación mínima) la primera etapa de la exposición, la cual ocasiona el blanqueamiento de la piel, comezón, hormigueo y pérdida de la sensibilidad.

G

Garbage (Residuos) desperdicios o alimentos echados a perder, así como otros tipos de desecho doméstico.

Green lumber (Madera verde) madera que ha sido cortada de un tronco recientemente.

Grid-connected system (Sistema conectado a la red) sistema que utiliza energía de la red eléctrica y de la energía solar.

Ground fault circuit interrupter (GFCI) (Interruptor de circuito por falla de conexión a tierra (GFCI)) dispositivo eléctrico diseñado para detectar pequeñas fugas de corriente a tierra y desactivar el circuito antes de que se produzca una lesión.

Groundskeeper (Encargado de mantenimiento de terrenos) técnico que mantiene la funcionalidad y la apariencia de jardines y zonas ajardinadas.

Groundskeeping (Mantenimiento de terrenos) actividad de cuidar de un área con fines estéticos o funcionales. Esta incluye cortar el césped, podar setos, quitar hierbas, plantar flores, etc.

H

Hardwood (Madera dura) árboles caducos.

Heating element (Elemento calefactor) dispositivo utilizado para transformar la electricidad en calor a través de la resistencia eléctrica.

Heat of transmission (Calor de transmisión) la cantidad de calor transferida a través de los materiales de construcción.

Honesty (Honestidad) actuar de manera sincera con sus clientes.

I

Indirect gain (Ganancia indirecta) proceso en el cual el calor se acumula en un dispositivo de almacenamiento y luego se distribuye por medio de conducción, convección y radiación.

Infiltration (Infiltración) aire que entra en una estructura a través de grietas, ventanas, puertas u otras aberturas, debido a que la presión de aire interna es más baja que la presión de aire externa.

Illumination (Iluminación) alumbramiento.

Impact noise (Ruido por impacto) sonido causado por la caída de un objeto en el suelo.

Inorganic mulch (Abono orgánico) constituye grava, piedritas, plástico negro y cubiertas para jardín.

Insecticide (Insecticida) producto químico utilizado para matar insectos.

Insulator (Aislante) material que tiene una estructura atómica estable y no permite el flujo libre de electrones de un átomo a otro.

Integer (Entero) cualquier número entero, incluido el cero que es positivo o negativo, que pueda o no factorizarse.

K

Kill spot ("Kill spot" (sitio inicial/final)) área discreta utilizada para comenzar o terminar el papel tapiz en una sala.

L

Landscaper (Paisajista) persona que modifica la característica de un área (terreno) como paredes de piedra de construcción, cercas de madera, caminos de ladrillo, estatuas, fuentes, banquillos, árboles, flores, arbustos y césped.

Law of centrifugal force (Ley de fuerza centrífuga) afirma que el objeto giratorio tiene una tendencia a alejarse de su punto central y que, a medida que gire más rápido, mayor será la fuerza centrífuga.

Law of charges (Ley de cargas) afirma que las cargas similares se repelen y que las cargas opuestas se atraen.

Light (Luz) rango de radiación electromagnética que el ojo humano puede detectar.

Linear measurement (Medida de longitud) medida de dos puntos a lo largo de una línea recta.

Litter (Basura) desechos que se eliminan indebidamente en el exterior.

M

Mask (Máscara) un falso acabado utilizado para aplicar un nuevo color sobre una capa de base seca, para crear una imagen o una forma.

Matter (Materia) sustancia que ocupa espacio y tiene peso.

Megger (Megger) otro nombre para un megóhmetro.

Megohmmeter (Megóhmetro) instrumento utilizado para medir la resistencia eléctrica elevada.

Metering device (Dispositivo de medición) el componente del sistema que elimina de éste el calor restante y la presión introducida anteriormente por el compresor.

Metric system of measure (Sistema métrico de medición) sistema de medición que utiliza una sola unidad para cualquier cantidad física.

Mold (Moho) planta que pertenece a la familia de los hongos.

Multispur bits (Brocas de puntas múltiples) broca mecánica guiada por un patrón para hacer orificios, que se usa para hacer un agujero en donde se inserta la cerradura.

Mural (Mural) falso acabado utilizado para dar la ilusión de paisaje o elementos arquitectónicos.

O

Occupants (Ocupantes) (ascensor) personas que viajan en un ascensor.

Occupational Safety and Health Administration (OSHA) (Administración de Salud y Seguridad Ocupacional (OSHA)) rama del Departamento de Trabajo de EE. UU. que procura reducir las lesiones y las muertes en el lugar de trabajo.

Off-grid system (Sistema fuera de red) sistema autónomo no vinculado a la red en el cual debe considerarse la demanda pico porque el sistema no está conectado a la red eléctrica.

Ohm's law (Ley de Ohm) la relación entre el voltaje y la corriente, y la capacidad de un material de conducir electricidad.

Organic mulch (Abono orgánico) constituye sustancias naturales como corteza, trozos de madera, hojas de pino o recortes de hierbas.

P

Passive solar design (Diseño solar pasivo) el uso de la iluminación natural, masa térmica para paisajes y orientación adecuada de edificaciones para proporcionar un diseño eficaz de la construcción.

Personal protective equipment (PPE) (Equipo de protección personal (EPP)) cualquier equipo que proporciona protección personal ante una posible lesión.

Pesticide (Pesticida) producto químico utilizado para eliminar plagas (como roedores o insectos).

Plies (Capas) capas de madera utilizadas para construir un producto como madera laminada.

Photons (Fotones) partícula pequeña que compone la luz.

Photovoltaics (Fotovoltaicos) la producción de electricidad a partir de la luz solar.

Potable water (Agua potable) agua que es segura para el consumo humano.

Power (Potencia) el trabajo eléctrico que se realiza en un circuito determinado, que se mide en voltaje (vatios) para un circuito puramente resistivo, o en voltios-amperios (VA) para un circuito inductivo/capacitivo.

Power tool (Herramienta eléctrica) herramienta que contiene un motor.

Priority (Prioridad) dar precedencia a una tarea por encima de otras.

Q

Qualified technician (Técnico cualificado) persona que cuenta con el conocimiento y las habilidades necesarias que se relacionan directamente con el funcionamiento del equipo eléctrico.

R

Radiation (Radiación) calor que viaja a través del aire, calentando los objetos sólidos que a su vez calientan el área circundante.

Real number (Número real) cualquier número (incluido el cero) que sea positivo o negativo, racional o irracional.

Reliability (Confiabilidad/ fiabilidad) la calidad de ser confiable.

Renewable energy (Energía renovable) cualquier forma de energía obtenida de fuentes que pueden regenerarse.

Resistance (Resistencia) la oposición al flujo de corriente en un circuito eléctrico determinado, que se mide en ohmios.

Return on investment (ROI) (Retorno de la inversión (ROI)) la tasa a la cual se recupera la inversión total a partir de los ahorros de la conservación de la energía.

S

Safety interlock (Enclavamiento de seguridad) dispositivo utilizado para impedir que una máquina cause daños a sí misma o lesiones al operador al desconectar el dispositivo cuando se activa el enclavamiento.

Screws (Tornillos) se utilizan cuando se requiere mayor fuerza de unión.

Self-talk (Hablar con uno mismo) lo que se dice en silencio para uno mismo a medida que transcurre el día o al enfrentarse a situaciones peligrosas.

Shoulder months (Meses "shoulder" (de transición entre épocas del año)) meses en los cuales no se utilizan los sistemas de calefacción y refrigeración.

Softboard (Cartón comprimido) cartón de fibra de baja densidad.

Softwood (Madera blanda) árboles cuyos frutos tienen forma de cono.

Solar array (Arreglo solar) varios paneles solares que están conectados entre sí.

Solar collector (Colector solar) dispositivo que utiliza el efecto invernadero y dirige la luz solar para calentar una plaza de cobre que contiene tubos de transferencia de calor (intercambiador de calor).

Solar constant (Constante solar) cantidad de energía disponible del sol.

Sound transmission class (STC) (Clasificación de la transmisión del sonido (STC)) una clasificación utilizada para describir la resistencia de una sección de un edificio al paso del sonido.

Spark ignition (Encendido por chispa) dispositivo utilizado para encender una estufa a gas.

Stand-alone grid-connected system (Sistema autónomo conectado a la red) básicamente igual que un sistema conectado a la red, con la excepción de que tiene un dispositivo de almacenamiento adicional, por lo general en la forma de un sistema de batería.

Stand-alone off-grid hybrid system (Sistema híbrido autónomo fuera de red) sistema autónomo que incorpora un generador diésel.

Striker plate (Placa del cerradero) placa instalada en la jamba de la puerta en la que encaja el pestillo cuando se cierra la puerta.

T

Task (Tarea) actividad que debe ser realizada para completar un proyecto.

Thermodynamics (Termodinámica) rama de la física que se ocupa de la conversión de la energía.

Thermostat (Termostato) dispositivo utilizado para controlar la temperatura del agua al controlar la fuente de calor.

Troubleshooting (Resolución de problemas) el proceso de realizar una búsqueda sistemática para resolver un problema técnico.

U

Unconditioned space (Espacio sin aire acondicionado) espacio de diseño que no tiene aire acondicionado.

U factor (Factor U) el recíproco del valor R (medida de resistencia de un material al flujo de calor).

V

Valence shell (Capa de valencia) la capa más exterior de un átomo.

Valves (Válvulas) dispositivos utilizados para controlar el flujo de agua, desechos, gas, etc.

Ventilation (Ventilación) inducción planificada del aire externo para crear una leve presión positiva en la estructura.

Voltage (Voltaje) la diferencia eléctrica potencial para el flujo de electrones de una línea a otra en un circuito eléctrico.

W

Watt (Vatio) unidad de potencia aplicada al flujo de electrones. Un vatio equivale a 3,414 Btu.

Wave–particle duality (Dualidad onda-partícula) el hecho de que un fotón exhibe las propiedades de una onda y de una partícula.

Whole numbers (Números enteros) números de conteo, incluido el cero.

Index

CPSIA information can be obtained
at www.ICGtesting.com
Printed in the USA
FFHW010530050119
50066442-54888FF